T0178080

MARKOV CHAINS and DECISION PROCESSES for ENGINEERS and MANAGERS

MARKOV CHAINS and DECISION PROCESSES for ENGINEERS and MANAGERS

Theodore J. Sheskin

CRC Press

Taylor & Francis Group

Boca Raton London New York

CRC Press is an imprint of the
Taylor & Francis Group, an **informa** business

CRC Press
Taylor & Francis Group
6000 Broken Sound Parkway NW, Suite 300
Boca Raton, FL 33487-2742

First issued in paperback 2019

© 2011 by Taylor and Francis Group, LLC
CRC Press is an imprint of Taylor & Francis Group, an Informa business

No claim to original U.S. Government works

ISBN-13: 978-1-4200-5111-7 (hbk)
ISBN-13: 978-0-367-38343-5 (pbk)

Visit the Taylor & Francis Web site at
http://www.taylorandfrancis.com

and the CRC Press Web site at
http://www.crcpress.com

Contents

Preface

The goal of this book is to provide engineers and managers with a unified treatment of Markov chains and Markov decision processes. The unified treatment of both subjects distinguishes this book from many others. The prerequisites are matrix algebra and elementary probability. In addition, linear programming is used in several sections of Chapter 5 as an alternative procedure for finding an optimal policy for a Markov decision process. These sections may be omitted without loss of continuity. The book will be of interest to seniors and beginning graduate students in quantitative disciplines including engineering, science, applied mathematics, operations research, management, and economics. Although written as a textbook, the book is also suitable for self-study by engineers, managers, and other quantitatively educated individuals. People who study this book will be prepared to construct and solve Markov models for a variety of random processes.

Many books on Markov chains or decision processes are either highly theoretical, with few examples, or else highly prescriptive, with little justification for the steps of the algorithms used to solve Markov models. This book balances both algorithms and applications. Engineers and quantitatively trained managers will be reluctant to use a formula or execute an algorithm without an explanation of the logical relationships on which they are based. On the other hand, they are not interested in proving theorems. In this book, formulas and algorithms are derived informally, occasionally in a section labeled as an optional insight. The validity of a formula is often justified by applying it to a small generic Markov model for which transition probabilities or rewards are expressed as symbols. The validity of other relationships is demonstrated by applying them to larger Markov models with numerical transition probabilities or rewards. Informal derivations and demonstrations of the validity of formulas are carried out in considerable detail.

Since engineers and managers are interested in applications, considerable attention is devoted to the construction of Markov models. A large number of simplified Markov models are constructed for a wide assortment of processes important to engineers and managers including the weather, gambling, diffusion of gases, a waiting line, inventory, component replacement, machine maintenance, selling a stock, a charge account, a career path, patient flow in a hospital, marketing, and a production line. The book is distinguished by the high level of detail with which the construction and solution of Markov models are described. Many of these Markov models have numerical transition probabilities and rewards. Descriptions of the step-by-step calculations made by the algorithms implemented for solving these models will facilitate the student's ability to apply the algorithms.

The text is organized around a Markov chain structure. Chapter 1 describes Markov chain states transitions, structure, and models. Chapter 2 discusses steady-state distributions and passage to a target state in a regular Markov chain. A regular chain has the simplest structure because all of its states communicate. Chapter 3 treats canonical forms and passage to target states or to classes of target states for reducible Markov chains. Reducible chains have more complex structures because they contain one or more closed classes of communicating states plus transient states. Chapter 4 adds an economic dimension to a Markov chain by associating rewards with states, thereby linking a Markov chain to a Markov decision process. The measure of effectiveness for a Markov chain with rewards is an expected reward. Chapter 5 adds decisions to create a Markov decision process, enabling an analyst to choose among alternative Markov chains with rewards so as to maximize an expected reward. Last, Chapter 6 introduces two interesting special topics that are rarely included in a senior and beginning graduate level text: state reduction and hidden Markov chains.

Markov chains and Markov decision processes are commonly treated in separate books, which use different conventions for numbering the time periods in a planning horizon. When a Markov chain is analyzed, time periods are numbered forward, starting with epoch 0 at the beginning of the horizon. However, in books in which value iteration is executed for a Markov decision process over an infinite planning horizon, time periods are numbered backward, starting with epoch 0 at the end of the horizon. In this book, the time periods are always numbered forward to establish a consistent convention for both Markov chains and decision processes. When value iteration is executed for a Markov decision process over an infinite horizon, time periods are numbered as consecutive negative integers, starting with epoch 0 at the end of the horizon.

I thank Professor David Sheskin of Western Connecticut State University for helpful suggestions and for encouraging me to write this book. I also thank Eugene Sheskin for his support. I am grateful to Cindy Carelli, editor of industrial engineering and operations research at CRC Press, for her enormous patience. I also thank the reviewers: Professors Isaac Sonin of University of North Carolina at Charlotte, M. Jeya Chandra of Pennsylvania State University, and Henry Lewandowski of Baldwin Wallace College.

Theodore J. Sheskin

Author

Theodore J. Sheskin is professor emeritus of industrial engineering at Cleveland State University. He earned a B.S. in electrical engineering from the Massachusetts Institute of Technology, an M.S. in electrical engineering from Syracuse University, and a Ph.D. in industrial engineering and operations research from Pennsylvania State University. Professor Sheskin is the sole author of 21 papers published in peer-reviewed journals of engineering and mathematical methods. He is a registered professional engineer.

1

Markov Chain Structure and Models

The goal of this book is to prepare students of engineering and management science to construct and solve Markov models of physical and social systems that may produce random outcomes. Chapter 1 is an introduction to Markov chain models and structure. A Markov chain is an indexed collection of random variables used to model a sequence of dependent events such that the probability of the next event depends only on the present event. In other words, a Markov chain assumes that, given the present event, the probability of the next event is independent of the past events. A Markov decision process is a sequential decision process for which the decisions produce a sequence of Markov chains with rewards. In this book, closed form solutions will be obtained algebraically for small generic Markov models. Larger models will be solved numerically.

1.1 Historical Note

The Markov chain model was created by A. A. Markov, a professor of mathematics at St. Petersburg University in Russia. He lived from 1856 to 1922, made significant contributions to the theory of probability, and also participated in protests against the Czarist government. In the United States, modern treatments of Markov chains were initiated by W. Feller of Princeton University in his 1950 book, *An Introduction to Probability Theory and its Applications*, and by J. G. Kemeny and J. L. Snell of Dartmouth College in their 1960 book, *Finite Markov Chains* [5]. This author follows the approach of Kemeny and Snell.

Many of the basic concepts of a Markov decision process were developed by mathematicians at the RAND Corporation in California in the late 1940s and early 1950s. The policy iteration algorithm used to determine an optimal set of decisions was created by R. A. Howard. His pioneering and eloquent 1960 book, *Dynamic Programming and Markov Processes*, based on his doctoral dissertation at M.I.T., stimulated great interest in Markov decision processes, and motivated this author's treatment of that subject.

1.2 States and Transitions

A random process, which is also called a stochastic process, is an indexed sequence of random variables. Assume that the sequence is observed at equally spaced points in time called epochs. The time interval between successive epochs is called a period or step. At each epoch, the system is observed to be in one of a finite number of mutually exclusive categories or states. A state is one of the possible values that the random variable can have. A Markov chain is the simplest kind of random process because it has the Markov property. The Markov property assumes the simplest kind of dependency, namely, that given the present state of a random process, the conditional probability of the next state depends only on the present state, and is independent of the past history of the process. This simplification has enabled engineers and managers to develop mathematically tractable models that can be used to analyze a variety of physical, economic, and social systems. A random process that lacks the Markov property is one for which knowledge of its past history is needed in order to probabilistically model its future behavior.

A Markov chain has N states designated $1, 2, \ldots, N$. All Markov chains treated in this book are assumed to have a finite number of states, so that N is a finite integer. The set of all possible states, called the state space, is denoted by $E = \{1, 2, \ldots, N\}$. The state of the chain is observed at equally spaced points in time called epochs. An epoch is denoted by $n = 0, 1, 2, \ldots$. Epoch n designates the end of time period n, which is also the beginning of period $n + 1$. A sequence of n consecutive time periods, each marked by an end-of-period epoch, is shown in Figure 1.1.

The random variable X_n represents the state of the chain at epoch n. A Markov chain is an indexed sequence of random variables, $\{X_0, X_1, X_2, \ldots\}$, which has the Markov property. The index, n, is the epoch at which the state of the random variable is observed. Thus, a Markov chain may be viewed as a sequence of states, which are observed at consecutive epochs, as shown in Figure 1.2.

While the index n most commonly represents an epoch or time of observation, it can also represent other parameters, such as the order of an observation. For example, the index n can indicate the nth item inspected, or the nth customer served, or the nth trial in a contest.

If at epoch n the chain is in state i, then $X_n = i$. The probability that the chain is in state i at epoch n, which may represent the present epoch, is denoted by $P(X_n = i)$. Then the probability that the chain is in state j at

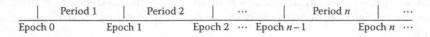

	Period 1		Period 2	$\cdots$		Period n	$\cdots$
Epoch 0		Epoch 1		Epoch 2 $\cdots$	Epoch $n-1$		Epoch n $\cdots$

FIGURE 1.1
Sequence of consecutive epochs.

X_0		X_1		X_2	$\cdots$		X_n	$\cdots$	State
	Period 1		Period 2		$\cdots$ Period n			$\cdots$	
0		1		2	$\cdots$		n	$\cdots$	Epoch

FIGURE 1.2
Sequence of states.

epoch $n+1$, the next epoch, is denoted by $P(X_{n+1} = j)$. The conditional probability that the chain will be in state j at epoch $n + 1$, given that it is in state i at epoch n, is denoted by $P(X_{n+1} = j | X_n = i)$. This conditional probability is called a transition probability, and is denoted by p_{ij}. Thus, the transition probability,

$$p_{ij} = P(X_{n+1} = j | X_n = i), \tag{1.1}$$

represents the conditional probability that if the chain is in state i at the current epoch, then it will be in state j at the next epoch. The transition probability, p_{ij}, represents a one-step transition because it is the probability of making a transition in one time period or step. The Markov property ensures that a transition probability depends only on the present state of the process, denoted by X_n. In other words, the history of the process, represented by the sequence of states, $\{X_0, X_1, X_2, \ldots, X_{n-1}\}$, occupied prior to the present epoch, can be ignored. Transition probabilities are assumed to be stationary over time, or time homogeneous. That is, they do not change over time. Therefore, the transition probability,

$$p_{ij} = P(X_{n+1} = j | X_n = i) = P(X_1 = j | X_0 = i), \tag{1.2}$$

is constant, independent of the epoch n.

The transition probabilities for a Markov chain with N states are collected in an $N \times N$ square matrix, called a one-step transition probability matrix, or more simply, a transition probability matrix, which is denoted by P. To avoid abstraction, and without loss of generality, consider a Markov chain with $N = 4$ states. The transition probability matrix is given by

$$
P =
\begin{array}{c|cccc}
X_n \backslash X_{n+1} & 1 & 2 & 3 & 4 \\
\hline
1 & p_{11} & p_{12} & p_{13} & p_{14} \\
2 & p_{21} & p_{22} & p_{23} & p_{24} \\
3 & p_{31} & p_{32} & p_{33} & p_{34} \\
4 & p_{41} & p_{42} & p_{43} & p_{44}
\end{array}. \tag{1.3}
$$

Each row of P represents the present state, at epoch n. Each column represents the next state, at epoch $n + 1$. Since the probability of a transition is conditioned on the present state, the entries in every row of P sum to one. In addition, all entries are nonnegative, and no entry is greater than one. The matrix P is called a stochastic matrix.

When the number of states in a Markov chain is small, the chain can be represented by a graph, which consists of nodes connected by directed arcs. In a transition probability graph, a node i denotes a state. An arc directed from node i to node j denotes a transition from state i to node j with transition probability p_{ij}. For example, consider a two-state generic Markov chain with the following transition probability matrix:

$$P = \begin{array}{c|cc} \text{State} & 1 & 2 \\ \hline 1 & p_{11} & p_{12} \\ 2 & p_{21} & p_{22} \end{array}. \tag{1.4}$$

The corresponding transition probability graph is shown in Figure 1.3.

A Markov chain evolves over time as it moves from state to state in accordance with its transition probabilities. Suppose the probability that a Markov chain is in a particular state after n transitions is of interest. Then, in addition to the one-step transition probabilities, the initial state probabilities are also needed. The initial state probability for a state j is the probability that at epoch 0 the process starts in state j. Let $p_j^{(0)} = P(X_0 = j)$ denote the initial state probability for state j. The initial state probabilities for all states in a Markov chain with four states are arranged in a row vector of initial state probabilities, designated by

$$\begin{aligned} p^{(0)} &= [p_1^{(0)} \quad p_2^{(0)} \quad p_3^{(0)} \quad p_4^{(0)}] \\ &= [P(X_0 = 1) \quad P(X_0 = 2) \quad P(X_0 = 3) \quad P(X_0 = 4)]. \end{aligned} \tag{1.5}$$

The initial state probability vector specifies the probability distribution of the initial state. The elements of every state probability vector, including the

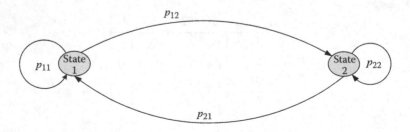

FIGURE 1.3
Transition probability graph for a two-state Markov chain.

$$X_0 = a \;\rightarrow\; X_1 = b \;\rightarrow\; X_2 = c \;\rightarrow\; X_3 = d \;\rightarrow\; X_4 = e \qquad \text{State}$$

$p_a^{(0)}$	p_{ab}		p_{bc}		p_{cd}		p_{de}		Transition probability
0		1		2		3		4	Epoch

FIGURE 1.4
State transitions for a sample path.

elements of an initial state probability vector, must sum to one. Hence,

$$p_1^{(0)} + p_2^{(0)} + p_3^{(0)} + p_4^{(0)} = 1. \tag{1.6}$$

A sequence of states visited by a Markov chain as it evolves over time is called a sample path. The state sequence $\{X_0, X_1, X_2, X_3, X_4\}$ is a sample path during the first four periods. State transitions for a sample path are shown in Figure 1.4.

The probability of a sample path is a joint probability, which can be calculated by invoking the Markov property. For example, the probability of the sample path is shown in Figure 1.4 is

$$\left.\begin{aligned}
&P(X_0 = a,\ X_1 = b,\ X_2 = c,\ X_3 = d,\ X_4 = e) \\
&= P(X_0 = a)P(X_1 = b | X_0 = a)\,P(X_2 = c | X_1 = b) \\
&\quad \times P(X_3 = d | X_2 = c)\,P(X_4 = e | X_3 = d) \\
&= p_a^{(0)} p_{ab} p_{bc} p_{cd} p_{de}.
\end{aligned}\right\} \tag{1.7}$$

If the initial state is known, so that $P(X_0 = a) = p_a^{(0)} = 1$, the probability of a particular sample path, given the initial state, is calculated below by using the Markov property once again

$$\left.\begin{aligned}
&P(X_1 = b, X_2 = c, X_3 = d, X_4 = e | X_0 = a) \\
&= P(X_1 = b | X_0 = a)P(X_2 = c | X_1 = b)\,P(X_3 = d | X_2 = c)\,P(X_4 = e | X_3 = d) \\
&= p_{ab} p_{bc} p_{cd} p_{de}.
\end{aligned}\right\} \tag{1.8}$$

1.3 Model of the Weather

As an example of a Markov chain model, suppose that weather in Cleveland, Ohio, can be classified as either raining, snowing, cloudy, or sunny [5]. Observations of the weather are made at the same time every day. The daily weather is assumed to have the Markov property, which means that the weather tomorrow depends only on the weather today. That is, the weather yesterday or on prior days will not affect the weather tomorrow. Since the daily weather is assumed to have the Markov property, the weather will be modeled as a Markov chain with four states. The state X_n denotes the weather on day n for $n = 0, 1, 2, 3, \dots$. The states are indexed below:

State, X_n	Description
1	Raining
2	Snowing
3	Cloudy
4	Sunny

The state space is $E = \{1, 2, 3, 4\}$.

   Transition probabilities are based on the following observations. If it is raining today, the probabilities that tomorrow will bring rain, snow, clouds, or sun are 0.3, 0.1, 0.4, and 0.2, respectively. If it is snowing today, the probabilities of rain, snow, clouds, or sun tomorrow are 0.2, 0.5, 0.2, and 0.1, respectively. If today is cloudy, the probabilities that rain, snow, clouds, or sun will appear tomorrow are 0.3, 0.2, 0.1, and 0.4, respectively. Finally, if today is sunny, the probabilities that tomorrow it will be snowing, cloudy, or sunny are 0.6, 0.3, and 0.1, respectively. (A sunny day is never followed by a rainy day.) Transition probabilities are obtained in the following manner. If day n designates today, then day $n + 1$ designates tomorrow. If X_n designates the state today, then X_{n+1} designates the state tomorrow. If it is raining today, then $X_n = 1$. If it is cloudy tomorrow, then $X_{n+1} = 3$. The observation that if it is raining today, then the conditional probability that tomorrow will be cloudy is 0.4, is expressed as the following transition probability:

$$P(\text{Cloudy tomorrow}|\text{Raining today}) = P(X_{n+1} = 3 | X_n = 1) = p_{13} = 0.4.$$

The 15 remaining transition probabilities for the four-state Markov chain are obtained in similar fashion to produce the following transition probability matrix:

$X_n \backslash X_{n+1}$	1	2	3	4	State	1	2	3	4
1	p_{11}	p_{12}	p_{13}	p_{14}	1	0.3	0.1	0.4	0.2
$P =$ 2	p_{21}	p_{22}	p_{23}	p_{24} $=$	2	0.2	0.5	0.2	0.1
3	p_{31}	p_{32}	p_{33}	p_{34}	3	0.3	0.2	0.1	0.4
4	p_{41}	p_{42}	p_{43}	p_{44}	4	0	0.6	0.3	0.1

$$(1.9)$$

Observe that the entries in every row of P sum to one.

Suppose that an initial state probability vector for the four-state Markov chain model of the weather is

$$
\left.
\begin{aligned}
p^{(0)} &= \begin{bmatrix} 0.2 & 0.3 & 0.4 & 0.1 \end{bmatrix} \\
&= \begin{bmatrix} p_1^{(0)} & p_2^{(0)} & p_3^{(0)} & p_4^{(0)} \end{bmatrix} \\
&= [P(X_0 = 1) \quad P(X_0 = 2) \quad P(X_0 = 3) \quad P(X_0 = 4)].
\end{aligned}
\right\}
\tag{1.10}
$$

Since $p_3^{(0)} = P(X_0 = 3) = 0.4$, the chain has a probability of 0.4 of starting in state 3, representing a cloudy day. This vector also indicates that the chain has a probability of 0.2 of starting in state 1, denoting a rainy day, a probability of 0.3 of starting in state 2, indicating a snowy day, and a 0.1 probability of starting in state 4, designating a sunny day. Note that the entries of $p^{(0)}$ sum to one.

Suppose the chain starts in state 3, indicating a cloudy day, so that $P(X_0 = 3) = p_3^{(0)} = 1$. The probability of a particular sample path during the next 5 days, given that the initial state is $X_0 = 3$, is calculated below:

$$
P(X_1 = 1, X_2 = 4, X_3 = 2, X_4 = 3, X_5 = 4 | X_0 = 3) = p_{31}p_{14}p_{42}p_{23}p_{34} = (0.3)(0.2)(0.6)(0.2)(0.4)
$$
$$
= 0.00288
$$

1.4 Random Walks

As a second example of a Markov chain model, suppose that a particle moves among positions on a line by one position at a time. The positions on the line are marked by consecutive integers. At each epoch, the particle moves one position to the right with probability p, or one position to the left with probability $1 - p$. The particle continues to move, either right or left, until it reaches one of two end positions called barriers. Since the movement of the particle depends only on its present position and not on any of its previous positions, the process can be modeled as a Markov chain. This chain is called a random walk because it may describe the unsteady movement of an intoxicated individual [5]. The states, represented by consecutive integers, are the possible positions of the particle on the line. When the particle is in position i, the random walk is in state i. Hence, the state $X_n = i$ denotes the position of the particle after n moves. Consider a five-state random walk shown in Figure 1.5 with the state space $E = \{0, 1, 2, 3, 4\}$.

FIGURE 1.5
Random walk with five states.

For a state i such that $0 < i < 4$, the probability of moving one position to the right is

$$p_{i,i+1} = P(X_{n+1} = i+1 \mid X_n = i) = p.$$

Similarly, the probability of moving one position to the left is

$$p_{i,i-1} = P(X_{n+1} = i-1 \mid X_n = i) = 1-p.$$

The remaining transition probabilities are zero. That is

$$p_{i,j} = 0 \quad \text{for } j \neq i-1 \text{ and for } j \neq i+1.$$

The transition probabilities for $i = 0$ and $i = 4$, the terminal states, depend on the nature of the barriers represented by these terminal states.

1.4.1 Barriers

Random walks may have various types of barriers. Three common types of barriers are absorbing, reflecting, and partially reflecting.

1.4.1.1 Absorbing Barriers

When a particle reaches an absorbing barrier, it remains there. Thus, an absorbing barrier is represented by a state, which is called an absorbing state. The transition probability matrix for a five-state random walk with two absorbing barriers, represented by absorbing states 0 and 4, is

$$P = [p_{ij}] = \begin{array}{c|ccccc} X_n \backslash X_{n+1} & 0 & 1 & 2 & 3 & 4 \\ \hline 0 & 1 & 0 & 0 & 0 & 0 \\ 1 & 1-p & 0 & p & 0 & 0 \\ 2 & 0 & 1-p & 0 & p & 0 \\ 3 & 0 & 0 & 1-p & 0 & p \\ 4 & 0 & 0 & 0 & 0 & 1 \end{array}. \quad (1.11)$$

As Section 1.8 indicates, when a state i is an absorbing state, $p_{ii} = 1$. A Markov chain cannot leave an absorbing state.

1.4.1.2 Gambler's Ruin

The gambler's ruin model is an interesting application of a random walk with two absorbing barriers [1, 3]. Suppose that a game between two gamblers, called player A and player B, is a sequence of independent bets or trials. Each player starts with $2. Both players bet $1 at a time. The loser pays $1 to the winner. They agree to continue betting until one player wins all the money, a total of $4. The gambler who loses all her money is ruined. To model this game as a random walk, let the state, X_n, denote the amount of money player A has after n trials. Since the total money is $4, the state space, in units of dollars, is $E = \{0, 1, 2, 3, 4\}$. Since player A starts with $2, $X_0 = 2$. If player A wins each $1 bet with probability p, and loses each $1 bet with probability $1 - p$, the five-state gambler's ruin model has the transition probability matrix given for the random walk in Section 1.4.1.1. States 0 and 4 are both absorbing states, which indicate that the game is over. If the game ends in state 0, then player A has lost all her initial money, $2. If the game ends in state 4, then player A has won all of her opponent's money, increasing her total to $4. The transition probability matrix is shown in Equation (1.11). This model of a gambler's ruin is treated in Section 3.3.3.2.1.

1.4.1.3 Reflecting Barriers

When a particle reaches a reflecting barrier, it enters the adjacent state on the next transition. The transition probability matrix for a five-state random walk with two reflecting barriers, represented by states 0 and 4, is

$$P = [p_{ij}] = \begin{array}{c|ccccc} X_n \backslash X_{n+1} & 0 & 1 & 2 & 3 & 4 \\ \hline 0 & 0 & 1 & 0 & 0 & 0 \\ 1 & 1-p & 0 & p & 0 & 0 \\ 2 & 0 & 1-p & 0 & p & 0 \\ 3 & 0 & 0 & 1-p & 0 & p \\ 4 & 0 & 0 & 0 & 1 & 0 \end{array}. \qquad (1.12)$$

1.4.2 Circular Random Walk

A circular five-state random walk represents a particle which moves on a circle through the state space $E = \{0, 1, 2, 3, 4\}$. The probability of moving clockwise to an adjacent state is p, and the probability of moving counterclockwise to an adjacent state is $1 - p$. Since states 0 and 4 are adjacent on the circle, a circular random walk, has no barriers. The transition probability matrix is

$$P = [p_{ij}] = \begin{array}{c|ccccc} X_n \backslash X_{n+1} & 0 & 1 & 2 & 3 & 4 \\ \hline 0 & 0 & p & 0 & 0 & 1-p \\ 1 & 1-p & 0 & p & 0 & 0 \\ 2 & 0 & 1-p & 0 & p & 0 \\ 3 & 0 & 0 & 1-p & 0 & p \\ 4 & p & 0 & 0 & 1-p & 0 \end{array}. \qquad (1.13)$$

1.5 Estimating Transition Probabilities

In developing a Markov chain model, an engineer must assume but cannot prove that a process possesses the Markov property. Transition probabilities can be estimated by counting the transitions among states over a long period of time [1]. A procedure for estimating transition probabilities is demonstrated by the following example of a sequential inspection process. Suppose that every hour a quality inspector either accepts a product, sends it back to be reworked, or rejects it. The inspector's goal is to construct a Markov chain model of product quality.

1.5.1 Conditioning on the Present State

As a first approximation, the inspector assumes that a prediction of the outcome of the next inspection depends solely on the outcome of the present inspection. By making this assumption, a Markov chain model can be constructed. The state, X_n, denotes the outcome of the nth inspection. Three outcomes are defined:

State, X_n	Outcome
A	Product is accepted
W	Product is sent back to be reworked
R	Product is reiected

The state space has three entries given by $E = \{A, W, R\}$. After 61 inspections, suppose that the following sequence of states has been observed: {A, A, A, W, A, R, A, A, A, W, W, A, A, R, A, A, A, W, R, A, A, W, A, W, A, A, R, A, A, A, R, W, A, R, A, A, A, W, W, A, A, A, A, R, A, A, W, A, A, R, A, A, R, W, A, A, R, A, A, W, A}.

A transition occurs when one inspection follows another. A transition probability, p_{ij}, is estimated by the proportion of inspections in which state i is followed by state j. Let n_{ij} denote the number of transitions from state i to state j. Let n_i denote the number of transitions made from state i. By counting transitions, Table 1.1 of transition counts is obtained.

Note that the 61st or last outcome is not counted in the sum of transitions because it is not followed by a transition to another outcome. The estimated transition probability is

$$p_{ij} = n_{ij} / n_i = P(X_{n+1} = j | X_n = i). \tag{1.14}$$

The matrix of estimated transition probabilities appears in Table 1.2.

TABLE 1.1

Transition Counts

State	A	W	R	Row Sum of Transitions
A	$n_{AA} = 21$	$n_{AW} = 8$	$n_{AR} = 9$	$n_A = 38$
W	$n_{WA} = 9$	$n_{WW} = 2$	$n_{WR} = 1$	$n_W = 12$
R	$n_{RA} = 8$	$n_{RW} = 2$	$n_{RR} = 0$	$n_R = 10$
		Sum of transitions		$\sum = 60$

TABLE 1.2

Matrix of Estimated Transition Probabilities

$X_n \backslash X_{n+1}$	A	W	R
A	$p_{AA} = 21/38$	$p_{AW} = 8/38$	$p_{AR} = 9/38$
W	$p_{WA} = 9/12$	$p_{WW} = 2/12$	$p_{WR} = 1/12$
R	$p_{RA} = 8/10$	$p_{RW} = 2/10$	$p_{RR} = 0/10$

$P = $ (applies to the rows A, W, R above)

1.5.2 Conditioning on the Present and Previous States

Now suppose that knowledge of the outcome of the present inspection is no longer considered sufficient to predict the outcome of the next inspection. Instead, this prediction will be based on the outcomes of both the present inspection and the previous inspection. If the current definition of a state as the outcome of the most recent inspection is retained, the process is no longer a Markov chain. However, the process will become a Markov chain if a new state is defined as the pair of outcomes of the two most recent inspections.

To obtain a better approximation, the model has been enlarged, which makes it less tractable computationally. Since the original model has three states, and all pairs of consecutive outcomes may be possible, the new model has $(3)^2 = 9$ states. In the enlarged model, the state, $Y_n = X_{n-1}, X_n$, denotes the pair of outcomes of the $(n-1)$th and nth inspections. The state space has nine entries given by $E = \{AA, AW, AR, WA, WW, WR, RA, RW, RR\}$. For example, states WA and RW for the enlarged process are defined below:

WA: the previous product was sent back to be reworked, and the present product was accepted,

RW: the previous product was rejected, and the present product was sent back to be reworked.

TABLE 1.3

Transition Counts for the Enlarged Process

State	AA	AW	AR	WA	WW	WR	RA	RW	RR	Row Sum
AA	7	7	7	0	0	0	0	0	0	21
AW	0	0	0	5	2	1	0	0	0	8
AR	0	0	0	0	0	0	7	2	0	9
WA	5	1	2	0	0	0	0	0	0	8
WW	0	0	0	2	0	0	0	0	0	2
WR	0	0	0	0	0	0	1	0	0	1
RA	8	0	0	0	0	0	0	0	0	8
RW	0	0	0	2	0	0	0	0	0	2
RR	0	0	0	0	0	0	0	0	0	0
										$\Sigma = 59$

The 61 outcomes for the sequence of inspections indicate that the process has moved from state AA to AA to AW to WA to AR to RA to AA to AA to AW to WW to WA to AA to AR to RA to AA to AA to AW to WR to RA to AA to AW to WA to AW to WA to AA to AR to RA to AA to AA to AR to RW to WA to AR to RA to AA to AA to AW to WW to WA to AA to AA to AA to AR to RA to AA to AW to WA to AA to AR to RA to AA to AR to RW to WA to AA to AR to RA to AA to AW, and lastly to WA. Observe that the first letter of the state to which the process moves must agree with the second letter of the state from which it moves, since they both refer to the outcome of the same inspection. Since the new model has nine states, the new transition matrix has $(9)^2 = 81$ entries. Table 1.3 of transition counts is obtained for the enlarged process.

Note that unless additional products are inspected, state RR is not a possible outcome. Hence, the number of states can be reduced by one. A transition probability, $p_{ij,jk}$, is estimated by the proportion of inspections in which the pair of outcomes i, j is followed by the pair of outcomes j, k. The estimated transition probability is

$$p_{ij,jk} = P(X_n = j, X_{n+1} = k | X_{n-1} = i, X_n = j). \tag{1.15}$$

The matrix of estimated transition probabilities for the enlarged process appears in Table 1.4.

1.6 Multiple-Step Transition Probabilities

As Section 1.2 indicates, a one-step transition probability matrix for a Markov chain is denoted by P. Suppose that transition probabilities after more than

TABLE 1.4

Matrix of Estimated Transition Probabilities for the Enlarged Process

$X_{n-1}=i, X_n=j \backslash X_n=j,$ $X_{n+1}=k$		AA	AW	AR	WA	WW	WR	RA	RW
	AA	7/21	7/21	7/21	0	0	0	0	0
	AW	0	0	0	5/8	2/8	1/8	0	0
	AR	0	0	0	0	0	0	7/9	2/9
$P=$	WA	5/8	1/8	2/8	0	0	0	0	0
	WW	0	0	0	2/2	0	0	0	0
	WR	0	0	0	0	0	0	1/1	0
	RA	8/8	0	0	0	0	0	0	0
	RW	0	0	0	2/2	0	0	0	0
	RA	8/8	0	0	0	0	0	0	0
	RW	0	0	0	2/2	0	0	0	0

one epoch are of interest. Initially, suppose that transition probabilities at every second epoch are desired. Such transition probabilities are called two-step transition probabilities. They are denoted by $p_{ij}^{(2)}$. Since transition probabilities are stationary in time,

$$p_{ij}^{(2)} = P(X_{n+2} = j \mid X_n = i) = P(X_2 = j \mid X_0 = i) \tag{1.16}$$

is the conditional probability of moving from state i to state j in two steps. The two-step transition probabilities are arranged in a matrix designated by $P^{(2)} = [p_{ij}^{(2)}]$. To see how to calculate two-step transition probabilities, consider the one-step transition probability matrix shown in Equation (1.9) for the four-state Markov chain model of the weather.

The conditional probability of going from state i to state j in two steps is equal to the probability of going from state i to an intermediate state k on the first step multiplied by the probability of going from state k to state j on the second step, summed over all states k. The state k can be any state of the process, including i and j. Therefore, $k = 1, 2, 3,$ and 4, and

$$p_{ij}^{(2)} = p_{i1}p_{1j} + p_{i2}p_{2j} + p_{i3}p_{3j} + p_{i4}p_{4j} = \sum_{k=1}^{4} p_{ik}p_{kj}. \tag{1.17}$$

By the rules of matrix multiplication, $p_{ij}^{(2)}$, the (i, j)th element of the two-step transition matrix, $P^{(2)}$, is also the (i, j)th element of the product matrix, P^2, obtained by multiplying the matrix of one-step transition probabilities by itself. That is,

$$P^{(2)} = P \cdot P = P^2. \tag{1.18}$$

For the model of the weather,

$$P^{(2)} = P \cdot P = P^2 = \begin{array}{c} 1 \\ 2 \\ 3 \\ 4 \end{array}\begin{bmatrix} 0.3 & 0.1 & 0.4 & 0.2 \\ 0.2 & 0.5 & 0.2 & 0.1 \\ 0.3 & 0.2 & 0.1 & 0.4 \\ 0 & 0.6 & 0.3 & 0.1 \end{bmatrix}\begin{bmatrix} 0.3 & 0.1 & 0.4 & 0.2 \\ 0.2 & 0.5 & 0.2 & 0.1 \\ 0.3 & 0.2 & 0.1 & 0.4 \\ 0 & 0.6 & 0.3 & 0.1 \end{bmatrix}$$

$$= \begin{array}{c} 1 \\ 2 \\ 3 \\ 4 \end{array}\begin{bmatrix} 0.23 & 0.28 & 0.24 & 0.25 \\ 0.22 & 0.37 & 0.23 & 0.18 \\ 0.16 & 0.39 & 0.29 & 0.16 \\ 0.21 & 0.42 & 0.18 & 0.19 \end{bmatrix}.$$

(1.19)

Observe that

$$p_{34}^{(2)} = p_{31}p_{14} + p_{32}p_{24} + p_{33}p_{34} + p_{34}p_{44} = \sum_{k=1}^{4} p_{3k}p_{k4}$$

$$= (0.3)(0.2) + (0.2)(0.1) + (0.1)(0.4) + (0.4)(0.1) = 0.16,$$

which is element $(3, 4)$ of P^2. Thus,

$$p_{34}^{(2)} = P(X_2 = 4 \mid X_0 = 3) = P(X_{n+2} = 4 \mid X_n = 3) = 0.16$$

$$= P(\text{Sunny 2 days from now} \mid \text{Cloudy today}).$$

An n-step transition probability is denoted by

$$p_{ij}^{(n)} = P(X_n = j \mid X_0 = i) = P(X_{r+n} = j \mid X_r = i).$$

(1.20)

The result for a two-step transition matrix can be generalized to show that for an N-state Markov chain, a matrix of n-step transition probabilities is obtained by raising the matrix of one-step transition probabilities to the nth power. That is,

$$P^{(n)} = P^n = PP^{n-1} = P^{n-1}P.$$

(1.21)

The matrix of $(n + m)$-step transition probabilities is equal to

$$P^{(n+m)} = P^{n+m} = P^n P^m = P^{(n)} P^{(m)}.$$

(1.22)

By the rules of matrix multiplication, for a Markov chain with N states, element (i, j) of $P^{(n+m)}$ is given by

$$p_{ij}^{(n+m)} = \sum_{k=1}^{N} p_{ik}^{(n)} p_{kj}^{(m)}, \quad \text{for } n = 0, 1, 2, \ldots; \ m = 0, 1, 2, \ldots;$$

$$i = 1, 2, \ldots, N; \ j = 1, 2, 3, \ldots, N. \tag{1.23}$$

These expressions are called the Chapman–Kolmogorov equations. They can be justified informally by the following algebraic argument:

$$p_{ij}^{(n+m)} = \sum_{i=1}^{N} P(X_{n+m} = j \mid X_0 = i)$$

$$= \sum_{i=1}^{N} P(X_{n+m} = j, \ X_n = k \mid X_0 = i)$$

$$= \sum_{i=1}^{N} P(X_{n+m} = j, \ X_n = k, \ X_0 = i) / P(X_0 = i)$$

$$= \sum_{i=1}^{N} P(X_{n+m} = j \mid X_n = k, X_0 = i) P(X_n = k, X_0 = i) / P(X_0 = i)$$

$$= \sum_{i=1}^{N} P(X_{n+m} = j \mid X_n = k, X_0 = i) P(X_n = k \mid X_0 = i) P(X_0 = i) / P(X_0 = i)$$

$$= \sum_{i=1}^{N} P(X_{n+m} = j \mid X_n = k) P(X_n = k \mid X_0 = i)$$

$$= \sum_{i=1}^{N} P(X_n = k \mid X_0 = i) P(X_{n+m} = j \mid X_n = k)$$

$$= \sum_{i=1}^{N} p_{ik}^{(n)} p_{kj}^{(m)}.$$

1.7 State Probabilities after Multiple Steps

An n-step transition probability, $p_{ij}^{(n)} = P(X_n = j \mid X_0 = i)$, is a conditional probability, conditioned on the process starting in a given state i. In contrast, an unconditional probability that the process is in state j after n steps, denoted by $P(X_n = j)$, requires a probability distribution for the starting state. This unconditional probability is called a state probability, and is also denoted by $p_j^{(n)}$. Thus,

$$p_j^{(n)} = P(X_n = j). \tag{1.24}$$

The state probabilities for an N-state Markov chain after n steps are collected in a $1 \times N$ row vector, $p^{(n)}$, termed a vector of state probabilities. The vector of state probabilities specifies the probability distribution of the state after n steps. For a four-state Markov chain, the vector of state probabilities is represented by

$$
\begin{aligned}
p^{(n)} &= [p_1^{(n)} \quad p_2^{(n)} \quad p_3^{(n)} \quad p_4^{(n)}] \\
&= [P(X_n = 1) \quad P(X_n = 2) \quad P(X_n = 3) \quad P(X_n = 4)].
\end{aligned}
\quad\quad\text{(1.25)}
$$

The vector of state probabilities after n transitions, $p^{(n)}$, can be determined only if the matrix of one-step transition probabilities, P, is accompanied by a vector of initial state probabilities. The vector of initial state probabilities, introduced in Section 1.2, is a special case of the vector of state probabilities, when $n = 0$. As Equation (1.5) indicates, the initial state probability vector specifies the probability distribution of the initial state.

A state probability can be calculated by multiplying the probability of starting in a particular state by an n-step transition probability, and summing over all states. For a four-state Markov chain,

$$
P(X_n = j) = \sum_{i=1}^{4} P(X_0 = i)\, P(X_n = j \mid X_0 = i), \text{ or } \quad p_j^{(n)} = \sum_{i=1}^{4} p_i^{(0)} p_{ij}^{(n)}. \quad\text{(1.26)}
$$

In matrix form,

$$
p^{(n)} = p^{(0)} P^{(n)}. \quad\quad\text{(1.27)}
$$

Note that

$$
\begin{aligned}
p^{(1)} &= p^{(0)} P^{(1)} = p^{(0)} P \\
p^{(2)} &= p^{(0)} P^{(2)} = p^{(0)}(PP) = (p^{(0)}P)P = p^{(1)}P \\
p^{(3)} &= p^{(0)} P^{(3)} = p^{(0)}(P^{(2)}P) = (p^{(0)}P^{(2)})P = p^{(2)}P.
\end{aligned}
\quad\quad\text{(1.28)}
$$

By induction,

$$
p^{(n)} = p^{(0)} P^{(n)} = p^{(n-1)} P.
$$

Thus, the vector of state probabilities is equal to the initial state probability vector post multiplied by the matrix of n-step transition probabilities, and is also equal to the vector of state probabilities after $n - 1$ transitions post multiplied by the matrix of one-step transition probabilities.

Consider the four-state Markov chain model of the weather for which the transition probability matrix is given in Equation (1.9). Suppose that the initial state probability vector is given by Equation (1.10). After 1 day, the vector of state probabilities is

$$p^{(1)} = p^{(0)}P = \begin{bmatrix} 0.2 & 0.3 & 0.4 & 0.1 \end{bmatrix} \begin{bmatrix} 0.3 & 0.1 & 0.4 & 0.2 \\ 0.2 & 0.5 & 0.2 & 0.1 \\ 0.3 & 0.2 & 0.1 & 0.4 \\ 0 & 0.6 & 0.3 & 0.1 \end{bmatrix}$$

$$= \begin{bmatrix} 0.24 & 0.31 & 0.21 & 0.24 \end{bmatrix}. \tag{1.29}$$

Thus, after 1 day, the process has a probability of 0.31 of being in state 2, and a 0.24 probability of being in state 4. The matrix of two-step transition probabilities is given by

$$P^{(2)} = P^2 = \begin{matrix} 1 \\ 2 \\ 3 \\ 4 \end{matrix} \begin{bmatrix} 0.3 & 0.1 & 0.4 & 0.2 \\ 0.2 & 0.5 & 0.2 & 0.1 \\ 0.3 & 0.2 & 0.1 & 0.4 \\ 0 & 0.6 & 0.3 & 0.1 \end{bmatrix} \begin{bmatrix} 0.3 & 0.1 & 0.4 & 0.2 \\ 0.2 & 0.5 & 0.2 & 0.1 \\ 0.3 & 0.2 & 0.1 & 0.4 \\ 0 & 0.6 & 0.3 & 0.1 \end{bmatrix}$$

$$= \begin{bmatrix} 0.23 & 0.28 & 0.24 & 0.25 \\ 0.22 & 0.37 & 0.23 & 0.18 \\ 0.16 & 0.39 & 0.29 & 0.16 \\ 0.21 & 0.42 & 0.18 & 0.19 \end{bmatrix}. \tag{1.30}$$

Thus, the probability of moving from state 1 to state 4 in 2 days is 0.25. After 2 days the vector of state probabilities is equal to

$$p^{(2)} = p^{(0)}P^{(2)} = \begin{bmatrix} 0.2 & 0.3 & 0.4 & 0.1 \end{bmatrix} \begin{bmatrix} 0.23 & 0.28 & 0.24 & 0.25 \\ 0.22 & 0.37 & 0.23 & 0.18 \\ 0.16 & 0.39 & 0.29 & 0.16 \\ 0.21 & 0.42 & 0.18 & 0.19 \end{bmatrix}$$

$$= \begin{bmatrix} 0.197 & 0.365 & 0.251 & 0.187 \end{bmatrix}. \tag{1.31}$$

Alternatively, the vector of state probabilities after 2 days is equal to

$$p^{(2)} = p^{(1)}P = \begin{bmatrix} 0.24 & 0.31 & 0.21 & 0.24 \end{bmatrix} \begin{bmatrix} 0.3 & 0.1 & 0.4 & 0.2 \\ 0.2 & 0.5 & 0.2 & 0.1 \\ 0.3 & 0.2 & 0.1 & 0.4 \\ 0 & 0.6 & 0.3 & 0.1 \end{bmatrix} \tag{1.32}$$

$$= \begin{bmatrix} 0.197 & 0.365 & 0.251 & 0.187 \end{bmatrix}.$$

After 2 days the probability of being in state 2 is 0.365. Finally, the matrix of three-step transition probabilities is given by

$$P^{(3)} = P^3 = P^{(2)}P = \begin{matrix} 1 \\ 2 \\ 3 \\ 4 \end{matrix} \begin{bmatrix} 0.23 & 0.28 & 0.24 & 0.25 \\ 0.22 & 0.37 & 0.23 & 0.18 \\ 0.16 & 0.39 & 0.29 & 0.16 \\ 0.21 & 0.42 & 0.18 & 0.19 \end{bmatrix} \begin{bmatrix} 0.3 & 0.1 & 0.4 & 0.2 \\ 0.2 & 0.5 & 0.2 & 0.1 \\ 0.3 & 0.2 & 0.1 & 0.4 \\ 0 & 0.6 & 0.3 & 0.1 \end{bmatrix}$$

$$= \begin{bmatrix} 0.197 & 0.361 & 0.247 & 0.195 \\ 0.209 & 0.361 & 0.239 & 0.191 \\ 0.213 & 0.365 & 0.219 & 0.203 \\ 0.201 & 0.381 & 0.243 & 0.175 \end{bmatrix}. \tag{1.33}$$

Thus, the probability of moving from state 1 to state 4 in 3 days is 0.195. After 3 days the vector of state probabilities is equal to

$$p^{(3)} = p^{(0)}P^{(3)} = \begin{bmatrix} 0.2 & 0.3 & 0.4 & 0.1 \end{bmatrix} \begin{bmatrix} 0.197 & 0.361 & 0.247 & 0.195 \\ 0.209 & 0.361 & 0.239 & 0.191 \\ 0.213 & 0.365 & 0.219 & 0.203 \\ 0.201 & 0.381 & 0.243 & 0.175 \end{bmatrix} \tag{1.34}$$

$$= \begin{bmatrix} 0.2074 & 0.3646 & 0.2330 & 0.1950 \end{bmatrix}.$$

Alternatively, the vector of state probabilities after 3 days is equal to

$$p^{(3)} = p^{(2)}P = \begin{bmatrix} 0.197 & 0.365 & 0.251 & 0.187 \end{bmatrix} \begin{bmatrix} 0.3 & 0.1 & 0.4 & 0.2 \\ 0.2 & 0.5 & 0.2 & 0.1 \\ 0.3 & 0.2 & 0.1 & 0.4 \\ 0 & 0.6 & 0.3 & 0.1 \end{bmatrix} \tag{1.35}$$

$$= \begin{bmatrix} 0.2074 & 0.3646 & 0.2330 & 0.1950 \end{bmatrix}.$$

After 3 days the probability of being in state 2 is 0.3646.

Now suppose that the process starts day 1 in state 2, denoting snow. The initial state probability vector is

$$p^{(0)} = \begin{bmatrix} 0 & 1 & 0 & 0 \end{bmatrix}. \tag{1.36}$$

After 3 days the vector of state probabilities is equal to

$$p^{(3)} = p^{(0)}P^{(3)} = \begin{bmatrix} 0 & 1 & 0 & 0 \end{bmatrix} \begin{bmatrix} 0.197 & 0.361 & 0.247 & 0.195 \\ 0.209 & 0.361 & 0.239 & 0.191 \\ 0.213 & 0.365 & 0.219 & 0.203 \\ 0.201 & 0.381 & 0.243 & 0.175 \end{bmatrix} \tag{1.37}$$

$$= \begin{bmatrix} 0.209 & 0.361 & 0.239 & 0.191 \end{bmatrix}.$$

Observe that when $p^{(0)} = [0\ 1\ 0\ 0]$, the entries of vector $p^{(3)}$ are identical to the entries in row 2 of matrix $p^{(3)}$. One can generalize this observation to conclude that when a Markov chain starts in state i, then after n steps, the entries of vector $p^{(n)}$ are identical to those in row i of matrix $P^{(n)}$.

1.8 Classification of States

A state j is said to be accessible from a state i if $p_{ij}^{(n)} > 0$ for some $n \geq 0$. That is, state j is accessible from state i if starting from state i, it is possible to eventually reach state j. If state j is accessible from state i, and if state i is also accessible from state j, then both states are said to communicate. A set of states is called a closed set if no state outside the set is accessible from a state inside the set. When the process enters a closed set of states, it will remain forever inside the closed set. If each pair of states inside a closed set communicates, then the closed set is also called a closed communicating class. If all states in a Markov chain communicate, then all states belong to a single closed communicating class. Such a Markov chain is called irreducible. Thus, an irreducible chain is one in which it is possible to go from every state to every other state, not necessarily in one step. That is, all states in an irreducible chain communicate.

Consider the four-state Markov chain model of the weather for which the transition probability matrix is given in Equation (1.9). All states belong to the same closed communicating class denoted by $R = \{1, 2, 3, 4\}$. Since $p_{41} = 0$, the chain cannot move from state 4 to state 1 in one step. Instead, the chain can move from state 4 to state 2 in one step, or from state 4 to state 3 in one step, and then enter state 1 from either of those states after one or more additional steps. The chain can also move from state 4 to state 4 in one step, and then enter state 1 after two or more additional steps. Note that the two-step transition probability for going from state 4 to state 1 is positive. That is,

$$p_{41}^{(2)} = p_{41}p_{11} + p_{42}p_{21} + p_{43}p_{31} + p_{44}p_{41}$$
$$= (0)(0.3) + (0.6)(0.2) + (0.3)(0.3) + (0.1)(0) = 0.21 > 0,$$

so that the chain can move from state 4 to state 1 in two or more steps.

Two different types of states can be distinguished: recurrent states and transient states. A recurrent state is one to which eventual return is certain. A transient state is one to which the process may not eventually return. To quantify this distinction, let f_{ii} denote the probability that, starting in state i, the chain will ever return to state i. State i is recurrent if $f_{ii} = 1$, and transient if $f_{ii} < 1$. An absorbing state is a special case of a recurrent state i for which $p_{ii} = f_{ii} = 1$. A Markov chain that enters an absorbing state will never leave it because the chain will always return to it on every transition. A chain that enters an absorbing state is said to be absorbed in that state.

In a finite-state Markov chain, any state that belongs to a closed communicating class of states is recurrent. Any state that does not belong to a closed communicating class is transient. When the state space is finite, not all states can be transient. The reason is that eventually the process will never return to a transient state. Since the process must always be in one state, at least one state must be recurrent. All the states in a finite-state, irreducible Markov chain are recurrent. For that reason an irreducible chain is also called a recurrent chain.

A recurrent state i is called periodic, with period $d > 1$, if the process will always return to state i after a multiple of d transitions, with d an integer. If any state in a closed communicating class of states is periodic, with period d, then all states in the closed communicating class are periodic, with the same period d. For example, the random walk with reflecting barriers for which the transition matrix is shown in Equation (1.12) is an irreducible Markov chain because it is possible to go from every state to every other state. Observe that every transition from an even numbered state (0, 2, 4) is to an odd numbered state (1, 3), and every transition from an odd numbered state (1, 3) is to an even numbered state (0, 2, 4). Starting in state 0, the chain will alternately be in odd and even numbered states. That is, starting in state 0, the chain can enter an even numbered state i only at epochs 0, 2, 4, ... The chain is therefore periodic, with period $d = 2$, because it is possible to return to a state only in an even number of steps. A state that is not periodic is called aperiodic. With the exception of Sections 1.4.1.3, 1.9.1.1, and 1.10.1.1.1.2, only aperiodic chains are considered in this book.

1.9 Markov Chain Structure

A Markov chain can be classified as unichain or multichain. A unichain or multichain that has transient states is called reducible. Examples of reducible unichains and multichains are given in Sections 1.10.1.2, 1.10.2, and 3.1. All Markov chains treated in this book have a finite number of states.

1.9.1 Unichain

A Markov chain is termed unichain if it consists of a single closed set of recurrent states plus a possibly empty set of transient states.

1.9.1.1 *Irreducible*

A unichain with no transient states is called an irreducible or recurrent chain. All states in an irreducible chain are recurrent states, which belong to a single closed communicating class. In an irreducible chain, it is possible to move from every state to every other state, not necessarily in one step [4].

An irreducible or recurrent Markov chain is called a regular chain if some power of the transition matrix has only positive elements. The irreducible four-state Markov chain model of weather in Section 1.3 is a regular chain. Another example of a regular Markov chain is the model for diffusion of two gases treated in Section 1.10.1.1.1.1 for which

$$P = \begin{array}{c|cccc} & 0 & 1 & 2 & 3 \\ \hline 0 & 0 & 1 & 0 & 0 \\ 1 & \dfrac{1}{9} & \dfrac{4}{9} & \dfrac{4}{9} & 0 \\ 2 & 0 & \dfrac{4}{9} & \dfrac{4}{9} & \dfrac{1}{9} \\ 3 & 0 & 0 & 1 & 0 \end{array}. \tag{1.38}$$

The easiest way to check regularity is to keep track of whether the entries in the powers of P are positive. This can be done without computing numerical values by putting an x in the entry if it is positive and a 0 otherwise. To check regularity, let

$$P = \begin{array}{c} 0 \\ 1 \\ 2 \\ 3 \end{array}\begin{bmatrix} 0 & x & 0 & 0 \\ x & x & x & 0 \\ 0 & x & x & x \\ 0 & 0 & x & 0 \end{bmatrix}$$

$$P^2 = PP = \begin{array}{c} 0 \\ 1 \\ 2 \\ 3 \end{array}\begin{bmatrix} 0 & x & 0 & 0 \\ x & x & x & 0 \\ 0 & x & x & x \\ 0 & 0 & x & 0 \end{bmatrix}\begin{bmatrix} 0 & x & 0 & 0 \\ x & x & x & 0 \\ 0 & x & x & x \\ 0 & 0 & x & 0 \end{bmatrix} = \begin{bmatrix} x & x & x & 0 \\ x & x & x & x \\ x & x & x & x \\ 0 & x & x & x \end{bmatrix}$$

$$P^4 = P^2 P^2 = \begin{array}{c} 0 \\ 1 \\ 2 \\ 3 \end{array}\begin{bmatrix} x & x & x & 0 \\ x & x & x & x \\ x & x & x & x \\ 0 & x & x & x \end{bmatrix}\begin{bmatrix} x & x & x & 0 \\ x & x & x & x \\ x & x & x & x \\ 0 & x & x & x \end{bmatrix} = \begin{bmatrix} x & x & x & x \\ x & x & x & x \\ x & x & x & x \\ x & x & x & x \end{bmatrix}.$$

Since all entries in P^4 are positive, the chain is regular. Note that the test for regularity is made faster by squaring the result each time.

The irreducible Markov chain constructed in Section 1.10.1.1.1.2 for the Ehrenfest model of diffusion for which

$$
P = \begin{array}{c|cccc}
 & 0 & 1 & 2 & 3 \\
\hline
0 & 0 & 1 & 0 & 0 \\
1 & \dfrac{1}{3} & 0 & \dfrac{2}{3} & 0 \\
2 & 0 & \dfrac{2}{3} & 0 & \dfrac{1}{3} \\
3 & 0 & 0 & 1 & 0
\end{array} . \tag{1.39}
$$

can be interpreted as a four-state random walk with reflecting barriers introduced in Section 1.4.1.3 in which $p = 2/3$. Since the random walk with reflecting barriers is a periodic chain, the Ehrenfest model is also a periodic chain. To check regularity, let

$$
P = \begin{array}{c}
0 \\ 1 \\ 2 \\ 3
\end{array}
\begin{bmatrix}
0 & x & 0 & 0 \\
x & 0 & x & 0 \\
0 & x & 0 & x \\
0 & 0 & x & 0
\end{bmatrix}
$$

$$
P^2 = PP = \begin{array}{c}
0 \\ 1 \\ 2 \\ 3
\end{array}
\begin{bmatrix}
0 & x & 0 & 0 \\
x & 0 & x & 0 \\
0 & x & 0 & x \\
0 & 0 & x & 0
\end{bmatrix}
\begin{bmatrix}
0 & x & 0 & 0 \\
x & 0 & x & 0 \\
0 & x & 0 & x \\
0 & 0 & x & 0
\end{bmatrix}
=
\begin{bmatrix}
x & 0 & x & 0 \\
0 & x & 0 & x \\
x & 0 & x & 0 \\
0 & x & 0 & x
\end{bmatrix}
$$

$$
P^4 = P^2 P^2 = \begin{array}{c}
0 \\ 1 \\ 2 \\ 3
\end{array}
\begin{bmatrix}
x & 0 & x & 0 \\
0 & x & 0 & x \\
x & 0 & x & 0 \\
0 & x & 0 & x
\end{bmatrix}
\begin{bmatrix}
x & 0 & x & 0 \\
0 & x & 0 & x \\
x & 0 & x & 0 \\
0 & x & 0 & x
\end{bmatrix}
=
\begin{bmatrix}
x & 0 & x & 0 \\
0 & x & 0 & x \\
x & 0 & x & 0 \\
0 & x & 0 & x
\end{bmatrix} .
$$

Observe that even powers of P will have 0s in the odd numbered entries of row 0. Furthermore,

$$P^3 = P^2 P = \begin{matrix} 0 \\ 1 \\ 2 \\ 3 \end{matrix} \begin{bmatrix} x & 0 & x & 0 \\ 0 & x & 0 & x \\ x & 0 & x & 0 \\ 0 & x & 0 & x \end{bmatrix} \begin{bmatrix} 0 & x & 0 & 0 \\ x & 0 & x & 0 \\ 0 & x & 0 & x \\ 0 & 0 & x & 0 \end{bmatrix} = \begin{bmatrix} 0 & x & 0 & x \\ x & 0 & x & 0 \\ 0 & x & 0 & x \\ x & 0 & x & 0 \end{bmatrix}$$

$$P^5 = P^2 P^3 = \begin{matrix} 0 \\ 1 \\ 2 \\ 3 \end{matrix} \begin{bmatrix} x & 0 & x & 0 \\ 0 & x & 0 & x \\ x & 0 & x & 0 \\ 0 & x & 0 & x \end{bmatrix} \begin{bmatrix} 0 & x & 0 & x \\ x & 0 & x & 0 \\ 0 & x & 0 & x \\ x & 0 & x & 0 \end{bmatrix} = \begin{bmatrix} 0 & x & 0 & x \\ x & 0 & x & 0 \\ 0 & x & 0 & x \\ x & 0 & x & 0 \end{bmatrix}.$$

Note that odd powers of P will have 0s in the even numbered entries of row 0. This chain is not regular because no power of the transition matrix has only positive elements. This example has demonstrated that a periodic chain cannot be regular. Hence, a regular Markov chain is irreducible and aperiodic. Regular Markov chains are the subject of Chapter 2.

1.9.1.2 Reducible Unichain

A unichain with transient states is called a reducible unichain. Thus, a reducible unichain consists of one recurrent chain plus one or more transient states. Transient states are those states that do not belong to the recurrent chain. The state space for a reducible unichain can be partitioned into a recurrent chain plus one or more transient states. For brevity, a reducible unichain is often called a unichain.

The transition probability matrix for a generic four-state reducible unichain is shown below. The transition matrix is partitioned to show that states 1 and 2 belong to the recurrent chain denoted by $R = \{1, 2\}$, and states 3 and 4 are transient. The set of transient states is denoted by $T = \{3, 4\}$.

$$P = [p_{ij}] = \begin{matrix} 1 \\ 2 \\ 3 \\ 4 \end{matrix} \begin{bmatrix} p_{11} & p_{12} & 0 & 0 \\ p_{21} & p_{22} & 0 & 0 \\ p_{31} & p_{32} & p_{33} & p_{34} \\ p_{41} & p_{42} & p_{43} & p_{44} \end{bmatrix}. \tag{1.40}$$

If the recurrent chain consists of a single absorbing state, the reducible unichain is called an absorbing unichain or an absorbing Markov chain. The transition probability matrix for a generic three-state absorbing unichain is shown below. The transition matrix is partitioned to show that state 1 is absorbing, and states 2 and 3 are transient. The recurrent chain is denoted by $R = \{1\}$. The transient set of states is denoted by $T = \{2, 3\}$.

$$P = [p_{ij}] = \begin{matrix} 1 \\ 2 \\ 3 \end{matrix} \begin{bmatrix} 1 & 0 & 0 \\ p_{21} & p_{22} & p_{23} \\ p_{31} & p_{32} & p_{33} \end{bmatrix}. \qquad (1.41)$$

1.9.2 Multichain

A Markov chain is termed multichain if it consists of two or more closed sets of recurrent states plus a possibly empty set of transient states. Transient states are those states that do not belong to any of the recurrent closed classes. A multichain with transient states is called a reducible multichain. The state space for a reducible multichain can be partitioned into two or more mutually exclusive closed communicating classes of recurrent states plus one or more transient states. The mutually exclusive closed sets of recurrent states are called recurrent chains. There is no interaction among the recurrent chains. Hence, each recurrent chain, which may consist of only a single absorbing state, may be analyzed separately by treating it as an irreducible Markov chain. If every recurrent chain consists solely of one absorbing state, then the reducible multichain is called an absorbing multichain or an absorbing Markov chain. For brevity, a reducible multichain is often called a multichain. A multichain with no transient states is called a recurrent multichain.

The transition probability matrix for a generic five-state reducible multichain is shown below. The transition matrix is partitioned to show that the chain has two recurrent chains plus two transient states.

$$P = \begin{matrix} 1 \\ 2 \\ 3 \\ 4 \\ 5 \end{matrix} \begin{bmatrix} 1 & 0 & 0 & 0 & 0 \\ 0 & p_{22} & p_{23} & 0 & 0 \\ 0 & p_{32} & p_{33} & 0 & 0 \\ p_{41} & p_{42} & p_{43} & p_{44} & p_{45} \\ p_{51} & p_{52} & p_{53} & p_{54} & p_{55} \end{bmatrix}. \qquad (1.42)$$

State 1 is an absorbing state, and constitutes the first recurrent chain, denoted by $R_1 = \{1\}$. Recurrent states 2 and 3 belong to the second recurrent chain, denoted by $R_2 = \{2, 3\}$. States 4 and 5 are transient. The set of transient states is denoted by $T = \{4, 5\}$.

1.9.3 Aggregated Canonical Form of the Transition Matrix

Suppose that the states in a reducible Markov chain are reordered so that the recurrent states come first, followed by the transient states. The states belonging to each recurrent chain in a reducible multichain are grouped together and numbered consecutively. All the recurrent chains are combined into one recurrent chain with a transition matrix denoted by S. The transient states are also numbered consecutively. Then the transition probability matrix, P, for a reducible Markov chain can be partitioned into four submatrices labeled S, 0, D, and Q, and represented in the following aggregated canonical or standard form:

$$P = \begin{bmatrix} S & 0 \\ D & Q \end{bmatrix}. \tag{1.43}$$

For example, the aggregated canonical form of the transition matrix for the generic five-state reducible multichain shown in Equation (1.42) appears below:

$$P = [p_{ij}] = \begin{matrix} 1 \\ 2 \\ 3 \\ 4 \\ 5 \end{matrix} \begin{bmatrix} 1 & 0 & 0 & 0 & 0 \\ 0 & p_{22} & p_{23} & 0 & 0 \\ 0 & p_{32} & p_{33} & 0 & 0 \\ p_{41} & p_{42} & p_{43} & p_{44} & p_{45} \\ p_{51} & p_{52} & p_{53} & p_{54} & p_{55} \end{bmatrix} = \begin{bmatrix} S & 0 \\ D & Q \end{bmatrix}, \tag{1.44}$$

where

$$S = \begin{matrix} 1 \\ 2 \\ 3 \end{matrix} \begin{bmatrix} 1 & 0 & 0 \\ 0 & p_{22} & p_{23} \\ 0 & p_{32} & p_{33} \end{bmatrix}, \quad D = \begin{matrix} 4 \\ 5 \end{matrix} \begin{bmatrix} p_{41} & p_{42} & p_{43} \\ p_{51} & p_{52} & p_{53} \end{bmatrix}, \quad Q = \begin{matrix} 4 \\ 5 \end{matrix} \begin{bmatrix} p_{44} & p_{45} \\ p_{54} & p_{55} \end{bmatrix}, \quad \text{and}$$

$$0 = \begin{matrix} 1 \\ 2 \\ 3 \end{matrix} \begin{bmatrix} 0 & 0 \\ 0 & 0 \\ 0 & 0 \end{bmatrix}.$$

Additional examples of transition matrices represented in canonical form, with and without aggregation, are given in Sections 1.10.1.2, 1.10.2, 3.1, and in Chapter 3.

1.10 Markov Chain Models

A variety of random processes can be modeled as Markov chains. Examples of unichain and multichain models are given in this section.

1.10.1 Unichain

Unichain models include irreducible chains, which have no transient states, and unichains, which have transient states.

1.10.1.1 Irreducible

The model of the weather in Section 1.3, and the circular random walk in Section 1.4.2 are both irreducible chains. Since these two Markov chains are aperiodic, they are also regular chains. Several additional models of regular Markov chains will be constructed, plus one model of an irreducible, periodic chain in Section 1.10.1.1.2.

1.10.1.1.1 Diffusion

This section will describe two simplified Markov chain models for diffusion. In the first model, two different gases are diffused. The second model, which produces a periodic chain, is for the diffusion of one gas between two containers [4, 5].

1.10.1.1.1.1 Two Gases
Consider a collection of $2k$ molecules of two gases, of which k are molecules of gas U, and k are molecules of gas V. The molecules are placed in two containers, labeled A and B, such that there are k molecules in each container. A single transition consists of choosing a molecule at random from each container, moving the molecule obtained from container A to container B, and moving the molecule obtained from container B to container A. The state of the process, denoted by X_n, is the number of molecules of gas V in container A after n transitions. Suppose that $X_n = i$. Then there are i molecules of gas V in container A, $k - i$ molecules of gas U in container A, $k - i$ molecules of gas V in container B, and i molecules of gas U in container B. The number of molecules of each type of gas in each container when the system is in state i is summarized in Table 1.5.

To compute transition probabilities, assume the probability that a molecule changes containers is proportional to the number of molecules in the container that the molecule leaves. Suppose that the process is in state i. If a

TABLE 1.5

Number of Molecules in Each Container
When the System Is in State i

Gas/Container	A	B
Gas U	$k - i$	i
Gas V	i	$k - i$

molecule of the same gas is chosen from each container, the system remains in state i. Hence,

$$p_{ii} = P(\text{a gas V molecule moves from A to B})$$
$$P(\text{a gas V molecule moves from B to A})$$
$$+ P(\text{a gas U molecule moves from A to B})$$
$$P(\text{a gas U molecule moves from B to A})$$

$$p_{ii} = \left(\frac{i}{k}\right)\left(\frac{k-i}{k}\right) + \left(\frac{k-i}{k}\right)\left(\frac{i}{k}\right) = 2\left(\frac{i}{k}\right)\left(\frac{k-i}{k}\right) = \frac{2i(k-i)}{k^2}.$$

If a molecule of gas V is chosen from container A and a molecule of gas U is chosen from container B, the system moves from state i to state $i - 1$. Hence,

$$p_{i,i-1} = P(\text{a gas V molecule moves from A to B})$$
$$P(\text{a gas U molecule moves from B to A})$$

$$= \left(\frac{i}{k}\right)\left(\frac{i}{k}\right) = \left(\frac{i}{k}\right)^2.$$

Finally, If a molecule of gas U is chosen from container A and a molecule of gas V is chosen from container B, the system moves from state i to state $i + 1$. Hence,

$$p_{i,i+1} = P(\text{a gas U molecule moves from A to B})$$
$$P(\text{a gas V molecule moves from B to A})$$

$$= \left(\frac{k-i}{k}\right)\left(\frac{k-i}{k}\right) = \left(\frac{k-i}{k}\right)^2.$$

No other transitions from state i are possible.

To gain insight into this model, consider the special case in which $k = 3$ molecules. The transition probabilities are

$$p_{ii} = \frac{2i(k-i)}{k^2} = \frac{2i(3-i)}{3^2} = \frac{2i(3-i)}{9}, \quad \text{for } i = 0,1,2,3.$$

$$p_{i,i-1} = \left(\frac{i}{k}\right)^2 = \left(\frac{i}{3}\right)^2 = \frac{i^2}{9}, \quad \text{for } i = 1,2,3$$

$$p_{i,i+1} = \left(\frac{k-i}{k}\right)^2 = \left(\frac{3-i}{3}\right)^2 = \frac{(3-i)^2}{9}, \quad \text{for } i = 0,1,2.$$

The transition probability matrix is shown below:

$$P = \begin{array}{c|cccc} & 0 & 1 & 2 & 3 \\ \hline 0 & 0 & 1 & 0 & 0 \\ 1 & \dfrac{1}{9} & \dfrac{4}{9} & \dfrac{4}{9} & 0 \\ 2 & 0 & \dfrac{4}{9} & \dfrac{4}{9} & \dfrac{1}{9} \\ 3 & 0 & 0 & 1 & 0 \end{array}. \tag{1.38}$$

This model for the diffusion of two gases is a regular Markov chain.

1.10.1.1.1.2 Ehrenfest Model   In the second model of diffusion, developed by the physicist T. Ehrenfest, k molecules of a single gas are stored in a container that is divided by a permeable membrane into two compartments, labeled A and B. At each transition, a molecule is chosen at random from the set of molecules and moved from the compartment in which it currently resides to the other compartment. The state of the process, denoted by X_n, is the number of molecules of gas in compartment A after n transitions. Therefore, if the process is in state i, then compartment A contains i molecules of gas, and compartment B contains $k - i$ molecules of gas. If the process is in state i, then with probability i/k a molecule moves from compartment A to B, and with probability $(k - i)/k$ a molecule moves from compartment B to A. If $X_n = i$, then at each transition, X_n either decreases by one molecule, so that $X_{n+1} = i - 1$, or increases by one molecule, so that $X_{n+1} = i + 1$. Thus, the transition probabilities are

$$p_{i,i-1} = \frac{i}{k}$$

$$p_{i,i+1} = \frac{k-i}{k}.$$

No other transitions from state i are possible.

For the special case in which $k = 3$ molecules, the transition probabilities are

$$p_{i,i-1} = \frac{i}{k} = \frac{i}{3}, \quad \text{for } i = 1, 2, 3$$

$$p_{i,i+1} = \frac{k-i}{k} = \frac{3-i}{3}, \quad \text{for } i = 0, 1, 2.$$

The transition probability matrix is shown below:

$$P = \begin{array}{c|cccc} & 0 & 1 & 2 & 3 \\ \hline 0 & 0 & 1 & 0 & 0 \\ 1 & \dfrac{1}{3} & 0 & \dfrac{2}{3} & 0 \\ 2 & 0 & \dfrac{2}{3} & 0 & \dfrac{1}{3} \\ 3 & 0 & 0 & 1 & 0 \end{array}. \tag{1.39}$$

As Section 1.9.1.1 has indicated, this Ehrenfest model of diffusion can be interpreted as a four-state random walk with reflecting barriers in which $p = 2/3$. Hence, this Ehrenfest model is a periodic Markov chain with a period of 2.

1.10.1.1.2 Waiting Line Inside an Office

A recruiter of engineers interviews applicants inside her office for positions in engineering management [1, 2]. The recruiter's office has a capacity of three candidates including the one being interviewed. Applicants who arrive when the office is full are not admitted. All interviews last 30 min. The number of applicants who arrive during the nth 30 min time period needed to conduct an interview is an independent and identically distributed random variable, which is denoted by A_n. The random variable A_n has the following probability distribution, which is stationary in time:

Number A_n of arrivals in period n, $A_n = k$	0	1	2	3	4	5 or more
Probability, $p_k = P(A_n = k)$	0.30	0.25	0.20	0.15	0.10	0

Let X_n represent the number of applicants in the recruiter's office immediately after the completion of the nth interview. Since the 30-min period required to conduct an interview is the same length as the interval during which new applicants arrive, the following relationships between X_n and X_{n+1} can be established. When $X_n = 0$, no candidates are in the recruiter's office so that X_n cannot be decreased by the completion of an interview. Furthermore, when $X_n = 0$, X_n can be increased by the arrival of A_{n+1} new applicants during the $(n + 1)$th 30 min period. Although new candidates may arrive, X_n cannot exceed three candidates, the capacity of the recruiter's office. Hence,

$$X_{n+1} = \min(3, A_{n+1}), \quad \text{if } X_n = 0.$$

When $X_n > 0$, one candidate is being interviewed and $X_n - 1$ candidates are in the recruiter's office waiting to be interviewed. In this case, X_n will be decreased by one at the completion of an interview at the end of the nth 30 min period. In addition, X_n can be increased by the arrival of A_{n+1} new applicants during the $(n + 1)$th 30 min period. Hence,

$$X_{n+1} = \min(3, X_n - 1 + A_{n+1}) \quad \text{if } X_n > 0.$$

These relationships show that X_{n+1} is a function only of X_n and A_{n+1}, an independent random variable. Therefore, $\{X_0, X_1, X_2 \dots\}$ is a Markov chain. Since the recruiter's office has a capacity of three candidates, the state space is $E = \{0, 1, 2, 3\}$. The transition probabilities are computed below:

If $X_n = i = 0$, the next state is $X_{n+1} = j = \min(3, A_{n+1})$.

The transition probabilities for state 0 are

$$p_{0j} = P(X_{n+1} = j | X_n = 0) = P(X_{n+1} = j = \min(3, A_{n+1}) | X_n = 0) = P(j = \min(3, A_{n+1}))$$

$$p_{00} = P(j = \min(3, A_{n+1})) = P(0 = \min(3, A_{n+1})) = P(A_{n+1} = 0) = 0.30$$

$$p_{01} = P(j = \min(3, A_{n+1})) = P(1 = \min(3, A_{n+1})) = P(A_{n+1} = 1) = 0.25$$

$$p_{02} = P(j = \min(3, A_{n+1})) = P(2 = \min(3, A_{n+1})) = P(A_{n+1} = 2) = 0.20$$

$$p_{03} = P(j = \min(3, A_{n+1})) = P(3 = \min(3, A_{n+1})) = P(A_{n+1} \geq 3)$$

$$= P(A_{n+1} = 3) + P(A_{n+1} = 4) = 0.15 + 0.10 = 0.25.$$

If $X_n = i = 1$, the next state is

$$X_{n+1} = j = \min(3, X_n - 1 + A_{n+1}) = \min(3, 1 - 1 + A_{n+1}) = \min(3, A_{n+1}).$$

Thus, the transition probabilities for state 1 are the same as those for state 0. The transition probabilities for state 1 are

$$p_{1j} = P(X_{n+1} = j | X_n = 1) = P(X_{n+1} = j = \min(3, X_n - 1 + A_{n+1}) | X_n = 1)$$

$$= P(j = \min(3, X_n - 1 + A_{n+1})) = P(j = \min(3, 1 - 1 + A_{n+1})) = P(j = \min(3, A_{n+1}))$$

$$p_{10} = P(j = \min(3, A_{n+1})) = P(0 = \min(3, A_{n+1})) = P(A_{n+1} = 0) = 0.30$$

$$p_{11} = P(j = \min(3, A_{n+1})) = P(1 = \min(3, A_{n+1})) = P(A_{n+1} = 1) = 0.25$$

$$p_{12} = P(j = \min(3, A_{n+1})) = P(2 = \min(3, A_{n+1})) = P(A_{n+1} = 2) = 0.20$$

$$p_{13} = P(j = \min(3, A_{n+1})) = P(3 = \min(3, A_{n+1})) = P(A_{n+1} \geq 3)$$

$$= P(A_{n+1} = 3) + P(A_{n+1} = 4) = 0.15 + 0.10 = 0.25.$$

If $X_n = i = 2$, the next state is

$$X_{n+1} = j = \min(3, X_n - 1 + A_{n+1})$$
$$= \min(3, 2 - 1 + A_{n+1}) = \min(3, 1 + A_{n+1}).$$

The transition probabilities for state 2 are

$$p_{2j} = P(X_{n+1} = j | X_n = 2) = P(X_{n+1} = j = \min(3, X_n - 1 + A_{n+1}) | X_n = 2)$$
$$= P(j = \min(3, X_n - 1 + A_{n+1})) = P(j = \min(3, 2 - 1 + A_{n+1}))$$
$$= P(j = \min(3, 1 + A_{n+1}))$$

$$p_{20} = P(j = \min(3, 1 + A_{n+1})) = P(0 = \min(3, 1 + A_{n+1})) = P(1 + A_{n+1} = 0)$$
$$= P(A_{n+1} = -1) = 0$$

$$p_{21} = P(j = \min(3, 1 + A_{n+1})) = P(1 = \min(3, 1 + A_{n+1})) = P(1 + A_{n+1} = 1)$$
$$= P(A_{n+1} = 0) = 0.30$$

$$p_{22} = P(j = \min(3, 1 + A_{n+1})) = P(2 = \min(3, 1 + A_{n+1})) = P(1 + A_{n+1} = 2)$$
$$= P(A_{n+1} = 1) = 0.25$$

$$p_{23} = P(j = \min(3, 1 + A_{n+1})) = P(3 = \min(3, 1 + A_{n+1})) = P(1 + A_{n+1} \geq 3)$$
$$= P(A_{n+1} \geq 2)$$
$$= P(A_{n+1} = 2) + P(A_{n+1} = 3) + P(A_{n+1} = 4) = 0.20 + 0.15 + 0.10 = 0.45.$$

If $X_n = i = 3$, the next state is

$$X_{n+1} = j = \min(3, X_n - 1 + A_{n+1}) = \min(3, 3 - 1 + A_{n+1}) = \min(3, 2 + A_{n+1}).$$

The transition probabilities for state 3 are

$$p_{3j} = P(X_{n+1} = j | X_n = 3) = P(X_{n+1} = j = \min(3, X_n - 1 + A_{n+1}) | X_n = 3)$$
$$= P(j = \min(3, X_n - 1 + A_{n+1})) = P(j = \min(3, 3 - 1 + A_{n+1}))$$
$$= P(j = \min(3, 2 + A_{n+1}))$$

$$p_{30} = P(j = \min(3, 2 + A_{n+1})) = P(0 = \min(3, 2 + A_{n+1})) = P(2 + A_{n+1} = 0)$$
$$= P(A_{n+1} = -2) = 0$$

$$p_{31} = P(j = \min(3, 2 + A_{n+1})) = P(1 = \min(3, 2 + A_{n+1})) = P(2 + A_{n+1} = 1)$$
$$= P(A_{n+1} = -1) = 0$$

$$p_{32} = P(j = \min(3, 2 + A_{n+1})) = P(2 = \min(3, 2 + A_{n+1})) = P(2 + A_{n+1} = 2)$$
$$= P(A_{n+1} = 0) = 0.30$$

$$p_{33} = P(j = \min(3, 2 + A_{n+1})) = P(3 = \min(3, 2 + A_{n+1})) = P(2 + A_{n+1} \geq 3)$$
$$= P(A_{n+1} \geq 1) = P(A_{n+1} = 1) + P(A_{n+1} = 2) + P(A_{n+1} = 3) + P(A_{n+1} = 4)$$
$$= 0.25 + 0.20 + 0.15 + 0.10 = 0.70.$$

The transition probabilities for the four-state Markov chain model of the waiting line inside the recruiter's office are collected to construct the following transition probability matrix:

$$
P = \begin{array}{c|cccc}
\text{State} & 0 & 1 & 2 & 3 \\
\hline
0 & P(A_n = 0) & P(A_n = 1) & P(A_n = 2) & P(A_n \geq 3) \\
1 & P(A_n = 0) & P(A_n = 1) & P(A_n = 2) & P(A_n \geq 3) \\
2 & 0 & P(A_n = 0) & P(A_n = 1) & P(A_n \geq 2) \\
3 & 0 & 0 & P(A_n = 0) & P(A_n \geq 1)
\end{array}
\tag{1.45}
$$

$$
= \begin{array}{c|cccc}
\text{State} & 0 & 1 & 2 & 3 \\
\hline
0 & 0.30 & 0.25 & 0.20 & 0.25 \\
1 & 0.30 & 0.25 & 0.20 & 0.25 \\
2 & 0 & 0.30 & 0.25 & 0.45 \\
3 & 0 & 0 & 0.30 & 0.70
\end{array}
\tag{1.46}
$$

1.10.1.1.3 Inventory System

A retailer who sells personal computers can order the computers from a manufacturer at the beginning of every period [1, 2]. All computers that are ordered are delivered immediately. The demand for computers during each period is an independent, identically distributed random variable with a known probability distribution. The retailer has a limited storage capacity sufficient to accommodate a maximum inventory of three computers. The number of computers in inventory at epoch n, the end of period n, is denoted by X_n. Hence, the inventory on hand at epoch $n - 1$, the beginning of period n, is denoted by X_{n-1}. The number of computers ordered and delivered immediately at the beginning of period n is denoted by c_{n-1}. The number of computers demanded by customers during period n is an independent, identically distributed random variable denoted by d_n. The demand for computers in every period has the following stationary probability distribution.

Demand d_n in period n	0	1	2	3
Probability, $p(d_n)$	0.3	0.4	0.1	0.2

Note that the number of computers demanded in a period cannot exceed three.

The retailer observes the inventory level at the beginning of every period. Computers are ordered and delivered immediately at the beginning of a period according to the following inventory ordering policy:

If $X_{n-1} < 2$, the retailer orders $c_{n-1} = 3 - X_{n-1}$ computers, which are delivered immediately to increase the beginning inventory to three computers.

If $X_{n-1} \geq 2$, the retailer does not order, so that the beginning inventory remains X_{n-1} computers.

This policy has the form (s, S), and is called an (s, S) inventory ordering policy. The quantity $s = 2$ is called the reorder point, and the quantity $S = 3$ is called the reorder level. Observe that under this $(2, 3)$ policy,

If $X_{n-1} = 0$, the retailer orders $c_{n-1} = 3 - 0 = 3$ computers, which increase the beginning inventory level to three computers.

If $X_{n-1} = 1$, the retailer orders $c_{n-1} = 3 - 1 = 2$ computers, which increase the beginning inventory level to three computers.

If $X_{n-1} = 2$, the retailer does not order, so that $c_{n-1} = 0$, and the beginning inventory level remains two computers.

If $X_{n-1} = 3$, the retailer does not order, so that $c_{n-1} = 0$, and the beginning inventory level remains three computers.

Observe that the inventory level X_n at the end of period n is a random variable, which is dependent on the inventory level X_{n-1} at the beginning of the period, on the quantity c_{n-1} ordered and delivered at the beginning of the period, and on the demand d_n, during the period. The demand, d_n, is governed by a probability distribution, which is known and is independent of the inventory on hand. Therefore, the sequence $\{X_0, X_1, X_2, ...\}$ forms a Markov chain, which can be used to model the retailer's inventory system. The state X_{n-1} is the inventory level or number of computers on hand at the beginning of period n. The Markov chain has four states. The state space is $E = \{0, 1, 2, 3\}$.

When the inventory on hand at the beginning of a period plus the number of computers ordered exceeds the number demanded by customers during the period, the number of unsold computers left over at the end of the period is $X_{n-1} + c_{n-1} - d_n > 0$. However, when the demand exceeds the beginning inventory plus the quantity ordered, sales are lost because customers who are unable to buy computers in the current period will not wait until the next period to purchase them. When sales are lost, the ending inventory level will be zero. In this case, the number of lost sales of computers, or the shortage of computers, at the end of the period is $d_n - X_{n-1} - c_{n-1} > 0$.

Transition probabilities are calculated by evaluating the relationship between beginning and ending inventory levels during a period.

In state 0, three computers are ordered. When $X_{n-1} = i = 0$ and $c_{n-1} = 3$, the next state is

$$X_n = X_{n-1} + c_{n-1} - d_n = j$$

$$= j = i + c_{n-1} - d_n = 0 + 3 - d_n = 3 - d_n.$$

The transition probabilities for state 0 are

$$p_{0j} = P(X_n = j | X_{n-1} = 0) = P(X_n = 3 - d_n | X_{n-1} = 0) = P(j = 3 - d_n).$$

Hence,

When $j = 0$,

$$p_{00} = P(X_n = 0 | X_{n-1} = 0) = P(j = 3 - d_n) = P(0 = 3 - d_n) = P(d_n = 3) = 0.2$$

When $j = 1$,

$$p_{01} = P(X_n = 1 | X_{n-1} = 0) = P(j = 3 - d_n) = P(1 = 3 - d_n) = P(d_n = 2) = 0.1$$

When $j = 2$,

$$p_{02} = P(X_n = 2 | X_{n-1} = 0) = P(j = 3 - d_n) = P(2 = 3 - d_n) = P(d_n = 1) = 0.4$$

When $j = 3$,

$$p_{03} = P(X_n = 3 | X_{n-1} = 0) = P(j = 3 - d_n) = P(3 = 3 - d_n) = P(d_n = 0) = 0.3.$$

In state 1, two computers are ordered. When $X_{n-1} = i = 1$ and $c_{n-1} = 2$, the next state is

$$X_n = X_{n-1} + c_{n-1} - d_n = j$$
$$= j = i + c_{n-1} - d_n = 1 + 2 - d_n = 3 - d_n.$$

The transition probabilities for state 1 are

$$p_{1j} = P(X_n = j | X_{n-1} = 1) = P(X_n = 3 - d_n | X_{n-1} = 1) = P(j = 3 - d_n).$$

Hence,

When $j = 0$,

$$p_{10} = P(X_n = 0 | X_{n-1} = 1) = P(j = 3 - d_n) = P(0 = 3 - d_n) = P(d_n = 3) = 0.2$$

When $j = 1$,

$$p_{11} = P(X_n = 1 | X_{n-1} = 1) = P(j = 3 - d_n) = P(1 = 3 - d_n) = P(d_n = 2) = 0.1$$

When $j = 2$,

$$p_{12} = P(X_n = 2|X_{n-1} = 1) = P(j = 3 - d_n) = P(2 = 3 - d_n) = P(d_n = 1) = 0.4$$

When $j = 3$,

$$p_{13} = P(X_n = 3|X_{n-1} = 1) = P(j = 3 - d_n) = P(3 = 3 - d_n) = P(d_n = 0) = 0.3.$$

Observe that the transition probabilities in state 1 are identical to those in state 0. In state 2, zero computers are ordered. When $X_{n-1} = i = 2$ and $c_{n-1} = 0$, the next state is $X_n = X_{n-1} + c_{n-1} - d_n = j$, provided that $X_{n-1} + c_{n-1} - d_n \geq 0$.

$X_n = j = i + c_{n-1} - d_n = 2 + 0 - d_n = 2 - d_n$, provided that $2 - d_n \geq 0$ or $d_n \leq 2$.

When $d_n = 3$, the demand exceeds the beginning inventory plus the quantity ordered, and the sale of one computer is lost. To ensure that the ending inventory is nonnegative, the equation for the next state, which represents the ending inventory, is expressed in the form

$$X_n = j = \max(2 - d_n, 0).$$

For example,

When $d_n = 0$, $X_n = \max (2-d_n, 0) = \max (2-0, 0) = \max (2, 0) = 2$.
When $d_n = 1$, $X_n = \max (2-d_n, 0) = \max (2-1, 0) = \max (1, 0) = 1$.
When $d_n = 2$, $X_n = \max (2-d_n, 0) = \max (2-2, 0) = \max (0, 0) = 1$.
When $d_n = 3$, $X_n = \max (2-d_n, 0) = \max (2-3, 0) = \max (-1, 0) = 0$.

The transition probabilities for state 2 are

$$p_{2j} = P(X_n = j|X_{n-1} = 2) = P(X_n = \max(2 - d_n, 0)|X_{n-1} = 2) = P(j = \max(2 - d_n, 0)).$$

Hence,

When $j = 0$,

$$p_{20} = P(X_n = 0|X_{n-1} = 2) = P(j = \max(2 - d_n, 0)) = P(0 = \max(2 - d_n, 0)) = P(d_n \geq 2)$$
$$= P(d_n = 2) + P(d_n = 3) = 0.1 + 0.2 = 0.3$$

When $j = 1$,

$$p_{21} = P(X_n = 1|X_{n-1} = 2) = P(j = \max(2 - d_n, 0)) = P(1 = \max(2 - d_n, 0)) = P(d_n = 1) = 0.4$$

When $j = 2$,

$$p_{22} = P(X_n = 2 | X_{n-1} = 2) = P(j = \max(2 - d_n, 0)) = P(2 = \max(2 - d_n, 0)) = P(d_n = 0) = 0.3$$

When $j = 3$,

$$p_{23} = P(X_n = 3 | X_{n-1} = 2) = P(j = \max(2 - d_n, 0)) = P(3 = \max(2 - d_n, 0)) = P(d_n = -1) = 0.$$

In state 3, zero computers are ordered. When $X_{n-1} = i = 3$ and $c_{n-1} = 0$, the next state is

$$X_n = X_{n-1} + c_{n-1} - d_n = j$$
$$= j = i + c_{n-1} - d_n = 3 + 0 - d_n = 3 - d_n.$$

The transition probabilities for state 3 are

$$p_{3j} = P(X_n = j | X_{n-1} = 3) = P(X_n = 3 - d_n | X_{n-1} = 3) = P(j = 3 - d_n).$$

Hence,

When $j = 0$,

$$p_{30} = P(X_n = 0 | X_{n-1} = 3) = P(j = 3 - d_n) = P(0 = 3 - d_n) = P(d_n = 3) = 0.2$$

When $j = 1$,

$$p_{31} = P(X_n = 1 | X_{n-1} = 3) = P(j = 3 - d_n) = P(1 = 3 - d_n) = P(d_n = 2) = 0.1$$

When $j = 2$,

$$p_{32} = P(X_n = 2 | X_{n-1} = 3) = P(j = 3 - d_n) = P(2 = 3 - d_n) = P(d_n = 1) = 0.4$$

When $j = 3$,

$$p_{33} = P(X_n = 3 | X_{n-1} = 3) = P(j = 3 - d_n) = P(3 = 3 - d_n) = P(d_n = 0) = 0.3.$$

The transition probabilities in state 3 are identical to those in states 0 and 1.

The transition probabilities for the four-state Markov chain model of the retailer's inventory system under a (2, 3) policy are collected to construct the following transition probability matrix:

Beginning Inventory, X_{n-1}	Order, c_{n-1}	$X_{n-1}+c_{n-1}$		State	0	1	2	3
0	3	3		0	$P(d_n=3)$	$P(d_n=2)$	$P(d_n=1)$	$P(d_n=0)$
1	2	3	$P=$	1	$P(d_n=3)$	$P(d_n=2)$	$P(d_n=1)$	$P(d_n=0)$
2	0	2		2	$P(d_n\geq 2)$	$P(d_n=1)$	$P(d_n=0)$	0
3	0	3		3	$P(d_n=3)$	$P(d_n=2)$	$P(d_n=1)$	$P(d_n=0)$

$$(1.47)$$

State	0	1	2	3
0	0.2	0.1	0.4	0.3
1	0.2	0.1	0.4	0.3
2	0.3	0.4	0.3	0
3	0.2	0.1	0.4	0.3

$$P=[p_{ij}]=$$

$$(1.48)$$

Observe that the transition probabilities in states 0, 1, and 3 are the same. The inventory system is enlarged to create a Markov chain with rewards in Section 4.2.3.4.1 and a Markov decision process in Sections 5.1.3.3.1 and 5.2.2.4.1.

1.10.1.1.4 Machine Breakdown and Repair

A factory has two machines and one repair crew [4]. Only one machine is used at any given time. A machine breaks down at the end of a day with probability p. The repair crew can work on only one machine at a time. When a machine has broken down at the end of the previous day, a repair can be completed in 1 day with probability r or in 2 days with probability $1 - r$. All repairs are completed at the end of the day, and no repair takes longer than 2 days. All breakdowns and repairs are independent events. To model this process as a Markov chain, let the state $X_n = (u, v)$ be a vector consisting of the pairs (u, v). Element u is the number of machines in operating condition at the end of day n. Element v is the number of days' work expended on a machine not yet repaired. The state space is given by $E = \{(2, 0) (1, 0) (1, 1) (0, 1)\}$. For example, state $(0, 1)$ indicates that neither machine is in operating condition, and 1 days' work was expended on a machine not yet repaired. If the process is in states $(1, 1)$ or $(0, 1)$, a repair is certain to be completed in 1 day. To see how to compute transition probabilities, suppose, for example, that the process is in state $(1, 0)$, which indicates that one machine is in operating condition and that 0 days' work was expended repairing the other machine. If with probability $1 - p$ the machine in operating condition does not fail, and with probability r the repair of the other machine is completed at the end of the day, then the process moves to state $(2, 0)$ with transition probability $(1 - p)r$. If with probability p the machine in operating condition breaks down, and with probability r the repair of the other machine is completed at the end of the day, then the process remains in state $(1, 0)$ with transition probability pr. If with probability $1 - p$ the machine in operating condition does not fail, and with probability $1 - r$ the repair of the other machine is not completed at the end of the day, then the process moves to state $(1, 1)$ with transition probability $(1 - p)$ $(1 - r)$. Finally, if with probability p the machine in operating condition breaks down, and with probability $1 - r$ the repair of the other machine is not completed at the end of the day, then the process moves to state $(0, 1)$ with transition probability $p(1 - r)$. The transition probability matrix is

$$
P = \begin{array}{c|cccc}
 & (2,0) & (1,0) & (1,1) & (0,1) \\
\hline
(2,0) & 1-p & p & 0 & 0 \\
(1,0) & (1-p)r & pr & (1-p)(1-r) & p(1-r) \\
(1,1) & 1-p & p & 0 & 0 \\
(0,1) & 0 & 1 & 0 & 0
\end{array}. \tag{1.49}
$$

1.10.1.1.5 Component Replacement

Consider an electronic device, which contains a number of interchangeable components, that operates independently [2]. Each component is subject to random failure, which makes it completely inoperable. All components are inspected at the end of every week. At the end of a week, a component that is observed to have failed is immediately replaced with an identical new component. When the new component fails, it is again replaced by an identical one, and so forth. Table 1.6 contains mortality data collected for a group of 100 of these components used over a 4-week period.

Observe that at the end of the 4-week period, none of the 100 components has survived, that is, all have failed. Therefore, the mortality data indicates that the service life of an individual component is at most 4 weeks. The replacement policy is to replace a component at the end of the week in which

TABLE 1.6

Mortality Data for 100 Components Used Over a 4-Week Period

End of Week	# of Survivors	# of Failures	P(Failures)	Cond. P(Failure)
n	S_n	F_n	$=F_n/100$	$=F_n/S_{n-1}$
0	100	0	0	–
1	80	20	0.20	$0.2 = 20/100$
2	50	30	0.30	$0.375 = 30/80$
3	10	40	0.40	$0.8 = 40/50$
4	0	10	0.10	$1 = 10/10$

it has failed. Since no component survives longer than 4 weeks, a component may be replaced every 1, 2, 3, or 4 weeks.

Note that 0.20 of new components fail during their first week of life, 0.30 of new components fail during their second week of life, 0.40 fail during their third week, and the remaining 0.10 fail during their fourth week. In addition, 0.375 of 1-week-old components fail during their second week of life, 0.80 of 2-week-old components fail during their third week of life, and all 3-week-old components fail during their fourth week of life. Let S_n denote the number of components, which survive until the end of week n. Let $F_n = S_{n-1} - S_n$. Thus, F_n denotes the number of components, which fail during week n of life. The probability that a component fails during its nth week of life is equal to $F_n/100$. The conditional probability that an individual component fails during week n of its life, given that it has survived for $n - 1$ weeks, is equal to $F_n/S_{n-1} = (S_{n-1} - S_n)/S_{n-1}$.

The mortality data indicates that the conditional probability that an individual component will fail depends only on its age. Let i represent the age of a component in weeks. The conditional probability that a component of age i will fail during week $i + 1$ of its life, given that it has survived for i weeks, is indicated in Table 1.7.

The process of component failure and replacement can be modeled as a four-state recurrent Markov chain. Let the state X_{n-1} denote the age of a component when it is observed at the end of week $n - 1$. The state space is $E = \{0, 1, 2, 3\}$. If a component of age i fails during week n, then it is replaced with a new component of age 0 at the end of week n, so that $X_n = 0$. Thus, if a component of age i fails during week n, the Markov chain model makes a transition from the state $X_{n-1} = i$ at the end of week $n - 1$ to the state $X_n = 0$ at the end of week n. The associated transition probability is

$$p_{i0} = P(X_n = 0 | X_{n-1} = i) = F_{i+1}/S_i = (S_i - S_{i+1})/S_i.$$

The mortality data indicates that $p_{00} = 0.2$, $p_{10} = 0.375$, and $p_{20} = 0.8$. Since a component of age 3 is certain to fail during the current week and be replaced at the end of the week, the transition probability for a 3-week-old component is $p_{30} = P(X_n = 0 | X_{n-1} = 3) = F_4/S_3 = 1$.

The age of the component in use during week $n - 1$ is $X_{n-1} = i$. For any $n - 1$, $X_n = 0$ if the component failed during week n. If the component survived during week n, then the component is 1 week older, so that $X_n = X_{n-1} + 1 = i + 1$. The conditional probabilities that a component of age

TABLE 1.7

Conditional Probability That a Component of Age i Will Fail During Week $i + 1$

Age i of a component in week, i	0	1	2	3
Conditional probability of	$0.2 = 20/100$	$0.375 = 30/80$	$0.8 = 40/50$	$1 = 10/10$
Failing during week $i + 1$ of life	$= F_1/S_0$	$= F_2/S_1$	$= F_3/S_2$	$= F_4/S_3$

i survives one additional week are calculated below:

$$p_{i,i+1} = P(X_n = i+1 | X_{n-1} = i)$$

$$= P(\text{Component survives 1 additional week} | \text{Component has survived } i \text{ weeks})$$

$$= S_{i+1}/S_i$$

$$= 1 - p_{10} = 1 - F_{i+1}/S_i = 1 - (S_i - S_{i+1})/S_i$$

$$p_{01} = P(X_n = 1 | X_{n-1} = 0) = S_1/S_0 = 80/100 = 0.8$$

$$p_{12} = P(X_n = 2 | X_{n-1} = 1) = S_2/S_1 = 50/80 = 0.625$$

$$p_{23} = P(X_n = 3 | X_{n-1} = 2) = S_3/S_2 = 10/50 = 0.2$$

$$p_{34} = P(X_n = 4 | X_{n-1} = 3) = S_4/S_3 = 0/10 = 0.$$

The four-state transition probability matrix is given below:

$$
P =
\begin{array}{c|cccc}
\text{State} & 0 & 1 & 2 & 3 \\
\hline
0 & 0.2 & 0.8 & 0 & 0 \\
1 & 0.375 & 0 & 0.625 & 0 \\
2 & 0.8 & 0 & 0 & 0.2 \\
3 & 1 & 0 & 0 & 0
\end{array}
\tag{1.50}
$$

The component replacement problem is enlarged to create a Markov chain with rewards in Section 4.2.3.4.2 and a Markov decision process in Section 5.1.3.3.2.

1.10.1.1.6 Independent Trials Process: Evaluating Candidates for a Secretarial Position

An independent trials process is a sequence of independent experiments or trials such that the outcome of any one trial does not affect the outcome of any other trial [2, 4]. Each trial is assumed to have a finite number of outcomes. The set of possible outcomes and the probability distribution for this set of outcomes are the same for every trial. To treat this process as a special case of a Markov chain, let the state X_n denote the outcome of the nth trial. The Markov property holds because given the outcome of the present trial, the outcome of the next trial is independent of the outcomes of the previous trials (and is also independent of the outcome of the present trial). Therefore, a transition probability is given by

$$p_{ij} = P(X_{n+1} = j | X_n = i) = P(X_{n+1} = j) = P(X_n = j) = P(X_{n-1} = j) = \cdots = P(X_1 = j).$$

Because the trials are independent, the joint probability of a sequence of n outcomes is equal to the product of the n marginal probabilities. That is, after n trials,

$$P(X_1 = j_1, X_2 = j_2, \ldots, X_n = j_n) = P(X_1 = j_1)P(X_2 = j_2) \cdots P(X_n = j_n).$$

Consider the following example of an independent trials process involving the evaluation of candidates or applicants for a secretarial position. Suppose that an executive must hire a secretary. The executive will interview one candidate per day. After each interview the executive will assign one of the following four numerical scores to the current candidate:

Candidate	Poor	Fair	Good	Excellent
Score	15	20	25	30

The scores assigned to the candidates, who arrive independently to be interviewed, are expected to vary according to the following stationary probability distribution:

Candidate	Poor	Fair	Good	Excellent
Score	15	20	25	30
Probability	0.3	0.4	0.2	0.1

The possible scores and the probabilities of these scores are the same for every applicant. The process of interviewing and assigning scores to candidates is an independent trials process because the score assigned to one candidate does not affect the score assigned to any other candidate.

To formulate this independent trials process as an irreducible Markov chain, let the state X_n denote the score assigned to the nth candidate, for $n = 1, 2, 3, \ldots$. The state space is $E = \{15, 20, 25, 30\}$. The sequence $\{X_1, X_2, X_3, \ldots\}$ is a collection of independent, identically distributed random variables. The probability distribution of X_n is shown in Table 1.8.

The score or state X_{n+1} of the next applicant is independent of the state X_n of the current applicant. Hence, $p_{ij} = P(X_{n+1} = j \mid X_n = i) = P(X_{n+1} = j)$. The

TABLE 1.8

Probability Distribution of Candidate Scores

Candidate	Poor	Fair	Good	Excellent
State $X_n = i$	15	20	25	30
$P(X_n = i)$	0.3	0.4	0.2	0.1

sequence $\{X_1, X_2, X_3,...\}$ for this independent trials process forms a Markov chain. The transition probability matrix is

$$
P = \begin{array}{c|cccc}
\text{State} & 15 & 20 & 25 & 30 \\
\hline
15 & P(X_{n+1}=15) & P(X_{n+1}=20) & P(X_{n+1}=25) & P(X_{n+1}=30) \\
20 & P(X_{n+1}=15) & P(X_{n+1}=20) & P(X_{n+1}=25) & P(X_{n+1}=30) \\
25 & P(X_{n+1}=15) & P(X_{n+1}=20) & P(X_{n+1}=25) & P(X_{n+1}=30) \\
30 & P(X_{n+1}=15) & P(X_{n+1}=20) & P(X_{n+1}=25) & P(X_{n+1}=30)
\end{array}
$$

$$
= \begin{array}{c|cccc}
\text{State} & 15 & 20 & 25 & 30 \\
\hline
15 & 0.3 & 0.4 & 0.2 & 0.1 \\
20 & 0.3 & 0.4 & 0.2 & 0.1 \\
25 & 0.3 & 0.4 & 0.2 & 0.1 \\
30 & 0.3 & 0.4 & 0.2 & 0.1
\end{array}
$$

(1.51)

Note that all the rows of P are identical. This property holds for every independent trials process. Two extended forms of this problem, called the secretary problem, will be formulated as Markov decision processes in Sections 5.1.3.3.3 and 5.2.2.4.2.

1.10.1.1.7 Birth and Death Process

A birth and death process is a Markov chain in which a transition can be made to an adjacent state or else leave the present state unchanged [6]. If at epoch n the chain is in state i, then at epoch $n + 1$ the chain is in state $i + 1$, $i - 1$, or i. A birth and death process can model a service facility by letting X_n denote the number of customers inside the facility at epoch n. Suppose the facility has a capacity of three customers including the one receiving service. The state space is $E = \{0, 1, 2, 3\}$. A birth represents the arrival of a customer. A customer who arrives when the server is not busy and the facility is not full goes directly into service. If the server is busy, the customer waits for service. Any customer who arrives when the facility is full is not admitted. When $X_n = i < 3$, the probability of a birth is b_i. A death, which occurs with probability d_i when $X_n = i > 0$, represents the departure of a customer who has completed service. A birth and death cannot occur simultaneously.

A birth and death process is a Markov chain because the number of customers at the next epoch depends only on the number at the present epoch. The transition probabilities are given below:

$$
P(X_{n+1} = i+1 \mid X_n = i) = \begin{cases} b_i, & i = 0,1,2 \\ 0, & i = 3. \end{cases}
$$

$$P(X_{n+1} = i-1 \mid X_n = i) = \begin{cases} d_i, & i = 1,2,3 \\ 0, & i = 0 \end{cases}$$

$$P(X_{n+1} = i \mid X_n = i) = 1 - b_i - d_i.$$

The chain has the following transition probability matrix:

$$P = \begin{array}{c|cccc} \text{State} & 0 & 1 & 2 & 3 \\ \hline 0 & 1-b_0 & b_0 & 0 & 0 \\ 1 & d_1 & 1-b_1-d_1 & b_1 & 0 \\ 2 & 0 & d_2 & 1-b_2-d_2 & b_2 \\ 3 & 0 & 0 & d_3 & 1-d_3 \end{array}. \qquad (1.52)$$

1.10.1.2 Reducible Unichain

A reducible unichain consists of one closed communicating class of recurrent states plus one or more transient states. When the recurrent chain is a single absorbing state, the reducible unichain is called an absorbing unichain.

1.10.1.2.1 Absorbing Markov Chain

An absorbing unichain contains one absorbing state plus a set of transient states. Generally, an absorbing unichain is called an absorbing Markov chain.

1.10.1.2.1.1 Selling a Stock for a Target Price Suppose that at the end of a month a woman buys one share of a certain stock for $10. The share price, rounded to the nearest $10, has been varying among the prices $0, $10, and $20 from month to month. She plans to sell the stock at the end of first month in which the share price rises to $20. She believes that the price of the stock can be modeled as a Markov chain in which the state, X_n, denotes the share price at the end of month n. The state space for the stock price is $E = \{\$0, \$10, \$20\}$. The state $X_n = \$20$ is an absorbing state, reached when the stock is sold. The two remaining states, which are entered when the stock is held, are transient. She models her investment as an absorbing unichain with the following transition probability matrix represented in the canonical form of Equation (1.43).

$$P = \begin{array}{c|ccc} X_n \backslash X_{n+1} & 20 & 10 & 0 \\ \hline 20 & 1 & 0 & 0 \\ 10 & 0.1 & 0.6 & 0.3 \\ 0 & 0.4 & 0.2 & 0.4 \end{array} = \begin{bmatrix} 1 & 0 \\ D & Q \end{bmatrix}. \qquad (1.53)$$

A model for selling a stock with two target prices will be treated in Sections 4.2.5.4.2 and 4.2.5.5.

1.10.1.2.1.2 Machine Deterioration   Consider a machine used in a production process [6]. Suppose that the condition of the machine deteriorates over time. The machine is observed at the beginning of each day. As Table 1.9 indicates, the condition of the machine can be represented by one of the four states.

TABLE 1.9

States of a Machine

State	Description
1	Not Working (NW), Inoperable
2	Working, with a Major Defect (WM)
3	Working, with a Minor Defect (Wm)
4	Working Properly (WP)

(Note that the states are labeled so that as the index of the state increases, the condition of the machine improves.) The state of the machine at the start of tomorrow depends only on its state at the start of today, and is independent of its past history. Hence, the condition of the machine can be modeled as a four-state Markov chain. Let X_{n-1} denote the state of the machine when it is observed at the start of day n. The state space is $E = \{1, 2, 3, 4\}$. Assume that at the start of each day, the engineer in charge of the production process does nothing to respond to the deterioration of the machine, that is, she does not perform maintenance. Therefore, the condition of the machine will either deteriorate by one or more states or remain unchanged. In other words, if the engineer responsible for the machine always does nothing, then at the start of tomorrow, the condition of the machine will be worse than or equal to the condition today. All state transitions caused by deterioration are assumed to occur at the end of the day. A transition probability matrix for the machine when it is left alone for one day is given below in the canonical form of Equation (1.43):

$$
P = \begin{array}{c} \text{1 Not Working} \\ \text{2 Major Defect} \\ \text{3 Minor Defect} \\ \text{4 Working Properly} \end{array}
\left[
\begin{array}{c|ccc}
1 & 0 & 0 & 0 \\
\hline
0.6 & 0.4 & 0 & 0 \\
0.2 & 0.3 & 0.5 & 0 \\
0.3 & 0.2 & 0.1 & 0.4
\end{array}
\right]
= \begin{bmatrix} 1 & \mathbf{0} \\ D & \mathbf{Q} \end{bmatrix}.
\qquad (1.54)
$$

The transition probability matrix for a machine, which is left alone, represents an absorbing Markov chain. As the machine deteriorates daily, it will eventually enter state 1, an absorbing state, where it will remain, not working. For example, if today a machine is in state 3, working, with a minor defect, then tomorrow, with transition probability $p_{32} = P(X_n = 2|X_{n-1} = 3) = 0.3$, the machine will be in state 2, working, with a major defect. One day later, with transition probability $p_{21} = P(X_n = 1|X_{n-1} = 2) = 0.6$, the chain will be absorbed in state 1, where the machine will remain, not working. The model of machine deterioration will be revisited in Sections 3.3.2.1 and 3.3.3.1.

1.10.1.2.2 Unichain with Recurrent States

A reducible unichain with recurrent states and no absorbing states consists of one recurrent chain plus transient states.

1.10.1.2.2.1 Machine Maintenance Consider the absorbing Markov chain model of machine deterioration introduced in Section 1.10.1.2.1.2 [6]. Suppose that, at the start of each day, the engineer in charge of the production process can respond to the deterioration of the machine either by doing nothing (as in Section 1.10.1.2.1.2) or by choosing among the following four alternative maintenance actions:

Decision, k	Maintenance Action
1	Do Nothing
2	Overhaul
3	Repair
4	Replace

The success or failure of a maintenance action depends only on the present state of the machine, and does not depend on its past behavior. All maintenance actions will take exactly 1 day to complete. However, decisions to overhaul or repair the machine are not certain to succeed. If the engineer has the machine overhauled, the overhaul will be completed, either successfully or unsuccessfully, in 1 day. An overhaul may be successful, with probability 0.8, or unsuccessful, with probability 0.2. If an overhaul is successful, then at the start of the next day, the condition of the machine will be improved by one state if that is possible. That is, if an overhaul at the start of today is successful for a machine in state i, where $i = 1, 2$, or 3, then with probability 0.8, the machine will be in state $i + 1$ at the start of tomorrow. If the overhaul is not successful, the machine will remain in state i with probability 0.2. If a machine in state 4 is overhauled, the machine will remain in state 4 with probability 1. If the engineer has the machine repaired, the repair will also be completed in 1 day, either successfully or unsuccessfully. A repair may be successful, with probability 0.7, or unsuccessful, with probability 0.3. If a repair is successful, then at the start of the next day, the condition of the machine will be improved by two states if that is possible. That is, if a repair at the start of today is successful for a machine in state i, where $i = 1$ or 2, then with probability 0.7, the machine will be in state $i + 2$ at the start of tomorrow. If the repair is not successful, the machine will remain in state i with probability 0.3. If a machine in state 3 is repaired successfully, then with probability 0.7, the condition of the machine will improve to state 4. If a machine in state 4 is repaired, the machine will remain in state 4 with probability 1. If a machine in any state i is replaced with a new machine at the start of today, then tomorrow, with probability 1, the machine will be in state 4, working properly. The replacement process takes 1 day to complete. The four possible maintenance actions are summarized in Table 1.10.

The engineer who manages the production process has implemented the following maintenance policy, called the original maintenance policy, for the machine. In state 1, when the machine is not working, it is always overhauled. In state 2, when the machine is working, with a major defect, the engineer always does nothing. In state 3, when the machine is working, with a minor defect, it is always overhauled. Finally, in state 4, when the machine is working properly, the engineer always does nothing. The original maintenance policy is summarized in Table 1.11.

The transition probability matrix associated with this original maintenance policy appears in Equation (1.55).

TABLE 1.10

Four Possible Maintenance Actions

Decision	Action	Outcome
1	Do nothing (DN)	The condition tomorrow will be worse than or equal to the condition today
2	Overhaul (OV)	If, with probability 0.8, an overhaul in states 1, 2, or 3, is successful, the condition tomorrow will be superior by one state to the condition today. If unsuccessful, the condition tomorrow will be unchanged
3	Repair (RP)	If, with probability 0.7, a repair in state 1 or 2 is successful, the condition tomorrow will be superior by two states to the condition today. If unsuccessful, the condition tomorrow will be unchanged
4	Replace (RL)	The machine will work properly tomorrow

TABLE 1.11

Original Maintenance Policy

State, i	Description	Decision, k	Maintenance Action
1	Not Working, Inoperable	2	Overhaul
2	Working, with a major defect	1	Do Nothing
3	Working, with a minor defect	2	Overhaul
4	Working Properly	1	Do Nothing

TABLE 1.12

Modified Maintenance Policy

State, i	Description	Decision, k	Maintenance Action
1	Not Working, Inoperable	2	Overhaul
2	Working, with a major defect	1	Do Nothing
3	Working, with a minor defect	1	Do Nothing
4	Working Properly	1	Do Nothing

$$
\begin{array}{cccc}
\text{State, } X_{n-1} = i & \text{Decision, } k & & \text{State} \\
1 & 2, \text{ Overhaul} & & 1 \\
2 & 1, \text{ Do Nothing} & P = & 2 \\
3 & 2, \text{ Overhaul} & & 3 \\
4 & 1, \text{ Do Nothing} & & 4
\end{array}
\quad
\begin{array}{c|cccc}
 & 1 & 2 & 3 & 4 \\
\hline
 & 0.2 & 0.8 & 0 & 0 \\
 & 0.6 & 0.4 & 0 & 0 \\
 & 0 & 0 & 0.2 & 0.8 \\
 & 0.3 & 0.2 & 0.1 & 0.4
\end{array}
= \begin{bmatrix} S & 0 \\ D & Q \end{bmatrix}. \quad (1.55)
$$

Observe that this Markov chain is unichain as states 1 and 2 form a recurrent closed class while states 3 and 4 are transient.

Now suppose that the engineer who manages the production process modifies the original maintenance policy by always doing nothing when the machine is in state 3 instead of overhauling it in state 3. The modified maintenance policy, under which the engineer always overhauls the machine in state 1 and does nothing in the other three states, is summarized in Table 1.12.

The transition probability matrix associated with the modified maintenance policy appears below in Equation (1.56).

$$
\begin{array}{cccc}
\text{State, } X_{n-1} = i & \text{Decision, } k & & \text{State} \\
1 & 2, \text{ Overhaul} & & 1 \\
2 & 1, \text{ Do Nothing} & P = & 2 \\
3 & 1, \text{ Do Nothing} & & 3 \\
4 & 1, \text{ Do Nothing} & & 4
\end{array}
\quad
\begin{array}{c|cccc}
 & 1 & 2 & 3 & 4 \\
\hline
 & 0.2 & 0.8 & 0 & 0 \\
 & 0.6 & 0.4 & 0 & 0 \\
 & 0.2 & 0.3 & 0.5 & 0 \\
 & 0.3 & 0.2 & 0.1 & 0.4
\end{array}
= \begin{bmatrix} S & 0 \\ D & Q \end{bmatrix}. \quad (1.56)
$$

Observe that this Markov chain under the modified maintenance policy is also unichain as states 1 and 2 form a recurrent closed class while states 3 and 4 are transient. The machine maintenance model will be revisited in Sections 3.3.2.1, 3.3.3.1, 3.5.4.2, 4.2.4.1, 4.2.4.2, 4.2.4.3, and 5.1.4.4.

1.10.1.2.2.2 Career Path with Lifetime Employment Consider a company owned by its employees that offers both lifetime employment and career flexibility. Employees are so well treated that no one ever leaves the company or retires. New employees begin their careers in engineering. Eventually, all engineers will be promoted to management positions, but managers never return to engineering. Engineers may have their engineering job assignments changed among the three areas of product design, systems integration, and systems testing. Managers may have their supervisory assignments changed among the three areas of hardware, software, and marketing. All changes in job assignment and all promotions occur monthly.

By assigning a state to represent each job assignment, the career path of an employee is modeled as a six-state Markov chain. The state X_n denotes an employee's job assignment at the end of month n. The states are indexed as follows:

State, X_n	Description
1	Management of Hardware
2	Management of Software
3	Management of Marketing
4	Engineering Product Design
5	Engineering Systems Integration
6	Engineering Systems Testing

Since people in management never return to engineering, states 1, 2, and 3 form a closed communicating class of recurrent states, denoted by $R = \{1, 2, 3\}$. States 4, 5, and 6 are transient because all engineers will eventually be promoted to management. The set of transient states is denoted by $T = \{4, 5, 6\}$. Thus, the model is a reducible unichain, which has one closed class of three recurrent states, and three transient states. After many years, the company has recorded sufficient data to construct the following transition probability matrix, displayed in the canonical form of Equation (1.43):

$$
P = \begin{array}{l}
\text{1 Management of Hardware} \\
\text{2 Management of Software} \\
\text{3 Management of Marketing} \\
\text{4 Engineering Product Design} \\
\text{5 Engineering Systems Integration} \\
\text{6 Engineering Systems Testing}
\end{array}
\left[\begin{array}{ccc|ccc}
0.30 & 0.20 & 0.50 & 0 & 0 & 0 \\
0.40 & 0.25 & 0.35 & 0 & 0 & 0 \\
0.50 & 0.10 & 0.40 & 0 & 0 & 0 \\
\hline
0.05 & 0.15 & 0.10 & 0.30 & 0.16 & 0.24 \\
0.04 & 0.07 & 0.05 & 0.26 & 0.40 & 0.18 \\
0.08 & 0.06 & 0.12 & 0.14 & 0.32 & 0.28
\end{array}\right] = \begin{bmatrix} S & 0 \\ D & Q \end{bmatrix}.
$$

$$(1.57)$$

The model of a career path is revisited in Section 4.2.4.4.2.

1.10.2 Reducible Multichain

A reducible multichain can be partitioned into two or more mutually exclusive closed communicating classes of recurrent states plus one or more transient states.

1.10.2.1 Absorbing Markov Chain

An absorbing multichain has two or more absorbing states plus transient states. Generally, an absorbing multichain is called an absorbing Markov chain.

1.10.2.1.1 Charge Account

Suppose that a store classifies a customer charge account into one of the following five states [1]:

State	Description
0	0 months (1–30 days) old
1	1 month (31–60 days) old
2	2 months (61–90 days) old
P	Paid in full
B	Bad debt

A charge account is classified according to the oldest unpaid debt, starting from the billing date. (Assume that all debt includes interest and finance charges.) For example, if a customer has one unpaid bill, which is 1 month old, and a second unpaid bill, which is 2 months old, then the account is classified as 2 months old. If she makes a payment less than the 2-month-old bill, the account remains classified as 2 months old. However, if she makes a payment greater than the 2-month-old bill but less than the sum of the 1-month-old and 2-month-old bills, the account is reclassified as 1 month old. If at any time she pays the entire balance owed, the account is labeled as paid in full. When an account becomes 3 months old, it is labeled as a bad debt and sent to a collection agency.

Assume that the change of status of a charge account depends only on its present classification. Then the process can be modeled as a five-state Markov chain. Let X_n denote the state of an account at the nth month since the account was opened. The state space is $E = \{0, 1, 2, P, B\}$. The states 0, 1, and 2 indicate the age of an account, in months. An account that is 0 months old is a new account with only current charges. State P indicates that an account is paid in full, and state B indicates a bad debt. When the account is in states 0, 1, or 2, it may stay in its present state. When the account is in state 0, it may move to state 1. When the account is in state 1, it may move to states 0 or 2. When the account is in state 2, it may move to states 0 or 1. Because an account can be paid in full at any time, transitions are possible from states 0, 1, and 2 to state P. Since an account is reclassified as a bad debt only when it becomes 3 months old, state B is reached only by a transition from state 2. States 0, 1, and 2 are transient states because eventually the charge account will either be paid in full or labeled as a bad debt. States P and B are absorbing states because once one of these states is entered, the account is settled and no further activity is possible. The process is an absorbing multichain.

After rearranging the five states so that the two absorbing states appear first, the transition probability matrix is given below in the canonical form of Equation (1.43):

$$P = \begin{array}{c|cc|ccc} \text{State} & P & B & 0 & 1 & 2 \\ \hline P & 1 & 0 & 0 & 0 & 0 \\ B & 0 & 1 & 0 & 0 & 0 \\ \hline 0 & p_{0P} & 0 & p_{00} & p_{01} & 0 \\ 1 & p_{1P} & 0 & p_{10} & p_{11} & p_{12} \\ 2 & p_{2P} & p_{2B} & p_{20} & p_{21} & p_{22} \end{array} = \begin{bmatrix} I_1 & 0 & 0 \\ 0 & I_2 & 0 \\ D_1 & D_2 & Q \end{bmatrix} = \begin{bmatrix} I & 0 \\ D & Q \end{bmatrix}.$$

Suppose that observations over a period of time have produced the following transition probability matrix, also displayed in the canonical form of Equation (1.43):

$$P = \begin{array}{c|cc|ccc} \text{State} & P & B & 0 & 1 & 2 \\ \hline P & 1 & 0 & 0 & 0 & 0 \\ B & 0 & 1 & 0 & 0 & 0 \\ \hline 0 & 0.5 & 0 & 0.2 & 0.3 & 0 \\ 1 & 0.2 & 0 & 0.1 & 0.4 & 0.3 \\ 2 & 0.1 & 0.2 & 0.4 & 0.2 & 0.1 \end{array} = \begin{bmatrix} I & 0 \\ D & Q \end{bmatrix}. \tag{1.58}$$

1.10.2.1.2 Patient Flow in a Hospital

Suppose that the patients in a hospital may be treated in one of six departments, which are indexed by the following states:

State	Department
0	Discharged
1	Diagnostic
2	Outpatient
3	Surgery
4	Physical Therapy
5	Morgue

During a given day, 40% of all diagnostic patients will not be moved, 10% will become outpatients, 20% will enter surgery, and 30% will begin physical therapy. Also, 15% of all outpatients will be discharged, 5% will die, 10%

will be moved to the diagnostic department, 20% will remain as outpatients, 30% will undergo surgery, and 20% will start physical therapy. In addition, 7% of all patients in surgery will be discharged, 3% will die, 20% will be transferred to the diagnostic unit, 10% will become outpatients, 40% will remain in surgery, and 20% will begin physical therapy. Furthermore, 30% of all patients in physical therapy will be transferred to the diagnostic department, 40% will become outpatients, 20% will enter surgery, and 10% will remain in physical therapy. Assume that discharged patients will never reenter the hospital.

Assume that the daily movement of a patient depends only on the department in which the patient currently resides. Then the daily movement of patients in the hospital can be modeled as an absorbing multichain in which states 0 and 5 are absorbing states, and the other four states are transient. Let the state X_n denote the department in which a patient resides on day n. The state space is $E = \{0, 1, 2, 3, 4, 5\}$. After rearranging the six states so that the two absorbing states appear first, the transition probability matrix is given below in the canonical form of Equation (1.43):

$$
P = \begin{array}{c|cc|cccc}
\text{State} & 0 & 5 & 1 & 2 & 3 & 4 \\
\hline
0 & 1 & 0 & 0 & 0 & 0 & 0 \\
5 & 0 & 1 & 0 & 0 & 0 & 0 \\
1 & 0 & 0 & 0.4 & 0.1 & 0.2 & 0.3 \\
2 & 0.15 & 0.05 & 0.1 & 0.2 & 0.3 & 0.2 \\
3 & 0.07 & 0.03 & 0.2 & 0.1 & 0.4 & 0.2 \\
4 & 0 & 0 & 0.3 & 0.4 & 0.2 & 0.1
\end{array} = \begin{bmatrix} S & 0 \\ D & Q \end{bmatrix} = \begin{bmatrix} I & 0 \\ D & Q \end{bmatrix}. \qquad (1.59)
$$

The model of patient flow is revisited in Chapter 3.

1.10.2.2 Eight-State Multichain Model of a Production Process

A multichain model may have recurrent states as well as absorbing states. Consider, for example, a production process that consists of three manufacturing stages in series. When the work at a stage is completed, the output of the stage is inspected. Output from stage 1 or 2 that is not defective is passed on to the next stage. Output with a minor defect is reworked at the current stage. Output with a major defect is scrapped. Nondefective but blemished output from stage 3 is sent to a training center where it is used to train engineers, technicians, and technical writers. Output from stage 3 that is neither defective nor blemished is sold.

By assigning a state to represent each operation of the production process, the following eight states are identified:

State	Operation
1	Scrapped
2	Sold
3	Training Engineers
4	Training Technicians
5	Training Technical Writers
6	Stage 3
7	Stage 2
8	Stage 1

The next operation on an item depends only on the outcome of the current operation. Therefore, the production process can be modeled as a Markov chain. An epoch is the instant at which an item passes through a production stage and is inspected, or is transferred to an employee in the training center, or is transferred within the training center. Let the state X_n denote the operation on an item at epoch n. Since output that is scrapped will not be reused, and output that is sold will not be returned, states 1 and 2 are absorbing states. Output sent to the training center will remain there permanently, and will therefore not rejoin the production stages or be scrapped or sold. Output received by the training center is dedicated exclusively to training engineers, technicians, and technical writers, and will be shared by these employees. Hence, states 3, 4, and 5 form a closed communicating class of recurrent states. States 6, 7, and 8 are transient because all output must eventually leave the production stages to be scrapped, sold, or sent to the training center. Thus, the model is a reducible multichain, which has two absorbing states, one closed class of three recurrent states, and three transient states. Observe that production stage i is represented by transient state $(9 - i)$. An item enters the production process at stage 1, which is transient state 8.

The following transition probabilities for the transient states are expressed in terms of the probabilities of producing output, which has a major defect, a minor defect, is blemished, or has no defect:

$p_{88} = 0.75 = P$(output from stage 1 has a minor defect and is reworked)

$p_{77} = 0.65 = P$(output from stage 2 has a minor defect and is reworked)

$p_{66} = 0.55 = P$(output from stage 3 has a minor defect and is reworked)

$p_{87} = 0.15 = P$(output from stage 1 has no defect and is passed to stage 2)

$p_{76} = 0.20 = P(\text{output from stage 2 has no defect and is passed to stage 3})$

$p_{62} = 0.16 = P(\text{output from stage 3 has no defect or blemish and is sold})$

$p_{65} = 0.02 = \begin{cases} P(\text{output from stage 3 has no defect but is blemished, and} \\ \text{is sent to train technical writers}) \end{cases}$

$p_{64} = 0.03 = \begin{cases} P(\text{output from stage 3 has no defect but is blemished, and} \\ \text{is sent to train technicians}) \end{cases}$

$p_{63} = 0.04 = \begin{cases} P(\text{output from stage 3 has no defect but is blemished, and} \\ \text{is sent to train engineers}) \end{cases}$

$p_{81} = 0.10 = P(\text{output from stage 1 has a major defect and is scrapped})$

$p_{71} = 0.15 = P(\text{output from stage 2 has a major defect and is scrapped})$

$p_{61} = 0.20 = P(\text{output from stage 3 has a major defect and is scrapped}).$

The transition probability matrix is shown below in the canonical form of Equation (1.43):

		1	2	3	4	5	6	7	8
Scrapped	1	1	0	0	0	0	0	0	0
Sold	2	0	1	0	0	0	0	0	0
Training Engineers	3	0	0	0.50	0.30	0.20	0	0	0
Training Technicians	4	0	0	0.30	0.45	0.25	0	0	0
Training Tech. Writers	5	0	0	0.10	0.35	0.55	0	0	0
Stage 3	6	0.20	0.16	0.04	0.03	0.02	0.55	0	0
Stage 2	7	0.15	0	0	0	0	0.20	0.65	0
Stage 1	8	0.10	0	0	0	0	0	0.15	0.75

$$P = \begin{bmatrix} P_1 & 0 & 0 & 0 \\ 0 & P_2 & 0 & 0 \\ 0 & 0 & P_3 & 0 \\ D_1 & D_2 & D_3 & Q \end{bmatrix} = \begin{bmatrix} I & 0 & 0 & 0 \\ 0 & I & 0 & 0 \\ 0 & 0 & P_3 & 0 \\ D_1 & D_2 & D_3 & Q \end{bmatrix} = \begin{bmatrix} S & 0 \\ D & Q \end{bmatrix}. \tag{1.60}$$

The passage of an item through the production process is shown in Figure 1.6.

The multichain model of a production process is revisited in Chapters 3 and 4.

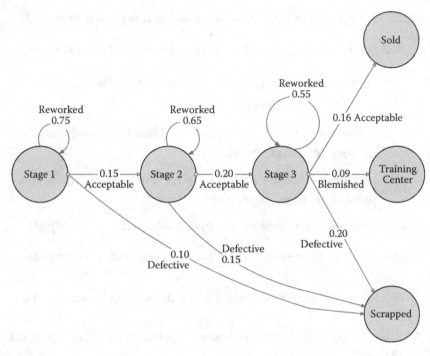

FIGURE 1.6
Passage of an item through a three-stage production process.

PROBLEMS

1.1   The condition of a machine, which is observed at the beginning of each day, can be represented by one of the following four states:

State	Description
1 (NW)	Not Working, Under Repair
2 (MD)	Working, with a major defect
3 (mD)	Working, with a minor defect
4 (WP)	Working Properly

The next state of the machine depends only on the present state and is independent of its past history. Hence, the condition of the machine can be modeled as a four-state Markov chain. Let X_{n-1} denote the state of the machine when it is observed at the start of day n. The state space is $E = \{1, 2, 3, 4\}$. Assume that at the start of each day, the engineer in charge of the machine does nothing to respond to the condition of the machine when it is in states 2, 3, or 4. A machine in states 2, 3, or 4 is allowed to fail and enter state 1, which represents a repair process. Therefore, a machine in state 1, not working, is assumed to be under repair.

The repair process carried out when the machine is in state 1 (NW) takes 1 day to complete. When the machine is in state 1 (NW), the repair process will be completely successful with probability 0.5 and transfer the machine to state 4 (WP), or largely successful with probability 0.2 and transfer the machine to state 3 (mD), or marginally successful with probability 0.2 and transfer the machine to state 2 (MD), or unsuccessful with probability 0.1 and leave the machine in state 1 (NW). A machine is not repaired when it is in states 2, 3, or 4. A machine in state 2 (MD) will remain in state 2 with probability 0.7, or fail with probability 0.3 and enter state 1 (NW). A machine in state 3 (mD) will remain in state 3 with probability 0.3, or acquire a major defect with probability 0.5 and enter state 2, or fail with probability 0.2 and enter state 1 (NW). Finally, a machine in state 4 (WP) will remain in state 4 with probability 0.4, or acquire a minor defect with probability 0.3 and enter state 3, or acquire a major defect with probability 0.2 and enter state 2, or fail with probability 0.1 and enter state 1 (NW).

Construct the transition probability matrix for this four-state Markov chain model under which a machine is left alone in states 2, 3, and 4, and repaired in state 1.

1.2 Suppose that the machine modeled in Problem 1.1 is modified by making the repair time last 2 days. Two states are needed to represent both days of a 2-day repair process. When the machine fails, it goes to state 1 (NW1), which denotes the first day of the repair process. When the first day of repair ends, the machine must enter state 2 (NW2), which denotes the second day of repair. The condition of the machine, which is observed at the beginning of each day, can be represented by one of the following five states:

State	Description
1 (NW1)	Not Working, in first day of repair
2 (NW2)	Not Working, in second day of repair
3 (MD)	Working, with a major defect
4 (mD)	Working, with a minor defect
5 (WP)	Working Properly

The use of two states to distinguish the 2 days of the repair process allows the next state of the machine to depend only on the present state and to be independent of its past history. Hence, the condition of the machine can be modeled as a five-state Markov chain. Let X_{n-1} denote the state of the machine when it is observed at the start of day n. The state space is $E = \{1, 2, 3, 4, 5\}$. Assume that at the start of each day, the engineer in charge of the machine does nothing to respond to the condition of the machine when it is in states 3, 4, or 5.

A machine in states 3, 4, or 5 that fails will enter state 1 (NW1). One day later the machine will move from state 1 to state 2 (NW2). When the machine is in state 2 (NW2), the repair process will be completely successful with probability 0.5 and transfer the machine to state 5 (WP), or largely successful with probability 0.2 and transfer the machine to state 4 (mD), or marginally successful with probability 0.2 and transfer the machine to state 3 (MD), or unsuccessful with probability 0.1 and transfer the machine to state 1 (NW1). A machine in state 3 (MD) will remain in state 3 with probability 0.7, or fail with probability 0.3 and enter state 1 (NW1). A machine in state 4 (mD) will remain in state 4 with probability 0.3, or acquire a major defect with probability 0.5 and enter state 3, or fail with probability 0.2 and enter state 1 (NW1). Finally, a machine in state 5 (WP) will remain in state 5 with probability 0.4, or acquire a minor defect with probability 0.3 and enter state 4, or acquire a major defect with probability 0.2 and enter state 3, or fail with probability 0.1 and enter state 1 (NW1).

Construct the transition probability matrix for this five-state Markov chain model under which a machine is left alone in states 3, 4, and 5.

1.3　Consider a machine that breaks down on a given day with probability p. A repair can be completed in 1, 2, 3, or 4 days with respective probabilities q_1, q_2, q_3, and q_4. The condition of the machine can be represented by one of the following five states:

State	Description
0 (W)	Working
1 (D1)	Not Working, in first day of repair
2 (D2)	Not Working, in second day of repair
3 (D3)	Not Working, in third day of repair
4 (D4)	Not Working, in fourth day of repair

Since the next state of the machine depends only on its present state, the condition of the machine can be modeled as a five-state Markov chain.

Construct the transition probability matrix.

1.4　Many products are classified as either acceptable or defective. Such products are often shipped in lots, which may contain a large number of individual items. A purchaser wants assurance that the proportion of defective items in a lot is not excessive. Instead of inspecting each item in a lot, a purchaser may follow an acceptance sampling plan under which a random sample selected from the lot is inspected. An acceptance sampling plan is said to be sequential if, after each item is inspected, one of the following decisions is made: accept a lot, reject it, or inspect another item. Suppose that the proportion of defective items in a lot is denoted by p. Consider the following sequential inspection plan: accept the lot if four acceptable items are found, reject the lot if two defective items are found, or inspect another item if neither four acceptable items nor two defective items have been found.

The sequential inspection plan represents an independent trials process in which the condition of the nth item to be inspected is independent of the condition of its predecessors. Hence, the sequential inspection plan can be modeled as a Markov chain. A state is represented by a pair of numbers. The first number in the pair is the number of acceptable items inspected, and the second is the number of defective items inspected. The model is an absorbing Markov chain because when the lot is accepted or rejected, the inspection process stops in an absorbing state. The transient states indicate that the inspection process will continue. The states are indexed and identified in the table below:

State	Number Pair	Number of Acceptable Items	Number of Defective Items	State Classification
1	(0, 0)	0	0	Transient
2	(1, 0)	1	0	Transient
3	(2, 0)	2	0	Transient
4	(3, 0)	3	0	Transient
5	(4, 0)	4, stop, accept lot	0	Absorbing
6	(0, 1)	0	1	Transient
7	(1, 1)	1	1	Transient
8	(2, 1)	2	1	Transient
9	(3, 1)	3	1	Transient
10	(4, 1)	4, stop, accept lot	1	Absorbing
11	(0, 2)	0	2, stop, reject lot	Absorbing
12	(1, 2)	1	2, stop, reject lot	Absorbing
13	(2, 2)	2	2, stop, reject lot	Absorbing
14	(3, 2)	3	2, stop, reject lot	Absorbing

Construct the transition probability matrix for this 14-state absorbing Markov chain.

1.5 A credit counselor accepts customers who arrive at her office without appointments. Her office consists of a private room in which she counsels one customer, plus a small waiting room, which can hold up to two additional customers. Any customer who arrives when the waiting room is full does not enter. Each credit counseling session for one customer lasts 45 min. If no customers are present when she has finished counseling a customer, she takes a 45 min break until the beginning of the next 45 min period. The number of customers who arrive during a 45 min counseling period is an independent, identically distributed random variable. If A_n is the number of arrivals during the nth 45 min counseling period, then A_n has the following probability distribution, which is stationary in time:

Number A_n of arrivals in period n, $A_n = k$	0	1	2	3	4 or more
Probability, $p_k = P(A_n = k)$	p_0	p_1	p_2	p_3	p_4

To model this problem as a regular Markov chain, let the state X_n represent the number of customers in the counselor's office at the end of the nth 45 min counseling period, immediately after the departure of the nth customer.

Construct the transition probability matrix for the Markov chain.

1.6 Consider a small clinic staffed by one nurse who injects patients with an antiflu vaccine. The clinic has space for only four chairs. One chair is reserved for the patient being vaccinated. The other three chairs are reserved for patients who wait. The arrival process is deterministic. One new patient arrives every 20 min. A single patient arrives at the clinic at precisely on the hour, at precisely 20 min past the hour, and at precisely 40 min past the hour. Any patient who arrives to find all four chairs occupied does not enter the clinic.

If the clinic is empty, the nurse is idle. If the clinic is not empty, the number of patients that she vaccinates during a 20 min interval is an independent, identically distributed random variable. If V_n is the number of vaccinations that the nurse provides during the nth 20 min interval, then V_n has the following conditional probability distribution, which is stationary in time.

Number V_n of vaccinations in period n, $V_n = k$	0	1	2	3	4	5 or more
Probability, $p_k = P(V_n = k)$	p_0	p_1	p_2	p_3	p_4	p_5

The quantity $p_k = P(V_n = k)$ is a conditional probability, conditioned on the presence of at least k patients inside the clinic available to be vaccinated.

To model this problem as a regular Markov chain, let the state X_n represent the number of patients in the clinic at the end of the nth 20 min interval, immediately before the arrival of the nth patient.

Construct the transition probability matrix for the Markov chain.

1.7 Consider a small convenience store that is open 24 h a day, 7 days a week. The store is staffed by one checkout clerk. The convenience store has space for only three customers including the one who is being served. Customers may arrive at the store every 30 min, immediately before the hour and immediately before the half-hour. Any customer who arrives to find three customers already inside the store does not enter the store. The probability of an arrival at the end of a consecutive 30 min interval is p. If the store is empty, the clerk is idle. If the store is not empty, the probability of a service completion at the end of a consecutive 30 min interval is r.

To model this problem as a regular Markov chain, let the state X_n represent the number of customers in the store at the end of the nth consecutive 30 min interval.

Construct the transition probability matrix for the Markov chain.

1.8 Consider the following inventory system controlled by a (1, 3) ordering policy. An office supply store sells laptops. The state of the system is the inventory on hand at the beginning of the day. If the number of laptops on hand at the start of the day is less than 1 (in other words, equal to 0), then the store places an order, which is delivered immediately to raise the beginning inventory level to three laptops. If the store starts the day with 1 or more laptops in stock, no laptops are ordered. The number d_n of laptops demanded by customers during day n is an independent, identically distributed random variable which has the following stationary probability distribution:

Demand d_n on day n, $d_n = k$	0	1	2	3
$P(d_n = k)$	$P(d_n = 0)$	$P(d_n = 1)$	$P(d_n = 2)$	$P(d_n = 3)$

Construct the transition probability matrix for a four-state recurrent Markov chain model of this inventory system under a (1, 3) policy.

1.9 The following mortality data has been collected for a group of S_0 integrated circuit chips (ICs) over a 4-year period:

Age i of an IC in years, i	0	1	2	3	4
Number surviving to age i	S_0	S_1	S_2	S_3	$S_4 = 0$

Observe that at the end of the 4-year period, none of the S_0 ICs has survived, that is, all have failed. Every IC that fails is replaced with a new one at the end of the year in which it has failed. The life of an IC can be modeled as a four-state recurrent Markov chain. Let the state X_n denote the age of an IC at the end of year n.

Construct the transition probability matrix for the Markov chain.

1.10 A small refinery produces one barrel of gasoline per hour. Each barrel of gasoline has an octane rating of either 120, 110, 100, or 90. The refinery engineer models the octane rating of the gasoline as a Markov chain in which the state X_n represents the octane rating of the nth barrel of gasoline. The states of the Markov chain, $\{X_0, X_1, X_2, \ldots\}$, are shown below:

State	Octane Rating
1	120
2	110
3	100
4	90

Suppose that the Markov chain, $\{X_0, X_1, X_2, \ldots\}$, has the following transition probability matrix:

$$P = \begin{array}{c|cccc} \text{State} & 1 & 2 & 3 & 4 \\ \hline 1 & p_{11} & p_{12} & b & k-b \\ 2 & p_{21} & p_{22} & k-d & d \\ 3 & p_{31} & p_{32} & e & h-e \\ 4 & p_{41} & p_{42} & h-u & u \end{array}$$

Observe that in columns 3 and 4 of row 1 of the transition probability matrix,

$$p_{13} + p_{14} = b + (k-b) = k.$$

In columns 3 and 4 of row 2,

$$p_{23} + p_{24} = (k-d) + d = k.$$

Hence,

$$p_{11} + p_{12} = p_{21} + p_{22} = 1 - k.$$

Note that in columns 3 and 4 of row 3 of the transition probability matrix,

$$p_{33} + p_{34} = e + (h-e) = h.$$

Similarly, in columns 3 and 4 of row 4,

$$p_{43} + p_{44} = (h-u) + u = h.$$

Hence,

$$p_{31} + p_{32} = p_{41} + p_{42} = 1 - h.$$

The initial state probability vector for the four-state Markov chain is

$$p^{(0)} = \begin{bmatrix} p_1^{(0)} & p_2^{(0)} & p_3^{(0)} & p_4^{(0)} \end{bmatrix}$$
$$= [P(X_0 = 1) \quad P(X_0 = 2) \quad P(X_0 = 3) \quad P(X_0 = 4)].$$

The refinery sells only 2 grades of gasoline: premium and economy. Premium has an octane rating of 120 or 110, and economy has an octane rating of 100 or 90. The refinery engineer measures the octane rating of each barrel of gasoline produced, and classifies it as either premium (P) or economy (E). Suppose that Y_n denotes the classification given to the nth barrel of gasoline, where

$$Y_n = \begin{cases} P, & \text{if } X_n = 1 \text{ or } 2 \\ E, & \text{if } X_n = 3 \text{ or } 4 \end{cases}$$

Let G be the transition matrix for the two-state process $\{Y_0, Y_1, Y_2, ...\}$, where

$$G = \begin{array}{c|cc} \text{State} & P & E \\ \hline P & g_{PP} & g_{PE} \\ E & g_{EP} & g_{EE} \end{array}$$

The two-state process $\{Y_0, Y_1, Y_2, ...\}$ is termed partially observable because the premium grade (state P) does not distinguish between the octane ratings of 120 (state 1) and 110 (state 2). Similarly, the economy grade (state E) does not distinguish between the octane ratings of 110 (state 3) and 90 (state 4).

(a) Construct the transition probabilities, g_{ij}, for the process $\{Y_0, Y_1, Y_2, ...\}$.

(b) A four-state Markov chain, with the state space, $S = \{1, 2, 3, 4\}$, which has the relationships, $p_{11} + p_{12} = p_{21} + p_{22}$ and $p_{31} + p_{32} = p_{41} + p_{42}$, is said to be lumpable with respect to the partition $P = \{1, 2\}$ and $E = \{3, 4\}$. The lumped process $\{Y_0, Y_1, Y_2, ...\}$ is also a Markov chain. Show that, as a consequence of the lumpability of the Markov chain $\{X_0, X_1, X_2, ...\}$, the transition probabilities, g_{ij}, for the lumped process $\{Y_0, Y_1, Y_2, ...\}$ are independent of the choice of the initial state probability vector, $p^{(0)} = \begin{bmatrix} p_1^{(0)} & p_2^{(0)} & p_3^{(0)} & p_4^{(0)} \end{bmatrix}$.

Now suppose that the Markov chain $\{X_0, X_1, X_2, ...\}$ has the following transition probability matrix, such that $p_{11} + p_{12} \neq p_{21} + p_{22}$ and $p_{31} + p_{32} \neq p_{41} + p_{42}$.

$$P = \begin{array}{c|cccc} \text{State} & 1 & 2 & 3 & 4 \\ \hline 1 & p_{11} & p_{12} & p_{13} & p_{14} \\ 2 & p_{21} & p_{22} & p_{23} & p_{24} \\ 3 & p_{31} & p_{32} & p_{33} & p_{34} \\ 4 & p_{41} & p_{42} & p_{43} & p_{44} \end{array}$$

Since $p_{11} + p_{12} \neq p_{21} + p_{22}$ and $p_{31} + p_{32} \neq p_{41} + p_{42}$, the Markov chain is not lumpable with respect to the partition $P = \{1, 2\}$ and $E = \{3, 4\}$. The process $\{Y_0, Y_1, Y_2, ...\}$ is not a Markov chain.

(c) With respect to this partition, calculate the transition probabilities, g_{ij}, for the process $\{Y_0, Y_1, Y_2, \ldots\}$.

(d) Show that the transition probabilities, g_{ij}, depend
on the choice of the initial state probability vector,
$p^{(0)} = \begin{bmatrix} p_1^{(0)} & p_2^{(0)} & p_3^{(0)} & p_4^{(0)} \end{bmatrix}$.

1.11 A dam is used for generating electricity, controlling floods,
and irrigating land. The dam has a capacity of 4 units of water.
Assume that the volume of water stored in the dam is always
an integer. Let X_n be the volume of water in the dam at the end
of week n. During week n, a volume of water denoted by W_n
flows into the dam. The probability distribution of W_n, which
is an independent, identically distributed, and integer random
variable, is given below:

Volume W_n of water flowing into dam in week n, $W_n = k$	0	1	2	3	4	5 or more
Probability, $w_k = P(W_n = k)$	w_0	w_1	w_2	w_3	w_4	w_5

At the end of each week, if the dam contains one or more
units of water, exactly one unit of water is released. Whenever
the inflow of water to the dam exceeds its capacity of 4 units, the
surplus water is released over the spillway and lost.
Construct the transition probability matrix.

1.12 A certain engineering college has three academic ranks for its
faculty: assistant professor, associate professor, and professor.
The first two ranks are tenure-track. Only professors have tenure. A tenure-track faculty member may be discharged, remain
at her present rank, or be promoted to the next higher rank. A
tenure-track faculty member who is discharged is never rehired.
Only tenure-track faculty members can be promoted or discharged. A professor may remain at her present rank or retire.
All changes in rank occur at the end of an academic year.

By assigning a state to represent each academic rank, the
career path of a faculty member of this college is modeled as a
five-state absorbing multichain. The state X_n denotes a faculty
member's rank at the end of academic year n. The states are
indexed as follows:

State, X_n	Academic Rank
1	Assistant professor
2	Associate professor
3	Professor
4	Discharged
5	Retired

A tenure-track faculty member of rank i will be discharged
with probability d_i, remain at her present rank with probability
r_i, or be promoted to the next higher rank with probability p_i.

A professor may remain at her present rank with probability r_3 or retire with probability $1 - r_3$.

(a) Construct the transition probability matrix.

(b) Classify the states.

(c) Represent the transition probability matrix in the canonical form of Equation (1.43).

1.13 A certain college of management has five academic ranks for its faculty: instructor, assistant professor, associate professor, research professor, and applications professor. The first three ranks are tenure-track. Only research professors and applications professors have tenure. A tenure-track faculty member may be discharged, remain at her present rank, or be promoted to the next higher rank. Only tenure-track faculty members can be promoted or discharged. A tenure-track faculty member who is discharged is never rehired. An associate professor may apply for promotion with tenure to the rank of research professor or applications professor. A research professor may remain a research professor or switch to the position of applications professor. Similarly, an applications professor may remain an applications professor or switch to the position of research professor.

By assigning a state to represent each academic rank, the career path of a faculty member of this college is modeled as a six-state multichain. The state X_n denotes a faculty member's rank at the end of academic year n. The states are indexed as follows:

State, X_n	Academic Rank
1	Instructor
2	Assistant professor
3	Associate professor
4	Research professor
5	Applications professor
6	Discharged

An instructor or assistant professor of rank i will be discharged with probability d_i, remain at her present rank with probability r_i, or be promoted to the next higher rank with probability p_i. An associate professor may be discharged with probability d_3, remain an associate professor with probability r_3, be promoted to the rank of research professor with probability p_R, or be promoted to the rank of applications professor with probability p_A. A research professor may remain a research professor with probability r_4 or switch to the position of applications professor with probability $1 - r_4$. Similarly, an applications professor may remain an applications professor with probability r_5 or switch to the position of research professor with probability $1 - r_5$.

(a) Construct the transition probability matrix.

(b) Classify the states.

(c) Represent the transition probability matrix in canonical form.

1.14 A small optical shop is staffed by an optometrist and an optician. Customers may arrive at the shop every 30 min, immediately after the hour and immediately after the half-hour. Two stages of service are provided in the following sequence. A customer is given an eye examination by the optometrist at stage 1 followed by the fitting of glasses by the optician at stage 2. Only one chair is provided by the optometrist for a customer at stage 1, and only one chair is provided by the optician for a customer at stage 2. Both the optometrist and the optician may be idle if they have no customer, or busy if they are working with a customer. In addition, the optometrist may be blocked if she has completed work on a customer at stage 1 before the optician has completed work on a customer at stage 2.

A person who arrives at the beginning of a 30 min interval during which the optometrist is busy or blocked does not enter the shop. A customer enters the shop with probability p at the beginning of a 30 min interval during which the optometrist is idle. The optometrist begins her eye examination promptly but may not complete it during the current 30 min. interval. If the optometrist is giving an eye examination at the beginning of a 30 min interval, she has a probability r of completing it and passing the customer to the optician at the beginning of the next 30 min interval if the optician is idle. Otherwise, the customer remains with the optometrist who is blocked until the optician completes work with a customer at stage 2. If the optician is working with a customer at the beginning of a 30 min interval, she has a probability q of completing her work during the current 30 min interval so that the customer departs.

To model the operation of the optical shop as a Markov chain, let the state X_n be the pair (u, v). The quantity u is I if the optometrist is idle at the beginning of the nth 30 min interval, W if the optometrist is working at the beginning of the nth 30 min interval, and B if the optometrist is blocked at the beginning of the nth 30 min interval. The quantity v is I if the optician is idle at the beginning of the nth 30 min interval, and W if the optician is working at the beginning of the nth 30 min interval. The five states are indexed below:

State, X_n	Optometrist	Optician
1 = (I, I)	Idle	Idle
2 = (I, W)	Idle	Working
3 = (W, I)	Working	Idle
4 = (W, W)	Working	Working
5 = (B, W)	Blocked	Working

Construct the transition probability matrix.

References

1. Bhat, N., *Elements of Applied Stochastic Processes*, 2nd ed., Wiley, New York, 1985.
2. Cinlar, E., *Introduction to Stochastic Processes*, Prentice-Hall, Englewood Cliffs, NJ, 1975.
3. Clarke, A. B. and Disney R. L., *Probability and Random Processes: A First Course with Applications*, 2nd ed., Wiley, New York, 1985.
4. Kemeny, J. G., Mirkil, H., Snell, J. L., and Thompson, G. L., *Finite Mathematical Structures*, Prentice-Hall, Englewood Cliffs, NJ, 1959.
5. Kemeny, J. G. and Snell, J. L., *Finite Markov Chains*, Van Nostrand, Princeton, NJ, 1960. Reprinted by Springer-Verlag, New York, 1976.
6. Heyman, D. P. and Sobel, M. J., *Stochastic Models in Operations Research*, vol. 1, McGraw Hill, New York, 1982.

2

Regular Markov Chains

Recall from Section 1.9.1.1 that an irreducible or recurrent Markov chain is one in which all the states communicate. Therefore, all states in an irreducible chain are recurrent states. An irreducible chain has no transient states. A Markov chain is termed a regular chain if some power of the transition matrix has only positive elements. A regular Markov chain is irreducible and aperiodic. Regular chains with a finite number of states are the subject of this chapter.

2.1 Steady-State Probabilities

When a Markov chain has made a small number of transitions, its behavior is called transient, or short term, or time-dependent. The set of consecutive time periods over which a Markov chain is analyzed is called the planning horizon. Under transient conditions, the planning horizon is finite, of length T periods. As Section 1.7 demonstrates, when the planning horizon is finite, the probability state vector after n steps is completely determined by the initial state probability vector and the one-step transition probability matrix. As the number of steps or transitions increases, the affect of the initial state probability vector decreases. After a large number of transitions, the behavior of the chain changes from transient to what is called steady state, or long term, or time-independent. In the steady state, the planning horizon approaches infinity, and the state probability vector becomes independent of the initial state probability vector. In the steady state, the state probability vector becomes a fixed or stationary probability vector, which does not change over time. The fixed or stationary probability vector is called a steady-state probability vector. The entries of a steady-state probability vector are called steady-state probabilities. In the steady state, a Markov chain continues to make transitions indefinitely among its various states. Hence, a steady-state probability can be interpreted as the long run proportion of time that a particular state is occupied [1–5].

To see how to compute the steady-state probability vector for a regular Markov chain, it is instructive to begin by examining the rows of the n-step transition matrix, $P^{(n)}$, as n grows larger. It will be seen that all of the rows approach the same stationary probability vector, namely, the steady-state probability vector. Consider the four-state regular Markov chain model of the weather introduced in Section 1.3. The one-step transition probability matrix is given in Equation (1.9). The two-step transition probability matrix is calculated in Equation (1.19). The four-step transition probability matrix is calculated as

$$
P^{(4)} = P^4 = P^{(2)}P^{(2)} =
\begin{matrix} 1 \\ 2 \\ 3 \\ 4 \end{matrix}
\begin{bmatrix}
0.23 & 0.28 & 0.24 & 0.25 \\
0.22 & 0.37 & 0.23 & 0.18 \\
0.16 & 0.39 & 0.29 & 0.16 \\
0.21 & 0.42 & 0.18 & 0.19
\end{bmatrix}
\begin{bmatrix}
0.23 & 0.28 & 0.24 & 0.25 \\
0.22 & 0.37 & 0.23 & 0.18 \\
0.16 & 0.39 & 0.29 & 0.16 \\
0.21 & 0.42 & 0.18 & 0.19
\end{bmatrix}
$$

$$
=
\begin{matrix} 1 \\ 2 \\ 3 \\ 4 \end{matrix}
\begin{bmatrix}
0.2054 & 0.3666 & 0.2342 & 0.1938 \\
0.2066 & 0.3638 & 0.2370 & 0.1926 \\
0.2026 & 0.3694 & 0.2410 & 0.1870 \\
0.2094 & 0.3642 & 0.2334 & 0.1930
\end{bmatrix}. \tag{2.1}
$$

The eight-step transition probability matrix is computed as follows:

$$
P^{(8)} = P^8 = P^{(4)}P^{(4)}
$$

$$
=
\begin{matrix} 1 \\ 2 \\ 3 \\ 4 \end{matrix}
\begin{bmatrix}
0.2054 & 0.3666 & 0.2342 & 0.1938 \\
0.2066 & 0.3638 & 0.2370 & 0.1926 \\
0.2026 & 0.3694 & 0.2410 & 0.1870 \\
0.2094 & 0.3642 & 0.2334 & 0.1930
\end{bmatrix}
\begin{bmatrix}
0.2054 & 0.3666 & 0.2342 & 0.1938 \\
0.2066 & 0.3638 & 0.2370 & 0.1926 \\
0.2026 & 0.3694 & 0.2410 & 0.1870 \\
0.2094 & 0.3642 & 0.2334 & 0.1930
\end{bmatrix}
$$

$$
=
\begin{matrix} 1 \\ 2 \\ 3 \\ 4 \end{matrix}
\begin{bmatrix}
0.2060 & 0.3658 & 0.2367 & 0.1916 \\
0.2059 & 0.3658 & 0.2367 & 0.1916 \\
0.2059 & 0.3658 & 0.2367 & 0.1916 \\
0.2060 & 0.3658 & 0.2367 & 0.1916
\end{bmatrix}. \tag{2.2}
$$

Observe that as the exponent n increases from 1 to 2, from 2 to 4, and from 4 to 8, the entries of $P^{(n)}$ approach limiting values. When $n = 8$, all the rows of $P^{(8)}$ are almost identical. One may infer that as n becomes very large, all the rows of $P^{(n)}$ approach the same stationary probability vector, namely,

$$
p^{(n)} = [p_1^{(n)} \quad p_2^{(n)} \quad p_3^{(n)} \quad p_4^{(n)}] = [0.2059 \quad 0.3658 \quad 0.2367 \quad 0.1916]. \tag{2.3}
$$

That is, after n transitions, as n becomes very large, the n-step transition probability $p_{ij}^{(n)}$ approaches a limiting probability, $p_j^{(n)}$, irrespective of the starting state i. If π_j denotes the limiting probability for state j in an N-state Markov chain, then the limiting probability is defined by the formula

$$\pi_j = \lim_{n \to \infty} p_{ij}^{(n)}, \quad \text{for } j = 1, \ldots, N. \tag{2.4}$$

The limiting probability π_j is called a steady-state probability. The vector of steady-state probabilities for an N-state Markov chain is a $1 \times N$ row vector denoted by

$$\pi = [\pi_1 \quad \pi_2 \quad \ldots \quad \pi_N]. \tag{2.5}$$

Since π is a probability vector, the entries of π must sum to one. Thus,

$$\sum_{j=1}^{N} \pi_j = 1. \tag{2.6}$$

Equation (2.6) is called the normalizing equation.

The behavior of $P^{(n)}$ for the four-state regular Markov chain suggests that as $n \to \infty$, $P^{(n)}$ will converge to a matrix Π with identical rows. Each row of Π is equal to the steady-state probability vector, π. To see why this is true for all regular Markov chains, note that the n-step transition probability, $p_{ij}^{(n)}$, is the (i, j)th element of the n-step transition probability matrix, $P^{(n)} = [p_{ij}^{(n)}]$. Since $\lim_{n \to \infty} p_{ij}^{(n)} = \pi_j$, the steady-state probability for state j, it follows that the limiting transition probability matrix is

$$\lim_{n \to \infty} P^{(n)} = \lim_{n \to \infty} P^n = \Pi = \begin{matrix} 1 \\ 2 \\ \vdots \\ N \end{matrix} \begin{bmatrix} \pi \\ \pi \\ \vdots \\ \pi \end{bmatrix} = \begin{matrix} 1 \\ 2 \\ \vdots \\ N \end{matrix} \begin{bmatrix} \pi_1 & \pi_2 & \ldots & \pi_N \\ \pi_1 & \pi_2 & \ldots & \pi_N \\ \ldots & \ldots & \ldots & \ldots \\ \pi_1 & \pi_2 & \ldots & \pi_N \end{bmatrix}. \tag{2.7}$$

Thus, Π is a matrix with each row π, the steady-state probability vector.

As Section 1.9.1.1 indicates, if a Markov chain is regular, then some power of the transition matrix has only positive elements. Thus, if P is the transition probability matrix for a regular Markov chain, then after some number of steps denoted by a positive integer K, P^K has no zero entries. If P raised to the power K has only positive entries, then P raised to all powers higher than K also has only positive entries. In particular, $\Pi = \lim_{n \to \infty} P^{(n)} = \lim_{n \to \infty} P^n$ is a matrix with only positive elements. Therefore, all entries of the matrix Π are also positive. Since Π is a matrix with each row π, all the entries of π are positive. Hence, the entries of the probability vector π for a regular Markov chain are strictly positive and sum to one. For an N-state regular Markov chain,

$$\pi_j > 0 \quad \text{for } j = 1, \ldots, N, \quad \text{and} \quad \sum_{j=1}^{N} \pi_j = 1. \tag{2.8}$$

For the four-state Markov chain model of the weather, the rows of $P^{(8)}$ calculated in Equation (2.2) indicate that

$$\pi \approx \begin{bmatrix} 0.20595 & 0.3658 & 0.2367 & 0.1916 \end{bmatrix}. \tag{2.9}$$

For large n, the state probability $p_j^{(n)}$ approaches the limiting probability π_j. That is,

$$\pi_j = \lim_{n\to\infty} p_j^{(n)} = \lim_{n\to\infty} p_{ij}^{(n)}, \quad \text{for } j = 1, \dots, N, \tag{2.10}$$

and does not depend on the starting state. Thus, the vector π of steady-state probabilities is equal to the limit, as the number of transitions approaches infinity, of the vector $p^{(n)}$ of state probabilities. That is,

$$\pi = \lim_{n\to\infty} p^{(n)}. \tag{2.11}$$

An informal derivation of the equations used to calculate the steady-state probability π_j begins by conditioning on the state at epoch n.

$$P(X_{n+1} = j) = \sum_{i=1}^{N} P(X_n = i) \, P(X_{n+1} = j \mid X_n = i)$$

$$p_j^{(n+1)} = \sum_{i=1}^{N} p_i^{(n)} p_{ij}.$$

Letting $n \to \infty$,

$$\lim_{n\to\infty} p_j^{(n+1)} = \lim_{n\to\infty} \sum_{i=1}^{N} p_i^{(n)} p_{ij}.$$

Interchanging the limit and the summation,

$$\lim_{n\to\infty} p_j^{(n+1)} = \sum_{i=1}^{N} \lim_{n\to\infty} p_i^{(n)} p_{ij}$$

$$\pi_j = \sum_{i=1}^{N} \pi_i p_{ij}.$$

In matrix form,

$$\pi = \pi P$$
$$\pi I = \pi P$$
$$\pi(I - P) = 0.$$

Hence, $\pi = \pi P$ is a homogeneous system of linear equations, which has infinitely many solutions.

Since the homogeneous system $\pi = \pi P$ has infinitely many solutions, one of the equations is redundant because it can be expressed as a linear combination of the others. A unique solution for π is obtained by dropping one of the homogeneous equations and replacing it with the normalizing equation (2.6). For a regular Markov chain with N states, the quantities π_j are the unique positive solution of the following steady-state equations in algebraic form:

$$\left. \begin{aligned} \pi_j &= \sum_{i=1}^{N} \pi_i p_{ij}, \quad \text{for } j = 1, 2, 3, \dots, N \\ \sum_{i=1}^{N} \pi_j &= 1 \\ \pi_i &> 0, \quad \text{for } i = 1, 2, 3, \dots, N. \end{aligned} \right\}$$

(2.12)

The matrix form of the steady-state equations is

$$\left. \begin{aligned} \pi &= \pi P \\ \pi e &= 1 \\ \pi &> 0 \end{aligned} \right\},$$

(2.13)

where e is a column vector with all entries one.

In summary, after a large number of transitions, the behavior of a Markov chain changes from short term, or transient, to long term, or steady state. In the steady state the process continues to make transitions among the various states in accordance with the stationary one-step transition probabilities, p_{ij}. The steady-state probability, π_j, represents the long run proportion of time that the process will spend in state j. For an N-state irreducible Markov chain, the steady-state probabilities, π_j, are the components of a steady-state probability vector, π, which is a $1 \times N$ row vector. For an irreducible Markov chain, the components of π are nonnegative. For a regular Markov chain, which is both irreducible and aperiodic, all the components of π are positive.

Two approaches can be followed to solve the homogeneous system of equations $\pi = \pi P$ and the normalizing equation (2.6). The first approach is based on the observation that the homogeneous system augmented by the normalizing equation produces a linear system of $N + 1$ equations in N unknowns. The homogeneous system contains one redundant equation. In order for the augmented system to have a unique solution, one of the homogeneous equations is dropped. The equation that is dropped is replaced by the normalizing equation (2.6) to avoid the trivial solution and thereby ensure that π is a probability vector.

The second approach is to first solve the homogeneous system of N equations in N unknowns for $\pi_1, \pi_2, \ldots, \pi_{N-1}$ interms of π_N. These values are then substituted into the normalizing equation (2.6) to determine π_N. The remaining quantities π_j are equal to constants times π_N. This approach for computing the steady-state probability vector is followed by the Markov chain partitioning algorithm, which is described in Section 6.1.1 as an example of a computational procedure called state reduction.

2.1.1 Calculating Steady-State Probabilities for a Generic Two-State Markov Chain

Both approaches for calculating steady-state probabilities will be illustrated by applying them initially to the simplest kind of generic regular Markov chain, a two-state chain for which the transition probability matrix is shown in Equation (1.4), The matrix form of the steady-state equations (2.13) to be solved is

$$\left.\begin{aligned} [\pi_1 \quad \pi_2] &= [\pi_1 \quad \pi_2] \begin{bmatrix} p_{11} & p_{12} \\ p_{21} & p_{22} \end{bmatrix} \\ [\pi_1 \quad \pi_2][1 \quad 1]^{T} &= 1. \end{aligned}\right\} \quad (2.14)$$

In algebraic form, the steady-state equations (2.12) consist of the following three equations in two unknowns, π_1 and π_2:

$$\left.\begin{aligned} \pi_1 &= p_{11}\pi_1 + p_{21}\pi_2 \\ \pi_2 &= p_{12}\pi_1 + p_{22}\pi_2 \\ \pi_1 + \pi_2 &= 1. \end{aligned}\right\} \quad (2.15)$$

Note that if the substitutions $p_{12} = 1 - p_{11}$ and $p_{22} = 1 - p_{21}$ are made in the second equation, the second equation is transformed into the first equation, as shown below:

$$\pi_2 = (1 - p_{11})\pi_1 + (1 - p_{21})\pi_2$$
$$\pi_2 = \pi_1 - p_{11}\pi_1 + \pi_2 - p_{21}\pi_2$$
$$\pi_1 = p_{11}\pi_1 + p_{21}\pi_2.$$

Thus, the system of equations $\pi = \pi P$ is linearly dependent. Any one of the equations is redundant and can be discarded.

When the first approach to solving the steady-state equations is followed, and the second equation is arbitrarily deleted, the resulting system of two equations in two unknowns is shown below:

$$\pi_1 = p_{11}\pi_1 + p_{21}\pi_2,$$

or

$$(1 - p_{11})\pi_1 - p_{21}\pi_2 = 0,$$
$$\pi_1 + \pi_2 = 1.$$

After multiplying both sides of the second equation by p_{21} and adding the result to the first equation, the solution for the vector of steady-state probabilities is

$$\pi = \begin{bmatrix} \pi_1 & \pi_2 \end{bmatrix} = \begin{bmatrix} p_{21}/(p_{12} + p_{21}) & p_{12}/(p_{12} + p_{21}) \end{bmatrix}. \qquad (2.16)$$

For example, the vector of steady-state probabilities for a two-state regular Markov chain with the transition probability matrix,

$$P = \begin{array}{c} 1 \\ 2 \end{array}\begin{bmatrix} p_{11} & p_{12} \\ p_{21} & p_{22} \end{bmatrix} = \begin{array}{c} 1 \\ 2 \end{array}\begin{bmatrix} 0.2 & 0.8 \\ 0.6 & 0.4 \end{bmatrix} \qquad (2.17)$$

is

$$\pi = \begin{bmatrix} \pi_1 & \pi_2 \end{bmatrix} = \begin{bmatrix} p_{21}/(p_{12} + p_{21}) & p_{12}/(p_{12} + p_{21}) \end{bmatrix} = \begin{bmatrix} 3/7 & 4/7 \end{bmatrix}. \qquad (2.18)$$

In the second approach to solving the steady-state equations, the normalizing equation is initially ignored. The resulting system, $\pi = \pi P$, of two linearly dependent equations in two unknowns is shown below:

$$\pi_1 = p_{11}\pi_1 + p_{21}\pi_2$$
$$\pi_2 = p_{12}\pi_1 + p_{22}\pi_2$$

The first equation is solved to express π_1 as the following constant times π_2:

$$\pi_1 = \frac{p_{21}}{1 - p_{11}} \quad \pi_2 = \frac{p_{21}}{p_{12}} \pi_2.$$

This expression is substituted into the normalizing equation (2.6) to solve for π_2.

$$\frac{p_{21}}{p_{12}}\pi_2 + \pi_2 = 1.$$

Once again, the solution for the vector of steady-state probabilities is given by Equation (2.16).

2.1.2 Calculating Steady-State Probabilities for a Four-State Model of Weather

Both approaches to solving the steady-state equations will be illustrated numerically by applying them to the four-state regular Markov chain model of the weather for which the transition probability matrix is shown in Equation (1.9) of Section 1.3. The matrix form of the steady-state equations (2.13) is

$$\left. \begin{array}{c} [\pi_1 \quad \pi_2 \quad \pi_3 \quad \pi_4] = [\pi_1 \quad \pi_2 \quad \pi_3 \quad \pi_4] \begin{bmatrix} p_{11} & p_{12} & p_{13} & p_{14} \\ p_{21} & p_{22} & p_{23} & p_{24} \\ p_{31} & p_{32} & p_{33} & p_{34} \\ p_{41} & p_{42} & p_{43} & p_{44} \end{bmatrix}, \\ [\pi_1 \quad \pi_2 \quad \pi_3 \quad \pi_4][1 \quad 1 \quad 1 \quad 1]^{\mathrm{T}} = 1. \end{array} \right\} \quad (2.19)$$

When the numerical values given in Equation (1.9) are substituted for the transition probabilities, the following system of five equations in four unknowns is produced:

$$\left. \begin{array}{c} [\pi_1 \quad \pi_2 \quad \pi_3 \quad \pi_4] = [\pi_1 \quad \pi_2 \quad \pi_3 \quad \pi_4] \begin{bmatrix} 0.3 & 0.1 & 0.4 & 0.2 \\ 0.2 & 0.5 & 0.2 & 0.1 \\ 0.3 & 0.2 & 0.1 & 0.4 \\ 0 & 0.6 & 0.3 & 0.1 \end{bmatrix} \\ [\pi_1 \quad \pi_2 \quad \pi_3 \quad \pi_4][1 \quad 1 \quad 1 \quad 1]^{\mathrm{T}} = 1. \end{array} \right\} \quad (2.20)$$

In algebraic form the system of steady-state equations (2.19) is

$$\left. \begin{array}{c} \pi_1 = (0.3)\pi_1 + (0.2)\pi_2 + (0.3)\pi_3 + (0)\pi_4 \\ \pi_2 = (0.1)\pi_1 + (0.5)\pi_2 + (0.2)\pi_3 + (0.6)\pi_4 \\ \pi_3 = (0.4)\pi_1 + (0.2)\pi_2 + (0.1)\pi_3 + (0.3)\pi_4 \\ \pi_4 = (0.2)\pi_1 + (0.1)\pi_2 + (0.4)\pi_3 + (0.1)\pi_4 \\ \pi_1 + \pi_2 + \pi_3 + \pi_4 = 1. \end{array} \right\} \quad (2.21)$$

When the first approach is followed, and the fourth equation is arbitrarily deleted, the resulting system of four equations in four unknowns is shown below:

$$\pi_1 = (0.3)\pi_1 + (0.2)\pi_2 + (0.3)\pi_3 + (0)\pi_4$$
$$\pi_2 = (0.1)\pi_1 + (0.5)\pi_2 + (0.2)\pi_3 + (0.6)\pi_4$$
$$\pi_3 = (0.4)\pi_1 + (0.2)\pi_2 + (0.1)\pi_3 + (0.3)\pi_4$$
$$\pi_1 + \pi_2 + \pi_3 + \pi_4 = 1.$$

The solution of this system is

$$\pi = [\pi_1 \quad \pi_2 \quad \pi_3 \quad \pi_4] = [1407/6832 \quad 2499/6832 \quad 1617/6832 \quad 1309/6832]$$
$$= [0.2059 \quad 0.3658 \quad 0.2367 \quad 0.1916]. \tag{2.22}$$

This solution almost matches the approximate one obtained in Equation (2.9) by calculating P^8.

In the second approach, the normalizing equation (2.6) is initially ignored. The first three equations contained in the system $\pi = \pi P$ are solved to express $\pi_1, \pi_2,$ and π_3 as the following constants times π_4:

$$\pi_1 = (1407/1309) \, \pi_4, \, \pi_2 = (2499/1309) \, \pi_4, \, \text{and} \, \pi_3 = (1617/1309) \, \pi_4. \tag{2.23}$$

These values for $\pi_1, \pi_2,$ and π_3 expressed in terms of π_4 are substituted into the normalizing equation to solve for π_4.

$$1 = \pi_1 + \pi_2 + \pi_3 + \pi_4$$
$$1 = (1407/1309)\pi_4 + (2499/1309)\pi_4 + (1617/1309)\pi_4 + (1309/1309)\pi_4.$$

The result is

$$\pi_4 = (1309/6832). \tag{2.24}$$

Substituting the result for π_4 to solve for the other steady-state probabilities gives the values obtained by following the first approach:

$$\pi_1 = (1407/6832), \quad \pi_2 = (2499/6832), \quad \text{and} \quad \pi_3 = (1617/6832). \tag{2.25}$$

The steady-state probability π_i represents the long run proportion of time that the weather will be represented by state i. For example, the long run proportion of cloudy days is equal to $\pi_3 = (1617/6832) = 0.2367$.

2.1.3 Steady-State Probabilities for Four-State Model of Inventory System

The model of the inventory system described in Section 1.10.1.1.3 for which the transition matrix is given in Equation (1.48) is another example of a regular four-state Markov chain. The system of matrix equations (2.13) for finding the steady-state probabilities is

$$\left.\begin{array}{c}\begin{bmatrix} \pi_1 & \pi_2 & \pi_3 & \pi_4 \end{bmatrix} = \begin{bmatrix} \pi_1 & \pi_2 & \pi_3 & \pi_4 \end{bmatrix}\begin{bmatrix} 0.2 & 0.1 & 0.4 & 0.3 \\ 0.2 & 0.1 & 0.4 & 0.3 \\ 0.3 & 0.4 & 0.3 & 0 \\ 0.2 & 0.1 & 0.4 & 0.3 \end{bmatrix} \\ \begin{bmatrix} \pi_1 & \pi_2 & \pi_3 & \pi_4 \end{bmatrix}\begin{bmatrix} 1 & 1 & 1 & 1 \end{bmatrix}^{T} = 1. \end{array}\right\} \qquad (2.26)$$

The solution of the system (2.26) is

$$\pi = \begin{bmatrix} \pi_0 & \pi_1 & \pi_2 & \pi_3 \end{bmatrix} = \begin{bmatrix} 0.4167 & 0.3333 & 0.2083 & 0.0417 \end{bmatrix}. \qquad (2.27)$$

The steady-state probability π_i represents the long run proportion of time that the retailer will have i computers in stock. For example, the long run proportion of time that the retailer will have no computers in stock is given by $\pi_0 = 0.2364$. Similarly, the long run proportion of time that she will have three computers in stock is equal to $\pi_3 = 0.1909$.

2.1.4 Steady-State Probabilities for Four-State Model of Component Replacement

The model of component replacement described in Section 1.10.1.1.5 is also a regular four-state Markov chain. The transition matrix is given in Equation (1.50). To obtain the age distribution of the components in an electronic device, which has been in operation for a long time, the steady-state probabilities are required. The matrix equations for finding the steady-state probabilities (2.13) are

$$\left.\begin{array}{c}\begin{bmatrix} \pi_1 & \pi_2 & \pi_3 & \pi_4 \end{bmatrix} = \begin{bmatrix} \pi_1 & \pi_2 & \pi_3 & \pi_4 \end{bmatrix}\begin{bmatrix} 0.2 & 0.8 & 0 & 0 \\ 0.375 & 0 & 0.625 & 0 \\ 0.8 & 0 & 0 & 0.2 \\ 1 & 0 & 0 & 0 \end{bmatrix} \\ \begin{bmatrix} \pi_1 & \pi_2 & \pi_3 & \pi_4 \end{bmatrix}\begin{bmatrix} 1 & 1 & 1 & 1 \end{bmatrix}^{T} = 1. \end{array}\right\} \qquad (2.28)$$

The solution of the system (2.28) is

$$\pi = \begin{bmatrix} \pi_0 & \pi_1 & \pi_2 & \pi_3 \end{bmatrix} = \begin{bmatrix} 0.4167 & 0.3333 & 0.2083 & 0.0417 \end{bmatrix}. \qquad (2.29)$$

The steady-state probability π_i represents the long run probability that a randomly selected component is i weeks old. For example, the long run probability that a component has just been replaced is given by $\pi_0 = 0.4167$. Similarly, the long run probability that a component has been in service 3 weeks is equal to $\pi_3 = 0.0417$.

2.2 First Passage to a Target State

The number of time periods, or steps, needed to move from a starting state i to a destination or target state j for the first time is called a first passage time.

2.2.1 Probability of First Passage in n Steps

The first passage time from state i to state j is a random variable denoted by T_{ij}. If $i = j$, then T_{ii} is called the recurrence time for state i. If $T_{ij} = n$, then the first passage time from state i to state j equals n steps. Let $f_{ij}^{(n)}$ denote the probability that the first passage time from state i to state j equals n steps [1–3]. That is,

$$f_{ij}^{(n)} = P(T_{ij} = n) \qquad (2.30)$$

Consider a generic four-state regular Markov chain with the transition probability matrix shown in Equation (1.3).

$$f_j^{(n)} = [f_{ij}^{(n)}] = \begin{bmatrix} f_{1j}^{(n)} \\ f_{2j}^{(n)} \\ f_{3j}^{(n)} \\ f_{4j}^{(n)} \end{bmatrix}, \quad \text{where } j = 1, 2, 3, \text{ or } 4 \qquad (2.31)$$

denote the column vector of n-step first passage time probabilities to a target state j. To obtain a formula for computing the probability distribution of the n-step first passage times, let $j = 1$. Suppose that $f_{41}^{(n)}$, the distribution of n-step first passage time probabilities from state 4 to a target state 1 is desired. To compute $f_{41}^{(n)}$, one may start with $n = 1$. The probability of going from state 4 to state 1 for the first time in one step, $f_{41}^{(1)}$, is simply the one-step transition probability, p_{41}. That is,

$$f_{41}^{(1)} = p_{41}^{(1)} = p_{41}. \tag{2.32}$$

To move from state 4 to state 1 for the first time in two steps, the chain must go from state 4 to any nontarget state k different from state 1 on the first step, and from that nontarget state k to state 1 on the second step. Therefore, for $k = 2, 3, 4,$

$$f_{41}^{(2)} = p_{42}p_{21} + p_{43}p_{31} + p_{44}p_{41} = p_{42}f_{21}^{(1)} + p_{43}f_{31}^{(1)} + p_{44}f_{41}^{(1)}. \tag{2.33}$$

To move from state 4 to state 1 for the first time in three steps, the chain must go from state 4 to any nontarget state k different from state 1 on the first step, and from that nontarget state k to state 1 for the first time after two additional steps. Hence, for $k = 2, 3, 4,$

$$f_{41}^{(3)} = p_{42}f_{21}^{(2)} + p_{43}f_{31}^{(2)} + p_{44}f_{41}^{(2)}. \tag{2.34}$$

By induction, for $n = 2, 3, ...,$ one can show that

$$f_{41}^{(n)} = p_{42}f_{21}^{(n-1)} + p_{43}f_{31}^{(n-1)} + p_{44}f_{41}^{(n-1)} = \sum_{k \neq 1} p_{4k}f_{k1}^{(n-1)}.$$

Thus, the n-step probability distribution of first passage times from a state i to a target state j can be computed recursively by using the algebraic formula

$$f_{ij}^{(n)} = \sum_{k \neq j} p_{ik}f_{kj}^{(n-1)}. \tag{2.35}$$

To express formula (2.35) for recursively calculating $f_{ij}^{(n)}$ in matrix form, let j denote the target state in a generic four-state regular Markov chain. Suppose that $j = 1$. Let column vector $f_1^{(1)}$ denote column 1 of P, so that

$$f_1^{(1)} = \begin{bmatrix} p_{11} \\ p_{21} \\ p_{31} \\ p_{41} \end{bmatrix} = \begin{bmatrix} f_{11}^{(1)} \\ f_{21}^{(1)} \\ f_{31}^{(1)} \\ f_{41}^{(1)} \end{bmatrix}. \tag{2.36}$$

Thus $f_1^{(1)}$ is the column vector of one-step transition probabilities from any state to state 1, the target state. Let the matrix Z be the matrix P with column j of the target state replaced by a column of zeroes. When $j = 1$,

$$Z = \begin{bmatrix} 0 & p_{12} & p_{13} & p_{14} \\ 0 & p_{22} & p_{23} & p_{24} \\ 0 & p_{32} & p_{33} & p_{34} \\ 0 & p_{42} & p_{43} & p_{44} \end{bmatrix}. \tag{2.37}$$

Matrix Z is the matrix of one-step transition probabilities from any state to any nontarget state, because the column of the target state has been replaced with zeroes. Hence, the chain has zero probability of entering the target state in one step. The recursive matrix equations corresponding to the algebraic equations are given below:

$$\begin{aligned} f_j^{(1)} &= f_j^{(1)} = Z^0 f_j^{(1)} \\ f_j^{(2)} &= Z f_j^{(1)} = Z^1 f_j^{(1)} \\ f_j^{(3)} &= Z f_j^{(2)} = Z(Z f_j^{(1)}) = Z^2 f_j^{(1)} \\ f_j^{(4)} &= Z f_j^{(3)} = Z(Z^2 f_j^{(1)}) = Z^3 f_j^{(1)}. \end{aligned} \tag{2.38}$$

After n steps

$$f_j^{(n)} = Z f_j^{(n-1)} = Z^{n-1} f_j^{(1)}.$$

Matrix Z is the matrix of probabilities of not entering the target state in one step because all entries in the column of the target state are zero. Therefore, Z^{n-1} is the matrix of probabilities of not entering the target state in $n-1$ steps. Vector $f_j^{(1)}$ is the vector of probabilities of entering the target state in one step. Hence, $f_j^{(n)} = Z^{n-1} f_j^{(1)}$ is the vector of probabilities of not entering the target state during the first $n-1$ steps, and then entering the target state for the first time on the nth step.

To see that these recursive matrix formulas are equivalent to the corresponding algebraic formulas, they will be applied to a generic regular four-state Markov chain to calculate the vectors $f_1^{(2)}$ and $f_1^{(3)}$ when state 1 is the target state. The column vector of probabilities of first passage to state 1 in two steps is represented by

$$f_1^{(2)} = Z f_1^{(1)} = \begin{bmatrix} 0 & p_{12} & p_{13} & p_{14} \\ 0 & p_{22} & p_{23} & p_{24} \\ 0 & p_{32} & p_{33} & p_{34} \\ 0 & p_{42} & p_{43} & p_{44} \end{bmatrix} \begin{bmatrix} p_{11} \\ p_{21} \\ p_{31} \\ p_{41} \end{bmatrix} = \begin{bmatrix} p_{12}p_{21} + p_{13}p_{31} + p_{14}p_{41} \\ p_{22}p_{21} + p_{23}p_{31} + p_{24}p_{41} \\ p_{32}p_{21} + p_{33}p_{31} + p_{34}p_{41} \\ p_{42}p_{21} + p_{43}p_{31} + p_{44}p_{41} \end{bmatrix} = \begin{bmatrix} f_{11}^{(2)} \\ f_{21}^{(2)} \\ f_{31}^{(2)} \\ f_{41}^{(2)} \end{bmatrix}. \tag{2.39}$$

Observe that $f_{41}^{(2)}$, the fourth entry in column vector $f_1^{(2)}$, is given by

$$f_{41}^{(2)} = p_{42} f_{21}^{(1)} + p_{43} f_{31}^{(1)} + p_{44} f_{41}^{(1)} = p_{42} p_{21} + p_{43} p_{31} + p_{44} p_{41},$$

in agreement with the algebraic formula (2.32) obtained earlier. The vector of probabilities of first passage to state 1 in three steps is given by

$$f_1^{(3)} = Z^2 f_1^{(1)} = \begin{bmatrix} 0 & p_{12} & p_{13} & p_{14} \\ 0 & p_{22} & p_{23} & p_{24} \\ 0 & p_{32} & p_{33} & p_{34} \\ 0 & p_{42} & p_{43} & p_{44} \end{bmatrix} \begin{bmatrix} 0 & p_{12} & p_{13} & p_{14} \\ 0 & p_{22} & p_{23} & p_{24} \\ 0 & p_{32} & p_{33} & p_{34} \\ 0 & p_{42} & p_{43} & p_{44} \end{bmatrix} \begin{bmatrix} f_{11}^{(1)} \\ f_{21}^{(1)} \\ f_{31}^{(1)} \\ f_{41}^{(1)} \end{bmatrix} = \begin{bmatrix} f_{11}^{(3)} \\ f_{21}^{(3)} \\ f_{31}^{(3)} \\ f_{41}^{(3)} \end{bmatrix}$$

$$= \begin{bmatrix} 0 & p_{12}p_{22} + p_{13}p_{32} + p_{14}p_{42} & p_{12}p_{23} + p_{13}p_{33} + p_{14}p_{43} & p_{12}p_{24} + p_{13}p_{34} + p_{14}p_{44} \\ 0 & p_{22}p_{22} + p_{23}p_{32} + p_{24}p_{42} & p_{22}p_{23} + p_{23}p_{33} + p_{24}p_{43} & p_{22}p_{24} + p_{23}p_{34} + p_{24}p_{44} \\ 0 & p_{32}p_{22} + p_{33}p_{32} + p_{34}p_{42} & p_{32}p_{23} + p_{33}p_{33} + p_{34}p_{43} & p_{32}p_{24} + p_{33}p_{34} + p_{34}p_{44} \\ 0 & p_{42}p_{22} + p_{43}p_{32} + p_{44}p_{42} & p_{42}p_{23} + p_{43}p_{33} + p_{44}p_{43} & p_{42}p_{24} + p_{43}p_{34} + p_{44}p_{44} \end{bmatrix} \begin{bmatrix} p_{11} \\ p_{21} \\ p_{31} \\ p_{41} \end{bmatrix}$$

$$= \begin{bmatrix} (p_{12}p_{22} + p_{13}p_{32} + p_{14}p_{42})p_{21} + (p_{12}p_{23} + p_{13}p_{33} + p_{14}p_{43})p_{31} + (p_{12}p_{24} + p_{13}p_{34} + p_{14}p_{44})p_{41} \\ (p_{22}p_{22} + p_{23}p_{32} + p_{24}p_{42})p_{21} + (p_{22}p_{23} + p_{23}p_{33} + p_{24}p_{43})p_{31} + (p_{22}p_{24} + p_{23}p_{34} + p_{24}p_{44})p_{41} \\ (p_{32}p_{22} + p_{33}p_{32} + p_{34}p_{42})p_{21} + (p_{32}p_{23} + p_{33}p_{33} + p_{34}p_{43})p_{31} + (p_{32}p_{24} + p_{33}p_{34} + p_{34}p_{44})p_{41} \\ (p_{42}p_{22} + p_{43}p_{32} + p_{44}p_{42})p_{21} + (p_{42}p_{23} + p_{43}p_{33} + p_{44}p_{43})p_{31} + (p_{42}p_{24} + p_{43}p_{34} + p_{44}p_{44})p_{41} \end{bmatrix}$$

$$(2.40)$$

Observe that $f_{41}^{(3)}$, the fourth entry in column vector $f_1^{(3)}$, is given by

$$\begin{aligned} f_{41}^{(3)} &= (p_{42}p_{22} + p_{43}p_{32} + p_{44}p_{42})p_{21} + (p_{42}p_{23} + p_{43}p_{33} + p_{44}p_{43})p_{31} \\ &\quad + (p_{42}p_{24} + p_{43}p_{34} + p_{44}p_{44})p_{41} \\ &= p_{42}(p_{22}p_{21} + p_{23}p_{31} + p_{24}p_{41}) + p_{43}(p_{32}p_{21} + p_{33}p_{31} + p_{34}p_{41}) \\ &\quad + p_{44}(p_{42}p_{21} + p_{43}p_{31} + p_{44}p_{41}) \\ &= p_{42} f_{21}^{(2)} + p_{43} f_{31}^{(2)} + p_{44} f_{41}^{(2)}, \end{aligned}$$

in agreement with the algebraic formula (2.33) obtained earlier.
    Alternatively,

$$f_1^{(3)} = Z f_1^{(2)} = \begin{bmatrix} 0 & p_{12} & p_{13} & p_{14} \\ 0 & p_{22} & p_{23} & p_{24} \\ 0 & p_{32} & p_{33} & p_{34} \\ 0 & p_{42} & p_{43} & p_{44} \end{bmatrix} \begin{bmatrix} f_{11}^{(2)} \\ f_{21}^{(2)} \\ f_{31}^{(2)} \\ f_{41}^{(2)} \end{bmatrix} = \begin{bmatrix} p_{12} f_{21}^{(2)} + p_{13} f_{31}^{(2)} + p_{14} f_{41}^{(2)} \\ p_{22} f_{21}^{(2)} + p_{23} f_{31}^{(2)} + p_{24} f_{41}^{(2)} \\ p_{32} f_{21}^{(2)} + p_{33} f_{31}^{(2)} + p_{34} f_{41}^{(2)} \\ p_{42} f_{21}^{(2)} + p_{43} f_{31}^{(2)} + p_{44} f_{41}^{(2)} \end{bmatrix} = \begin{bmatrix} f_{11}^{(3)} \\ f_{21}^{(3)} \\ f_{31}^{(3)} \\ f_{41}^{(3)} \end{bmatrix}.$$

$$(2.41)$$

Observe that $f_{41}^{(3)}$, the fourth entry in column vector $f_1^{(3)}$, is given by

$$f_{41}^{(3)} = p_{42} f_{21}^{(2)} + p_{43} f_{31}^{(2)} + p_{44} f_{41}^{(2)},$$

in agreement with the algebraic formula (2.33) obtained earlier.

To calculate numerical probabilities of first passage in n-steps, consider the four-state regular Markov chain model of the weather introduced in Section 1.3. The transition probability matrix is given in Equation (1.9). If state 1 (rain) is the target state, then

$$f_1^{(1)} = \begin{array}{c} 1 \\ 2 \\ 3 \\ 4 \end{array}\begin{bmatrix} 0.3 \\ 0.2 \\ 0.3 \\ 0 \end{bmatrix}, \quad \text{and} \quad Z = \begin{bmatrix} 0 & 0.1 & 0.4 & 0.2 \\ 0 & 0.5 & 0.2 & 0.1 \\ 0 & 0.2 & 0.1 & 0.4 \\ 0 & 0.6 & 0.3 & 0.1 \end{bmatrix}. \tag{2.42}$$

The vectors of n-step first passage probabilities, for $n = 2, 3$, and 4, are calculated below, along with Z^{n-1}:

$$f_1^{(2)} = Z f_1^{(1)} = \begin{array}{c} 1 \\ 2 \\ 3 \\ 4 \end{array}\begin{bmatrix} 0 & 0.1 & 0.4 & 0.2 \\ 0 & 0.5 & 0.2 & 0.1 \\ 0 & 0.2 & 0.1 & 0.4 \\ 0 & 0.6 & 0.3 & 0.1 \end{bmatrix}\begin{bmatrix} 0.3 \\ 0.2 \\ 0.3 \\ 0 \end{bmatrix} = \begin{bmatrix} 0.14 \\ 0.16 \\ 0.07 \\ 0.21 \end{bmatrix} = \begin{bmatrix} f_{11}^{(2)} \\ f_{21}^{(2)} \\ f_{31}^{(2)} \\ f_{41}^{(2)} \end{bmatrix} \tag{2.43}$$

$$f_1^{(3)} = Z f_1^{(2)} = \begin{array}{c} 1 \\ 2 \\ 3 \\ 4 \end{array}\begin{bmatrix} 0 & 0.1 & 0.4 & 0.2 \\ 0 & 0.5 & 0.2 & 0.1 \\ 0 & 0.2 & 0.1 & 0.4 \\ 0 & 0.6 & 0.3 & 0.1 \end{bmatrix}\begin{bmatrix} 0.14 \\ 0.16 \\ 0.07 \\ 0.21 \end{bmatrix} = \begin{bmatrix} 0.086 \\ 0.115 \\ 0.123 \\ 0.138 \end{bmatrix} = \begin{bmatrix} f_{11}^{(3)} \\ f_{21}^{(3)} \\ f_{31}^{(3)} \\ f_{41}^{(3)} \end{bmatrix}. \tag{2.44}$$

Alternatively,

$$f_1^{(3)} = Z^2 f_1^{(1)} = \begin{array}{c} 1 \\ 2 \\ 3 \\ 4 \end{array}\begin{bmatrix} 0 & 0.25 & 0.12 & 0.19 \\ 0 & 0.35 & 0.15 & 0.14 \\ 0 & 0.36 & 0.17 & 0.10 \\ 0 & 0.42 & 0.18 & 0.19 \end{bmatrix}\begin{bmatrix} 0.3 \\ 0.2 \\ 0.3 \\ 0 \end{bmatrix} = \begin{bmatrix} 0.086 \\ 0.115 \\ 0.123 \\ 0.138 \end{bmatrix} = \begin{bmatrix} f_{11}^{(3)} \\ f_{21}^{(3)} \\ f_{31}^{(3)} \\ f_{41}^{(3)} \end{bmatrix} \tag{2.45}$$

$$f_1^{(4)} = Zf_1^{(3)} = \begin{array}{c} 1 \\ 2 \\ 3 \\ 4 \end{array}\begin{bmatrix} 0 & 0.1 & 0.4 & 0.2 \\ 0 & 0.5 & 0.2 & 0.1 \\ 0 & 0.2 & 0.1 & 0.4 \\ 0 & 0.6 & 0.3 & 0.1 \end{bmatrix}\begin{bmatrix} 0.086 \\ 0.115 \\ 0.123 \\ 0.138 \end{bmatrix} = \begin{bmatrix} 0.0883 \\ 0.0959 \\ 0.0905 \\ 0.1197 \end{bmatrix} = \begin{bmatrix} f_{11}^{(4)} \\ f_{21}^{(4)} \\ f_{31}^{(4)} \\ f_{41}^{(4)} \end{bmatrix}. \qquad (2.46)$$

Alternatively,

$$f_1^{(4)} = Z^3 f_1^{(1)} = \begin{array}{c} 1 \\ 2 \\ 3 \\ 4 \end{array}\begin{bmatrix} 0 & 0.263 & 0.119 & 0.092 \\ 0 & 0.289 & 0.127 & 0.109 \\ 0 & 0.274 & 0.119 & 0.114 \\ 0 & 0.360 & 0.159 & 0.133 \end{bmatrix}\begin{bmatrix} 0.3 \\ 0.2 \\ 0.3 \\ 0 \end{bmatrix} = \begin{bmatrix} 0.0883 \\ 0.0959 \\ 0.0905 \\ 0.1197 \end{bmatrix} = \begin{bmatrix} f_{11}^{(4)} \\ f_{21}^{(4)} \\ f_{31}^{(4)} \\ f_{41}^{(4)} \end{bmatrix}. \qquad (2.47)$$

The probability that the chain moves from state 4 to target state 1 for the first time in four steps is given by $f_{41}^{(4)} = 0.1197$. Therefore, the probability that the next rainy day (state 1) will appear for the first time 4 days after a sunny day (state 4) is 0.1197.

2.2.2 Mean First Passage Times

Recall from Section 2.2.1 that the first passage time from a state i to a target state j is a random variable denoted by T_{ij}. The mean first passage time from a state i to a state j is denoted by m_{ij} and is abbreviated as MFPT [1, 4, 5]. The MFPT is defined by the formula for mathematical expectation,

$$m_{ij} = E(T_{ij}) = \sum_{n=1}^{\infty} nP(T_{ij} = n) = \sum_{n=1}^{\infty} nf_{ij}^{(n)}, \qquad (2.48)$$

where, as Equation (2.30) indicates, $f_{ij}^{(n)} = P(T_{ij} = n)$ is the probability distribution of T_{ij}. However, calculating $f_{ij}^{(n)}$ for all n is not feasible because the number of steps in a first passage is unbounded. Hence, Equation (2.48) cannot be used to compute the MFPTs.

2.2.2.1 MFPTs for a Five-State Markov Chain

As this section will demonstrate, the MFPTs can be obtained by solving a system of linear equations in which the unknowns are the entries m_{ij} to a target state j. To see how to compute the MFPTs for a regular Markov chain, consider the following example of a generic Markov chain with five states, indexed j, 1, 2, 3, and 4. Suppose that the transition probability matrix, denoted by P, is

$$
P = \begin{matrix} j \\ 1 \\ 2 \\ 3 \\ 4 \end{matrix} \begin{bmatrix} p_{jj} & p_{j1} & p_{j2} & p_{j3} & p_{j4} \\ p_{1j} & p_{11} & p_{12} & p_{13} & p_{14} \\ p_{2j} & p_{21} & p_{22} & p_{23} & p_{24} \\ p_{3j} & p_{31} & p_{32} & p_{33} & p_{34} \\ p_{4j} & p_{41} & p_{42} & p_{43} & p_{44} \end{bmatrix}. \tag{2.49}
$$

The vector of MFPTs can be found by following a procedure, which is also used in Section 3.3.1. The procedure begins by making the target state j an absorbing state. (As Sections 1.4.1.1 and 1.8 indicate, when a state j is an absorbing state, $p_{jj} = 1$, and the chain never leaves state j.) All the remaining states become transient states. The modified transition matrix represents an absorbing Markov chain with a single absorbing state, j. In the terminology of Section 1.9.1.2, the Markov chain has been transformed into an absorbing unichain. The chain stops as soon as it enters state j for the first time. When the chain stops, it is said to have been absorbed in state j. Therefore, the MFPT from a state i to a target state j in the original chain is also the mean time to absorption from a transient state i to the absorbing state j in the modified chain. The positions of the target state and the first state may be interchanged, if necessary, to make the target state the first state. In this example the target state is already the first state, so that no interchange is necessary. The transition probability matrix, P_M, for the modified absorbing Markov chain is partitioned, as shown below, to put it in the canonical form of Equation (1.43):

$$
P_M = \begin{matrix} j \\ 1 \\ 2 \\ 3 \\ 4 \end{matrix} \left[\begin{array}{c|cccc} 1 & 0 & 0 & 0 & 0 \\ \hline p_{1j} & p_{11} & p_{12} & p_{13} & p_{14} \\ p_{2j} & p_{21} & p_{22} & p_{23} & p_{24} \\ p_{3j} & p_{31} & p_{32} & p_{33} & p_{34} \\ p_{4j} & p_{41} & p_{42} & p_{43} & p_{44} \end{array} \right] = \begin{bmatrix} 1 & 0 \\ D & Q \end{bmatrix}, \tag{2.50a}
$$

where

$$
D = \begin{matrix} 1 \\ 2 \\ 3 \\ 4 \end{matrix} \begin{bmatrix} p_{1j} \\ p_{2j} \\ p_{3j} \\ p_{4j} \end{bmatrix}, \quad Q = \begin{matrix} 1 \\ 2 \\ 3 \\ 4 \end{matrix} \begin{bmatrix} p_{11} & p_{12} & p_{13} & p_{14} \\ p_{21} & p_{22} & p_{23} & p_{24} \\ p_{31} & p_{32} & p_{33} & p_{34} \\ p_{41} & p_{42} & p_{43} & p_{44} \end{bmatrix}. \tag{2.50b}
$$

In this partition, the submatrix Q is formed by deleting row j and column j of the original transition probability matrix P.

Suppose that the MFPT from state 4 to state j, that is, $m_{4j} = E(T_{4j})$, is of interest. To obtain a formula for computing the MFPTs, one may condition on the outcome of the first step, multiply this outcome by the probability that it occurs, and sum these products over all possible outcomes. Given that the process is initially in state 4, then either (a) the first step, with probability p_{4j}, is to target state j, in which case the MFPT is exactly one step, or (b) the first step, with probability p_{4k}, is to a nontarget state, $k \neq j$, in which case the MFPT will be $(1 + m_{kj})$, equal to the one step already taken plus m_{kj}, the MFPT from nontarget state k to target state j. Weighting each outcome by the probability that it occurs produces the following formula for the MFPT from state 4 to state j:

$$
\begin{aligned}
m_{4j} &= p_{4j}(1) + p_{41}(1 + m_{1j}) + p_{42}(1 + m_{2j}) + p_{43}(1 + m_{3j}) + p_{44}(1 + m_{4j}) \\
&= (p_{4j} + p_{41} + p_{42} + p_{43} + p_{44}) + p_{41}m_{1j} + p_{42}m_{2j} + p_{43}m_{3j} + p_{44}m_{4j} \quad (2.51)\\
&= 1 + p_{41}m_{1j} + p_{42}m_{2j} + p_{43}m_{3j} + p_{44}m_{4j}
\end{aligned}
$$

$$
= 1 + \sum_{k=1}^{4} p_{4k}m_{kj}. \tag{2.52}
$$

Since m_{4j} in a five-state regular Markov chain is a linear function of the four unknowns, m_{1j}, m_{2j}, m_{3j}, and m_{4j}, m_{4j} cannot be found by solving a single equation. Instead, the following linear system of four equations in the four unknowns, m_{1j}, m_{2j}, m_{3j}, and m_{4j}, must be solved to find the MFPTs to a target state j:

$$
\left.
\begin{aligned}
m_{1j} &= 1 + p_{11}m_{1j} + p_{12}m_{2j} + p_{13}m_{3j} + p_{14}m_{4j} \\
m_{2j} &= 1 + p_{21}m_{1j} + p_{22}m_{2j} + p_{23}m_{3j} + p_{24}m_{4j} \\
m_{3j} &= 1 + p_{31}m_{1j} + p_{32}m_{2j} + p_{33}m_{3j} + p_{34}m_{4j} \\
m_{4j} &= 1 + p_{41}m_{1j} + p_{42}m_{2j} + p_{43}m_{3j} + p_{44}m_{4j}
\end{aligned}
\right\}. \tag{2.53}
$$

In the general case of an $(N + 1)$-state regular Markov chain, given a starting state i and a target state j, the algebraic formula for computing the MFPTs is the system of N linear equations

$$
m_{ij} = 1 + \sum_{k \neq j} p_{ik}m_{kj}. \tag{2.54}
$$

For the five-state regular Markov chain, the matrix form of the system of algebraic equations (2.53) for calculating the vector of MFPTs to a target state j is

$$
\begin{bmatrix} m_{1j} \\ m_{2j} \\ m_{3j} \\ m_{4j} \end{bmatrix} = \begin{bmatrix} 1 \\ 1 \\ 1 \\ 1 \end{bmatrix} + \begin{bmatrix} p_{11} & p_{12} & p_{13} & p_{14} \\ p_{21} & p_{22} & p_{23} & p_{24} \\ p_{31} & p_{32} & p_{33} & p_{34} \\ p_{41} & p_{42} & p_{43} & p_{44} \end{bmatrix} \begin{bmatrix} m_{1j} \\ m_{2j} \\ m_{3j} \\ m_{4j} \end{bmatrix}. \tag{2.55}
$$

For an $(N + 1)$-state regular Markov chain, the concise form of the matrix equation for computing the vector of MFPTs to a target state j is

$$m_j = e + Qm_j, \qquad (2.56)$$

where

$$m_j = \begin{bmatrix} m_{1j} & m_{2j} & \cdots & m_{Nj} \end{bmatrix}^{\mathrm{T}}$$

is an $N \times 1$ column vector of MFPTs to a target state j, and e is an $N \times 1$ column vector with all entries one. Equation (2.56) is the matrix form of Equation (2.54).

To see how to calculate the MFPTs for a regular Markov chain with numerical transition probabilities, consider the following example of a Markov chain with five states, indexed j, 1, 2, 3, and 4. The transition probability matrix is

$$P = \begin{array}{c} j \\ 1 \\ 2 \\ 3 \\ 4 \end{array} \begin{bmatrix} 0.3 & 0.1 & 0.2 & 0.4 & 0 \\ 0 & 0.4 & 0.1 & 0.2 & 0.3 \\ 0.2 & 0.1 & 0.2 & 0.3 & 0.2 \\ 0.1 & 0.2 & 0.1 & 0.4 & 0.2 \\ 0 & 0.3 & 0.4 & 0.2 & 0.1 \end{bmatrix}. \qquad (2.57)$$

Suppose that the vector of MFPTs to target state j is desired. When target state j is made an absorbing state, the modified transition probability matrix, P_M, is partitioned in the following manner to put it in canonical form of Equation (1.43):

$$P_M = \begin{array}{c|c|cccc} \text{State} & j & 1 & 2 & 3 & 4 \\ \hline j & 1 & 0 & 0 & 0 & 0 \\ 1 & 0 & 0.4 & 0.1 & 0.2 & 0.3 \\ 2 & 0.2 & 0.1 & 0.2 & 0.3 & 0.2 \\ 3 & 0.1 & 0.2 & 0.1 & 0.4 & 0.2 \\ 4 & 0 & 0.3 & 0.4 & 0.2 & 0.1 \end{array} = \begin{bmatrix} 1 & 0 \\ D & Q \end{bmatrix}, \qquad (2.58)$$

where

$$D = \begin{array}{c} 1 \\ 2 \\ 3 \\ 4 \end{array} \begin{bmatrix} 0 \\ 0.2 \\ 0.1 \\ 0 \end{bmatrix}, \quad Q = \begin{array}{c} 1 \\ 2 \\ 3 \\ 4 \end{array} \begin{bmatrix} 0.4 & 0.1 & 0.2 & 0.3 \\ 0.1 & 0.2 & 0.3 & 0.2 \\ 0.2 & 0.1 & 0.4 & 0.2 \\ 0.3 & 0.4 & 0.2 & 0.1 \end{bmatrix}, \quad e = \begin{array}{c} 1 \\ 2 \\ 3 \\ 4 \end{array} \begin{bmatrix} 1 \\ 1 \\ 1 \\ 1 \end{bmatrix}. \qquad (2.59)$$

In algebraic form, the Equations (2.53) or (2.54) for the MFPTs are

$$\left.\begin{aligned}
m_{1j} &= 1 + (0.4)m_{1j} + (0.1)m_{2j} + (0.2)m_{3j} + (0.3)m_{4j} \\
m_{2j} &= 1 + (0.1)m_{1j} + (0.2)m_{2j} + (0.3)m_{3j} + (0.2)m_{4j} \\
m_{3j} &= 1 + (0.2)m_{1j} + (0.1)m_{2j} + (0.4)m_{3j} + (0.2)m_{4j} \\
m_{4j} &= 1 + (0.3)m_{1j} + (0.4)m_{2j} + (0.2)m_{3j} + (0.1)m_{4j}
\end{aligned}\right\}. \qquad (2.60)$$

The matrix form of these Equations (2.54) or (2.55) is

$$\begin{bmatrix} m_{1j} \\ m_{2j} \\ m_{3j} \\ m_{4j} \end{bmatrix} = \begin{bmatrix} 1 \\ 1 \\ 1 \\ 1 \end{bmatrix} + \begin{bmatrix} 0.4 & 0.1 & 0.2 & 0.3 \\ 0.1 & 0.2 & 0.3 & 0.2 \\ 0.2 & 0.1 & 0.4 & 0.2 \\ 0.3 & 0.4 & 0.2 & 0.1 \end{bmatrix} \begin{bmatrix} m_{1j} \\ m_{2j} \\ m_{3j} \\ m_{4j} \end{bmatrix}. \qquad (2.61)$$

The solution for the vector m_j of MFPTs to target state j, is

$$m_j = \begin{bmatrix} m_{1j} & m_{2j} & m_{3j} & m_{4j} \end{bmatrix}^{\mathrm{T}} = \begin{bmatrix} 15.7143 & 12.1164 & 13.8624 & 14.8148 \end{bmatrix}^{\mathrm{T}}. \qquad (2.62)$$

For example, the mean number of steps to move from state 2 to state j for the first time is given by $m_{2j} = 12.1164$. Observe that matrix Q in Equation (2.59) is identical to matrix Q in Equation (3.20). However, the entries in the vector of MFPTs in Equation (2.62) differ slightly from those calculated by different methods in Equations (3.22) and (6.68). Discrepancies are due to roundoff error because only the first four significant decimal digits were stored.

2.2.2.2 MFPTs for a Four-State Model of Component Replacement

Consider the Markov chain model of component replacement introduced in Section 1.10.1.1.5. The transition probability matrix is given in Equation (1.50). Each time the Markov chain enters state 0, a component is replaced with a new component of age 0. The expected number of weeks until a component of age i is replaced is equal to m_{i0}, the MFPT from state i to state 0. To calculate m_0, the vector of MFPTs to target state 0, state 0 is made an absorbing state. The modified transition probability matrix is partitioned below to put it in canonical form:

$$P_M = \begin{matrix} 0 \\ 1 \\ 2 \\ 3 \end{matrix} \begin{bmatrix} 1 & 0 & 0 & 0 \\ 0.375 & 0 & 0.625 & 0 \\ 0.8 & 0 & 0 & 0.2 \\ 1 & 0 & 0 & 0 \end{bmatrix} = \begin{bmatrix} 1 & 0 \\ D & Q \end{bmatrix}. \qquad (2.63)$$

The matrix form of the equations (2.56) to be solved to find the vector m_0 of MFPTs is

$$\begin{bmatrix} m_{10} \\ m_{20} \\ m_{30} \end{bmatrix} = \begin{bmatrix} 1 \\ 1 \\ 1 \end{bmatrix} + \begin{bmatrix} 0 & 0.625 & 0 \\ 0 & 0 & 0.2 \\ 0 & 0 & 0 \end{bmatrix} \begin{bmatrix} m_{10} \\ m_{20} \\ m_{30} \end{bmatrix}. \tag{2.64}$$

The solution for the vector m_0 is

$$m_0 = \begin{bmatrix} m_{10} & m_{20} & m_{30} \end{bmatrix}^T = \begin{bmatrix} 1.75 & 1.2 & 1 \end{bmatrix}^T. \tag{2.65}$$

Thus, the expected length of time until a 1-week-old component will be replaced is 1.75 weeks. On average, a 2-week-old component will be replaced every 1.2 weeks, and a 3-week-old component will be replaced weekly.

2.2.2.3 MFPTs for a Four-State Model of Weather

Consider the Markov chain model of weather introduced in Section 1.3. The transition probability matrix is given in Equation (1.9). The vector of MFPTs to target state 1, rain, will be computed. After making state 1 an absorbing state, the modified transition probability matrix in canonical form is

$$P_M = \begin{array}{c|cccc} \text{State} & 1 & 2 & 3 & 4 \\ \hline 1 & 1 & 0 & 0 & 0 \\ 2 & 0.2 & 0.5 & 0.2 & 0.1 \\ 3 & 0.3 & 0.2 & 0.1 & 0.4 \\ 4 & 0 & 0.6 & 0.3 & 0.1 \end{array} = \begin{bmatrix} 1 & 0 \\ D & Q \end{bmatrix}. \tag{2.66}$$

The matrix equation (2.56) for computing m_1, the vector of MFPTs to target state 1, is

$$\begin{bmatrix} m_{21} \\ m_{31} \\ m_{41} \end{bmatrix} = \begin{bmatrix} 1 \\ 1 \\ 1 \end{bmatrix} + \begin{bmatrix} 0.5 & 0.2 & 0.1 \\ 0.2 & 0.1 & 0.4 \\ 0.6 & 0.3 & 0.1 \end{bmatrix} \begin{bmatrix} m_{21} \\ m_{31} \\ m_{41} \end{bmatrix}. \tag{2.67}$$

The solution for m_1 is

$$m_1 = \begin{bmatrix} m_{21} & m_{31} & m_{41} \end{bmatrix}^T = \begin{bmatrix} 5.3234 & 5.1244 & 6.3682 \end{bmatrix}^T. \tag{2.68}$$

Snow, state 2, is followed by rain after a mean interval of 5.3234 days, and a sunny day, state 4, is followed by rain after a mean interval of 6.3682 days.

2.2.3 Mean Recurrence Time

If $i = j$, then the starting state is also the target state. When $i = j$, T_{jj} is called the recurrence time for state j, and $m_{jj} = E(T_{jj})$ denotes the mean recurrence time for state j [1, 4, 5]. The mean recurrence time for state j can be calculated by two different methods. The first method is to first compute the MFPTs to state j by solving the system of equations (2.54) or (2.56) for the case in which $i \neq j$, and then to solve the additional equation

$$m_{jj} = 1 + \sum_{k \neq j} p_{jk} m_{kj} \tag{2.69}$$

for the case in which $i = j$. The second method is to calculate the steady-state probabilities for the original N-state regular Markov chain, The mean recurrence times are the reciprocals of the steady-state probabilities [1, 4, 5].

2.2.3.1 Mean Recurrence Time for a Five-State Markov Chain

Consider the numerical example of a five-state regular Markov chain for which the transition probability matrix is given in Equation (2.57). The MFPTs to a target state j, m_{1j}, m_{2j}, m_{3j}, and m_{4j}, have already been calculated by solving the system of equations (2.60) or (2.61) . The vector of MFPTs to state j is shown in Equation (2.62). The following additional equation can be solved to calculate m_{jj}, the mean recurrence time for state j:

$$\left.\begin{aligned}
m_{jj} &= 1 + p_{j1}m_{1j} + p_{j2}m_{2j} + p_{j3}m_{3j} + p_{j4}m_{4j} \\
&= 1 + (0.1)m_{1j} + (0.2)m_{2j} + (0.4)m_{3j} + (0)m_{4j} \\
&= 1 + (0.1)(15.7143) + (0.2)(12.1164) + (0.4)(13.8624) + (0)(14.8148) \\
&= 10.5397.
\end{aligned}\right\} \tag{2.70}$$

Alternatively, if the steady probabilities are known, the mean recurrence time m_{jj} for a target state j is simply the reciprocal of the steady-state probability π_j for the target state j. The steady-state probability vector for the five-state regular Markov chain for which the transition probability matrix is given in Equation (2.57) is obtained by solving the system of equations (2.13).

$$\begin{bmatrix} \pi_j & \pi_1 & \pi_2 & \pi_3 & \pi_4 \end{bmatrix} = \begin{bmatrix} \pi_j & \pi_1 & \pi_2 & \pi_3 & \pi_4 \end{bmatrix} \begin{matrix} j \\ 1 \\ 2 \\ 3 \\ 4 \end{matrix} \begin{bmatrix} 0.3 & 0.1 & 0.2 & 0.4 & 0 \\ 0 & 0.4 & 0.1 & 0.2 & 0.3 \\ 0.2 & 0.1 & 0.2 & 0.3 & 0.2 \\ 0.1 & 0.2 & 0.1 & 0.4 & 0.2 \\ 0 & 0.3 & 0.4 & 0.2 & 0.1 \end{bmatrix}$$

$$\begin{bmatrix} \pi_j & \pi_1 & \pi_2 & \pi_3 & \pi_4 \end{bmatrix} \begin{bmatrix} 1 & 1 & 1 & 1 & 1 \end{bmatrix}^T = 1. \tag{2.71}$$

The solution is

$$\pi = [\pi_j \quad \pi_1 \quad \pi_2 \quad \pi_3 \quad \pi_4] = [0.0949 \quad 0.2385 \quad 0.1837 \quad 0.2967 \quad 0.1862].$$

(2.72)

Observe that $\pi_j = 0.0949$. Hence, the mean recurrence time for state j is,

$$m_{jj} = 1/\pi_j = 1/0.0949 = 10.5374,$$

which is close to the result obtained by the first method. Discrepancies are due to roundoff error.

2.2.3.2 Mean Recurrence Times for a Four-State Model of Component Replacement

Consider the Markov chain model of component replacement for which the vector of MFPTs to target state 0 was computed in Equation (2.65). The mean recurrence time for target state 0 can be computed by adding the equation

$$\left.\begin{array}{l}m_{00} = 1 + p_{01}m_{10} + p_{02}m_{20} + p_{03}m_{30} \\ \quad = 1 + (0.8)m_{10} + (0)m_{20} + (0)m_{30} \\ \quad = 1 + (0.8)(1.75) + (0)(1.2) + (0)(1) = 2.4.\end{array}\right\}$$

(2.73)

Alternatively, the steady-state probabilities for the model of component replacement were calculated in Equation (2.29). The steady-state probability that a component has just been replaced is given by $\pi_0 = 0.4167$. Therefore the mean recurrence time, m_{00}, is again 2.4 weeks, equal to the reciprocal of π_0. This indicates that a new component, of age 0 weeks, will be replaced about every 2.4 weeks. In other words, the mean life of a new component is 2.4 weeks.

2.2.3.3 Optional Insight: Mean Recurrence Time as the Reciprocal of the Steady-State Probability for a Two-State Markov Chain

The reciprocal relationship between the mean recurrence time for a state and its associated steady-state probability will be demonstrated in two different ways for the special case of a generic two-state regular Markov chain. The first demonstration depends on knowledge of the steady-state probability vector. The transition probability matrix is given in Equation (1.4). The steady-state probability vector was computed in Equation (2.16). First, the MFPTs to target states 1 and 2, respectively, are calculated by solving the system of equations (2.54).

$$
\left.\begin{array}{c}
m_{21} = 1 + p_{22}m_{21} \\
(1 - p_{22})m_{21} = 1 = p_{21}m_{21} \\
m_{21} = 1/p_{21},
\end{array}\right\}
$$

and

$$
\left.\begin{array}{c}
m_{12} = 1 + p_{11}m_{12} \\
(1 - p_{11})m_{12} = 1 = p_{12}m_{12} \\
m_{12} = 1/p_{12}
\end{array}\right\}
\tag{2.74}
$$

Next, the mean recurrence times for states 1 and 2, respectively, are calculated by solving Equation (2.69) in terms of the MFPTs previously calculated.

$$
\left.\begin{array}{l}
m_{11} = 1 + p_{12}m_{21} = 1 + p_{12}/p_{21} = (p_{12} + p_{21})/p_{21} = 1/\pi_1 \\
m_{22} = 1 + p_{21}m_{12} = 1 + p_{21}/p_{12} = (p_{12} + p_{21})/p_{12} = 1/\pi_2
\end{array}\right\}
\tag{2.75}
$$

Hence, for a regular two-state Markov chain,

$$
m_{jj} = 1/\pi_j.
\tag{2.76}
$$

An alternative demonstration of the reciprocal relationship between the mean recurrence time for a state and the associated steady-state probability is given below. This demonstration does not depend on knowledge of the steady-state probabilities. Consider once again, for simplicity, a two-state generic Markov chain for which the transition probability matrix is shown in Equation (1.4). A matrix of MFPTs and mean recurrence times is denoted by

$$
M = \begin{matrix} 1 \\ 2 \end{matrix}\begin{bmatrix} m_{11} & m_{12} \\ m_{21} & m_{22} \end{bmatrix}.
\tag{2.77}
$$

The systems of Equations (2.54) and (2.69) used to compute the MFPTs and the mean recurrence times can be written in the following form:

$$
\left.\begin{array}{l}
m_{11} = 1 + p_{12}m_{21} = p_{11}(m_{11} - m_{11}) + p_{12}m_{21} + 1 \\
m_{12} = 1 + p_{11}m_{12} = p_{12}(m_{22} - m_{22}) + p_{11}m_{12} + 1 \\
m_{21} = 1 + p_{22}m_{21} = p_{21}(m_{11} - m_{11}) + p_{22}m_{21} + 1 \\
m_{22} = 1 + p_{21}m_{12} = p_{22}(m_{22} - m_{22}) + p_{21}m_{12} + 1.
\end{array}\right\}
\tag{2.78}
$$

To express the final form of this system as a matrix equation, let

$$
M_D = \begin{bmatrix} m_{11} & 0 \\ 0 & m_{22} \end{bmatrix}, \quad E = \begin{bmatrix} 1 & 1 \\ 1 & 1 \end{bmatrix}, \quad \pi = \begin{bmatrix} \pi_1 & \pi_2 \end{bmatrix}.
$$

Then the matrix form of the system of equations (2.78) is

$$
\begin{bmatrix} m_{11} & m_{12} \\ m_{21} & m_{22} \end{bmatrix} = \begin{bmatrix} p_{11} & p_{12} \\ p_{21} & p_{22} \end{bmatrix} \begin{bmatrix} m_{11} - m_{11} & m_{12} \\ m_{21} & m_{22} - m_{22} \end{bmatrix} + \begin{bmatrix} 1 & 1 \\ 1 & 1 \end{bmatrix}
$$

$$
= \begin{bmatrix} p_{11} & p_{12} \\ p_{21} & p_{22} \end{bmatrix} \left(\begin{bmatrix} m_{11} & m_{12} \\ m_{21} & m_{22} \end{bmatrix} - \begin{bmatrix} m_{11} & 0 \\ 0 & m_{22} \end{bmatrix} \right) + \begin{bmatrix} 1 & 1 \\ 1 & 1 \end{bmatrix}.
$$

In symbolic form the matrix equation is

$$
M = P(M - M_D) + E.
$$

Premultiplying both sides by π gives

$$
\pi M = \pi P(M - M_D) + \pi E
$$
$$
= \pi P(M - M_D) + E, \quad \text{since } \pi E = E.
$$
$$
= \pi(M - M_D) + E, \quad \text{as } \pi P = \pi
$$
$$
= \pi M - \pi M_D + E
$$
$$
\pi M_D = E
$$
$$
\begin{bmatrix} \pi_1 & \pi_2 \end{bmatrix} \begin{bmatrix} m_{11} & 0 \\ 0 & m_{22} \end{bmatrix} = \begin{bmatrix} 1 & 1 \\ 1 & 1 \end{bmatrix}.
$$

In algebraic form the equations are

$$
\pi_1 m_{11} = 1
$$
$$
\pi_2 m_{22} = 1.
$$

Hence, $m_{ii} = 1/\pi_i$.

The conclusion that the mean recurrence time is the reciprocal of the steady-state probability can be generalized to apply to any regular Markov chain with a finite number of states.

PROBLEMS

2.1 This problem refers to Problem 1.1.
  (a) If the machine is observed today operating in state 4 (WP), what is the probability that it will enter state 2 (MD) after 4 days?
  (b) If the machine is observed today operating in state 4 (WP), what is the probability that it will enter state 2 (MD) for the first time after 4 days?

(c) If the machine is observed today operating in state 4 (WP), what is the expected number of days until the machine enters state 1 (NW) for the first time?

(d) If the machine is observed today operating in state 3 (mD), what is the expected number of days until the machine returns to state 3 (mD) for the first time?

(e) Find the long run proportion of time that the machine will occupy each state.

2.2   This problem refers to Problem 1.2.

(a) If the machine is observed today operating in state 5 (WP), what is the probability that it will enter state 3 (MD) after 4 days?

(b) If the machine is observed today operating in state 5 (WP), what is the probability that it will enter state 3 (MD) for the first time after 4 days?

(c) If the machine is observed today operating in state 5 (WP), what is the expected number of days until the machine enters state 1 (NW1) for the first time?

(d) If the machine is observed today operating in state 4 (mD), what is the expected number of days until the machine returns to state 4 (mD) for the first time?

(e) Find the long run proportion of time that the machine will occupy each state.

2.3   Consider the following modification of Problem 1.3 in which a machine breaks down on a given day with probability p. A repair can be completed in 1 day with probability q_1, or in 2 days with probability q_2, or in 3 days with probability q_3. The condition of the machine can be represented by one of the following four states:

State	Description
0 (W)	Working
1 (D1)	Not working, in first day of repair
2 (D2)	Not working, in second day of repair
3 (D3)	Not working, in third day of repair

Since the next state of the machine depends only on its present state, the condition of the machine can be modeled as a four-state Markov chain. Answer the following questions if $p = 0.1$, $q_1 = 0.5$, $q_2 = 0.3$, and $q_3 = 0.2$:

(a) Construct the transition probability matrix for this four state recurrent Markov chain.

(b) If the machine is observed today in state 3 (D3), what is the probability that it will enter state 2 (D2) after 3 days?

(c) If the machine is observed today in state 1 (D1), what is the probability that it will enter state 3 (D3) for the first time after 3 days?

(d) If the machine is observed today in state 2 (D2), what is the expected number of days until the machine enters state 0 (W) for the first time?

(e) If the machine is observed today in state 0 (W), what is the expected number of days until the machine returns to state 0 (W) for the first time?

(f) Find the long run proportion of time that the machine will occupy each state.

2.4 Consider the following modification of Problem 1.8 in which an inventory system is controlled by a (1, 3) ordering policy. An office supply store sells laptops. The state of the system is the inventory on hand at the beginning of the day. If the number of laptops on hand at the start of the day is less than one (in other words, equal to 0), the store places an order, which is delivered immediately, to raise the beginning inventory level to three laptops. If the store starts the day with one or more laptops in stock, no laptops are ordered. The number d_n of laptops demanded by customers during the day is an independent, identically distributed random variable, which has a stationary Poisson probability distribution with parameter $\lambda = 0.5$ laptops per day.

(a) Construct the transition probability matrix for a four-state recurrent Markov chain model of this inventory system under a (1, 3) policy.

(b) If one laptop is on hand at the beginning of today, what is the probability distribution of the inventory level 3 days from today?

(c) If two laptops are on hand at the beginning of today, what is the probability that no orders are placed tomorrow?

(d) If the inventory level is 2 at the beginning of today, what is the expected number of days before the inventory level is 0 for the first time?

(e) If the inventory level is 1 at the beginning of today, what is the expected number of days until the inventory level returns to 1 for the first time?

(f) Find the long run probability distribution of the inventory level.

2.5 In Problem 1.9, let

$$S_0 = 1{,}000, \quad S_1 = 700, \quad S_2 = 300, \quad S_3 = 100, \quad \text{and} \quad S_4 = 0.$$

(a) Construct the transition probability matrix.

(b) Starting with a new IC in a single location, find the probability distribution for the age of an IC after 3 years.

(c) Find the steady-state probability distribution for the age of an IC.

(d) Find the expected length of time until a component of age i years is replaced, for $i = 0, 1, 2, 3$.

2.6 In Problem 1.5, let

$$p_0 = 0.30, \quad p_1 = 0.10, \quad p_2 = 0.20, \quad p_3 = 0.25, \quad \text{and} \quad p_4 = 0.15.$$

(a) Construct the transition probability matrix.
(b) If the credit counselor starts the day with one customer in her office, find the probability distribution for the number of customers in her office at the end of the third 45-min counseling period.
(c) Find the steady-state probability distribution for the number of customers in the credit counselor's office.
(d) If the credit counselor starts the day with two customers in her office, find the expected length of time until her office is empty.

2.7   In Problem 1.6, let

$$p_0 = 0.24, \quad p_1 = 0.08, \quad p_2 = 0.18, \quad p_3 = 0.16, \quad p_4 = 0.22, \quad \text{and} \quad p_5 = 0.12.$$

(a) Construct the transition probability matrix.
(b) If the nurse starts the day with two patients in the clinic, find the probability distribution for the number of patients in the clinic at the end of the third 20-min. interval.
(c) Find the steady-state probability distribution for the number of patients in the clinic.
(d) If the nurse starts the day with four patients in the clinic, find the expected length of time until the clinic has one patient.

2.8   This problem refers to Problem 1.7.
(a) Verify that the following four-element vector is the steady-state probability vector:

$$\frac{r-p}{r(1-r)-p(1-p)s^3}[1 - r \quad s \quad s^2 \quad (1-p)s^3], \quad \text{where } s = \frac{p(1-r)}{(1-p)r}.$$

(b) What is the long run proportion of time that the checkout clerk is idle?
(c) What is the long run proportion of time that an arriving customer will be denied admission to the store?
   In Problem 1.7, let $p = 0.24$ and $r = 0.36$.
(d) Construct the transition probability matrix.
(e) If the checkout clerk starts the day with one customer inside the convenience store, find the probability distribution for the number of customers inside the store at the end of the third 30-min interval.
(f) Find the steady-state probability distribution for the number of customers inside the convenience store.
(g) If the checkout clerk starts the day with two customers inside the convenience store, find the expected length of time until the store has three customers.

2.9   In Problem 1.11, let

$$w_0 = 0.14, \quad w_1 = 0.10, \quad w_2 = 0.20, \quad w_3 = 0.18, \quad w_4 = 0.24, \quad \text{and} \quad w_5 = 0.14.$$

(a) Construct the transition probability matrix.

(b) If the dam starts the week with three units of water, find the probability distribution for the volume of water stored in the dam after 3 weeks.

(c) Find the steady-state probability distribution for the volume of water stored in the dam.

(d) If the dam starts the week with one unit of water, find the expected length of time until the dam has four units of water.

2.10 In Problem 1.14, let $p = 0.6$, $r = 0.8$, and $q = 0.7$.

(a) Construct the transition probability matrix.

(b) If the optical shop starts the day with both the optometrist and the optician idle, find the probability state vector after three 30-min intervals.

(c) Find the steady-state probability vector for the optical shop.

(d) If the optical shop starts the day with the optometrist blocked and the optician working, find the expected length of time until the optometrist is working and the optician is idle.

References

1. Bhat, N., *Elements of Applied Stochastic Processes*, 2nd ed., Wiley, New York, 1985.
2. Cinlar, E., *Introduction to Stochastic Processes*, Prentice-Hall, Englewood Cliffs, NJ, 1975.
3. Clarke, A. B. and Disney R. L., *Probability and Random Processes: A First Course with Applications*, 2nd ed., Wiley, New York, 1985.
4. Kemeny, J. G., Mirkil, H., Snell, J. L., and Thompson, G. L., *Finite Mathematical Structures*, Prentice-Hall, Englewood Cliffs, NJ, 1959.
5. Kemeny, J. G. and Snell, J. L., *Finite Markov Chains*, Van Nostrand, Princeton, NJ, 1960. Reprinted by Springer-Verlag, New York, 1976.

3

Reducible Markov Chains

All Markov chains treated in this book have a finite number of states. Recall from Section 1.9 that a reducible Markov chain has both recurrent states and transient states. The state space for a reducible chain can be partitioned into one or more mutually exclusive closed communicating classes of recurrent states plus a set of transient states. A closed communicating class of recurrent states is often called a recurrent class or a recurrent chain. The state space for a reducible unichain can be partitioned into one recurrent chain plus a set of transient states. The state space for a reducible multichain can be partitioned into two or more mutually exclusive recurrent chains plus a set of transient states. There is no interaction among different recurrent chains within a reducible multichain. Hence, each recurrent chain can be analyzed separately by treating it as an irreducible Markov chain. A reducible chain, which starts in a transient state, will eventually leave the set of transient states to enter a recurrent chain, within which it will continue to make transitions indefinitely.

3.1 Canonical Form of the Transition Matrix

As Section 1.9.3 indicates, when the state space of a reducible Markov chain is partitioned into one or more disjoint closed communicating classes of recurrent states plus a set of transient states, the transition probability matrix is said to be represented in a canonical or standard form. The canonical form of the transition matrix without aggregation is shown for unichain models in Section 1.10.1.2 and for multichain models in Section 1.10.2. The canonical form of the transition matrix with aggregation is shown in Sections 1.9.3 and 3.1.3 [1, 4, 5].

3.1.1 Unichain

Suppose that the states in a unichain are rearranged so that the recurrent states come first, numbered consecutively, followed by the transient states, which are also numbered consecutively. The closed class of recurrent states is denoted by R, and the set of transient states is denoted by T. This rearrangement of states can be used to partition the transition probability matrix for the unichain into a recurrent chain plus a set of transient states. As Equation (1.43) indicates, the transition probability matrix, P, is represented in the following canonical form:

$$P = \begin{bmatrix} S & 0 \\ D & Q \end{bmatrix}. \tag{3.1}$$

The square submatrix S governs transitions within the closed class of recurrent states. Rectangular submatrix 0 consists entirely of zeroes indicating that no transitions from recurrent states to transient states are possible. Rectangular submatrix D governs transitions from transient states to recurrent states. The square submatrix Q governs transitions among transient states. Submatrix Q is substochastic, which means that at least one row sum is less than one.

Consider the generic four-state reducible unichain described in Section 1.9.1.2. The transition probability matrix is partitioned in Equation (1.40) to show that states 1 and 2 belong to the recurrent chain denoted by $R = \{1, 2\}$, and states 3 and 4 are transient. The set of transient states is denoted by $T = \{3, 4\}$. The canonical form of the transition matrix is

$$P = [p_{ij}] = \begin{matrix} 1 \\ 2 \\ 3 \\ 4 \end{matrix} \begin{bmatrix} p_{11} & p_{12} & 0 & 0 \\ p_{21} & p_{22} & 0 & 0 \\ p_{31} & p_{32} & p_{33} & p_{34} \\ p_{41} & p_{42} & p_{43} & p_{44} \end{bmatrix} = \begin{bmatrix} S & 0 \\ D & Q \end{bmatrix},$$

where

$$S = \begin{matrix} 1 \\ 2 \end{matrix} \begin{bmatrix} p_{11} & p_{12} \\ p_{21} & p_{22} \end{bmatrix}, \quad D = \begin{matrix} 3 \\ 4 \end{matrix} \begin{bmatrix} p_{31} & p_{32} \\ p_{41} & p_{42} \end{bmatrix}, \quad \text{and} \quad Q = \begin{matrix} 3 \\ 4 \end{matrix} \begin{bmatrix} p_{33} & p_{34} \\ p_{43} & p_{44} \end{bmatrix} \tag{3.2}$$

A special case of a reducible unichain is an absorbing unichain in which the recurrent chain, denoted by $R = \{1\}$, is a single absorbing state. That is, $S = I = [1]$. Thus, in an absorbing unichain, which is often called an absorbing Markov chain, the only recurrent state is an absorbing state. All the remaining states are transient. The canonical form of the transition probability matrix shown in Equation (1.41) for a generic three-state absorbing unichain is

$$P = [p_{ij}] = \begin{matrix} 1 \\ 2 \\ 3 \end{matrix} \begin{bmatrix} 1 & 0 & 0 \\ p_{21} & p_{22} & p_{23} \\ p_{31} & p_{32} & p_{33} \end{bmatrix}, \quad P = \begin{bmatrix} 1 & 0 \\ D & Q \end{bmatrix},$$

where

$$D = \frac{2}{3}\begin{bmatrix} p_{21} \\ p_{31} \end{bmatrix}, \quad \text{and} \quad Q = \frac{2}{3}\begin{bmatrix} p_{22} & p_{23} \\ p_{32} & p_{33} \end{bmatrix}. \tag{3.3}$$

If passage from a transient state to the recurrent chain, R, in a reducible unichain is of interest, rather than passage to the individual states within R, then the recurrent states within R may be lumped together to be replaced by a single absorbing state. This procedure, which is used in Section 3.5.1, will transform a reducible unichain with recurrent states into an absorbing unichain with the same set of transient states.

3.1.2 Multichain

Consider a reducible multichain that has M recurrent chains, denoted by $R_1, \ldots, R_M$, with the respective transition probability matrices $P_1, \ldots, P_M$, plus a set of transient states denoted by T. The states can be rearranged if necessary so that the transition probability matrix, P, can be represented in the following canonical form:

$$P = \begin{bmatrix} P_1 & 0 & \cdots & 0 & 0 \\ 0 & P_2 & \cdots & 0 & 0 \\ \vdots & \vdots & \ddots & \vdots & \vdots \\ 0 & 0 & \cdots & P_M & 0 \\ D_1 & D_2 & \cdots & D_M & Q \end{bmatrix}. \tag{3.4}$$

The M square submatrices, $P_1, \ldots, P_M$, are the transition probability matrices, which specify the transitions after the chain has entered the corresponding closed class of recurrent states. The M rectangular submatrices, $D_1, \ldots, D_M$, are the transition probability matrices, which govern transitions from transient states to the corresponding closed classes of recurrent states. The square submatrix Q is the transition probability matrix, which governs transitions among the transient states. The submatrix 0 consists entirely of zeroes.

The transition probability matrix for a generic five-state reducible multichain is shown in Equation (1.42). As Section 1.9.2 indicates, the multichain has $M = 2$ recurrent closed sets, one of which is an absorbing state, plus two transient states. The two recurrent sets of states are $R_1 = \{1\}$ and $R_2 = \{2, 3\}$. The set of transient states is denoted by $T = \{4, 5\}$.

The canonical form of the transition matrix is

$$P = \begin{matrix} 1 \\ 2 \\ 3 \\ 4 \\ 5 \end{matrix} \begin{bmatrix} 1 & 0 & 0 & 0 & 0 \\ 0 & p_{22} & p_{23} & 0 & 0 \\ 0 & p_{32} & p_{33} & 0 & 0 \\ p_{41} & p_{42} & p_{43} & p_{44} & p_{45} \\ p_{51} & p_{52} & p_{53} & p_{54} & p_{55} \end{bmatrix} = \begin{bmatrix} P_1 & 0 & 0 \\ 0 & P_2 & 0 \\ D_1 & D_2 & Q \end{bmatrix},$$

where

$$P_1 = [1], \quad P_2 = \begin{array}{c}2\\3\end{array}\begin{bmatrix} p_{22} & p_{23} \\ p_{32} & p_{33} \end{bmatrix}, \quad D_1 = \begin{array}{c}4\\5\end{array}\begin{bmatrix} p_{41} \\ p_{51} \end{bmatrix},$$

$$D_2 = \begin{array}{c}4\\5\end{array}\begin{bmatrix} p_{42} & p_{43} \\ p_{52} & p_{53} \end{bmatrix}, \quad \text{and} \quad Q = \begin{array}{c}4\\5\end{array}\begin{bmatrix} p_{44} & p_{45} \\ p_{54} & p_{55} \end{bmatrix}. \tag{3.5}$$

3.1.3 Aggregation of the Transition Matrix in Canonical Form

The canonical form of the transition probability matrix, P, for a reducible multichain can be represented in an aggregated format. As Section 1.9.3 indicates, this is accomplished by combining all the closed communicating classes of recurrent states into a single recurrent closed class with a transition matrix denoted by S.

Since S in a multichain is formed by combining the transition probability matrices for several recurrent chains, Equation (3.1) represents the transition matrix of a reducible multichain in aggregated canonical form. Equation (1.44) in Section 1.9.3 shows the aggregated canonical form of the transition matrix for a generic five-state reducible multichain. Now consider a reducible multichain that has $M = 3$ closed communicating classes of recurrent states denoted by R_1, R_2, and R_3 with the corresponding transition probability matrices P_1, P_2, and P_3, plus a set of transient states. The transition probability matrix, P, in aggregated canonical form is

$$P = \begin{bmatrix} P_1 & 0 & 0 & 0 \\ 0 & P_2 & 0 & 0 \\ 0 & 0 & P_3 & 0 \\ D_1 & D_2 & D_3 & Q \end{bmatrix} = \left[\begin{array}{ccc|c} P_1 & 0 & 0 & 0 \\ 0 & P_2 & 0 & 0 \\ 0 & 0 & P_3 & 0 \\ \hline D_1 & D_2 & D_3 & Q \end{array}\right] = \begin{bmatrix} S & 0 \\ D & Q \end{bmatrix}.$$

where

$$S = \begin{bmatrix} P_1 & 0 & 0 \\ 0 & P_2 & 0 \\ 0 & 0 & P_3 \end{bmatrix}, \quad 0 = \begin{bmatrix} 0 \\ 0 \\ 0 \end{bmatrix}, \quad D = \begin{bmatrix} D_1 & D_2 & D_3 \end{bmatrix}. \tag{3.6}$$

The Markov chain model of a production process constructed in Section 1.10.2.2 is an example of an eight-state reducible multichain with three closed classes of recurrent states. States 1 and 2 are absorbing states, which represent the first two recurrent chains, respectively. States 3, 4, and 5 are recurrent states, which represent the third recurrent chain. States 6, 7, and 8 are the transient states. The aggregated transition matrix in canonical form is shown in Equation (1.60) and is repeated below with the submatrices identified individually:

$$
P = \begin{array}{r}
\text{Scrapped} \\
\text{Sold} \\
\text{Training engineers} \\
\text{Training technicians} \\
\text{Training tech. writers} \\
\text{Stage 1} \\
\text{Stage 2} \\
\text{Stage 3}
\end{array}
\begin{array}{r}
1 \\ 2 \\ 3 \\ 4 \\ 5 \\ 6 \\ 7 \\ 8
\end{array}
\left[
\begin{array}{cc|ccc|ccc}
1 & 0 & 0 & 0 & 0 & 0 & 0 & 0 \\
0 & 1 & 0 & 0 & 0 & 0 & 0 & 0 \\
0 & 0 & 0.50 & 0.30 & 0.20 & 0 & 0 & 0 \\
0 & 0 & 0.30 & 0.45 & 0.25 & 0 & 0 & 0 \\
0 & 0 & 0.10 & 0.35 & 0.55 & 0 & 0 & 0 \\
0.20 & 0.16 & 0.04 & 0.03 & 0.02 & 0.55 & 0 & 0 \\
0.15 & 0 & 0 & 0 & 0 & 0.20 & 0.65 & 0 \\
0.10 & 0 & 0 & 0 & 0 & 0 & 0.15 & 0.75
\end{array}
\right]
$$

$$
= \begin{bmatrix}
P_1 & 0 & 0 & 0 \\
0 & P_2 & 0 & 0 \\
0 & 0 & P_3 & 0 \\
D_1 & D_2 & D_3 & Q
\end{bmatrix}
= \begin{bmatrix} S & 0 \\ D & Q \end{bmatrix},
$$

where

$$
S = \begin{array}{r} 1 \\ 2 \\ 3 \\ 4 \\ 5 \end{array}
\left[
\begin{array}{cc|ccc}
1 & 0 & 0 & 0 & 0 \\
0 & 1 & 0 & 0 & 0 \\
0 & 0 & 0.50 & 0.30 & 0.20 \\
0 & 0 & 0.30 & 0.45 & 0.25 \\
0 & 0 & 0.10 & 0.35 & 0.55
\end{array}
\right]
= \begin{bmatrix}
P_1 & 0 & 0 \\
0 & P_2 & 0 \\
0 & 0 & P_3
\end{bmatrix},
\quad
Q = \begin{array}{r} 6 \\ 7 \\ 8 \end{array}
\begin{bmatrix}
0.55 & 0 & 0 \\
0.20 & 0.65 & 0 \\
0 & 0.15 & 0.75
\end{bmatrix},
$$

$$
D = \begin{array}{r} 6 \\ 7 \\ 8 \end{array}
\left[
\begin{array}{cc|ccc}
0.20 & 0.16 & 0.04 & 0.03 & 0.02 \\
0.15 & 0 & 0 & 0 & 0 \\
0.10 & 0 & 0 & 0 & 0
\end{array}
\right]
= [D_1 \quad D_2 \quad D_3],
$$

$$
P_1 = P_2 = [1], \quad
P_3 = \begin{array}{r} 3 \\ 4 \\ 5 \end{array}
\begin{bmatrix}
0.50 & 0.30 & 0.20 \\
0.30 & 0.45 & 0.25 \\
0.10 & 0.35 & 0.55
\end{bmatrix},
\quad
D_1 = \begin{array}{r} 6 \\ 7 \\ 8 \end{array}
\begin{bmatrix}
0.20 \\ 0.15 \\ 0.10
\end{bmatrix},
\quad
D_2 = \begin{array}{r} 6 \\ 7 \\ 8 \end{array}
\begin{bmatrix}
0.16 \\ 0 \\ 0
\end{bmatrix},
$$

$$
D_3 = \begin{array}{r} 6 \\ 7 \\ 8 \end{array}
\begin{bmatrix}
0.04 & 0.03 & 0.02 \\
0 & 0 & 0 \\
0 & 0 & 0
\end{bmatrix}.
$$

(3.7)

The transition probability matrices, P_1, P_2, and P_3, correspond to the three closed classes of recurrent states, denoted by $R_1 = \{1\}$, $R_2 = \{2\}$, and $R_3 = \{3, 4, 5\}$, respectively. Note that $P_1 = [1]$ and $P_2 = [1]$ are each associated with a single absorbing state. Each absorbing state forms its own closed communicating class. The set of transient states is denoted by $T = \{6, 7, 8\}$.

As Section 1.10.2.1 indicates, a special case of a reducible multichain is an absorbing multichain in which the submatrix S is an identity matrix, I. Thus, in an absorbing multichain, which is generally called an absorbing Markov chain, all the recurrent states are absorbing states. All the remaining states are transient states. The canonical form of the transition matrix for an absorbing multichain is

$$P = \begin{bmatrix} I & 0 \\ D & Q \end{bmatrix}. \tag{3.8}$$

If passage from a transient state to a particular recurrent chain, R_i, in a reducible multichain is of interest, rather than passage to an individual state within the recurrent chain, R_i, then the states within each recurrent chain may be lumped together to be replaced by a single absorbing state. This procedure, which is used in Section 3.4.1, will transform a reducible multichain into an absorbing multichain with the same set of transient states. Of course, if passage from a transient state to any recurrent state is of interest, then submatrix S may be replaced by a single absorbing state to create an absorbing unichain, for which the transition matrix is

$$P = \begin{bmatrix} 1 & 0 \\ D & Q \end{bmatrix}. \tag{3.9}$$

3.2 The Fundamental Matrix

A reducible Markov chain model can be used to answer the following three questions concerning the eventual passage from a transient state to a recurrent state, assuming that the chain starts in a given transient state [1, 4, 5].

1. What is the expected number of times the chain may enter a particular transient state before it eventually enters a recurrent state?
2. What is the expected total number of times the chain may enter all transient states before it eventually enters a recurrent state?
3. What is the probability of eventual passage to a recurrent state?

Questions 1, 2, and 3 are answered in Sections 3.2.2, 3.2.3, and 3.3.3, respectively.

3.2.1 Definition of the Fundamental Matrix

To answer these three questions, a matrix inverse called the fundamental matrix is defined. Assume that the transition matrix, P, for a reducible unichain, which may be produced by aggregating all the recurrent closed classes in a reducible multichain, is represented in the canonical form.

$$P = \begin{bmatrix} S & 0 \\ D & Q \end{bmatrix}. \tag{3.1}$$

The inverse of the matrix $(I-Q)$ exists, and is called the fundamental matrix, denoted by U. That is, the fundamental matrix, U, is defined to be

$$U = (I-Q)^{-1}. \tag{3.10}$$

To see how the fundamental matrix can be used to answer the three questions of Section 3.2, it is instructive to consider, for simplicity, the canonical form of the transition matrix for the reducible generic four-state unichain shown in Equation (3.2). Observe that

$$I-Q = \begin{array}{c} 3 \\ 4 \end{array}\begin{bmatrix} 1 & 0 \\ 0 & 1 \end{bmatrix} - \begin{array}{c} 3 \\ 4 \end{array}\begin{bmatrix} p_{33} & p_{34} \\ p_{43} & p_{44} \end{bmatrix} = \begin{array}{c} 3 \\ 4 \end{array}\begin{bmatrix} 1-p_{33} & -p_{34} \\ -p_{43} & 1-p_{44} \end{bmatrix}.$$

The fundamental matrix is

$$U = (I-Q)^{-1}$$

$$= \begin{array}{c} 3 \\ 4 \end{array}\begin{bmatrix} 1-p_{33} & -p_{34} \\ -p_{43} & 1-p_{44} \end{bmatrix}^{-1}$$

$$= \frac{1}{(1-p_{33})(1-p_{44})-p_{34}p_{43}} \begin{array}{cc} & 3 \\ \begin{bmatrix} 1-p_{44} & p_{34} \\ p_{43} & 1-p_{33} \end{bmatrix} & 4 \end{array}. \tag{3.11}$$

3.2.2 Mean Time in a Particular Transient State

To see how the fundamental matrix can be used to answer question one for the reducible four-state unichain for which the transition matrix is shown in Equation (3.2), let the random variable N_{ij} denote the number of times the chain enters transient state j before the chain eventually enters a recurrent state, given that the chain starts in transient state i. Let u_{ij} denote the mean number of times the chain enters transient state j before the chain eventually enters a recurrent state, given that the chain starts in transient state i. Then

$$u_{ij} = E(N_{ij}). \tag{3.12}$$

Suppose that i and j are different transient states, so that $i \neq j$. If the first transition from transient state i is to any transient state k, with probability p_{ik}, then state j will be entered a mean number of u_{kj} times. State j will be entered zero times if the first transition is to a recurrent state. For the case in which $i \neq j$,

$$u_{ij} = \sum_{k \in T} p_{ik} u_{kj}, \quad \text{if } i \neq j, \tag{3.13}$$

where the sum is over all transient states, k.

Now suppose that i and j are the same transient state, so that $i = j$. Then, by assumption, state i will be occupied once during the first time period. State i will be entered a mean number of u_{ki} additional times if, with probability p_{ik}, the first transition is to any transient state k. Thus,

$$u_{ii} = 1 + \sum_{k \in T} p_{ik} u_{ki}, \quad \text{if } i = j, \tag{3.14}$$

where the sum is over all transient states, k.

Since the reducible four-state unichain has two transient states, there are $2^2 = 4$ unknown quantities, u_{ij}, to be determined. The system of four algebraic equations used for calculating the u_{ij} for the reducible four-state unichain is given below:

$$u_{34} = \sum_{k=3}^{4} p_{3k} u_{k4} = 0 + p_{33} u_{34} + p_{34} u_{44}$$

$$u_{43} = \sum_{k=3}^{4} p_{4k} u_{k3} = 0 + p_{43} u_{33} + p_{44} u_{43}$$

$$u_{33} = 1 + \sum_{k=3}^{4} p_{3k} u_{k3} = 1 + p_{33} u_{33} + p_{34} u_{43}$$

$$u_{44} = 1 + \sum_{k=3}^{4} p_{4k} u_{k4} = 1 + p_{43} u_{34} + p_{44} u_{44}.$$

The matrix form of these equations is

$$\begin{bmatrix} u_{33} & u_{34} \\ u_{43} & u_{44} \end{bmatrix} = \begin{bmatrix} 1 & 0 \\ 0 & 1 \end{bmatrix} + \begin{bmatrix} p_{33} & p_{34} \\ p_{43} & p_{44} \end{bmatrix} \begin{bmatrix} u_{33} & u_{34} \\ u_{43} & u_{44} \end{bmatrix}.$$

In this matrix equation, let a matrix

$$U = \begin{bmatrix} u_{33} & u_{34} \\ u_{43} & u_{44} \end{bmatrix}.$$

Recall from Equation (3.2) that

$$Q = \begin{bmatrix} p_{33} & p_{34} \\ p_{43} & p_{44} \end{bmatrix}.$$

The compact form of the matrix equation is

$$U = I + QU$$

$$U - QU = I$$
$$IU - QU = I$$
$$(I - Q)U = I$$
$$U = (I - Q)^{-1}.$$

The following results, although obtained for a reducible four-state uni-chain, can be generalized to hold for any finite-state reducible Markov chain. The matrix form of Equations (3.13) and (3.14) is

$$U = I + QU. \tag{3.15}$$

An alternative matrix form of Equations (3.13) and (3.14) is

$$U = (I - Q)^{-1}. \tag{3.16}$$

Observe that the matrix $U = (I - Q)^{-1}$. Hence, U is the fundamental matrix defined in Equation (3.9). Since $U = [u_{ij}]$, u_{ij} is the (i, j)th entry of U. To answer question one of Section 3.2, u_{ij}, the (i, j)th entry of the fundamental matrix, specifies the mean number of times the chain enters transient state j before the chain eventually enters a recurrent state, given that the chain starts in transient state i. For example, the fundamental matrix for the reducible four-state unichain is shown in Equation (3.11). Suppose the unichain starts in transient state 4. Then

$$u_{43} = \frac{p_{43}}{(1 - p_{33})(1 - p_{44}) - p_{34}p_{43}}$$

is the mean number of time periods that a unichain, which is initially in transient state 4, spends in transient state 3 before the chain eventually enters recurrent state 1 or recurrent state 2, which together form a recurrent closed class.

3.2.3 Mean Time in All Transient States

To see how the fundamental matrix can be used to answer question two of Section 3.2, let the random variable N_i denote the total number of times the chain enters all transient states before the chain eventually enters a recurrent state, given that the chain starts in transient state i. Then

$$N_i = \sum_{j \in T} N_{ij},$$

where the sum is over all transient states, j. Let u_i denote the mean number of times the chain enters all transient states before the chain eventually enters a recurrent state, given that the chain starts in transient state i. It follows that

$$u_i = E(N_i) = E\left(\sum_{j \in T} N_{ij}\right) = \sum_{j \in T} E(N_{ij}) = \sum_{j \in T} u_{ij}, \tag{3.17}$$

where the sum is over all transient states, j. Thus, u_i is the sum of the elements in row i of U, the fundamental matrix. To answer question two, if the chain begins in transient state i, it will make a mean total number of u_i transitions before eventually entering a recurrent state.

This result can be expressed in the form of a matrix equation. Suppose that a reducible Markov chain has q transient states. Let e be a q-component column vector with all entries one. Then the ith component of the column vector

$$u = Ue \tag{3.18}$$

represents the mean total number of transitions to all transient states before the chain eventually enters a recurrent state, given that the chain started in transient state i. The column vector Ue is computed for the reducible four-state unichain for which the fundamental matrix is shown in Equation (3.11).

$$
u = \begin{bmatrix} u_3 \\ u_4 \end{bmatrix} = Ue = \begin{array}{c} 3 \\ 4 \end{array}\begin{bmatrix} u_{33} & u_{34} \\ u_{43} & u_{44} \end{bmatrix}\begin{bmatrix} 1 \\ 1 \end{bmatrix} = \begin{array}{c} 3 \\ 4 \end{array}\begin{bmatrix} u_{33} + u_{34} \\ u_{43} + u_{44} \end{bmatrix}
$$

$$
= \frac{1}{(1-p_{33})(1-p_{44})-p_{34}p_{43}}\begin{array}{c}\\ \end{array}\begin{bmatrix} 1-p_{44}+p_{34} \\ p_{43}+1-p_{33} \end{bmatrix}\begin{array}{c} 3 \\ 4 \end{array}. \tag{3.19}
$$

The sum

$$u_4 = u_{43} + u_{44} = \frac{p_{43}+1-p_{33}}{(1-p_{33})(1-p_{44})-p_{34}p_{43}},$$

which is the last entry of column vector Ue, represents the mean number of time periods that the reducible unichain, starting in transient state 4, spends in transient states 3 and 4 before the chain eventually enters recurrent state 1 or 2.

3.2.4 Absorbing Multichain Model of Patient Flow in a Hospital

As an illustration of the role of the fundamental matrix in analyzing a reducible Markov chain model, consider the example of the daily movement of patients within a hospital, which is modeled as a six-state absorbing multichain in Section 1.10.2.1.2. In this model, both recurrent states are absorbing states. The transition probability matrix for the absorbing multichain is given in canonical form in Equation (1.59). For this model,

$$Q = \begin{array}{c} 1 \\ 2 \\ 3 \\ 4 \end{array}\begin{bmatrix} 0.4 & 0.1 & 0.2 & 0.3 \\ 0.1 & 0.2 & 0.3 & 0.2 \\ 0.2 & 0.1 & 0.4 & 0.2 \\ 0.3 & 0.4 & 0.2 & 0.1 \end{bmatrix}, \quad (I-Q) = \begin{array}{c} 1 \\ 2 \\ 3 \\ 4 \end{array}\begin{bmatrix} 0.6 & -0.1 & -0.2 & -0.3 \\ -0.1 & 0.8 & -0.3 & -0.2 \\ -0.2 & -0.1 & 0.6 & -0.2 \\ -0.3 & -0.4 & -0.2 & 0.9 \end{bmatrix}. \quad (3.20)$$

The fundamental matrix is

$$U = (I-Q)^{-1} = \begin{array}{c|cccc} \text{State} & 1 & 2 & 3 & 4 \\ \hline 1 & 5.2364 & 2.8563 & 4.2847 & 3.3321 \\ 2 & 2.9263 & 3.2798 & 3.4383 & 2.4682 \\ 3 & 3.5078 & 2.4858 & 5.0253 & 2.8382 \\ 4 & 3.8254 & 2.9621 & 4.0731 & 3.9494 \end{array}. \quad (3.21)$$

To answer question one of Section 3.2, u_{ij}, the (i, j)th entry of the fundamental matrix, represents the mean number of days a patient spends in transient state j before the patient eventually enters an absorbing state, given that the patient starts in transient state i. Therefore, $u_{43} = 4.0731$ is the mean number of days that a patient who is initially in physical therapy (transient state 4) spends in surgery (transient state 3) before the patient is eventually discharged (absorbed in state 0) or dies (absorbed in state 5).

To answer question two, the column vector Ue has as components the mean total number of days a patient spends in each transient state before the patient eventually enters an absorbing state. In other words, component u_i of the column vector Ue denotes the mean time to absorption for a patient who starts in transient state i. The column vector Ue for the hospital model is

$$u = Ue = \begin{bmatrix} u_1 \\ u_2 \\ u_3 \\ u_4 \end{bmatrix} = \begin{array}{c} 1 \\ 2 \\ 3 \\ 4 \end{array}\begin{bmatrix} 5.2364 & 2.8563 & 4.2847 & 3.3321 \\ 2.9263 & 3.2798 & 3.4383 & 2.4682 \\ 3.5078 & 2.4858 & 5.0253 & 2.8382 \\ 3.8254 & 2.9621 & 4.0731 & 3.9494 \end{bmatrix}\begin{bmatrix} 1 \\ 1 \\ 1 \\ 1 \end{bmatrix} = \begin{array}{c} 1 \\ 2 \\ 3 \\ 4 \end{array}\begin{bmatrix} 15.7095 \\ 12.1126 \\ 13.8571 \\ 14.81 \end{bmatrix}. \quad (3.22)$$

The sum $u_4 = u_{41} + u_{42} + u_{43} + u_{44} = 14.81$, which is the fourth entry of column vector Ue, represents the mean number of days that a patient initially in physical therapy (transient state 4) spends in the diagnostic, outpatient, surgery, and physical therapy departments before the patient eventually is discharged or dies. Observe that matrix Q in Equation (3.20) is identical to matrix Q in Equation (2.58). However, the entries in the vector of mean first passage times (MFPTs) in Equation (3.22) differ slightly from those calculated by different methods in Equations (2.62) and (6.68). Discrepancies are due to roundoff error because only the first four significant decimal digits were stored.

3.3 Passage to a Target State

A reducible Markov chain contains a set of transient states plus one or more closed classes of recurrent states, some of which may be absorbing states. In many reducible chain models, passage from a transient state to a target recurrent state is of interest [1, 3–5].

3.3.1 Mean First Passage Times in a Regular Markov Chain Revisited

In this section, MFPTs in a regular Markov chain, calculated in Section 2.2.2, are revisited to show that they are equivalent to the mean times to absorption in an associated absorbing Markov chain created by changing the target recurrent state in the regular chain into an absorbing state.

Consider the generic five-state regular Markov chain which was introduced in Section 2.2.2.1. The five states are indexed j, 1, 2, 3, and 4. The transition probability matrix, P, is shown in Equation (2.49). As Section 2.2.2.1 indicates, the vector of MFPTs can be found by making the target state j an absorbing state. Equation (2.50) shows the canonical form of the transition probability matrix, P_M, for the modified absorbing Markov chain, obtained by deleting row j and column j of the original transition probability matrix. When Equation (2.56) is solved for the vector $\mathbf{m}_j$ of MFPTs to target state j, the result is

$$
\begin{aligned}
I\mathbf{m}_j &= e + Q\mathbf{m}_j \\
I\mathbf{m}_j - Q\mathbf{m}_j &= e \\
(I - Q)\mathbf{m}_j &= e \\
\mathbf{m}_j &= (I - Q)^{-1} e = Ue.
\end{aligned}
\tag{3.23}
$$

Observe that $(I - Q)^{-1} = U$ is the fundamental matrix of a reducible Markov chain. Formula (3.23) for the vector $\mathbf{m}_j$ indicates that the vector of MFPTs for a regular Markov chain is equal to the fundamental matrix of the associated absorbing Markov chain postmultiplied by a column vector of ones. By the rules of matrix multiplication, the MFPT from a state i to a target state j in the regular chain is equal to the sum of the entries in row i of the fundamental matrix. Therefore, the vector of MFPTs to the target state j is

$$m_j = \begin{bmatrix} m_{1j} \\ m_{2j} \\ m_{3j} \\ m_{4j} \end{bmatrix} = (I - Q)^{-1}e$$

$$= Ue$$

$$= \begin{bmatrix} u_{11} & u_{12} & u_{13} & u_{14} \\ u_{21} & u_{22} & u_{23} & u_{24} \\ u_{31} & u_{32} & u_{33} & u_{34} \\ u_{41} & u_{42} & u_{43} & u_{44} \end{bmatrix} \begin{bmatrix} 1 \\ 1 \\ 1 \\ 1 \end{bmatrix} = \begin{bmatrix} u_{11} & +u_{12} & +u_{13} & +u_{14} \\ u_{21} & +u_{22} & +u_{23} & +u_{24} \\ u_{31} & +u_{32} & +u_{33} & +u_{34} \\ u_{41} & +u_{42} & +u_{43} & +u_{44} \end{bmatrix}$$

$$= \begin{bmatrix} u_1 \\ u_2 \\ u_3 \\ u_4 \end{bmatrix}.$$

$$(3.24)$$

The column vector $Ue = m_j$ of mean times to absorption in absorbing state j is also the vector of MFPTs to the target state j in a regular Markov chain, which has been transformed into an associated absorbing unichain by changing the target state j into an absorbing state. For example, the sum $u_4 = u_{41} + u_{42} + u_{43} + u_{44}$, which is the fourth entry of column vector Ue in Equation (3.19), represents the mean time to absorption in absorbing state j, starting in transient state 4. The quantity $u_4 = m_{4j}$ is also the MFPT from recurrent state 4 to target state j in the original regular chain. Hence, an alternative way to compute MFPTs is to multiply the fundamental matrix by a vector of ones. However, this alternative procedure is recommended only if the fundamental matrix is already available because, for an N-state chain, inverting a matrix involves about N times as many arithmetic operations as solving a system of N linear equations.

For a numerical illustration of the alternative procedure, observe that the submatrix Q shown in Equation (2.58) for the modified absorbing unichain constructed for the regular five-state Markov chain is identical to the submatrix Q shown in Equation (3.20) for the absorbing multichain model of patient flow in a hospital. Hence, both models have the same fundamental matrix. The vector of MFPTs to the target state j was calculated in Equation (3.22) by summing the entries in the rows of the fundamental matrix. These values differ slightly from those obtained in Equation (2.62) by solving the matrix equation for the vector of MFPTs to target state j. These values also differ slightly from those obtained by state reduction in Equation (6.68). Discrepancies are caused by roundoff error because only the first four significant decimal digits were stored.

3.3.2 Probability of First Passage in *n* Steps

This section is focused on computing probabilities of first passage in n time periods, or steps, from a transient state to a target recurrent state in a reducible Markov chain. (The adjective "first" is redundant because a reducible chain allows only one passage from a transient state to a recurrent state.) For a reducible Markov chain, the first passage probabilities of interest involve transitions in one direction only, from transient states to recurrent states or to absorbing states. Recall that in a regular Markov chain all states are recurrent, and they all belong to the same closed communicating class of states. In Equation (2.35), $f_{ij}^{(n)}$ represents the probability that the first passage time from a recurrent state i to a target recurrent state j equals n steps, provided that states i and j belong to the same recurrent closed class. The n-step first passage probabilities in a recurrent chain can be computed recursively by using the algebraic formula (2.35) or the matrix formula (2.38).

3.3.2.1 Reducible Unichain

Consider a reducible unichain with a set of transient states plus one closed set of recurrent states. The canonical form of the transition probability matrix is shown in Equation (3.1). As in Equation (2.35), let $f_{ij}^{(n)}$ be the probability of first passage in n steps from a transient state i to a target recurrent state j. That is, for $n \geq 1$, $f_{ij}^{(n)}$ denotes the probability of moving from a transient state i to a target recurrent state j for the first time on the nth step.

Assume that the chain starts in a transient state i. To obtain a formula for recursively computing the n-step first passage probabilities from a transient state i to a recurrent state j in a reducible unichain, an argument similar to the one used in Section 2.2.1 for a regular Markov chain is followed. The argument begins by observing that

$$f_{ij}^{(1)} = p_{ij}^{(1)} = p_{ij}, \tag{3.25}$$

because the probability of entering state j for the first time after one step is simply the one-step transition probability. To enter recurrent state j for the first time after two steps, the chain must enter any transient state k after one step, and move from k to j after one additional step. Since each transition is conditioned only on the present state, these two transitions are independent, and

$$f_{ij}^{(2)} = \sum_{k \in T} p_{ik} f_{kj}^{(1)} = \sum_{k \in T} p_{ik} f_{kj}^{(2-1)}, \tag{3.26}$$

where the sum is over all transient states, k. To enter recurrent state j for the first time after three steps, the chain must enter any transient state k after one step, and move from k to j for the first time after two additional steps. Therefore,

$$f_{ij}^{(3)} = \sum_{k \in T} p_{ik} f_{kj}^{(2)} = \sum_{k \in T} p_{ik} f_{kj}^{(3-1)}, \qquad (3.27)$$

where, once again, the sum is over all transient states, k. By continuing in this manner, the following algebraic equation is obtained for recursively computing $f_{ij}^{(n)}$:

$$f_{ij}^{(n)} = \sum_{k \in T} p_{ik} f_{kj}^{(n-1)}, \qquad (3.28)$$

where the sum is over all transient states, k. Note that in order to reach the target recurrent state j for the first time on the nth step, the first step, with probability p_{ik}, must be to any transient state k. This probability is multiplied by the probability, $f_{kj}^{(n-1)}$, of moving from transient state k to the target recurrent state j for the first time in $n-1$ additional steps. These products are summed over all transient states k.

The recursive equation (3.28) can also be expressed in matrix form. Let $f_j^{(n)} = [f_{ij}^{(n)}]$ be the column vector of probabilities of first passage in n steps from transient states to recurrent state j. Then, the algebraic equation (3.25) has the matrix form

$$f_j^{(1)} = D_j, \qquad (3.29)$$

where D_j is the vector of one-step transition probabilities from transient states to recurrent state j, and $f_j^{(1)}$ is the vector of probabilities of first passage in one step from transient states to recurrent state j. Similarly, the algebraic equation (3.26) corresponds to the matrix equation

$$f_j^{(2)} = Q f_j^{(1)} = Q D_j, \qquad (3.30)$$

where Q is the matrix of one-step transition probabilities from transient states to transient states. The recursive algebraic equation (3.27) is equivalent to the matrix equation

$$f_j^{(3)} = Q f_j^{(2)} = Q(Q f_j^{(1)}) = Q^2 f_j^{(1)} = Q^2 D_j. \qquad (3.31)$$

Similarly, for $n = 4$, the matrix equation is

$$f_j^{(4)} = Q f_j^{(3)} = Q(Q f_j^{(2)}) = Q(Q(Q f_j^{(1)})) = Q^3 f_j^{(1)} = Q^3 D_j. \qquad (3.32)$$

By continuing in this manner, for successively higher values of n, the matrix equation for recursively computing $f_j^{(n)}$ is

$$f_j^{(n)} = Q^{n-1} D_j. \qquad (3.33)$$

3.3.2.1.1 *Probability of Absorption in n Steps in Absorbing Unichain Model of Machine Deterioration*

As Section 3.1.1 indicates, the simplest kind of reducible unichain is an absorbing unichain, which has one absorbing state plus transient states. In an absorbing unichain, the probability of first passage in n steps from a transient state i to the target absorbing state j, denoted by $f_{ij}^{(n)}$, is called a probability of absorption in n steps because the chain never leaves the absorbing state. The column vector of n-step probabilities of absorption in absorbing state j is denoted by $f_j^{(n)} = [f_{ij}^{(n)}]$.

Consider the four-state absorbing unichain model of machine deterioration introduced in Section 1.10.1.2.1.2. The canonical form of the transition probability matrix is

State		1	2	3	4
Not Working	1	1	0	0	0
$P = $ Working, with a Major Defect	2	0.6	0.4	0	0
Working, with a Minor Defect	3	0.2	0.3	0.5	0
Working Properly	4	0.3	0.2	0.1	0.4

$$= \begin{bmatrix} 1 & 0 \\ D_1 & Q \end{bmatrix}. \quad (1.54)$$

In this model, state 1, Not Working, is the absorbing state, and states 2, 3, and 4 are transient. The vector of probabilities of absorption in 1 day from the three transient states is

$$f_1^{(1)} = \begin{bmatrix} p_{21} \\ p_{31} \\ p_{41} \end{bmatrix} = \begin{bmatrix} f_{21}^{(1)} \\ f_{31}^{(1)} \\ f_{41}^{(1)} \end{bmatrix} = D_1 = \begin{matrix} 2 \\ 3 \\ 4 \end{matrix} \begin{bmatrix} 0.6 \\ 0.2 \\ 0.3 \end{bmatrix}. \quad (3.34)$$

The probability that a machine in state 3, which is working, with a minor defect, will be absorbed in state 1 and stop working after 1 day is given by $p_{31} = f_{31}^{(1)} = 0.2$. The vectors of probabilities of absorption in 2, 3, and 4 days, respectively, are computed below:

$$f_1^{(2)} = \begin{bmatrix} f_{21}^{(2)} \\ f_{31}^{(2)} \\ f_{41}^{(2)} \end{bmatrix} = Qf_1^{(1)} = QD_1 =$$

State	2	3	4	State 1		State	1
2	0.4	0	0	0.6		2	0.24
3	0.3	0.5	0	0.2	$=$	3	0.28
4	0.2	0.1	0.4	0.3		4	0.26

$$(3.35)$$

$$f_1^{(3)} = \begin{bmatrix} f_{21}^{(3)} \\ f_{31}^{(3)} \\ f_{41}^{(3)} \end{bmatrix} = Q^2 f_1^{(1)} = Q^2 D_1 = \begin{matrix} 2 \\ 3 \\ 4 \end{matrix} \begin{bmatrix} 0.4 & 0 & 0 \\ 0.3 & 0.5 & 0 \\ 0.2 & 0.1 & 0.4 \end{bmatrix} \begin{bmatrix} 0.4 & 0 & 0 \\ 0.3 & 0.5 & 0 \\ 0.2 & 0.1 & 0.4 \end{bmatrix} \begin{bmatrix} 0.6 \\ 0.2 \\ 0.3 \end{bmatrix}$$

$$= \begin{bmatrix} 0.16 & 0 & 0 \\ 0.27 & 0.25 & 0 \\ 0.19 & 0.09 & 0.16 \end{bmatrix} \begin{bmatrix} 0.6 \\ 0.2 \\ 0.3 \end{bmatrix} = \begin{bmatrix} 0.096 \\ 0.212 \\ 0.180 \end{bmatrix}$$

$$(3.36)$$

$$f_1^{(4)} = \begin{bmatrix} f_{21}^{(4)} \\ f_{31}^{(4)} \\ f_{41}^{(4)} \end{bmatrix} = Q^3 f_1^{(1)} = Q^3 D_1 = \begin{matrix} 2 \\ 3 \\ 4 \end{matrix} \begin{bmatrix} 0.16 & 0 & 0 \\ 0.27 & 0.25 & 0 \\ 0.19 & 0.09 & 0.16 \end{bmatrix} \begin{bmatrix} 0.4 & 0 & 0 \\ 0.3 & 0.5 & 0 \\ 0.2 & 0.1 & 0.4 \end{bmatrix} \begin{bmatrix} 0.6 \\ 0.2 \\ 0.3 \end{bmatrix}$$

$$= \begin{bmatrix} 0.064 & 0 & 0 \\ 0.183 & 0.125 & 0 \\ 0.135 & 0.061 & 0.064 \end{bmatrix} \begin{bmatrix} 0.6 \\ 0.2 \\ 0.3 \end{bmatrix} = \begin{bmatrix} 0.0384 \\ 0.1348 \\ 0.1124 \end{bmatrix}.$$

$$(3.37)$$

The probability that a machine in state 3, which is working, with a minor defect, will be absorbed in state 1 and stop working for the first time 4 days later is given by $f_{31}^{(4)} = 0.1348$ in Equation (3.37).

3.3.2.1.2 *Probability of First Passage in n Steps in a Regular Markov Chain Revisited*

In Sections 2.2.2.1 and 3.3.1, MFPTs for a regular Markov chain are computed by changing a target state in a regular Markov chain into an absorbing state to produce an absorbing unichain. In Section 2.2.1, probabilities of first passage in n steps are computed for a regular Markov chain. In this section, probabilities of first passage in n steps for a regular Markov chain will be computed by the alternative procedure of changing a target state in a regular Markov chain into an absorbing state. When a regular chain is converted into an absorbing unichain, the probability of first passage in n steps for the regular Markov chain is equal to the probability of absorption in n steps for the associated absorbing unichain. To demonstrate this alternative procedure for computing probabilities of first passage in n steps for a regular Markov chain, consider the four-state regular Markov chain model of the weather described in Section 1.3. The transition probability matrix is given by

$$P = \begin{matrix} 1 \\ 2 \\ 3 \\ 4 \end{matrix} \begin{bmatrix} 0.3 & 0.1 & 0.4 & 0.2 \\ 0.2 & 0.5 & 0.2 & 0.1 \\ 0.3 & 0.2 & 0.1 & 0.4 \\ 0 & 0.6 & 0.3 & 0.1 \end{bmatrix}. \qquad (1.9)$$

In Section 2.2.1, probabilities of first passage in n steps are computed for this regular Markov chain model with state 1 (rain) chosen as the target state. When state 1 is made an absorbing state, the probability transition matrix of the modified absorbing unichain model is shown below in canonical form:

$$P = \begin{array}{c|cccc} \text{State} & 1 & 2 & 3 & 4 \\ \hline \text{Rain} \quad 1 & 1 & 0 & 0 & 0 \\ \text{Snow} \quad 2 & 0.2 & 0.5 & 0.2 & 0.1 \\ \text{Cloudy} \quad 3 & 0.3 & 0.2 & 0.1 & 0.4 \\ \text{Sunny} \quad 4 & 0 & 0.6 & 0.3 & 0.1 \end{array} = \begin{bmatrix} 1 & 0 \\ D_1 & Q \end{bmatrix}. \tag{3.38}$$

The vector of probabilities of absorption or first passage in 1 day from the three transient states is

$$f_1^{(1)} = \begin{bmatrix} p_{21} \\ p_{31} \\ p_{41} \end{bmatrix} = \begin{bmatrix} f_{21}^{(1)} \\ f_{31}^{(1)} \\ f_{41}^{(1)} \end{bmatrix} = D_1 = \begin{array}{c} 2 \\ 3 \\ 4 \end{array} \begin{bmatrix} 0.2 \\ 0.3 \\ 0 \end{bmatrix}. \tag{3.39}$$

The vectors of n-step probabilities of absorption, or first passage probabilities, for $n = 2, 3,$ and 4, are calculated below:

$$f_1^{(2)} = \begin{bmatrix} f_{21}^{(2)} \\ f_{31}^{(2)} \\ f_{41}^{(2)} \end{bmatrix} = Qf_1^{(1)} = QD_1 = \begin{array}{c|ccc} \text{State} & 2 & 3 & 4 \\ \hline 2 & 0.5 & 0.2 & 0.1 \\ 3 & 0.2 & 0.1 & 0.4 \\ 4 & 0.6 & 0.3 & 0.1 \end{array} \begin{array}{c|c} \text{State} & 1 \\ \hline & 0.2 \\ & 0.3 \\ & 0 \end{array} = \begin{array}{c|c} \text{State} & 1 \\ \hline 2 & 0.16 \\ 3 & 0.07 \\ 4 & 0.21 \end{array}$$

$$\tag{3.40}$$

$$\begin{aligned} f_1^{(3)} \begin{bmatrix} f_{21}^{(3)} \\ f_{31}^{(3)} \\ f_{41}^{(3)} \end{bmatrix} &= Q^2 f_1^{(1)} = Q^2 D_1 = \begin{array}{c} 2 \\ 3 \\ 4 \end{array} \begin{bmatrix} 0.5 & 0.2 & 0.1 \\ 0.2 & 0.1 & 0.4 \\ 0.6 & 0.3 & 0.1 \end{bmatrix} \begin{bmatrix} 0.5 & 0.2 & 0.1 \\ 0.2 & 0.1 & 0.4 \\ 0.6 & 0.3 & 0.1 \end{bmatrix} \begin{bmatrix} 0.2 \\ 0.3 \\ 0 \end{bmatrix} \\ &= \begin{bmatrix} 0.35 & 0.15 & 0.14 \\ 0.36 & 0.17 & 0.10 \\ 0.42 & 0.18 & 0.19 \end{bmatrix} \begin{bmatrix} 0.2 \\ 0.3 \\ 0 \end{bmatrix} = \begin{bmatrix} 0.115 \\ 0.123 \\ 0.138 \end{bmatrix} \end{aligned}$$

$$\tag{3.41}$$

$$
f_1^{(4)} = \begin{bmatrix} f_{21}^{(4)} \\ f_{31}^{(4)} \\ f_{41}^{(4)} \end{bmatrix} = Q^3 f_1^{(1)} = Q^3 D_1 = \begin{matrix} 2 \\ 3 \\ 4 \end{matrix}\begin{bmatrix} 0.35 & 0.15 & 0.14 \\ 0.36 & 0.17 & 0.10 \\ 0.42 & 0.18 & 0.19 \end{bmatrix}\begin{bmatrix} 0.5 & 0.2 & 0.1 \\ 0.2 & 0.1 & 0.4 \\ 0.6 & 0.3 & 0.1 \end{bmatrix}\begin{bmatrix} 0.2 \\ 0.3 \\ 0 \end{bmatrix}
$$

$$
= \begin{bmatrix} 0.289 & 0.127 & 0.109 \\ 0.274 & 0.119 & 0.114 \\ 0.360 & 0.159 & 0.133 \end{bmatrix}\begin{bmatrix} 0.2 \\ 0.3 \\ 0 \end{bmatrix} = \begin{bmatrix} 0.0959 \\ 0.0905 \\ 0.1197 \end{bmatrix}.
$$

$$\tag{3.42}$$

The probability that a cloudy day, state 3, will be absorbed in state 1 and change to a rainy day for the first time 4 days later is given by $f_{31}^{(4)} = 0.0905$ in Equation (3.42). The same results are obtained in Equations (2.46) and (2.47).

Observe that the matrix Q in the treatment of the problem as an absorbing chain in this section is a submatrix of the matrix Z in the treatment as a regular chain in Section 2.2.1. The matrix Q is obtained by deleting the first row and first column of Z. As a result, the first entry, $f_{jj}^{(n)}$, of vector $f^{(n)}$ in the treatment as a regular chain is omitted from the vector $f^{(n)}$ in the treatment as an absorbing chain. Hence, the treatment as a regular chain in Section 2.2.1 yields additional information, namely, $f_{jj}^{(n)}$, the n-step first passage probability or n-step recurrence probability for the target state j in the original, regular Markov chain.

3.3.2.1.3 Unichain with Nonabsorbing Recurrent States

A reducible unichain may consist of a closed class of r recurrent states, none of which are absorbing states, plus one or more transient states. The recursive equation (3.33) for computing the vector of probabilities of first passage in n steps can be extended such that it has the form

$$
\begin{aligned}
F^{(n)} &= [f_1^{(n)} \quad f_2^{(n)} \quad \cdots \quad f_r^{(n)}] = Q^{n-1}D = Q^{n-1}[d_1 \quad d_2 \quad \cdots \quad d_r] \\
&= [Q^{n-1}d_1 \quad Q^{n-1}d_2 \quad \cdots \quad Q^{n-1}d_r],
\end{aligned}
\tag{3.43}
$$

where $F^{(n)}$ is the matrix of n-step first passage probabilities, vector $f_j^{(n)}$ is the jth column of matrix $F^{(n)}$, and vector d_j is the jth column of matrix D.

The probability of entering a recurrent closed set for the first time in n steps is the sum of the n-step first passage probabilities to the states, which belong to the recurrent closed set. If the recurrent closed set of states is denoted by R, and $f_{iR}^{(n)}$ denotes the n-step first passage probability from transient state i to the recurrent set R, then

$$
f_{iR}^{(n)} = \sum_{j \in R} f_{ij}^{(n)},
\tag{3.44}
$$

where the sum is over all recurrent states j, which belong to R.

Consider the generic four-state reducible unichain for which the transition matrix is shown below:

$$P = [p_{ij}] = \begin{array}{c} 1 \\ 2 \\ 3 \\ 4 \end{array} \left[\begin{array}{cc|cc} p_{11} & p_{12} & 0 & 0 \\ p_{21} & p_{22} & 0 & 0 \\ \hline p_{31} & p_{32} & p_{33} & p_{34} \\ p_{41} & p_{42} & p_{43} & p_{44} \end{array} \right] = \left[\begin{array}{cc} S & 0 \\ D & Q \end{array} \right]. \tag{3.2}$$

The matrix of one-step first passage probabilities is

$$F^{(1)} = Q^{1-1}D = Q^0 D = D = \left[\begin{array}{cc} p_{31} & p_{32} \\ p_{41} & p_{42} \end{array} \right] = \left[\begin{array}{cc} f_{31}^{(1)} & f_{32}^{(1)} \\ f_{41}^{(1)} & f_{42}^{(1)} \end{array} \right] = [f_1^{(1)} \quad f_2^{(1)}]. \tag{3.45}$$

The probability of first passage in one step from transient state 4 to recurrent state 2 is

$$f_{42}^{(1)} = p_{42}. \tag{3.46}$$

Using Equation (3.44), the probability of first passage in one step from transient state 4 to the recurrent set of states $R = \{1, 2\}$ is

$$f_{4R}^{(1)} = f_{41}^{(1)} + f_{42}^{(1)} = p_{41} + p_{42}. \tag{3.47}$$

The matrix of two-step first passage probabilities is

$$F^{(2)} = Q^{2-1}D = Q^1 D = QD = \left[\begin{array}{cc} p_{33} & p_{34} \\ p_{43} & p_{44} \end{array} \right] \left[\begin{array}{cc} p_{31} & p_{32} \\ p_{41} & p_{42} \end{array} \right]$$

$$= \left[\begin{array}{cc} (p_{33}p_{31} + p_{34}p_{41}) & (p_{33}p_{32} + p_{34}p_{42}) \\ (p_{43}p_{31} + p_{44}p_{41}) & (p_{43}p_{32} + p_{44}p_{42}) \end{array} \right] = \left[\begin{array}{cc} f_{31}^{(2)} & f_{32}^{(2)} \\ f_{41}^{(2)} & f_{42}^{(2)} \end{array} \right] = [f_1^{(2)} \quad f_2^{(2)}]. \tag{3.48}$$

The probability of first passage in two steps from transient state 4 to recurrent state 2 is

$$f_{42}^{(2)} = p_{43}p_{32} + p_{44}p_{42}.$$

The matrix of three-step first passage probabilities is

$$F^{(3)} = Q^{3-1}D = Q^2 D = \left[\begin{array}{cc} p_{33} & p_{34} \\ p_{43} & p_{44} \end{array} \right]^2 \left[\begin{array}{cc} p_{31} & p_{32} \\ p_{41} & p_{42} \end{array} \right]$$

$$= \begin{bmatrix} (p_{33}p_{33} + p_{34}p_{43}) & (p_{33}p_{34} + p_{34}p_{44}) \\ (p_{43}p_{33} + p_{44}p_{43}) & (p_{43}p_{34} + p_{44}p_{44}) \end{bmatrix} \begin{bmatrix} p_{31} & p_{32} \\ p_{41} & p_{42} \end{bmatrix}$$

$$= \begin{bmatrix} (p_{33}p_{33} + p_{34}p_{43})p_{31} + (p_{33}p_{34} + p_{34}p_{44})p_{41} & (p_{33}p_{33} + p_{34}p_{43})p_{32} + (p_{33}p_{34} + p_{34}p_{44})p_{42} \\ (p_{43}p_{33} + p_{44}p_{43})p_{31} + (p_{43}p_{34} + p_{44}p_{44})p_{41} & (p_{43}p_{33} + p_{44}p_{43})p_{32} + (p_{43}p_{34} + p_{44}p_{44})p_{42} \end{bmatrix}$$

$$= \begin{bmatrix} f_{31}^{(3)} & f_{32}^{(3)} \\ f_{41}^{(3)} & f_{42}^{(3)} \end{bmatrix} = [f_1^{(3)} \quad f_2^{(3)}].$$

The probability of first passage in three steps from transient state 4 to recurrent state 2 is

$$f_{42}^{(3)} = (p_{43}p_{33} + p_{44}p_{43})p_{32} + (p_{43}p_{34} + p_{44}p_{44})p_{42}.$$

Using Equation (3.44), the probability of first passage in three steps from transient state 4 to the recurrent set of states $R = \{1, 2\}$ is

$$f_{4R}^{(3)} = f_{41}^{(3)} + f_{42}^{(3)}. \tag{3.49}$$

3.3.2.1.4 Probability of First Passage in n Steps in a Unichain Model of Machine Maintenance Under a Modified Policy of Doing Nothing in State 3

The four-state Markov chain model of machine maintenance introduced in Section 1.10.1.2.2.1 provides a numerical example of a reducible unichain with recurrent states. The transition probability matrix associated with the modified maintenance policy, under which the engineer always overhauls the machine in state 1 and does nothing in all other states, is given below in canonical form.

State	1	2	3	4
1	0.2	0.8	0	0
2	0.6	0.4	0	0
3	0.2	0.3	0.5	0
4	0.3	0.2	0.1	0.4

$$P = \begin{bmatrix} S & 0 \\ D & Q \end{bmatrix}. \tag{1.56}$$

The matrix of one-step first passage probabilities is

$$F^{(1)} = Q^{1-1}D = Q^0 D = D = \begin{bmatrix} 0.2 & 0.3 \\ 0.3 & 0.2 \end{bmatrix} = \begin{bmatrix} f_{31}^{(1)} & f_{32}^{(1)} \\ f_{41}^{(1)} & f_{42}^{(1)} \end{bmatrix} = [f_1^{(1)} \quad f_2^{(1)}]. \tag{3.50}$$

The matrix of two-step first passage probabilities is

$$F^{(2)} = QD = \begin{bmatrix} 0.5 & 0 \\ 0.1 & 0.4 \end{bmatrix}\begin{bmatrix} 0.2 & 0.3 \\ 0.3 & 0.2 \end{bmatrix} = \begin{bmatrix} 0.10 & 0.15 \\ 0.14 & 0.11 \end{bmatrix} = \begin{bmatrix} f_{31}^{(2)} & f_{32}^{(2)} \\ f_{41}^{(2)} & f_{42}^{(2)} \end{bmatrix} = [f_1^{(2)} \quad f_2^{(2)}].$$

(3.51)

The matrix of three-step first passage probabilities is

$$F^{(3)} = Q^2 D = \begin{bmatrix} 0.5 & 0 \\ 0.1 & 0.4 \end{bmatrix}\begin{bmatrix} 0.5 & 0 \\ 0.1 & 0.4 \end{bmatrix}\begin{bmatrix} 0.2 & 0.3 \\ 0.3 & 0.2 \end{bmatrix} = \begin{bmatrix} 0.25 & 0 \\ 0.09 & 0.16 \end{bmatrix}\begin{bmatrix} 0.2 & 0.3 \\ 0.3 & 0.2 \end{bmatrix} = \begin{bmatrix} 0.05 & 0.075 \\ 0.066 & 0.059 \end{bmatrix}$$

$$= \begin{bmatrix} f_{31}^{(3)} & f_{32}^{(3)} \\ f_{41}^{(3)} & f_{42}^{(3)} \end{bmatrix} = [f_1^{(3)} \quad f_2^{(3)}].$$

(3.52)

Under the modified maintenance policy, the probability of first passage in 3 days from transient state 4, working properly, to recurrent state 2, working, with a major defect, is $f_{42}^{(3)} = 0.059$. Similarly, under this maintenance policy, the probability of first passage in 3 days from transient state 4, working properly, to the recurrent set of states, $R = \{1, 2\}$, which denotes either not working or working, with a major defect, is

$$f_{4R}^{(3)} = f_{41}^{(3)} + f_{42}^{(3)} = 0.066 + 0.059 = 0.125.$$

(3.53)

3.3.2.2 Reducible Multichain

Consider a reducible multichain with a set of transient states plus two or more closed sets of recurrent states. As Section 3.1.2 indicates, the canonical form of the transition probability matrix for a multichain with two recurrent chains is

$$P = \left[\begin{array}{cc|c} P_1 & 0 & 0 \\ 0 & P_2 & 0 \\ \hline D_1 & D_2 & Q \end{array}\right] = \begin{bmatrix} S & 0 \\ D & Q \end{bmatrix},$$

(3.4)

where

$$S = \begin{bmatrix} P_1 & 0 \\ 0 & P_2 \end{bmatrix}, \quad 0 = \begin{bmatrix} 0 \\ 0 \end{bmatrix}, \quad \text{and} \quad D = [D_1 \quad D_2].$$

The recursive equation (3.43) for computing the matrix of probabilities of first passage in n steps, extended to apply to a multichain with r recurrent chains, is

$$F^{(n)} = Q^{n-1}D = Q^{n-1}[D_1 \quad \cdots \quad D_r] = [Q^{n-1}D_1 \quad \cdots \quad Q^{n-1}D_r] = [F_1^{(n)} \quad \cdots \quad F_r^{(n)}].$$
(3.54)

Consider the canonical form of the following transition probability matrix for a generic five-state reducible multichain displayed in Section 3.1.2. The multichain has two recurrent closed sets, one of which is an absorbing state, plus two transient states.

$$P = \begin{matrix} 1 \\ 2 \\ 3 \\ 4 \\ 5 \end{matrix} \begin{bmatrix} 1 & 0 & 0 & 0 & 0 \\ 0 & p_{22} & p_{23} & 0 & 0 \\ 0 & p_{32} & p_{33} & 0 & 0 \\ p_{41} & p_{42} & p_{43} & p_{44} & p_{45} \\ p_{51} & p_{52} & p_{53} & p_{54} & p_{55} \end{bmatrix} = \begin{bmatrix} P_1 & 0 & 0 \\ 0 & P_2 & 0 \\ D_1 & D_2 & Q \end{bmatrix}.$$
(3.5)

Suppose that two-step first passage probabilities to $R_2 = \{2, 3\}$ are of interest. The matrix of two-step first passage probabilities is

$$F_2^{(2)} = Q^{2-1}D_2 = Q^1 D_2 = QD_2 = \begin{bmatrix} p_{44} & p_{45} \\ p_{54} & p_{55} \end{bmatrix} \begin{bmatrix} p_{42} & p_{43} \\ p_{52} & p_{53} \end{bmatrix}$$

$$= \begin{bmatrix} (p_{44}p_{42} + p_{45}p_{52}) & (p_{44}p_{43} + p_{45}p_{53}) \\ (p_{54}p_{42} + p_{55}p_{52}) & (p_{54}p_{43} + p_{55}p_{53}) \end{bmatrix} = \begin{bmatrix} f_{42}^{(2)} & f_{43}^{(2)} \\ f_{52}^{(2)} & f_{53}^{(2)} \end{bmatrix} = [F_2^{(2)} \quad F_3^{(2)}]. \quad (3.55)$$

Observe that the probability of first passage in two steps from transient state 4 to recurrent state 2 is

$$f_{42}^{(2)} = p_{44}p_{42} + p_{45}p_{52}.$$
(3.56)

Using Equation (3.44), the probability of first passage in two steps from state 4 to R_2 is

$$f_{4,R_2}^{(2)} = f_{42}^{(2)} + f_{43}^{(2)}.$$
(3.57)

If the target recurrent state j is an absorbing state, then the probability of absorption in two steps from transient state 4 to absorbing state 1 is

$$f_{41}^{(2)} = p_{44}p_{41} + p_{45}p_{51} = f_{4,R_1}^{(2)}.$$
(3.58)

3.3.2.2.1 Probability of Absorption in n Steps in an Absorbing Multichain Model of Hospital Patient Flow

Consider the absorbing multichain model of patient flow in a hospital introduced in Section 1.10.2.1.2, and for which the fundamental matrix is obtained in Section 3.2.4. The transition probability matrix in canonical form is shown in Equation (1.59).

State	0	5	1	2	3	4
0	1	0	0	0	0	0
5	0	1	0	0	0	0
1	0	0	0.4	0.1	0.2	0.3
2	0.15	0.05	0.1	0.2	0.3	0.2
3	0.07	0.03	0.2	0.1	0.4	0.2
4	0	0	0.3	0.4	0.2	0.1

$$P = \begin{bmatrix} S & 0 \\ D & Q \end{bmatrix} = \begin{bmatrix} I & 0 \\ D & Q \end{bmatrix}, \qquad (1.59a)$$

where

$$S = I = \begin{bmatrix} 1 & 0 \\ 0 & 1 \end{bmatrix}, \quad D = \begin{matrix} 1 \\ 2 \\ 3 \\ 4 \end{matrix}\begin{bmatrix} 0 & 0 \\ 0.15 & 0.05 \\ 0.07 & 0.03 \\ 0 & 0 \end{bmatrix}, \quad Q = \begin{matrix} 1 \\ 2 \\ 3 \\ 4 \end{matrix}\begin{bmatrix} 0.4 & 0.1 & 0.2 & 0.3 \\ 0.1 & 0.2 & 0.3 & 0.2 \\ 0.2 & 0.1 & 0.4 & 0.2 \\ 0.3 & 0.4 & 0.2 & 0.1 \end{bmatrix}. \qquad (1.59b)$$

The matrix of one-step absorption probabilities is

$$F^{(1)} = D = \begin{matrix} 1 \\ 2 \\ 3 \\ 4 \end{matrix}\begin{bmatrix} 0 & 0 \\ 0.15 & 0.05 \\ 0.07 & 0.03 \\ 0 & 0 \end{bmatrix} = \begin{bmatrix} f_{10}^{(1)} & f_{15}^{(1)} \\ f_{20}^{(1)} & f_{25}^{(1)} \\ f_{30}^{(1)} & f_{35}^{(1)} \\ f_{40}^{(1)} & f_{45}^{(1)} \end{bmatrix} \qquad (3.59)$$

After 1 day, a patient in state 3, surgery, will be in state 0, discharged, with probability $f_{30}^{(1)} = 0.07$, or in state 5, dead, with probability $f_{35}^{(1)} = 0.03$.

The matrix of two-step absorption probabilities is

$$F^{(2)} = QD = \begin{matrix} 1 \\ 2 \\ 3 \\ 4 \end{matrix}\begin{bmatrix} 0.4 & 0.1 & 0.2 & 0.3 \\ 0.1 & 0.2 & 0.3 & 0.2 \\ 0.2 & 0.1 & 0.4 & 0.2 \\ 0.3 & 0.4 & 0.2 & 0.1 \end{bmatrix}\begin{bmatrix} 0 & 0 \\ 0.15 & 0.05 \\ 0.07 & 0.03 \\ 0 & 0 \end{bmatrix} = \begin{bmatrix} 0.029 & 0.011 \\ 0.051 & 0.019 \\ 0.043 & 0.017 \\ 0.074 & 0.026 \end{bmatrix} = \begin{bmatrix} f_{10}^{(2)} & f_{15}^{(2)} \\ f_{20}^{(2)} & f_{25}^{(2)} \\ f_{30}^{(2)} & f_{35}^{(2)} \\ f_{40}^{(2)} & f_{45}^{(2)} \end{bmatrix}.$$

$$(3.60)$$

After 2 days, a patient in state 3, surgery, will be in state 0, discharged, with probability $f_{30}^{(2)} = 0.043$, or in state 5, dead, with probability $f_{35}^{(2)} = 0.017$.

3.3.2.2.2 Probability of First Passage in n Steps in a Multichain Model of a Production Process

Consider the multichain model of a production process introduced in Section 1.10.2.2. The transition probability matrix in canonical form is shown in Equation (3.5) of Section 3.1.3. Recall that production stage i is represented by transient state $(9 - i)$.

The matrix $F^{(n)}$ of n-step first passage probabilities has the form

$$F^{(n)} = [f_1^{(n)} \quad f_2^{(n)} \mid f_3^{(n)} \quad f_4^{(n)} \quad f_5^{(n)}], \tag{3.61}$$

where vector $f_j^{(n)}$ is the jth column of matrix $F^{(n)}$.

The matrix of one-step absorption probabilities is

$$F_A^{(1)} = [f_1^{(1)} \quad f_2^{(1)}] = [D_1 \quad D_2] = \begin{matrix} 6 \\ 7 \\ 8 \end{matrix} \begin{bmatrix} 0.20 & 0.16 \\ 0.15 & 0 \\ 0.10 & 0 \end{bmatrix} = \begin{bmatrix} f_{61}^{(1)} & f_{62}^{(1)} \\ f_{71}^{(1)} & f_{72}^{(1)} \\ f_{81}^{(1)} & f_{82}^{(1)} \end{bmatrix}. \tag{3.62}$$

After one step, an item in state 6, production stage 3, will be in state 1, scrapped, with probability $f_{61}^{(1)} = 0.20$, or in state 2, sold, with probability $f_{62}^{(1)} = 0.16$.

The matrix of one-step first passage probabilities is

$$F_R^{(1)} = [f_3^{(1)} \quad f_4^{(1)} \quad f_5^{(1)}] = D_3 = \begin{matrix} 6 \\ 7 \\ 8 \end{matrix} \begin{bmatrix} 0.04 & 0.03 & 0.02 \\ 0 & 0 & 0 \\ 0 & 0 & 0 \end{bmatrix} = \begin{bmatrix} f_{63}^{(1)} & f_{64}^{(1)} & f_{65}^{(1)} \\ f_{73}^{(1)} & f_{74}^{(1)} & f_{75}^{(1)} \\ f_{83}^{(1)} & f_{84}^{(1)} & f_{85}^{(1)} \end{bmatrix}.$$

$$\tag{3.63}$$

After one step, an item in state 6, production stage 3, will be in state 3, used to train engineers, with probability $f_{63}^{(1)} = 0.04$, or in state 4, used to train technicians, with probability $f_{64}^{(1)} = 0.03$, or in state 5, used to train technical writers, with probability $f_{65}^{(1)} = 0.02$. Therefore, after one step, an item in state 6, production stage 3, will be sent to the training center with probability $f_{63}^{(1)} + f_{64}^{(1)} + f_{65}^{(1)} = 0.04 + 0.03 + 0.02 = 0.09$, where it will remain.

The matrix of two-step absorption probabilities is

$$F_A^{(2)} = [f_1^{(2)} \quad f_2^{(2)}] = Q[D_1 \quad D_2]$$

$$= \begin{matrix} 6 \\ 7 \\ 8 \end{matrix} \begin{bmatrix} 0.55 & 0 & 0 \\ 0.20 & 0.65 & 0 \\ 0 & 0.15 & 0.75 \end{bmatrix} \begin{bmatrix} 0.20 & 0.16 \\ 0.15 & 0 \\ 0.10 & 0 \end{bmatrix} = \begin{bmatrix} 0.11 & 0.088 \\ 0.1375 & 0.032 \\ 0.0975 & 0 \end{bmatrix} = \begin{bmatrix} f_{61}^{(2)} & f_{62}^{(2)} \\ f_{71}^{(2)} & f_{72}^{(2)} \\ f_{81}^{(2)} & f_{82}^{(2)} \end{bmatrix}.$$

$$\tag{3.64}$$

After two steps, an item in state 6, production stage 3, will be in state 1, scrapped, with probability $f_{61}^{(2)} = 0.11$, or in state 2, sold, with probability $f_{62}^{(2)} = 0.088$.

The matrix of two-step first passage probabilities is

$$F_R^{(2)} = [f_3^{(2)} \quad f_4^{(2)} \quad f_5^{(2)}] = QD_3$$

$$
= \begin{matrix} 6 \\ 7 \\ 8 \end{matrix}
\begin{bmatrix} 0.55 & 0 & 0 \\ 0.20 & 0.65 & 0 \\ 0 & 0.15 & 0.75 \end{bmatrix}
\begin{bmatrix} 0.04 & 0.03 & 0.02 \\ 0 & 0 & 0 \\ 0 & 0 & 0 \end{bmatrix}
= \begin{bmatrix} 0.022 & 0.0165 & 0.011 \\ 0.008 & 0.006 & 0.004 \\ 0 & 0 & 0 \end{bmatrix}
$$

$$
= \begin{bmatrix} f_{63}^{(2)} & f_{64}^{(2)} & f_{65}^{(2)} \\ f_{73}^{(2)} & f_{74}^{(2)} & f_{75}^{(2)} \\ f_{83}^{(2)} & f_{84}^{(2)} & f_{85}^{(2)} \end{bmatrix}.
\tag{3.65}
$$

After two steps, an item in state 6, production stage 3, will, for the first time, be in state 3, used to train engineers, with probability $f_{63}^{(2)} = 0.022$, or in state 4, used to train technicians, with probability $f_{64}^{(2)} = 0.0165$, or in state 5, used to train technical writers, with probability $f_{65}^{(2)} = 0.011$. Therefore, after two steps, an item in state 6, production stage 3, will, for the first time, be sent to the training center with probability

$$f_{63}^{(2)} + f_{64}^{(2)} + f_{65}^{(2)} = 0.022 + 0.0165 + 0.011 = 0.0495. \tag{3.66}$$

3.3.3 Probability of Eventual Passage to a Recurrent State

In this section attention is focused on the probability of eventual passage (without counting the number of steps) from a transient state to a target recurrent state in a reducible unichain or multichain. Suppose that the transition probability matrix is represented in the aggregated canonical form of Equation (3.1).

$$P = \begin{bmatrix} S & 0 \\ D & Q \end{bmatrix}. \tag{3.1}$$

S may have been formed by aggregating the transition probability matrices for several recurrent closed classes of states belonging to a reducible multichain. Let f_{ij} denote the probability of eventual passage from a transient state i to a target recurrent state j. If the target state j is an absorbing state, then f_{ij} is called an absorption probability. Let $F = [f_{ij}]$ denote the matrix of probabilities of eventual passage from transient states to recurrent states. Since the number of steps required for eventual passage to a target recurrent state j is a random variable,

$$f_{ij} = f_{ij}^{(1)} + f_{ij}^{(2)} + f_{ij}^{(3)} + \cdots = \sum_{n=1}^{\infty} f_{ij}^{(n)}. \tag{3.67}$$

In matrix form,

$$F = \sum_{n=1}^{\infty} F^{(n)} = F^{(1)} + F^{(2)} + F^{(3)} + \cdots = D + F^{(2)} + F^{(3)} + \cdots$$

$$= D + \sum_{n=2}^{\infty} F^{(n)} = ID + \sum_{n=2}^{\infty} F^{(n)}$$

$$= ID + \sum_{n=2}^{\infty} Q^{n-1}D = ID + (Q + Q^2 + Q^3 + \cdots)D$$

$$= (I + Q + Q^2 + Q^3 + \cdots)D \tag{3.68}$$

The entries of Q^n give the probabilities of being in each of the transient states after n steps for each possible transient starting state. After zero steps the chain is in the transient state in which it started, so that $Q^0 = I$. As the chain is certain to eventually enter a closed set of recurrent states within which it will remain forever, the chain will eventually never return to a transient state. Therefore, the probability of being in the transient states after n steps approaches zero. In other words, if i and j are both transient states, then $\lim_{n\to\infty} p_{ij}^{(n)} = 0$, irrespective of the transient starting state i. It follows that every entry of Q^n must approach zero as $n\to\infty$. That is,

$$\lim_{n\to\infty} Q^n = 0, \tag{3.69}$$

the null matrix.

Let the matrix Y represent the following sum.

$$Y = I + Q + Q^2 + Q^3 + \cdots + Q^{n-1}. \tag{3.70}$$

Premultiplying both sides of Equation (3.69) by Q,

$$QY = Q + Q^2 + Q^3 + Q^4 + \cdots + Q^{n-1} + Q^n. \tag{3.71}$$

Subtracting Equation (3.70) from Equation (3.69),

$$Y - QY = IY - QY = (I - Q)Y = I - Q^n. \tag{3.72}$$

Now let $n \to \infty$. Since the $\lim_{n \to \infty} Q^n = 0$,

$$(I - Q)Y = I$$
$$Y = (I - Q)^{-1} = I + Q + Q^2 + Q^3 + \cdots, \tag{3.73}$$

where the sum of the infinite series, $Y = (I-Q)^{-1}$, is defined in Equation (3.10) as the fundamental matrix of a reducible Markov chain, denoted by U. Therefore, the fundamental matrix can be expressed as the sum of an infinite series of substochastic matrices, Q^n, as shown in Equation (3.74).

$$U = Y = (I-Q)^{-1} = I + Q + Q^2 + Q^3 + \cdots . \tag{3.74}$$

This result is analogous to the formula for the sum of an infinite geometric series, which indicates that

$$1 + q + q^2 + q^3 + \cdots = (1 - q)^{-1}, \tag{3.75}$$

where q is a number less than one in absolute value. Using this result in the calculation of F,

$$F = (I + Q + Q^2 + Q^3 + \cdots)D = (I - Q)^{-1}D = UD. \tag{3.76}$$

Therefore, the matrix F of the probabilities of eventual passage from transient states to recurrent states in a reducible Markov chain is equal to the fundamental matrix, $U = (I - Q)^{-1}$, postmultiplied by the matrix D of one-step transition probabilities from transient states to recurrent states. The (i, j)th entry, f_{ij}, of matrix F represents an answer to question 3 in Section 3.2 because f_{ij} is the probability of eventual passage from a transient state i to a target recurrent state j.

An alternative derivation of f_{ij}, the probability of eventual passage from a transient state i to a target recurrent state j, is instructive. Consider a reducible multichain that has been partitioned into two or more closed communicating classes of recurrent states plus a set of transient states. Suppose that the chain starts in a transient state i. On the first step, one of the following four mutually exclusive events may occur:

1. The chain enters target recurrent state j after one step with probability p_{ij}.
2. With probability p_{ih} the chain enters a nontarget recurrent state h, which communicates with the target recurrent state j. In this case the first step, with probability p_{ih}, does not contribute to f_{ij}, the probability of eventual passage from transient state i to the target state j, but

contributes instead to f_{ih}, the probability of eventual passage from transient state i to the nontarget state h.

3. The chain enters a recurrent state g, which belongs to a different closed set of recurrent states that does not contain the target state j. In this case the target state will never be reached because recurrent state g does not communicate with the target state j.

4. The chain enters a transient state k with probability p_{ik}. In this case the probability of eventual passage from transient state k to the target state j is f_{kj}.

By combining the two relevant events, (1) and (4), in which the target state j is reached after starting in a transient state i, the following system of algebraic equations is produced:

$$f_{ij} = p_{ij} + \sum_{k \in T} p_{ik} f_{kj}, \qquad (3.77)$$

where the sum is over all transient states k. The formula in Equation (3.77) represents a system of equations because the unknown f_{ij} is expressed in terms of all the unknowns f_{kj}. In matrix form, the system of equations (3.77) is

$$F = D + QF$$
$$F - QF = D$$
$$IF - QF = D \qquad (3.78)$$
$$(I - Q)F = D$$

$$F = (I - Q)^{-1} D = UD, \qquad (3.76)$$

which confirms the previous result. Thus, f_{ij}, the probability of eventual passage from a transient state i to a target recurrent state j, can be calculated by solving either the system (3.77) of algebraic equations, or the matrix equation (3.76). In Equation (3.76), $F = [f_{ij}]$ is the matrix of probabilities of eventual passage from transient states to recurrent states. Observe that two alternative matrix forms of the system of equations (3.77) are given in Equations (3.78) and (3.76).

3.3.3.1 Reducible Unichain

A reducible unichain has one closed class of recurrent states plus a set of transient states.

3.3.3.1.1 *Probability of Absorption in Absorbing Unichain Model of Machine Deterioration*

As Section 3.3.2.1.1 indicates, the simplest kind of reducible unichain is an absorbing unichain, which has one absorbing state plus transient states. In an absorbing unichain, the probability of eventual passage from a transient state i to the target absorbing state j, denoted by f_{ij}, is called a probability of absorption, or absorption probability, because the chain never leaves the absorbing state. The column vector of probabilities of absorption in absorbing state j is denoted by $F = f_j = [f_{ij}]$. Starting from every transient state in an absorbing unichain, absorption is certain. Therefore, $f_{ij} = 1$ for all transient states i, and $F = f_j$ is a vector of ones.

The four-state absorbing unichain model of machine deterioration introduced in Section 1.10.1.2.1.2 is an example of an absorbing unichain. The probability of absorption in n steps is computed in Section 3.3.2.1.1. The canonical form of the transition probability matrix for the unichain model of machine deterioration is shown in Equation (1.54).

State		1	2	3	4	
Not Working	1	1	0	0	0	
$P =$ Working, with a Major Defect	2	0.6	0.4	0	0	$= \begin{bmatrix} 1 & 0 \\ D & Q \end{bmatrix}.$
Working, with a Minor Defect	3	0.2	0.3	0.5	0	
Working Properly	4	0.3	0.2	0.1	0.4	

$$(1.54)$$

The vector of probabilities of absorption in absorbing state 1, starting from the three transient states, is calculated by Equation (3.76). For this example,

State		2	3	4
	2	0.6	0	0
$I - Q =$	3	-0.3	0.5	0
	4	-0.2	-0.1	0.6

State		2	3	4
	2	1.6667	0	0
$U = (I - Q)^{-1} =$	3	1	2	0
	4	0.7222	0.3333	1.6667

$$F = (I-Q)^{-1}D = UD = \begin{array}{c} 2 \\ 3 \\ 4 \end{array}\begin{bmatrix} 1.6667 & 0 & 0 \\ 1 & 2 & 0 \\ 0.7222 & 0.3333 & 1.6667 \end{bmatrix}\begin{bmatrix} 0.6 \\ 0.2 \\ 0.3 \end{bmatrix} = \begin{bmatrix} 1 \\ 1 \\ 1 \end{bmatrix}. \qquad (3.79)$$

As expected, $F = f_1 = e$ is a vector of ones. Regardless of the transient starting state, the machine will eventually end its life in absorbing state 1, not working.

3.3.3.1.2 Four-State Reducible Unichain

Consider the generic four-state reducible unichain for which the transition matrix is shown in canonical form in Equation (3.2).

$$P = [p_{ij}] = \begin{matrix} 1 \\ 2 \\ 3 \\ 4 \end{matrix} \begin{bmatrix} p_{11} & p_{12} & 0 & 0 \\ p_{21} & p_{22} & 0 & 0 \\ \hline p_{31} & p_{32} & p_{33} & p_{34} \\ p_{41} & p_{42} & p_{43} & p_{44} \end{bmatrix} = \begin{bmatrix} S & 0 \\ D & Q \end{bmatrix}. \tag{3.2}$$

Suppose that the probability f_{32} of eventual passage from transient state 3 to recurrent state 2 is of interest. Since Equation (3.77) for calculating f_{32} is expressed in terms of both f_{32} and f_{42}, the following system of two algebraic equations must be solved:

$$\left. \begin{aligned} f_{32} &= p_{32} + \sum_{k \in T} p_{3k} f_{k2} = p_{32} + p_{33} f_{32} + p_{34} f_{42} \\ f_{42} &= p_{42} + \sum_{k \in T} p_{4k} f_{k2} = p_{42} + p_{43} f_{32} + p_{44} f_{42} \end{aligned} \right\}. \tag{3.80a}$$

A matrix form of equations (3.80) based on Equation (3.78) is

$$\begin{bmatrix} f_{32} \\ f_{42} \end{bmatrix} = \begin{bmatrix} p_{32} \\ p_{42} \end{bmatrix} + \begin{bmatrix} p_{33} & p_{34} \\ p_{43} & p_{44} \end{bmatrix} \begin{bmatrix} f_{32} \\ f_{42} \end{bmatrix}. \tag{3.80b}$$

The solution of equations (3.80) for the probabilities of eventual passage to recurrent state 2 is

$$\left. \begin{aligned} f_{32} &= \frac{p_{32}(1 - p_{44}) + p_{34} p_{42}}{(1 - p_{33})(1 - p_{44}) - p_{34} p_{43}} \\ f_{42} &= \frac{p_{42}(1 - p_{33}) + p_{43} p_{32}}{(1 - p_{33})(1 - p_{44}) - p_{34} p_{43}} \end{aligned} \right\}. \tag{3.81}$$

Alternatively, the matrix F of probabilities of eventual passage from transient states to recurrent states in a reducible unichain can be calculated by solving Equation (3.76). The fundamental matrix for the generic four-state reducible unichain was obtained by Equation (3.11). Thus,

$$F = UD = (I - Q)^{-1} D = \frac{1}{(1 - p_{33})(1 - p_{44}) - p_{34} p_{43}} \begin{bmatrix} 1 - p_{44} & p_{34} \\ p_{43} & 1 - p_{33} \end{bmatrix} \begin{bmatrix} p_{31} & p_{32} \\ p_{41} & p_{42} \end{bmatrix}$$

$$= \frac{1}{(1-p_{33})(1-p_{44})-p_{34}p_{43}} \begin{bmatrix} (1-p_{44})p_{31}+p_{34}p_{41} & (1-p_{44})p_{32}+p_{34}p_{42} \\ p_{43}p_{31}+(1-p_{33})p_{41} & p_{43}p_{32}+(1-p_{33})p_{42} \end{bmatrix}$$

$$= \begin{bmatrix} f_{31} & f_{32} \\ f_{41} & f_{42} \end{bmatrix}. \tag{3.82}$$

These values agree with the results obtained in equations (3.81).

After algebraic simplification, observe that

$$f_{31}+f_{32} = \frac{p_{31}(1-p_{44})+p_{34}p_{41}}{(1-p_{33})(1-p_{44})-p_{34}p_{43}} + \frac{p_{32}(1-p_{44})+p_{34}p_{42}}{(1-p_{33})(1-p_{44})-p_{34}p_{43}} = 1 \tag{3.83}$$

$$f_{41}+f_{42} = 1. \tag{3.84}$$

These two sums both equal one because eventual passage from a transient state to the single recurrent closed class is certain. This result can be extended to apply to any reducible unichain. One may conclude that starting from any particular transient state, the sum of the probabilities of eventual passage to all of the recurrent states is one. For the generic four-state reducible unichain, note that the probabilities of eventual passage, $f_{31}, f_{32}, f_{41},$ and $f_{42},$ are independent of the transition probabilities for the recurrent closed class, $p_{11}, p_{12}, p_{21},$ and $p_{22}.$ Therefore, if states 1 and 2 were converted to absorbing states, making the chain a generic four-state absorbing multichain, the probabilities of eventual passage, which would be called absorption probabilities, would be unchanged. To illustrate this result, consider a generic four-state absorbing multichain, in which states 1 and 2 are absorbing states. The absorbing multichain has the following transition probability matrix in canonical form:

$$P = [p_{ij}] = \begin{matrix} 1 \\ 2 \\ 3 \\ 4 \end{matrix} \begin{bmatrix} 1 & 0 & 0 & 0 \\ 0 & 1 & 0 & 0 \\ p_{31} & p_{32} & p_{33} & p_{34} \\ p_{41} & p_{42} & p_{43} & p_{44} \end{bmatrix} = \begin{bmatrix} I & 0 \\ D & Q \end{bmatrix}. \tag{3.85}$$

The same result is obtained for F as the one obtained in Equation (3.82) for the matrix of probabilities of eventual passage to recurrent states for the generic four-state reducible unichain. However, in this case, as Section 3.5.2 indicates, F is called the matrix of absorption probabilities for the generic four-state absorbing multichain.

3.3.3.1.3 Reducible Unichain Model of Machine Maintenance Under a Modified Policy of Doing Nothing in State 3

Consider the unichain model of machine maintenance introduced in Section 1.10.1.2.2.1 and analyzed in Section 3.3.2.1.4. The transition probability matrix associated with the modified maintenance policy, under which the engineer always does nothing to a machine in state 3, is given in canonical form in Equation (1.56).

$$
P = \begin{array}{c|cccc}
\text{State} & 1 & 2 & 3 & 4 \\
\hline
1 & 0.2 & 0.8 & 0 & 0 \\
2 & 0.6 & 0.4 & 0 & 0 \\
3 & 0.2 & 0.3 & 0.5 & 0 \\
4 & 0.3 & 0.2 & 0.1 & 0.4
\end{array} = \begin{bmatrix} S & 0 \\ D & Q \end{bmatrix}. \tag{1.56}
$$

The recurrent set of states is $R = \{1,2\}$. The set of transient states is $T = \{3,4\}$. Solving Equation (3.84), the matrix of probabilities of eventual passage from transient states to recurrent states is

$$
F = \begin{bmatrix} f_{31} & f_{32} \\ f_{41} & f_{42} \end{bmatrix}
$$

$$
= \frac{1}{(1 - p_{33})(1 - p_{44}) - p_{34}p_{43}} \begin{bmatrix} (1 - p_{44})p_{31} + p_{34}p_{41} & (1 - p_{44})p_{32} + p_{34}p_{42} \\ p_{43}p_{31} + (1 - p_{33})p_{41} & p_{43}p_{32} + (1 - p_{33})p_{42} \end{bmatrix}
$$

$$
= \frac{1}{(1-0.5)(1-0.4) - (0)(0.1)} \begin{bmatrix} (1-0.4)(0.2) + (0)(0.3) & (1-0.4)(0.3) + (0)(0.2) \\ (0.1)(0.2) + (1-0.5)(0.3) & (0.1)(0.3) + (1-0.5)(0.2) \end{bmatrix}
$$

$$
= \begin{bmatrix} 0.4 & 0.6 \\ 0.5667 & 0.4333 \end{bmatrix}. \tag{3.86}
$$

When the machine is used under the modified maintenance policy prescribed in Section 1.10.1.2.2.1, the probability of eventual passage from transient state 3, working, with a minor defect, to recurrent state 1, not working, is $f_{31} = 0.4$. Also, under the modified maintenance policy, the probability of eventual passage from transient state 4, working properly, to recurrent state 2, working, with a major defect, is $f_{42} = 0.4333$.

Alternatively, using Equation (3.76), where

$$
Q = \begin{bmatrix} 0.5 & 0 \\ 0.1 & 0.4 \end{bmatrix}, \quad D = \begin{bmatrix} 0.2 & 0.3 \\ 0.3 & 0.2 \end{bmatrix}, \quad I - Q = \begin{bmatrix} 0.5 & 0 \\ -0.1 & 0.6 \end{bmatrix},
$$

$$
U = (I - Q)^{-1} = \frac{1}{0.3} \begin{bmatrix} 0.6 & 0 \\ 0.1 & 0.5 \end{bmatrix}.
$$

$F = UD = (I - Q)^{-1}D$

$$= \frac{1}{0.3}\begin{bmatrix} 0.6 & 0 \\ 0.1 & 0.5 \end{bmatrix}\begin{bmatrix} 0.2 & 0.3 \\ 0.3 & 0.2 \end{bmatrix} = \begin{bmatrix} 0.4 & 0.6 \\ 0.5667 & 0.4333 \end{bmatrix} = \begin{bmatrix} f_{31} & f_{32} \\ f_{41} & f_{42} \end{bmatrix}, \qquad (3.87)$$

confirming the results obtained previously.

3.3.3.2 Reducible Multichain

A reducible multichain has two or more closed classes of recurrent states plus a set of transient states.

3.3.3.2.1 Absorbing Multichain Model of a Gambler's Ruin

An absorbing multichain has two or more absorbing states plus one or more transient states. As Section 3.3.3.1.1 indicates, the probability of eventual passage from a transient state i to an absorbing state j, denoted by f_{ij}, is called an absorption probability. The matrix of probabilities of eventual passage from transient states to absorbing states, denoted by $F = [f_{ij}]$, is called the matrix of absorption probabilities.

In Section 1.4.1.2, a gambler's ruin is modeled as a five-state random walk with two absorbing barriers. This random walk is an absorbing multichain in which states 0 and 4 are absorbing states, and the other three states are transient states. Suppose that player A wants to compute the probability that she will eventually lose all her money and be ruined. This is the probability that the chain will be absorbed in state 0. The transition probability matrix for the absorbing multichain model of the gambler's ruin is given in canonical form in Equation (1.11).

$$P = [p_{ij}] = \begin{matrix} 0 \\ 4 \\ 1 \\ 2 \\ 3 \end{matrix}\begin{bmatrix} 1 & 0 & 0 & 0 & 0 \\ 0 & 1 & 0 & 0 & 0 \\ \hline 1-p & 0 & 0 & p & 0 \\ 0 & 0 & 1-p & 0 & p \\ 0 & p & 0 & 1-p & 0 \end{bmatrix} = \begin{bmatrix} S & 0 \\ D & Q \end{bmatrix} = \begin{bmatrix} I & 0 \\ D & Q \end{bmatrix}. \qquad (1.11)$$

Observe that

$$(I - Q) = \begin{array}{c|ccc} \text{State} & 1 & 2 & 3 \\ \hline 1 & 1 & -p & 0 \\ 2 & -(1-p) & 1 & -p \\ 3 & 0 & -(1-p) & 1 \end{array} \qquad (3.88)$$

The fundamental matrix is

$$U = (I-Q)^{-1} =$$

State	1	2	3
1	$\dfrac{p+(1-p)^2}{p^2+(1-p)^2}$	$\dfrac{p}{p^2+(1-p)^2}$	$\dfrac{p^2}{p^2+(1-p)^2}$
2	$\dfrac{(1-p)}{p^2+(1-p)^2}$	$\dfrac{1}{p^2+(1-p)^2}$	$\dfrac{p}{p^2+(1-p)^2}$
3	$\dfrac{(1-p)^2}{p^2+(1-p)^2}$	$\dfrac{(1-p)}{p^2+(1-p)^2}$	$\dfrac{p^2+(1-p)}{p^2+(1-p)^2}$

$$(3.89)$$

Note that when player A starts with \$2, the expected number of times that she will have \$2 is given by

$$u_{22} = \frac{1}{p^2+(1-p)^2}, \tag{3.90}$$

the entry in row 2 and column 2 of the fundamental matrix. The expected number of times that she will have \$2, given that she starts with \$2, varies between one and two. This expected value is one if $p = 0$ or $p = 1$. If player A starts with \$2 and $p = 0$, she will have \$1 remaining after the first bet, and lose the game with \$0 remaining after the second bet. If player A starts with \$2 and $p = 1$, she will have a total of \$3 after the first bet, and win the game with a total of \$4 after the second bet. Therefore, if $p = 0$ or 1, and player A starts in state 2, she will have \$2 only until she makes the first bet. If player A starts with \$2 and $p = 1/2$, the expected number of times that she will have \$2 is two.

Using Equation (3.76), the matrix of absorption probabilities is equal to

$$F = UD = (I-Q)^{-1}D$$

State	1	2	3
1	$\dfrac{p+(1-p)^2}{p^2+(1-p)^2}$	$\dfrac{p}{p^2+(1-p)^2}$	$\dfrac{p^2}{p^2+(1-p)^2}$
2	$\dfrac{(1-p)}{p^2+(1-p)^2}$	$\dfrac{1}{p^2+(1-p)^2}$	$\dfrac{p}{p^2+(1-p)^2}$
3	$\dfrac{(1-p)^2}{p^2+(1-p)^2}$	$\dfrac{(1-p)}{p^2+(1-p)^2}$	$\dfrac{p^2+(1-p)}{p^2+(1-p)^2}$

$$\begin{bmatrix} 1-p & 0 \\ 0 & 0 \\ 0 & p \end{bmatrix}$$

$$= \begin{array}{c|cc} \text{State} & 0 & 4 \\ \hline 1 & \dfrac{p+(1-p)+(1-p)^3}{p^2+(1-p)^2} & \dfrac{p^3}{p^2+(1-p)^2} \\[2mm] 2 & \dfrac{(1-p)^2}{p^2+(1-p)^2} & \dfrac{p^2}{p^2+(1-p)^2} \\[2mm] 3 & \dfrac{(1-p)^3}{p^2+(1-p)^2} & \dfrac{p(1-p)+p^3}{p^2+(1-p)^2} \end{array} . \qquad (3.91)$$

If player A starts with \$2, the probability that she will eventually lose all her money and be ruined is given by

$$f_{20} = \frac{(1-p)^2}{p^2+(1-p)^2}, \qquad (3.92)$$

the entry in row 2 and column 1 of F, the matrix of absorption probabilities.

3.3.3.2.2 Absorbing Multichain Model of Patient Flow in a Hospital

In Sections 1.10.2.1.2, 3.2.4, and 3.3.2.2.1, the movement of patients in a hospital is modeled as an absorbing multichain in which states 0 and 5 are absorbing states, and the other four states are transient. Suppose that the objective is to compute the probability that a patient will eventually be discharged. This is the probability that a patient in one of the four transient states will be absorbed in state 0. The transition probability matrix for the absorbing multichain is given in canonical form in Equation (1.59).

$$P = \begin{array}{c|cc|cccc} \text{State} & 0 & 5 & 1 & 2 & 3 & 4 \\ \hline 0 & 1 & 0 & 0 & 0 & 0 & 0 \\ 5 & 0 & 1 & 0 & 0 & 0 & 0 \\ \hline 1 & 0 & 0 & 0.4 & 0.1 & 0.2 & 0.3 \\ 2 & 0.15 & 0.05 & 0.1 & 0.2 & 0.3 & 0.2 \\ 3 & 0.07 & 0.03 & 0.2 & 0.1 & 0.4 & 0.2 \\ 4 & 0 & 0 & 0.3 & 0.4 & 0.2 & 0.1 \end{array} = \begin{bmatrix} S & 0 \\ D & Q \end{bmatrix} = \begin{bmatrix} I & 0 \\ D & Q \end{bmatrix}. \qquad (1.59)$$

The fundamental matrix was calculated in Equation (3.21). Using Equation (3.76), the matrix of absorption probabilities is equal to

$$F = UD = (I-Q)^{-1}D = \begin{array}{c} 1 \\ 2 \\ 3 \\ 4 \end{array} \begin{bmatrix} 5.2364 & 2.8563 & 4.2847 & 3.3321 \\ 2.9263 & 3.2798 & 3.4383 & 2.4682 \\ 3.5078 & 2.4858 & 5.0253 & 2.8382 \\ 3.8254 & 2.9621 & 4.0731 & 3.9494 \end{bmatrix} \begin{bmatrix} 0 & 0 \\ 0.15 & 0.05 \\ 0.07 & 0.03 \\ 0 & 0 \end{bmatrix} = \begin{array}{c} 1 \\ 2 \\ 3 \\ 4 \end{array} \begin{bmatrix} 0.7284 & 0.2714 \\ 0.7327 & 0.2671 \\ 0.7246 & 0.2750 \\ 0.7294 & 0.2703 \end{bmatrix} .$$

$$(3.93)$$

The entry $f_{30} = 0.7246$ shows that a surgery patient has a probability of 0.7246 of eventually being discharged, while the entry $f_{35} = 0.2750$ shows that a surgery patient has a probability of 0.2750 of dying. The entries in the first column of the matrix F indicate that the probabilities that a patient in the diagnostic, outpatient, surgery, and physical therapy departments, respectively, will eventually be discharged are 0.7284, 0.7327, 0.7246, and 0.7294, respectively. Of course, the absorption probabilities in each row of matrix F sum to one because eventually a patient must be discharged or die.

Now assume that the fundamental matrix for the absorbing multichain has not been calculated. As Section 3.3.3 indicates, f_{i0}, the probability of absorption in state 0, given that the chain starts in a transient state i, can also be calculated by solving the system of equations (3.77).

$$f_{i0} = p_{i0} + \sum_{k \in T} p_{ik} f_{k0}. \tag{3.77}$$

Equations (3.77) for the probabilities of eventual discharge are

$$\left. \begin{aligned} f_{10} &= 0 + (0.4)f_{10} + (0.1)f_{20} + (0.2)f_{30} + (0.3)f_{40} \\ f_{20} &= 0.15 + (0.1)f_{10} + (0.2)f_{20} + (0.3)f_{30} + (0.2)f_{40} \\ f_{30} &= 0.07 + (0.2)f_{10} + (0.1)f_{20} + (0.4)f_{30} + (0.2)f_{40} \\ f_{40} &= 0 + (0.3)f_{10} + (0.4)f_{20} + (0.2)f_{30} + (0.1)f_{40}. \end{aligned} \right\} \tag{3.94}$$

The solution of equations (3.94) for the vector of probabilities of absorption in absorbing state 0 is column one of matrix (3.93),

$$f_0 = [f_{10} \quad f_{20} \quad f_{30} \quad f_{40}]^T = [0.7284 \quad 0.7327 \quad 0.7246 \quad 0.7294]^T. \tag{3.95}$$

These probabilities of absorption in state 0 differ only slightly from those computed by state reduction in Equation (6.83). Discrepancies are due to roundoff error because only the first four significant digits after the decimal point were stored.

3.3.3.2.3 Five-State Reducible Multichain

Consider the canonical form of the transition probability matrix for a generic five-state reducible multichain, shown in Equation (3.5).

$$P = \begin{matrix} 1 \\ 2 \\ 3 \\ 4 \\ 5 \end{matrix} \begin{bmatrix} 1 & 0 & 0 & 0 & 0 \\ 0 & p_{22} & p_{23} & 0 & 0 \\ 0 & p_{32} & p_{33} & 0 & 0 \\ p_{41} & p_{42} & p_{43} & p_{44} & p_{45} \\ p_{51} & p_{52} & p_{53} & p_{54} & p_{55} \end{bmatrix} = \begin{bmatrix} P_1 & 0 & 0 \\ 0 & P_2 & 0 \\ D_1 & D_2 & Q \end{bmatrix}. \tag{3.5}$$

Suppose that the probability of eventual passage from transient state 4 to recurrent state 2, f_{42}, is of interest. The following system of equations (3.77) must be solved:

$$\left.\begin{aligned} f_{42} &= p_{42} + \sum_{k \varepsilon T} p_{4k} f_{k2} = p_{42} + p_{44} f_{42} + p_{45} f_{52} \\ f_{52} &= p_{52} + \sum_{k \varepsilon T} p_{5k} f_{k2} = p_{52} + p_{54} f_{42} + p_{55} f_{52}. \end{aligned}\right\} \tag{3.96a}$$

A matrix form of equations (3.96) based on Equation (3.78) is

$$\begin{bmatrix} f_{42} \\ f_{52} \end{bmatrix} = \begin{bmatrix} p_{42} \\ p_{52} \end{bmatrix} + \begin{bmatrix} p_{44} & p_{45} \\ p_{54} & p_{55} \end{bmatrix} \begin{bmatrix} f_{42} \\ f_{52} \end{bmatrix}. \tag{3.96b}$$

The solution of equations (3.96) for the probabilities of eventual passage to recurrent state 2 is

$$\left.\begin{aligned} f_{42} &= \frac{p_{42}(1-p_{55}) + p_{45} p_{52}}{(1-p_{44})(1-p_{55}) - p_{45} p_{54}} \\ f_{52} &= \frac{p_{52}(1-p_{44}) + p_{54} p_{42}}{(1-p_{44})(1-p_{55}) - p_{45} p_{54}}. \end{aligned}\right\} \tag{3.97}$$

Alternatively, Equation (3.96) for computing the matrix of probabilities of eventual passage from transient states to recurrent states, extended to apply to a multichain with r recurrent chains, is

$$F = UD = U[D_1 \quad \cdots \quad D_r] = [UD_1 \quad \cdots \quad UD_r] = [F_1 \quad \cdots \quad F_r]. \tag{3.98}$$

Since the five-state reducible multichain has $r = 2$ recurrent chains, and the target recurrent state 2 belongs to the second recurrent chain $R_2 = (2, 3)$, matrix F_2 will be calculated.

$$F_2 = UD_2 = (I - Q)^{-1} D_2. \tag{3.99}$$

The fundamental matrix is

$$U = (I - Q)^{-1} = \begin{bmatrix} 1 - p_{44} & -p_{45} \\ -p_{54} & 1 - p_{55} \end{bmatrix}^{-1} = \frac{1}{(1-p_{44})(1-p_{55}) - p_{45} p_{54}} \begin{bmatrix} 1 - p_{55} & p_{45} \\ p_{54} & 1 - p_{44} \end{bmatrix}. \tag{3.100}$$

Thus,

$$F_2 = \frac{1}{(1-p_{44})(1-p_{55})-p_{45}p_{54}} \begin{bmatrix} 1-p_{55} & p_{45} \\ p_{54} & 1-p_{44} \end{bmatrix} \begin{bmatrix} p_{42} & p_{43} \\ p_{52} & p_{53} \end{bmatrix}$$

$$= \frac{1}{(1-p_{44})(1-p_{55})-p_{45}p_{54}} \begin{bmatrix} (1-p_{55})p_{42}+p_{45}p_{52} & (1-p_{55})p_{43}+p_{45}p_{53} \\ p_{54}p_{42}+(1-p_{44})p_{52} & p_{54}p_{43}+(1-p_{44})p_{53} \end{bmatrix}$$

$$= \begin{bmatrix} f_{42} & f_{43} \\ f_{52} & f_{53} \end{bmatrix}. \tag{3.101}$$

These values agree with those obtained previously in Equation (3.97). As Section 3.3.3.1.2 has noted for the case of a reducible unichain, the probabilities f_{42}, f_{43}, f_{52}, and f_{53} of eventual passage are independent of the transition probabilities, p_{22}, p_{23}, p_{32}, and p_{33}, for the recurrent closed class. Thus, if states 2 and 3 were absorbing states, making the chain an absorbing multichain, the probabilities of eventual passage, which would be called absorption probabilities, would be unchanged.

3.3.3.2.4 Multichain Model of an Eight-State Serial Production Process

For another illustration of how to calculate the probability of eventual passage from a transient state to a target recurrent state, consider the eight-state reducible multichain model of a serial production process introduced in Section 1.10.2.2. The transition probability matrix in canonical form is shown in Equation (3.7).

The fundamental matrix is

$$U = (I-Q)^{-1} = \begin{array}{c} 6 \\ 7 \\ 8 \end{array} \begin{bmatrix} 0.45 & 0 & 0 \\ -0.20 & 0.35 & 0 \\ 0 & -0.15 & 0.25 \end{bmatrix}^{-1} = \begin{array}{c|ccc} & 6 & 7 & 8 \\ \hline 6 & 2.2222 & 0 & 0 \\ 7 & 1.2698 & 2.8571 & 0 \\ 8 & 0.7619 & 1.7143 & 4 \end{array}.$$

$$\tag{3.102}$$

In this model, production stage i is represented by transient state $(9-i)$. Recall that u_{ij}, the (i, j)th entry of the fundamental matrix, specifies the mean number of visits to transient state j before the chain eventually enters a recurrent state, given that the chain starts in transient state i. Hence, the entries in the bottom row of U specify the expected number of times that an entering item, starting at production stage 1 in transient state 8, will visit each of the three production stages prior to being sold, scrapped, or sent to the training center. On average, an entering item will make $u_{88} = 4$ visits to stage 1, $u_{87} = 1.7143$ visits to stage 2, and $u_{86} = 0.7619$ visits to stage 3. The expected total number of visits to all three production stages made by an entering item is 6.4762, equal to the sum of these three entries.

If the chain occupies a transient state, the matrix of probabilities of eventual passage to absorbing and recurrent states is calculated by applying Equation (3.98).

$$F = UD = U[D_1 \quad D_2 \quad D_R] = \begin{array}{c} \\ 6 \\ 7 \\ 8 \end{array} \begin{array}{|ccc|} \hline 6 & 7 & 8 \\ \hline 2.2222 & 0 & 0 \\ 1.2698 & 2.8571 & 0 \\ 0.7619 & 1.7143 & 4 \\ \hline \end{array} \begin{array}{|ccccc|} \hline 1 & 2 & 3 & 4 & 5 \\ \hline 0.20 & 0.16 & 0.04 & 0.03 & 0.02 \\ 0.15 & 0 & 0 & 0 & 0 \\ 0.10 & 0 & 0 & 0 & 0 \\ \hline \end{array}$$

$$= \begin{array}{c} \\ 6 \\ 7 \\ 8 \end{array} \begin{array}{|ccccc|} \hline 1 & 2 & 3 & 4 & 5 \\ \hline 0.4444 & 0.3556 & 0.0889 & 0.0667 & 0.0444 \\ 0.6825 & 0.2032 & 0.0508 & 0.0381 & 0.0254 \\ 0.8095 & 0.1219 & 0.0305 & 0.0229 & 0.0152 \\ \hline \end{array} = [F_1 \quad F_2 \quad F_R],$$

where

$$F_1 = UD_1 = \begin{array}{c} 6 \\ 7 \\ 8 \end{array} \begin{bmatrix} 2.2222 & 0 & 0 \\ 1.2698 & 2.8571 & 0 \\ 0.7619 & 1.7143 & 4 \end{bmatrix} \begin{bmatrix} 0.20 \\ 0.15 \\ 0.10 \end{bmatrix} = \begin{array}{c} 6 \\ 7 \\ 8 \end{array} \begin{bmatrix} 0.4444 \\ 0.6825 \\ 0.8095 \end{bmatrix} = \begin{array}{c} 6 \\ 7 \\ 8 \end{array} \begin{bmatrix} f_{61} \\ f_{71} \\ f_{81} \end{bmatrix}$$

$$F_2 = UD_2 = \begin{array}{c} 6 \\ 7 \\ 8 \end{array} \begin{bmatrix} 2.2222 & 0 & 0 \\ 1.2698 & 2.8571 & 0 \\ 0.7619 & 1.7143 & 4 \end{bmatrix} \begin{bmatrix} 0.16 \\ 0 \\ 0 \end{bmatrix} = \begin{array}{c} 6 \\ 7 \\ 8 \end{array} \begin{bmatrix} 0.3556 \\ 0.2032 \\ 0.1219 \end{bmatrix} = \begin{array}{c} 6 \\ 7 \\ 8 \end{array} \begin{bmatrix} f_{62} \\ f_{72} \\ f_{82} \end{bmatrix}$$

$$F_R = UD_R = \begin{array}{c} 6 \\ 7 \\ 8 \end{array} \begin{bmatrix} 2.2222 & 0 & 0 \\ 1.2698 & 2.8571 & 0 \\ 0.7619 & 1.7143 & 4 \end{bmatrix} \begin{bmatrix} 0.04 & 0.03 & 0.02 \\ 0 & 0 & 0 \\ 0 & 0 & 0 \end{bmatrix}$$

$$= \begin{array}{c} 6 \\ 7 \\ 8 \end{array} \begin{bmatrix} 0.0889 & 0.0667 & 0.0444 \\ 0.0508 & 0.0381 & 0.0254 \\ 0.0305 & 0.0229 & 0.0152 \end{bmatrix} = \begin{array}{c} 6 \\ 7 \\ 8 \end{array} \begin{bmatrix} f_{63} & f_{64} & f_{65} \\ f_{73} & f_{74} & f_{75} \\ f_{83} & f_{84} & f_{85} \end{bmatrix}. \qquad (3.103)$$

For an entering item, which starts in transient state 8, the probability of being scrapped, or absorbed in state 1, is $f_{81} = 0.8095$, while the probability of being sold, or absorbed in state 2, is $f_{82} = 0.1219$. The probability that an entering item will enter the training center to be used initially for training engineers is $f_{83} = 0.0305$. The probability that an entering item will eventually enter the training center, to be used for training engineers, technicians, and technical writers, is given by

$$f_{83} + f_{84} + f_{85} = 0.0305 + 0.0229 + 0.0152 = 0.0686. \tag{3.104}$$

Suppose that an order for 100 items is received from a customer. If exactly 100 items are started, then the firm can expect to sell only

$$100\,f_{82} = 100\,(0.1219) \approx 12 \tag{3.105}$$

items; the remainder will be scrapped or sent to the training center. Therefore, the expected number of entering items required to fill this order is equal to

$$100/f_{82} = 100/0.1219 \approx 821. \tag{3.106}$$

Alternatively, if the chain occupies a transient state, the probabilities of eventual passage to absorbing and recurrent states can be also computed by using Equation (3.77). Note that only output from production stage 3, represented by transient state 6, can reach the training center in one step. Suppose that the probability f_{84} that an entering item will enter the training center to be used initially for training technicians is of interest. States 6, 7, and 8 are transient. Recurrent state 4 communicates with recurrent states 3, 4, and 5. The following system of three linear equations (3.77) must be solved for the three unknowns, $f_{64}, f_{74},$ and f_{84}.

$$\left.\begin{aligned}
f_{64} &= p_{64} + \sum_{k=6,7,8} p_{6k} f_{k4} = p_{64} + p_{66} f_{64} + p_{67} f_{74} + p_{68} f_{84} \\
f_{74} &= p_{74} + \sum_{k=6,7,8} p_{7k} f_{k4} = p_{74} + p_{76} f_{64} + p_{77} f_{74} + p_{78} f_{84} \\
f_{84} &= p_{84} + \sum_{k=6,7,8} p_{8k} f_{k4} = p_{84} + p_{86} f_{64} + p_{87} f_{74} + p_{88} f_{84}.
\end{aligned}\right\} \tag{3.107a}$$

A matrix form of equations (3.107) based on Equation (3.78) is

$$\begin{bmatrix} f_{64} \\ f_{74} \\ f_{84} \end{bmatrix} = \begin{bmatrix} p_{64} \\ p_{74} \\ p_{84} \end{bmatrix} + \begin{bmatrix} p_{66} & p_{67} & p_{68} \\ p_{76} & p_{77} & p_{78} \\ p_{86} & p_{87} & p_{88} \end{bmatrix} \begin{bmatrix} f_{64} \\ f_{74} \\ f_{84} \end{bmatrix}. \tag{3.107b}$$

When numerical coefficients are inserted, the system of equations (3.107) is

$$\left.\begin{aligned}
f_{64} &= 0.03 + 0.55 f_{64} + 0 f_{74} + 0 f_{84} = 0.0667 \\
f_{74} &= 0 + 0.20 f_{64} + 0.65 f_{74} + 0 f_{84} = 0.0381 \\
f_{84} &= 0 + 0 f_{64} + 0.15 f_{74} + 0.75 f_{84} = 0.0229.
\end{aligned}\right\} \tag{3.108}$$

A matrix form and the solution of equations (3.108) is

$$
\begin{bmatrix} f_{64} \\ f_{74} \\ f_{84} \end{bmatrix} = \begin{bmatrix} 0.03 \\ 0 \\ 0 \end{bmatrix} + \begin{bmatrix} 0.55 & 0 & 0 \\ 0.20 & 0.65 & 0 \\ 0 & 0.15 & 0.75 \end{bmatrix} \begin{bmatrix} f_{64} \\ f_{74} \\ f_{84} \end{bmatrix} = \begin{bmatrix} 0.0667 \\ 0.0381 \\ 0.0229 \end{bmatrix}. \tag{3.109}
$$

Equation (3.109) produces the same solution for the vector $[f_{64}, f_{74}, f_{84}]^{\mathrm{T}}$ as does Equation (3.103).

3.4 Eventual Passage to a Closed Set within a Reducible Multichain

A reducible Markov chain can be partitioned into one or more disjoint closed communicating classes of recurrent states plus a set of transient states. In this section, passage from a transient state to a recurrent chain, R, in a reducible multichain is of interest, rather than passage to a particular state within R [1–3]. The probability of eventual passage from a transient state to the single closed class of recurrent states within a reducible unichain is one, because eventual passage to the recurrent closed class is certain. Two methods are presented for calculating the probability of eventual passage from a transient state to a closed set of recurrent states within a reducible multichain. Both methods verify that the probability of eventual passage from a transient state i to a closed class of recurrent states is equal to the sum of the probabilities of eventual passage from the transient state i to all of the states, which belong to the recurrent closed class.

3.4.1 Method One: Replacing Recurrent Sets with Absorbing States and Using the Fundamental Matrix

Method one, which requires calculation of the fundamental matrix, can be applied without first calculating the individual probabilities of eventual passage from transient states to recurrent states. To do this, all of the recurrent states, which belong to the same closed class, are lumped into a single absorbing state. When this is done for every recurrent closed class, the result is an absorbing multichain with as many absorbing states as there are recurrent closed classes in the original chain. The absorption probabilities for the absorbing multichain are equal to the probabilities of eventual passage to the corresponding recurrent closed classes in the original multichain. The set of transient states is unchanged.

3.4.1.1 Five-State Reducible Multichain

To demonstrate method one, consider, once again, the generic five-state reducible multichain, treated most recently in Section 3.3.3.2.3. The canonical form of the transition probability matrix shown in Equation (3.5).

$$P = \begin{matrix} 1 \\ 2 \\ 3 \\ 4 \\ 5 \end{matrix} \begin{bmatrix} 1 & 0 & 0 & 0 & 0 \\ 0 & p_{22} & p_{23} & 0 & 0 \\ 0 & p_{32} & p_{33} & 0 & 0 \\ p_{41} & p_{42} & p_{43} & p_{44} & p_{45} \\ p_{51} & p_{52} & p_{53} & p_{54} & p_{55} \end{bmatrix} = \begin{bmatrix} P_1 & 0 & 0 \\ 0 & P_2 & 0 \\ D_1 & D_2 & Q \end{bmatrix}. \tag{3.5}$$

Recall that the probabilities $f_{42}, f_{43}, f_{52},$ and f_{53} of eventual passage were computed in Section 3.3.3.2.3. To compute the probability of eventual passage from a transient state to the recurrent closed class denoted by $R_2 = \{2,3\}$, R_2 is replaced by an absorbing state. The following associated absorbing Markov chain is formed with transition probability matrix denoted by P_A.

$$P_A = \begin{matrix} 1 \\ R_2 \\ 4 \\ 5 \end{matrix} \begin{bmatrix} 1 & 0 & 0 & 0 \\ 0 & 1 & 0 & 0 \\ p_{41} & p_{42}+p_{43} & p_{44} & p_{45} \\ p_{51} & p_{52}+p_{53} & p_{54} & p_{55} \end{bmatrix} = \begin{bmatrix} 1 & 0 & 0 \\ 0 & 1 & 0 \\ D_1 & d_2 & Q \end{bmatrix}, \quad d_2 = \begin{bmatrix} p_{42}+p_{43} \\ p_{52}+p_{53} \end{bmatrix}.$$

$$\tag{3.110}$$

where the transition probability in each row of vector d_2 is equal to the sum of the transition probabilities in the same row of matrix D_2. As the matrix Q for the original multichain is unchanged, the fundamental matrix U for the absorbing multichain, calculated in Equation (3.100), is also unchanged. Applying Equation (3.99) to calculate f_2, which denotes the vector of probabilities of eventual passage from the transient states to the closed class of recurrent states, $R_2 = \{2,3\}$,

$$f_2 = Ud_2 = (I-Q)^{-1}d_2 = \frac{1}{(1-p_{44})(1-p_{55})-p_{45}p_{54}} \begin{bmatrix} 1-p_{55} & p_{45} \\ p_{54} & 1-p_{44} \end{bmatrix} \begin{bmatrix} p_{42}+p_{43} \\ p_{52}+p_{53} \end{bmatrix}$$

$$= \frac{1}{(1-p_{44})(1-p_{55})-p_{45}p_{54}} \begin{bmatrix} (1-p_{55})(p_{42}+p_{43})+p_{45}(p_{52}+p_{53}) \\ p_{54}(p_{42}+p_{43})+(1-p_{44})(p_{52}+p_{53}) \end{bmatrix}$$

$$= \begin{bmatrix} f_{42}+f_{43} \\ f_{52}+f_{53} \end{bmatrix} = \begin{bmatrix} f_{4R_2} \\ f_{5R_2} \end{bmatrix}, \tag{3.111}$$

where

$$\left. \begin{aligned} f_{4R_2} &= \frac{(1-p_{55})(p_{42}+p_{43})+p_{45}(p_{52}+p_{53})}{(1-p_{44})(1-p_{55})-p_{45}p_{54}} \\ f_{5R_2} &= \frac{p_{54}(p_{42}+p_{43})+(1-p_{44})(p_{52}+p_{53})}{(1-p_{44})(1-p_{55})-p_{45}p_{54}}. \end{aligned} \right\} \tag{3.112}$$

In this example,

$$f_{iR_2} = f_{i2} + f_{i3} \tag{3.113}$$

is the probability of eventual passage from a transient state i to the recurrent closed class $R_2 = \{2, 3\}$.

3.4.1.2 Multichain Model of an Eight-State Serial Production Process

As a demonstration of method one with numerical values given for transition probabilities, the procedure will be applied to the eight-state reducible multichain model of a serial production process for which probabilities of eventual passage to recurrent states are computed in Section 3.3.3.2.4. The model is a reducible multichain, which has two absorbing states, one closed set of three recurrent states, and three transient states. The transition probability matrix in canonical form is shown in Equation (3.7).

Recall that $R_3 = \{3, 4, 5\}$ denote the closed communicating class of three recurrent states, and $T = \{6, 7, 8\}$ denote the set of transient states. To compute the probabilities of eventual passage from transient states to the closed class of recurrent states represented by R_3, R_3 is replaced by an absorbing state. States 1 and 2 remain absorbing states. The following associated absorbing multichain is formed with transition probability matrix denoted by P_A.

$$
P_A =
\begin{array}{cc}
\begin{array}{c}
\text{Scrapped} \\
\text{Sold} \\
\text{Training Center} \\
\text{Stage 3} \\
\text{Stage 2} \\
\text{Stage 1}
\end{array}
&
\begin{array}{c}
1 \\
2 \\
R_3 \\
6 \\
7 \\
8
\end{array}
\left[
\begin{array}{ccc|ccc}
1 & 0 & 0 & 0 & 0 & 0 \\
0 & 1 & 0 & 0 & 0 & 0 \\
0 & 0 & 1 & 0 & 0 & 0 \\
\hline
0.20 & 0.16 & 0.09 & 0.55 & 0 & 0 \\
0.15 & 0 & 0 & 0.20 & 0.65 & 0 \\
0.10 & 0 & 0 & 0 & 0.15 & 0.75
\end{array}
\right]
\end{array}
$$

$$
=
\begin{array}{cc}
\begin{array}{c}
\text{Scrapped} \\
\text{Sold} \\
\text{Training Center} \\
\text{Production Stages}
\end{array}
&
\begin{array}{c}
1 \\
2 \\
R_3 \\
T
\end{array}
\left[
\begin{array}{cccc}
1 & 0 & 0 & 0 \\
0 & 1 & 0 & 0 \\
0 & 0 & 1 & 0 \\
D_1 & D_2 & d_3 & Q
\end{array}
\right]
\end{array}. \tag{3.114}
$$

where

$$
D_1 = \begin{array}{c} 6 \\ 7 \\ 8 \end{array}\begin{bmatrix} 0.20 \\ 0.15 \\ 0.10 \end{bmatrix}, \quad
D_2 = \begin{bmatrix} 0.16 \\ 0 \\ 0 \end{bmatrix}, \quad
d_3 = \begin{bmatrix} 0.04 + 0.03 + 0.02 = 0.09 \\ 0 \\ 0 \end{bmatrix} = \begin{bmatrix} 0.09 \\ 0 \\ 0 \end{bmatrix},
$$

$$
\text{and} \quad Q = \begin{bmatrix} 0.55 & 0 & 0 \\ 0.20 & 0.65 & 0 \\ 0 & 0.15 & 0.75 \end{bmatrix}.
$$

and the transition probability in each row of vector d_3 is equal to the sum of the transition probabilities in the same row of matrix D_3. As the matrix Q for the original multichain is unchanged, the fundamental matrix U for the absorbing multichain, calculated in Equation (3.102), is also unchanged.
Applying Equation (3.98),

$$F = UD = U[D_1 \quad D_2 \quad d_3] =$$

State	6	7	8	1	2	$R_3 = \{3,4,5\}$
6	2.2222	0	0	0.20	0.16	0.09
7	1.2698	2.8571	0	0.15	0	0
8	0.7619	1.7143	4	0.10	0	0

State	1	2	$R_3 = \{3,4,5\}$
6	0.4444	0.3556	0.2
7	0.6825	0.2032	0.1143
8	0.8095	0.1219	0.0686

$$= \begin{array}{c} 6 \\ 7 \\ 8 \end{array} \begin{bmatrix} f_{61} & f_{62} & f_{6R_3} \\ f_{71} & f_{72} & f_{7R_3} \\ f_{81} & f_{82} & f_{8R_3} \end{bmatrix} \qquad (3.115a)$$

$$\begin{bmatrix} f_{6R_3} \\ f_{7R_3} \\ f_{8R_3} \end{bmatrix} = \begin{array}{c} 6 \\ 7 \\ 8 \end{array} \begin{bmatrix} 0.2 \\ 0.1143 \\ 0.0686 \end{bmatrix}. \qquad (3.115b)$$

Thus, the probability of eventually reaching the closed set of recurrent states $R_3 = \{3, 4, 5\}$ from transient state 8 or production stage 1 is $f_{8R3} = 0.0686$.

As Equation (3.1.1.3) indicates, the probability of eventual passage from a transient state i to a recurrent closed set denoted by R is simply the sum of the probabilities of eventual passage from the transient state i to all of the recurrent states, which are members of R. Expressed as an equation, the probability f_{iR} of eventual passage from a transient state i to a recurrent closed set R is

$$f_{iR} = \sum_{j \in R} f_{ij}, \qquad (3.116)$$

where the recurrent states j belong to the recurrent closed set R. For this example, the probability that an entering item will eventually be sent to the training center is equal to

$$f_{8R} = f_{83} + f_{84} + f_{85} = 0.0305 + 0.0229 + 0.0152 = 0.0686, \qquad (3.117)$$

which confirms the result obtained in Equation (3.104) without replacing $R_3 = \{3, 4, 5\}$ with a single absorbing state.

3.4.2 Method Two: Direct Calculation without Using the Fundamental Matrix

In the second method for calculating the probability of eventual passage to a closed set, the fundamental matrix is not needed. Once again, let R represent a designated closed set of recurrent states. The probability of eventual passage from a transient state i to the designated closed set R is denoted by f_{iR}. Starting in transient state i, the chain may enter the designated recurrent set R in one or more steps. The probability of entering R on the first step is the sum of the one-step transition probabilities, p_{ik}, from transient state i to every recurrent state k, which belongs to R. If the chain does not enter R on the first step, the chain may move either to another closed set different from R, from which R will never be reached, or to a transient state, h. In the latter case, there is a probability f_{hR} of eventual passage to R from the transient state, h. Therefore, for a reducible multichain, the probabilities of eventual passage to a designated recurrent closed class R are obtained by solving the following system of linear equations:

$$f_{iR} = \sum_{k \in R} p_{ik} + \sum_{h \in T} p_{ih} f_{hR}, \tag{3.118}$$

where i is a transient starting state, the first sum is over all recurrent states, k, which belong to R, and the second sum is over all transient states, h, which belong to T, the set of all transient states.

Consider again the following transition matrix for the generic five-state reducible multichain treated by method one in Section 3.4.1.1. The chain has two recurrent closed sets, one of which is an absorbing state, plus two transient states. The transition probability matrix is represented on canonical form in Equation (3.5).

Method 2 will be applied to compute the probabilities of absorption in state 1 from the transient states 4 and 5. These absorption probabilities are denoted by f_{41} and f_{51}, respectively. Applying Equation (3.118),

$$\left.\begin{aligned} f_{41} &= p_{41} + p_{44} f_{41} + p_{45} f_{51} \\ f_{51} &= p_{51} + p_{54} f_{41} + p_{55} f_{51}. \end{aligned}\right\} \tag{3.119}$$

The solution of system (3.119) for the absorption probabilities is

$$\left.\begin{aligned} f_{41} &= \frac{p_{41}(1 - p_{55}) + p_{45} p_{51}}{(1 - p_{44})(1 - p_{55}) - p_{45} p_{54}} \\ f_{51} &= \frac{p_{51}(1 - p_{44}) + p_{54} p_{41}}{(1 - p_{44})(1 - p_{55}) - p_{45} p_{54}}. \end{aligned}\right\} \tag{3.120}$$

Method 2 will also be applied. to compute the probabilities of eventual passage from the transient states 4 and 5 to the closed set of recurrent states, $R_2 = \{2, 3\}$. The probability of eventual passage from a transient state i to a designated closed set R_2 is denoted by f_{iR_2}.

$$f_{4R_2} = (p_{42} + p_{43}) + (p_{44}f_{4R_2} + p_{45}f_{5R_2})$$
$$f_{5R_2} = (p_{52} + p_{53}) + (p_{54}f_{4R_2} + p_{55}f_{5R_2}).$$

(3.121)

The solution of system (3.121) for the probabilities of eventual passage to $R_2 = \{2, 3\}$ is

$$f_{4R_2} = \frac{(1-p_{55})(p_{42}+p_{43})+p_{45}(p_{52}+p_{53})}{(1-p_{44})(1-p_{55})-p_{45}p_{54}}$$
$$f_{5R_2} = \frac{(1-p_{44})(p_{52}+p_{53})+p_{54}(p_{42}+p_{43})}{(1-p_{44})(1-p_{55})-p_{45}p_{54}}.$$

(3.122)

These values agree with those obtained by using method 1 in Equation (3.112) of Section 3.4.1.1. Note that both $f_{41} + f_{4R_2} = 1$ and $f_{51} + f_{5R_2} = 1$ because eventual passage from a transient state to one of the recurrent closed classes in a multichain is certain.

3.5 Limiting Transition Probability Matrix

A one-step transition probability from state i to state j of a Markov chain is denoted by p_{ij}. An n-step transition probability is denoted by $p_{ij}^{(n)}$. Limiting transition probabilities govern the behavior of the chain after a large number of transitions, as $n \to \infty$, and the chain has entered a steady state. Thus, a limiting transition probability is designated by $\lim_{n \to \infty} p_{ij}^{(n)}$. The matrix of limiting transition probabilities is denoted by $\lim_{n \to \infty} P^{(n)}$, where P is the matrix of one-step transition probabilities. As Equation (2.7) indicates, the limiting transition probability matrix for an irreducible Markov chain is obtained by solving the equation $\lim_{n \to \infty} P^{(n)} = \Pi$, where Π is a matrix with each row π, the steady-state probability vector.

The matrix of limiting transition probabilities for a reducible Markov chain is also denoted by $\lim_{n \to \infty} P^{(n)}$. Prior to calculating the limiting transition probabilities for a reducible Markov chain, the transition probability matrix is arranged in the canonical form of Equation (3.1) [1, 2].

3.5.1 Recurrent Multichain

As Section 1.9.2 indicates, a multichain with no transient states is called a recurrent multichain. The state space for a recurrent multichain can be partitioned into two or more disjoint closed classes of recurrent states. The transition probability matrix for a recurrent multichain with M recurrent chains has the following canonical form:

$$P = \begin{bmatrix} P_1 & 0 & \cdots & 0 \\ 0 & P_2 & \cdots & 0 \\ \vdots & \vdots & \ddots & \vdots \\ 0 & 0 & \cdots & P_M \end{bmatrix},$$

(3.123)

where $P_1, \ldots, P_M$ are the transition probability matrices of the M separate recurrent chains denoted by $R_1, \ldots, R_M$, respectively. Once a chain enters a closed set of recurrent states, it never leaves that set. Suppose that i and j are two recurrent states belonging to the same recurrent closed class R_k, which contains N states. If $p(k)_{ij}^{(n)}$ is an n-step transition probability for P_k, let

$$\lim_{n \to \infty} p(k)_{ij}^{(n)} = \pi(k)_j,$$

(3.124)

where $\pi(k)_j$ is the steady-state probability for state j, which belongs to the recurrent chain R_k. Let

$$\pi(k) = [\pi(k)_1 \quad \pi(k)_2 \quad \cdots \quad \pi(k)_N]$$

(3.125)

denote the steady-state probability vector or limiting probability vector for P_k, so that

$$\pi(k) = \pi(k)P_k.$$

(3.126)

This result can be generalized to apply to all recurrent chains belonging to a recurrent multichain. The limiting transition probability matrix is

$$\lim_{n \to \infty} P^n = \begin{bmatrix} \lim_{n \to \infty} P_1^n & 0 & 0 \\ 0 & \ddots & 0 \\ 0 & 0 & \lim_{n \to \infty} P_M^n \end{bmatrix} = \begin{bmatrix} \Pi(1) & 0 & 0 \\ 0 & \ddots & 0 \\ 0 & 0 & \Pi(M) \end{bmatrix},$$

(3.127)

where $\Pi(k)$ is a matrix with each row $\pi(k)$, the steady-state probability vector for the transition probability matrix P_k.

The limiting transition probability matrix is computed for the following numerical example of a four-state recurrent multichain with $M = 2$ recurrent chains:

$$P = \begin{array}{c} 1 \\ 2 \\ 3 \\ 4 \end{array} \begin{bmatrix} 0.2 & 0.8 & 0 & 0 \\ 0.6 & 0.4 & 0 & 0 \\ \hline 0 & 0 & 0.7 & 0.3 \\ 0 & 0 & 0.5 & 0.5 \end{bmatrix} = P = \begin{bmatrix} P_1 & 0 \\ 0 & P_2 \end{bmatrix}$$

(3.128)

$$\lim_{n\to\infty} P^n = \begin{bmatrix} \lim\limits_{n\to\infty} P_1^n & 0 \\ 0 & \lim\limits_{n\to\infty} P_2^n \end{bmatrix} = \begin{bmatrix} \Pi(1) & 0 \\ 0 & \Pi(2) \end{bmatrix} = \begin{array}{c} 1 \\ 2 \\ 3 \\ 4 \end{array}\begin{bmatrix} \pi(1)_1 & \pi(1)_2 & 0 & 0 \\ \pi(1)_1 & \pi(1)_2 & 0 & 0 \\ 0 & 0 & \pi(2)_3 & \pi(2)_4 \\ 0 & 0 & \pi(2)_3 & \pi(2)_4 \end{bmatrix}$$

$$= \begin{bmatrix} 0.4286 & 0.5714 & 0 & 0 \\ 0.4286 & 0.5714 & 0 & 0 \\ 0 & 0 & 0.625 & 0.375 \\ 0 & 0 & 0.625 & 0.375 \end{bmatrix}.$$

$$(3.129)$$

3.5.2 Absorbing Markov Chain

As Equation (3.8) indicates, the canonical form of the two-step transition matrix for an absorbing Markov chain, including an absorbing multichain, is

$$P^2 = \begin{bmatrix} I & 0 \\ D & Q \end{bmatrix}\begin{bmatrix} I & 0 \\ D & Q \end{bmatrix} = \begin{bmatrix} I^2 & 0 \\ D+QD & Q^2 \end{bmatrix} = \begin{bmatrix} I & 0 \\ (I+Q)D & Q^2 \end{bmatrix}. \qquad (3.130)$$

The canonical form of the three-step transition matrix is

$$P^3 = PP^2 = \begin{bmatrix} I & 0 \\ D & Q \end{bmatrix}\begin{bmatrix} I^2 & 0 \\ (I+Q)D & Q^2 \end{bmatrix} = \begin{bmatrix} I^3 & 0 \\ (I+Q+Q^2)D & Q^3 \end{bmatrix} = \begin{bmatrix} I & 0 \\ (I+Q+Q^2)D & Q^3 \end{bmatrix}. \qquad (3.131)$$

By raising P to successively higher powers, one can show that the canonical form of the n-step transition matrix is

$$P^n = \begin{bmatrix} I & 0 \\ (I+Q+Q^2+\cdots+Q^{n-1})D & Q^n \end{bmatrix}. \qquad (3.132)$$

Hence,

$$\lim_{n\to\infty} P^n = \begin{bmatrix} I & 0 \\ \lim\limits_{n\to\infty}(I+Q+Q^2+\ldots+Q^{n-1})D & \lim\limits_{n\to\infty} Q^n \end{bmatrix}$$

$$= \begin{bmatrix} I & 0 \\ (I-Q)^{-1}D & \lim\limits_{n\to\infty} Q^n \end{bmatrix} = \begin{bmatrix} I & 0 \\ UD & \lim\limits_{n\to\infty} Q^n \end{bmatrix} = \begin{bmatrix} I & 0 \\ F & \lim\limits_{n\to\infty} Q^n \end{bmatrix}, \qquad (3.133)$$

where $U = (I - Q)^{-1}$ is the fundamental matrix defined in Equation (3.10), and $\mathbf{F} = UD$ is the matrix of eventual passage probabilities and absorption probabilities defined in Equation (3.76). As Equation (3.69) indicates, as n approaches infinity, Q^n approaches the null matrix. That is,

$$\lim_{n\to\infty} Q^n = 0. \tag{3.69}$$

Hence, the limiting transition probability matrix for an absorbing Markov chain, including an absorbing multichain, is

$$\lim_{n\to\infty} P^n = \begin{bmatrix} I & 0 \\ F & \lim_{n\to\infty} Q^n \end{bmatrix} = \begin{bmatrix} \mathbf{I} & \mathbf{0} \\ \mathbf{F} & \mathbf{0} \end{bmatrix}. \tag{3.134}$$

Consider the generic four-state absorbing multichain of Section 3.3.3.1.2, in which states 1 and 2 are absorbing states. The transition probability matrix in canonical form is shown below:

$$P = [p_{ij}] = \begin{matrix} 1 \\ 2 \\ 3 \\ 4 \end{matrix} \begin{bmatrix} 1 & 0 & 0 & 0 \\ 0 & 1 & 0 & 0 \\ p_{31} & p_{32} & p_{33} & p_{34} \\ p_{41} & p_{42} & p_{43} & p_{44} \end{bmatrix} = \begin{bmatrix} I & 0 \\ D & Q \end{bmatrix}. \tag{3.85}$$

The matrix of absorption probabilities is calculated as

$$F = \frac{1}{(1-p_{33})(1-p_{44})-p_{34}p_{43}} \begin{bmatrix} (1-p_{44})p_{31}+p_{34}p_{41} & (1-p_{44})p_{32}+p_{34}p_{42} \\ p_{43}p_{31}+(1-p_{33})p_{41} & p_{43}p_{32}+(1-p_{33})p_{42} \end{bmatrix}$$
$$= \begin{bmatrix} f_{31} & f_{32} \\ f_{41} & f_{42} \end{bmatrix}. \tag{3.82}$$

The limiting transition probability matrix for this absorbing multichain, is

$$\lim_{n\to\infty} P^n = \begin{bmatrix} \mathbf{I} & \mathbf{0} \\ \mathbf{F} & \mathbf{0} \end{bmatrix} = \begin{bmatrix} 1 & 0 & 0 & 0 \\ 0 & 1 & 0 & 0 \\ f_{31} & f_{32} & 0 & 0 \\ f_{41} & f_{42} & 0 & 0 \end{bmatrix}. \tag{3.135}$$

3.5.3 Absorbing Markov Chain Model of Patient Flow in a Hospital

Consider the absorbing multichain model of patient flow in a hospital for which the transition probability matrix is given in canonical form in Equation (1.59). The matrix of absorption probabilities is calculated in Equation (3.9.3). The matrix of limiting probabilities for this absorbing multichain is

$$\lim_{n\to\infty} P^n = \begin{bmatrix} I & 0 \\ F & 0 \end{bmatrix} = \begin{array}{c|ccc|cccc} \text{State} & 0 & 5 & 1 & 2 & 3 & 4 \\ \hline 0 & 1 & 0 & 0 & 0 & 0 & 0 \\ 5 & 0 & 1 & 0 & 0 & 0 & 0 \\ 1 & 0.7284 & 0.2714 & 0 & 0 & 0 & 0 \\ 2 & 0.7327 & 0.2671 & 0 & 0 & 0 & 0 \\ 3 & 0.7246 & 0.2750 & 0 & 0 & 0 & 0 \\ 4 & 0.7294 & 0.2703 & 0 & 0 & 0 & 0 \end{array}. \tag{3.136}$$

A patient receiving physical therapy, in state 4, has a probability of $f_{40} = 0.7294$ of eventually being discharged, and a probability of $f_{45} = 0.2703$ of eventually dying.

3.5.4 Reducible Unichain

As Section 1.9.1.2 indicates, a reducible unichain has a set of transient states plus one closed set of recurrent states. The canonical form of the transition probability matrix is given in Equation (3.1). The canonical form of the two-step transition probability matrix is

$$P^2 = PP = \begin{bmatrix} S & 0 \\ D & Q \end{bmatrix}\begin{bmatrix} S & 0 \\ D & Q \end{bmatrix} = \begin{bmatrix} S^2 & 0 \\ DS + QD & Q^2 \end{bmatrix} = \begin{bmatrix} S^2 & 0 \\ D_2 & Q^2 \end{bmatrix}, \tag{3.137}$$

where

$$D_2 = DS + QD. \tag{3.138}$$

The canonical form of the three-step transition matrix is

$$P^3 = PP^2 = \begin{bmatrix} S & 0 \\ D & Q \end{bmatrix}\begin{bmatrix} S^2 & 0 \\ D_2 & Q^2 \end{bmatrix} = \begin{bmatrix} S^3 & 0 \\ DS^2 + QD_2 & Q^3 \end{bmatrix} = \begin{bmatrix} S^3 & 0 \\ D_3 & Q^3 \end{bmatrix}, \tag{3.139}$$

where

$$D_3 = DS^2 + QD_2 = DS^2 + Q(DS + QD) = DS^2 + QDS + Q^2D = (DS + QD)S + Q^2D$$
$$= D_2S + Q^2D. \tag{3.140}$$

By raising P to successively higher powers, one can show that the canonical form of the n-step transition matrix is

$$P^n = \begin{bmatrix} S^n & 0 \\ D_n & Q^n \end{bmatrix}, \tag{3.141}$$

where

$$D_1 = D \tag{3.142}$$

and

$$D_n = D_{n-1}S + Q^{n-1}D. \tag{3.143}$$

Using Equation (3.69),

$$\lim_{n \to \infty} P^n = \begin{bmatrix} \lim\limits_{n \to \infty} S^n & 0 \\ \lim\limits_{n \to \infty} D_n & \lim\limits_{n \to \infty} Q^n \end{bmatrix} = \begin{bmatrix} \lim\limits_{n \to \infty} S^n & 0 \\ \lim\limits_{n \to \infty} D_n & 0 \end{bmatrix}. \tag{3.144}$$

If states i and j are both recurrent, then they belong to the same recurrent closed class, which acts as a separate irreducible chain with transition probability matrix S. The unichain has only one recurrent closed class, which is therefore certain to be reached from a transient state. Once the unichain has entered the recurrent closed class, its limiting behavior is governed by the steady-state probability vector for the recurrent closed class.

If $\pi = [\pi_j]$ is the steady-state probability vector for the recurrent closed class, then

$$\left. \begin{aligned} \pi &= \pi S, \\ \pi e &= 1, \\ \text{and} & \\ \pi_j &= \lim_{n \to \infty} p_{ij}^{(n)}. \end{aligned} \right\} \tag{3.145}$$

Thus,

$$\lim_{n \to \infty} S^{(n)} = \Pi, \tag{3.146}$$

where Π is a matrix with each row π.

Suppose that state i is transient and state j is recurrent. Then transitions from transient state i to recurrent state j are governed by submatrix D. Once again, $\lim_{n \to \infty} p_{ij}^{(n)} = \pi_j$ because the chain is certain to eventually enter the closed set of recurrent states. Thus,

$$\lim_{n \to \infty} D_n = \lim_{n \to \infty} S^{(n)} = \Pi. \tag{3.147}$$

In other words, if j is a recurrent state, then $\lim_{n \to \infty} p_{ij}^{(n)} = \pi_j$, irrespective of whether the starting state i is transient or recurrent. Therefore, the limiting transition probability matrix for a reducible unichain is

$$\lim_{n \to \infty} P^n = \begin{bmatrix} \lim\limits_{n \to \infty} S^n & 0 \\ \lim\limits_{n \to \infty} D_n & 0 \end{bmatrix} = \begin{bmatrix} \Pi & 0 \\ \Pi & 0 \end{bmatrix}. \tag{3.148}$$

Once the unichain has entered the recurrent closed set, which is certain to be reached, its limiting behavior is governed by the steady-state probability vector for the closed set.

3.5.4.1 Reducible Four-State Unichain

Consider the generic four-state reducible unichain for which the transition matrix is shown in canonical form in Equation (3.2).

$$P = [p_{ij}] = \begin{matrix} 1 \\ 2 \\ 3 \\ 4 \end{matrix} \begin{bmatrix} p_{11} & p_{12} & 0 & 0 \\ p_{21} & p_{22} & 0 & 0 \\ \hline p_{31} & p_{32} & p_{33} & p_{34} \\ p_{41} & p_{42} & p_{43} & p_{44} \end{bmatrix} = \begin{bmatrix} S & 0 \\ D & Q \end{bmatrix}. \tag{3.2}$$

The steady-state probability vector for the generic two-state transition matrix S was obtained in Equation (2.16). Using Equation (3.148), the limiting transition probability matrix for P is

$$\lim_{n \to \infty} P^n = \begin{bmatrix} \Pi & 0 \\ \Pi & 0 \end{bmatrix} = \begin{bmatrix} \pi_1 & \pi_2 & 0 & 0 \\ \pi_1 & \pi_2 & 0 & 0 \\ \pi_1 & \pi_2 & 0 & 0 \\ \pi_1 & \pi_2 & 0 & 0 \end{bmatrix} = \begin{bmatrix} p_{21}/(p_{12}+p_{21}) & p_{12}/(p_{12}+p_{21}) & 0 & 0 \\ p_{21}/(p_{12}+p_{21}) & p_{12}/(p_{12}+p_{21}) & 0 & 0 \\ p_{21}/(p_{12}+p_{21}) & p_{12}/(p_{12}+p_{21}) & 0 & 0 \\ p_{21}/(p_{12}+p_{21}) & p_{12}/(p_{12}+p_{21}) & 0 & 0 \end{bmatrix}. \tag{3.149}$$

3.5.4.2 Reducible Unichain Model of Machine Maintenance

The model of machine maintenance introduced in Section 1.10.1.2.2.1 is a four-state unichain. If the engineer who manages the production process follows the modified maintenance policy, under which she is always doing nothing when the machine is in state 3, the probability of first passage in n steps was computed in Section 3.3.2.1.4, and the probability of eventual passage was computed in Section 3.3.3.1.3. The transition probability matrix associated with the modified maintenance policy is shown in canonical form in Equation (1.56).

$$P = [p_{ij}] = \begin{matrix} 1 \\ 2 \\ 3 \\ 4 \end{matrix} \begin{bmatrix} 0.2 & 0.8 & 0 & 0 \\ 0.6 & 0.4 & 0 & 0 \\ \hline 0.2 & 0.3 & 0.5 & 0 \\ 0.3 & 0.2 & 0.1 & 0.4 \end{bmatrix} = \begin{bmatrix} S & 0 \\ D & Q \end{bmatrix}. \tag{1.56}$$

The transition probability matrix associated with the original maintenance policy, under which the engineer overhauls the machine when it is in state 3, is shown in canonical form in Equation (1.55).

$$
P = [p_{ij}] = \begin{matrix} 1 \\ 2 \\ 3 \\ 4 \end{matrix} \begin{bmatrix} 0.2 & 0.8 & 0 & 0 \\ 0.6 & 0.4 & 0 & 0 \\ \hline 0 & 0 & 0.2 & 0.8 \\ 0.3 & 0.2 & 0.1 & 0.4 \end{bmatrix} = \begin{bmatrix} S & 0 \\ D & Q \end{bmatrix}.
\tag{1.55}
$$

Observe that under both maintenance policies, states 1 and 2 form a recurrent closed class, while states 3 and 4 are transient. Submatrix S, which governs transitions for the recurrent chain, is identical under both maintenance policies. Using Equation (2.16), the vector of steady-state probabilities for submatrix S is

$$
\pi = [\pi_1 \quad \pi_2] = [0.6/(0.8+0.6) \quad 0.8/(0.8+0.6)] = [3/7 \quad 4/7].
\tag{3.150}
$$

Under both policies, using Equation (3.149), the limiting transition probability matrix is

$$
\lim_{n \to \infty} P^n = \begin{bmatrix} \Pi & 0 \\ \Pi & 0 \end{bmatrix} = \begin{bmatrix} \pi_1 & \pi_2 & 0 & 0 \\ \pi_1 & \pi_2 & 0 & 0 \\ \pi_1 & \pi_2 & 0 & 0 \\ \pi_1 & \pi_2 & 0 & 0 \end{bmatrix} = \begin{bmatrix} 3/7 & 4/7 & 0 & 0 \\ 3/7 & 4/7 & 0 & 0 \\ 3/7 & 4/7 & 0 & 0 \\ 3/7 & 4/7 & 0 & 0 \end{bmatrix}.
\tag{3.151}
$$

In the long run, under both maintenance policies, the machine will be in state 1, not working, 3/7 of the time, and in state 2, working, with a major defect, the remaining 4/7 of the time.

3.5.5 Reducible Multichain

Consider a reducible multichain that has M recurrent chains, denoted by $R_1, \ldots, R_M$, with the respective transition probability matrices $P_1, \ldots, P_M$, plus a set of transient states. The canonical form of the transition probability matrix is shown in Equation (3.4).

$$
P = \begin{bmatrix} P_1 & 0 & 0 & 0 \\ 0 & \ddots & 0 & 0 \\ 0 & 0 & P_M & 0 \\ D_1 & \cdots & D_M & Q \end{bmatrix} = \begin{bmatrix} S & 0 \\ D & Q \end{bmatrix},
\tag{3.4}
$$

where the transition probability matrix is also expressed in aggregated form, so that

$$S = \begin{bmatrix} P_1 & 0 & 0 \\ 0 & \ddots & 0 \\ 0 & 0 & P_M \end{bmatrix}, \quad D = \begin{bmatrix} D_1 & \cdots & D_M \end{bmatrix}. \tag{3.152}$$

By combining the results obtained in Equation (3.127) for a recurrent multi-chain and in Equation (3.144) for a reducible unichain, the limiting transition probability matrix for a reducible multichain has the form:

$$\lim_{n \to \infty} P^n = \begin{bmatrix} \lim_{n \to \infty} P_1^n & 0 & 0 & 0 \\ 0 & \ddots & 0 & 0 \\ 0 & \cdots & \lim_{n \to \infty} P_M^n & \vdots \\ \lim_{n \to \infty} D_{1,n} & \cdots & \lim_{n \to \infty} D_{M,n} & \lim_{n \to \infty} Q^n \end{bmatrix} = \begin{bmatrix} \lim_{n \to \infty} P_1^n & 0 & 0 & 0 \\ 0 & \ddots & 0 & 0 \\ 0 & \cdots & \lim_{n \to \infty} P_M^n & \vdots \\ \lim_{n \to \infty} D_{1,n} & \cdots & \lim_{n \to \infty} D_{M,n} & 0 \end{bmatrix}$$

$$= \begin{bmatrix} \Pi(1) & 0 & 0 & 0 \\ 0 & \ddots & 0 & 0 \\ 0 & \cdots & \Pi(M) & \vdots \\ \lim_{n \to \infty} D_{1,n} & \cdots & \lim_{n \to \infty} D_{M,n} & 0 \end{bmatrix}, \tag{3.153}$$

where $\Pi(j)$ is a matrix with each row $\pi(j)$, the steady-state probability vector for the transition probability matrix P_j, and $D_{j,n}$ is the matrix of n-step transition probabilities from transient states to the recurrent chain R_j.

As Section 3.4.1 and Equation (3.116) indicate, in a reducible multichain, the probability of eventual passage from a transient state i to a recurrent closed class is simply the sum of the probabilities of eventual passage from the transient state i to all of the recurrent states within the closed class. Suppose that R_k represents a recurrent chain for which the transition probabilities are governed by the matrix P_k. The probability f_{i,R_k} of eventual passage from a transient state i to the recurrent closed class R_k is given by

$$f_{iR_k} = \sum_{j \in R_k} f_{ij}, \tag{3.116}$$

where the recurrent states j belongs to the recurrent closed class R_k.

The limiting probability of a transition from a transient state i to a state j belonging to the recurrent closed class R_k is equal to the product of f_{i,R_k}, the probability of eventual passage from the transient state i to the recurrent chain R_k, and $\pi(k)_j$, the steady-state probability for state j, which belongs to the recurrent closed class R_k. That is,

$$\lim_{n\to\infty} p_{ij}^{(n)} = f_{i,R_k}\pi(k)_j, \tag{3.154}$$

represents the limiting probability of a transition from a transient state i to a state j within the recurrent closed class R_k.

3.5.5.1 Reducible Five-State Multichain

Consider the generic five-state reducible multichain, treated most recently in Section 3.4.1.1. The canonical form of the transition probability matrix is shown below:

$$\mathbf{P} = \begin{array}{c} 1 \\ 2 \\ 3 \\ 4 \\ 5 \end{array}\left[\begin{array}{c|cc|cc} 1 & 0 & 0 & 0 & 0 \\ \hline 0 & p_{22} & p_{23} & 0 & 0 \\ 0 & p_{32} & p_{33} & 0 & 0 \\ \hline p_{41} & p_{42} & p_{43} & p_{44} & p_{45} \\ p_{51} & p_{52} & p_{53} & p_{54} & p_{55} \end{array}\right] = \begin{bmatrix} P_1 & 0 & 0 \\ 0 & P_2 & 0 \\ D_1 & D_2 & Q \end{bmatrix}. \tag{3.5}$$

Recall that state 1 is an absorbing state. The two recurrent closed sets are denoted by $R_1 = \{1\}$ and $R_2 = \{2, 3\}$. The set of transient states is denoted by $T = \{4, 5\}$. The probabilities of eventual passage to the individual recurrent states, states 2 and 3, were computed in Section 3.3.3.2.3 in Equations (3.97) and (3.101). The probabilities of eventual passage to $R_2 = \{2, 3\}$ were computed in Section 3.4.1.1 in Equations (3.111) through (3.113), and in Section 3.4.2 in Equation (3.122). The limiting transition probability matrix for the reducible multichain has the form

$$\lim_{n\to\infty} \mathbf{P}^n = \lim_{n\to\infty} \begin{array}{c} 1 \\ 2 \\ 3 \\ 4 \\ 5 \end{array}\left[\begin{array}{c|cc|cc} 1 & 0 & 0 & 0 & 0 \\ \hline 0 & p_{22}^{(n)} & p_{23}^{(n)} & 0 & 0 \\ 0 & p_{32}^{(n)} & p_{33}^{(n)} & 0 & 0 \\ \hline p_{41}^{(n)} & p_{42}^{(n)} & p_{43}^{(n)} & p_{44}^{(n)} & p_{45}^{(n)} \\ p_{51}^{(n)} & p_{52}^{(n)} & p_{53}^{(n)} & p_{54}^{(n)} & p_{55}^{(n)} \end{array}\right]$$

$$= \begin{array}{c} 1 \\ 2 \\ 3 \\ 4 \\ 5 \end{array}\left[\begin{array}{c|cc|cc} 1 & 0 & 0 & 0 & 0 \\ \hline 0 & \pi_2 & \pi_3 & 0 & 0 \\ 0 & \pi_2 & \pi_3 & 0 & 0 \\ \hline f_{41} & \lim_{n\to\infty} p_{42}^{(n)} & \lim_{n\to\infty} p_{43}^{(n)} & 0 & 0 \\ f_{51} & \lim_{n\to\infty} p_{52}^{(n)} & \lim_{n\to\infty} p_{53}^{(n)} & 0 & 0 \end{array}\right], \tag{3.155}$$

where f_{41} and f_{51} are the absorption probabilities for transient states 4 and 5, respectively, and π_2 and π_3 are the steady-state probabilities for recurrent states 2 and 3, respectively. Using Equation (3.116), the probability f_{iR_2} of eventual passage from a transient state i to the recurrent chain $R_2 = \{2, 3\}$, is

$$f_{i,R_2} = f_{i2} + f_{i3}. \tag{3.156}$$

Using Equation (3.154), the limiting probability, $\lim\limits_{n \to \infty} p_{ij}^{(n)}$, of a transition from a transient state i to a state j belonging to the recurrent closed class $R_2 = \{2, 3\}$ is

$$\lim_{n \to \infty} p_{ij}^{(n)} = (f_{i2} + f_{i3})\pi_j = f_{iR_2}\pi_j. \tag{3.157}$$

Therefore, the limiting transition probability for the five-state reducible multichain is

$$\lim_{n \to \infty} P^n = \begin{array}{c} 1 \\ 2 \\ 3 \\ 4 \\ 5 \end{array}\left[\begin{array}{c|ccc|cc} 1 & 0 & & 0 & 0 & 0 \\ \hline 0 & \pi_2 & & \pi_3 & 0 & 0 \\ 0 & \pi_2 & & \pi_3 & 0 & 0 \\ \hline f_{41} & (f_{42}+f_{43})\pi_2 & (f_{42}+f_{43})\pi_3 & 0 & 0 \\ f_{51} & (f_{52}+f_{53})\pi_2 & (f_{52}+f_{53})\pi_3 & 0 & 0 \end{array}\right]$$

$$\tag{3.158}$$

$$= \begin{array}{c} 1 \\ 2 \\ 3 \\ 4 \\ 5 \end{array}\left[\begin{array}{c|cc|cc} 1 & 0 & 0 & 0 & 0 \\ \hline 0 & \pi_2 & \pi_3 & 0 & 0 \\ 0 & \pi_2 & \pi_3 & 0 & 0 \\ \hline f_{41} & f_{4R_2}\pi_2 & f_{4R_2}\pi_3 & 0 & 0 \\ f_{51} & f_{5R_2}\pi_2 & f_{5R_2}\pi_3 & 0 & 0 \end{array}\right].$$

3.5.5.2 Reducible Multichain Model of an Eight-State Serial Production Process

Consider the eight-state reducible multichain model of a serial production process for which probabilities of eventual passage to the recurrent closed set $R_3 = \{3, 4, 5\}$ are computed in Section 3.4.1.2. The transition probability matrix in canonical form is shown in Equation (3.7). The limiting transition probability matrix for the eight-state serial production process has the following form:

$$\lim_{n \to \infty} P^n =$$

$$
\begin{array}{c}
1 \\ 2 \\ 3 \\ 4 \\ 5 \\ 6 \\ 7 \\ 8
\end{array}
\left[
\begin{array}{cc|ccc|ccc}
1 & 0 & 0 & 0 & 0 & 0 & 0 & 0 \\
0 & 1 & 0 & 0 & 0 & 0 & 0 & 0 \\
0 & 0 & \pi_3 & \pi_4 & \pi_5 & 0 & 0 & 0 \\
0 & 0 & \pi_3 & \pi_4 & \pi_5 & 0 & 0 & 0 \\
0 & 0 & \pi_3 & \pi_4 & \pi_5 & 0 & 0 & 0 \\
f_{61} & f_{62} & (f_{63}+f_{64}+f_{65})\pi_3 & (f_{63}+f_{64}+f_{65})\pi_4 & (f_{63}+f_{64}+f_{65})\pi_5 & 0 & 0 & 0 \\
f_{71} & f_{72} & (f_{73}+f_{74}+f_{75})\pi_3 & (f_{73}+f_{74}+f_{75})\pi_4 & (f_{73}+f_{74}+f_{75})\pi_5 & 0 & 0 & 0 \\
f_{81} & f_{82} & (f_{83}+f_{84}+f_{85})\pi_3 & (f_{83}+f_{84}+f_{85})\pi_4 & (f_{83}+f_{84}+f_{85})\pi_5 & 0 & 0 & 0
\end{array}
\right]
$$

$$
=
\begin{array}{c}
1 \\ 2 \\ 3 \\ 4 \\ 5 \\ 6 \\ 7 \\ 8
\end{array}
\left[
\begin{array}{cc|ccc|ccc}
1 & 0 & 0 & 0 & 0 & 0 & 0 & 0 \\
0 & 1 & 0 & 0 & 0 & 0 & 0 & 0 \\
0 & 0 & \pi_3 & \pi_4 & \pi_5 & 0 & 0 & 0 \\
0 & 0 & \pi_3 & \pi_4 & \pi_5 & 0 & 0 & 0 \\
0 & 0 & \pi_3 & \pi_4 & \pi_5 & 0 & 0 & 0 \\
f_{61} & f_{62} & f_{6R_3}\pi_3 & f_{6R_3}\pi_4 & f_{6R_3}\pi_5 & 0 & 0 & 0 \\
f_{71} & f_{72} & f_{7R_3}\pi_3 & f_{7R_3}\pi_4 & f_{7R_3}\pi_5 & 0 & 0 & 0 \\
f_{81} & f_{82} & f_{8R_3}\pi_3 & f_{8R_3}\pi_4 & f_{8R_3}\pi_5 & 0 & 0 & 0
\end{array}
\right],
\tag{3.159}
$$

where, for, $i = 6, 7,$ and 8,

$$f_{iR_3} = f_{i3} + f_{i4} + f_{i5}. \tag{3.160}$$

The vectors of probabilities of absorption in absorbing states 1 and 2, starting from the set of transient states $T = \{6, 7, 8\}$, are computed in Equations (3.103) and (3.115). The matrix of probabilities of eventual passage starting from the set of transient states, $T = \{6, 7, 8\}$, to the three individual recurrent states, which belong to the recurrent closed class $R_3 = \{3, 4, 5\}$, are computed in Equation (3.103). The matrix of probabilities of eventual passage starting from the set of transient states, $T = \{6, 7, 8\}$, to the recurrent closed class $R_3 = \{3, 4, 5\}$ are computed in equations (3.115).

The steady-state probability vector for the transition probability matrix P_3, which governs transitions among the three recurrent states that belong to the recurrent chain $R_3 = \{3, 4, 5\}$, is calculated by solving equations (2.13).

$$
\left.
\begin{array}{c}
[\pi_3 \quad \pi_4 \quad \pi_5] = [\pi_3 \quad \pi_4 \quad \pi_5]
\begin{bmatrix}
0.50 & 0.30 & 0.20 \\
0.30 & 0.45 & 0.25 \\
0.10 & 0.35 & 0.55
\end{bmatrix} \\[2ex]
\pi_3 + \pi_4 + \pi_5 = 1
\end{array}
\right\}
\tag{3.161}
$$

The solution for the steady-state probability vector for P_3 is

$$\pi = [\pi_3 \quad \pi_4 \quad \pi_5] = [0.2909 \quad 0.3727 \quad 0.3364].$$ (3.162)

Using Equation (3.153) with $M = 3$, the limiting transition probability matrix for the eight-state serial production process has the form:

$$\lim_{n \to \infty} P^n = \begin{vmatrix} \Pi(1) & 0 & 0 & 0 \\ 0 & \Pi(2) & 0 & 0 \\ 0 & 0 & \Pi(3) & 0 \\ \lim_{n \to \infty} D_{1,n} & \lim_{n \to \infty} D_{2,n} & \lim_{n \to \infty} D_{3,n} & 0 \end{vmatrix}.$$ (3.153)

Using Equations (3.154) and (3.159), the matrix $\lim_{n \to \infty} D_{3,n}$ of the limiting probabilities of transitions from the set of transient states, $T = \{6, 7, 8\}$ associated with the three production stages, to the three recurrent states, which belong to the recurrent chain $R_3 = \{3, 4, 5\}$ associated with the training center, is given by

$$\lim_{n \to \infty} D_{3,n} = \begin{vmatrix} \lim_{n \to \infty} p_{6,3}^{(n)} & \lim_{n \to \infty} p_{64}^{(n)} & \lim_{n \to \infty} p_{65}^{(n)} \\ \lim_{n \to \infty} p_{7,3}^{(n)} & \lim_{n \to \infty} p_{74}^{(n)} & \lim_{n \to \infty} p_{75}^{(n)} \\ \lim_{n \to \infty} p_{8,3}^{(n)} & \lim_{n \to \infty} p_{84}^{(n)} & \lim_{n \to \infty} p_{85}^{(n)} \end{vmatrix} = \begin{vmatrix} f_{6R_3}\pi_3 & f_{6R_3}\pi_4 & f_{6R_3}\pi_5 \\ f_{7R_3}\pi_3 & f_{7R_3}\pi_4 & f_{7R_3}\pi_5 \\ f_{8R_3}\pi_3 & f_{8R_3}\pi_4 & f_{8R_3}\pi_5 \end{vmatrix}$$

$$= \begin{bmatrix} 0.2(0.2909) & 0.2(0.3727) & 0.2(0.3364) \\ 0.1143(0.2909) & 0.1143(0.3727) & 0.1143(0.3364) \\ 0.0686(0.2909) & 0.0686(0.3727) & 0.0686(0.3364) \end{bmatrix}$$

$$= \begin{bmatrix} 0.0582 & 0.0745 & 0.0673 \\ 0.0332 & 0.0426 & 0.0385 \\ 0.0199 & 0.0256 & 0.0231 \end{bmatrix}.$$

(3.163)

For example,

$$\lim_{n \to \infty} p_{84}^{(n)} = (f_{83} + f_{84} + f_{85})\pi_4 = f_{8R_3}\pi_4 = (0.0686)(0.3727) = 0.0256$$ (3.164)

represents the limiting probability or long run proportion of time that an entering item which has been sent to the training center will be used for training technicians.

By combining all of the results in Section 3.5.5.2, the limiting transition probability matrix for the reducible eight-state multichain model of the serial production process is

$$
\lim_{n \to \infty} P^{(n)} =
\begin{array}{c}
1 \\ 2 \\ 3 \\ 4 \\ 5 \\ 6 \\ 7 \\ 8
\end{array}
\begin{bmatrix}
1 & 0 & 0 & 0 & 0 & 0 & 0 & 0 \\
0 & 1 & 0 & 0 & 0 & 0 & 0 & 0 \\
0 & 0 & 0.2909 & 0.3727 & 0.3364 & 0 & 0 & 0 \\
0 & 0 & 0.2909 & 0.3727 & 0.3364 & 0 & 0 & 0 \\
0 & 0 & 0.2909 & 0.3727 & 0.3364 & 0 & 0 & 0 \\
0.4444 & 0.3556 & 0.0582 & 0.0745 & 0.0673 & 0 & 0 & 0 \\
0.6825 & 0.2032 & 0.0332 & 0.0426 & 0.0385 & 0 & 0 & 0 \\
0.8095 & 0.1219 & 0.0199 & 0.0256 & 0.0231 & 0 & 0 & 0
\end{bmatrix}.
\tag{3.165}
$$

3.5.5.3 Conditional Mean Time to Absorption

In this section, the conditional mean time to absorption will be calculated for an absorbing multichain when it is known that the process will end in a target absorbing state [5]. For example, consider the absorbing multichain model of a production process for which the transition probability matrix P is expressed in canonical form in Equation (3.114). Suppose that, for all items which will eventually be sold, management wishes to compute the mean number of transitions until an entering item is sold. That is, management wants to calculate the mean time to absorption in the target absorbing state 2 when the process starts in transient state 8. To calculate this number, the absorbing multichain, which has three absorbing states, is transformed into an absorbing unichain, which has only state 2 as an absorbing state. The absorbing unichain will have a modified transition probability matrix denoted by $\widehat{P} = [\hat{p}_{ij}]$. Recall that the matrix of absorption probabilities, $F = [f_{ij}]$, for the production process is calculated in Equation (3.115a). When the conditional process starts in a transient state i, the transition probabilities $\hat{p}_{ij}$ are computed in the following manner:

$$
\left.
\begin{aligned}
\hat{p}_{ij} &= P(i \to j \text{ in unichain}) \\
&= P(i \to j \text{ in multichain} \mid \text{multichain state } i \text{ absorbed in state 2}) \\
&= \frac{P(i \to j \text{ in multichain} \cap \text{multichain state } j \text{ absorbed in state 2})}{P(\text{multichain state } i \text{ absorbed in state 2})} \\
&= \frac{P(i \to j \text{ in multichain}) \, P(\text{multichain state } j \text{ absorbed in state 2})}{P(\text{multichain state } i \text{ absorbed in state 2})} \\
&= \frac{p_{ij} f_{j2}}{f_{i2}}.
\end{aligned}
\right\}
\tag{3.166}
$$

The transition probability matrix $\hat{P}$ for the absorbing unichain is shown below:

$$\hat{P} = \begin{array}{c} 2 \\ 6 \\ 7 \\ 8 \end{array} \begin{array}{c} \\ \end{array} \left[\begin{array}{c|ccc} 1 & 0 & 0 & 0 \\ \hline 0.45 & 0.55 & 0 & 0 \\ 0 & 0.35 & 0.65 & 0 \\ 0 & 0 & 0.25 & 0.75 \end{array} \right] = \left[\begin{array}{cc} I & 0 \\ \hat{D} & \hat{Q} \end{array} \right]. \tag{3.167}$$

The fundamental matrix $\hat{U}$ for the absorbing unichain is

$$\hat{U} = (I - \hat{Q})^{-1} = \begin{array}{c} 6 \\ 7 \\ 8 \end{array} \left[\begin{array}{ccc} 2.2222 & 0 & 0 \\ 2.2222 & 2.8571 & 0 \\ 2.2222 & 2.8571 & 4 \end{array} \right]. \tag{3.168}$$

Hence, the mean time to absorption in state 2 for an entering item is 9.0793 steps, the sum of the entries in the third row of $\hat{U}$.

PROBLEMS

3.1 Three basketball players compete in a basketball foul shooting contest. The eligible players are allowed one foul shot per round. After each round, all players who miss their foul shots are eliminated, and the remaining players participate in the next round. The contest ends when a single player who has not missed a foul shot remains, and is declared the winner, or when all players have been eliminated, and no one wins. In the past, player A has made an average of 80% of his foul shots, player B has made an average of 75%, and player C has made an average of 70 %.

This contest can be modeled as an absorbing multichain by choosing as states all sets of players who have not been eliminated. For example, if two players remain, the corresponding states are the three pairs (A, B), (A, C), and (B, C).

(a) Construct the transition probability matrix.

(b) What is the probability that the contest will end without a winner?

(c) What is the probability that player A will win the contest?

(d) If players B and C are the remaining contestants, what is the expected number of rounds needed before C wins?

3.2 Three hockey players compete in a defensive contest. In each round, each eligible player is allowed one shot at the goal of the remaining player who has the highest career scoring average. After each round, all players who allow their opponent to score a goal are eliminated, and the remaining players participate in the next round. The contest ends when a single player who has not allowed a goal remains, and is declared the winner, or when all players have been eliminated, and no one wins. The career scoring averages of players A, B, and C are 40%, 35%, and 20%, respectively.

The contest begins with all three players eligible to compete in the first round. Player A will shoot his puck toward the goal of B, A's competitor who has the highest career scoring average. Next, player B will shoot his puck toward the goal of A, B's competitor who has the highest career scoring average. Finally, player C will shoot his puck toward the goal of A, C's competitor who has the highest career scoring average. When the round ends, any player who has allowed a goal to be scored is eliminated. Note that after the first round, in which all three players compete, players A and B can both be eliminated if each scores a goal against the other. Player C, who has the lowest career scoring average, cannot be eliminated after the first round because no other player will shoot his puck toward the goal of C. The surviving players enter the next round.

This contest can be modeled as an absorbing multichain by choosing as states all sets of players who have not been eliminated. For example, if two players remain, the corresponding states are the three pairs (A, B), (A, C), and (B, C).
(a) Construct the transition probability matrix.
(b) What is the probability that the contest will end without a winner?
(c) What is the probability that player A will win the contest?
(d) If players B and C are the remaining contestants, what is the expected number of rounds needed before C wins?

3.3　A woman needs $5,000 for a down payment on a condominium. She will try to raise the money for the down payment by gambling. She will place a sequence of bets until she either accumulates $5,000 or loses all her money. She starts with $2,000, and will wager $1,000 on each bet. Each time that she bets $1,000, she will win $1,000 with probability 0.4, or lose $1,000 with probability 0.6.
(a) Model the woman's gambling experience as a six-state absorbing multichain. Let the state X_n denote the gambler's revenue after the nth bet. Construct the transition probability matrix.
(b) What is the expected number of bets that she will make?
(c) What is the probability that she will obtain her $5,000 down payment?

3.4　Suppose that the woman seeking a $5,000 down payment in Problem 3.3 has the option of betting either $1,000 or $2,000. She chooses the following aggressive strategy. If she has $1,000 or $4,000, she will bet $1,000, and will win $1,000 with probability 0.4 or lose $1,000 with probability 0.6. If she has $2,000 or $3,000, she will bet $2,000, and will either win $2,000 with probability 0.05, or win $1,000 with probability 0.15, or lose $2,000 with probability 0.8.

(a) Model the woman's gambling experience under the aggressive strategy as a six-state absorbing Markov chain. Let the state represent the amount of money that she has when she places a bet. Construct the transition probability matrix.

(b) What is the expected number of bets that she will make?

(c) What is the probability that she will obtain her $5,000 down payment?

3.5 A consumer electronics dealer sells new flat panel televisions for $1,000. The dealer offers customers a 4-year warranty. The warranty provides free replacement of a television that fails within the 4-year warranty period, but does not cover the cost of a replaced TV.

The dealer discloses that 5% of new flat panel televisions fail during their first year of life, 10% of 1-year-old televisions fail during their second year of life, 15% of 2-year-old televisions fail during their third year of life, and 20% of 3-year-old televisions fail during their fourth year of life. Suppose that the dealer sells a consumer a 4-year warranty for $140 along with the television.

(a) Model the 4-year warranty experience of the dealer as a six-state absorbing multichain with two absorbing states. Choose one absorbing state to represent the replacement of a TV, which has failed during the warranty period, and the other to represent the survival of the TV until the end of the warranty period. Choose the transient states to represent the age of the television. Construct the transition probability matrix.

(b) What is the probability that the TV will have to be replaced during the warranty period?

(c) What is the dealer's expected revenue from selling a 4-year warranty?

3.6 A production process contains three stages in series. An entering item starts in the first manufacturing stage. The output from each stage is inspected. An item of acceptable quality is passed on to the next stage, a defective item is scrapped, and an item of marginal quality is reworked at the current stage. An item at stage 1 has a 0.08 probability of being defective, a 0.12 probability of being of marginal quality, and a 0.80 probability of being of acceptable quality. An item at stage 2 has a 0.06 probability of being defective, a 0.09 probability of being of marginal quality, and a 0.85 probability of being acceptable. The probabilities that an item at stage 3 will be defective, marginal in quality, or acceptable, are 0.04, 0.06, and 0.90, respectively. All items of acceptable quality produced by stage 3 are sold.

(a) Model the production process as a five-state absorbing multichain with two absorbing states. Choose one absorbing state to represent scrapping a defective item, and the other to represent selling an item of acceptable quality produced by stage 3. Choose the transient states to represent the stages of the production process. Construct the transition probability matrix. Represent it in canonical form.

(b) Find the probability that an item will be scrapped, given that it is in stage 3.

(c) Find the probability that an item will be sold without being reworked.

(d) Find the probability that an item will be in stage 2 after three inspections.

(e) Find the probability that an item will be scrapped.

(f) Find the probability that an item will be sold.

(g) Find the mean number of inspections that an item will receive in stage 3.

(h) Given that an item is in stage 2, find the mean number of inspections that it will receive in stage 3.

3.7   An unmanned expendable rocket is launched to place a communications satellite in either high earth orbit (HEO) or low earth orbit (LEO). HEO is the preferred destination. As the rocket is tracked, a sequence of course correction signals is sent to it. The system has five states, which are labeled as follows:

State	Description
1	On course to HEO
2	On course to LEO
3	Minor deviation from course to HEO
4	Major deviation from course to HEO
5	Abort mission

Assume that the system can be modeled by an absorbing Markov chain in which states 1, 2, and 5 are absorbing, and states 3 and 4 are transient. The state of the system is observed after the nth course correction. Suppose that the five-state absorbing multi-chain has the following transition probability matrix:

State	1	2	3	4	5
1, On course to HEO	1	0	0	0	0
2, On course to LEO	0	1	0	0	0
3, Minor deviation	0.55	0.30	0.10	0.05	0
4, Major deviation	0	0.40	0.30	0.20	0.10
5, Abort	0	0	0	0	1

$P =$ (to the left of rows 2–4)

(a) Represent the transition matrix in canonical form.

(b) If, upon launch, the rocket is observed to start in state 3, find the probability that it eventually gets on course to LEO.

(c) If, upon launch, the rocket is observed to start in state 4, find the probability that it eventually gets on course to HEO.

(d) If, upon launch, the rocket is observed to start in state 3 with probability 0.8 or in state 4 with probability 0.2, find the probability that it eventually gets on course to HEO.

(e) If, upon launch, the rocket is observed to start in state 3 with probability 0.8 or in state 4 with probability 0.2, find the probability that the mission will eventually be aborted.

(f) If, upon launch, the rocket is observed to start in state 4, find the probability that it will be in state 3 after three course corrections.

(g) If, upon launch, the rocket is observed to start in state 3, find the mean number of course corrections before it reaches HEO or LEO or aborts.

3.8   In Problem 1.12, let $d_1 = d_2 = d$, $r_1 = r_2 = r_3 = r$, and $p_1 = p_2 = p$.

(a) Show that the following quantity is the fundamental matrix:

State	1	2	3
1 = Assistant professor	$\dfrac{1}{d+p}$	$\dfrac{p}{(d+p)^2}$	$\dfrac{p^2}{(d+p)^3}$
2 = Associate professor	0	$\dfrac{1}{d+p}$	$\dfrac{p}{(d+p)^2}$
3 = Professor	0	0	$\dfrac{1}{d+p}$

$$U = (I - Q)^{-1} =$$

(b) Find the average number of years that a newly hired assistant professor spends working for this college (in any academic rank).

(c) Find the average number of years that a newly hired assistant professor spends working for this college as an associate professor.

(d) Find the average number of years that a professor spends working for this college (as a professor).

(e) Find the probability that a newly hired assistant professor will eventually retire as a professor.

(f) Find the probability that an associate professor will eventually retire as a professor.

(g) Assuming that a faculty member at this college begin her career as assistant professor, find the probability that she will be an associate professor after 2 years.

3.9   In Problem 1.13, let $d_1 = 0.6$, $d_2 = 0.5$, $d_3 = 0.4$, $r_1 = 0.3$, $p_1 = 0.1$, $r_2 = 0.3$, $p_2 = 0.2$, $p_R = 0.1$, $p_A = 0.2$, $r_3 = 0.3$, $r_4 = 0.7$, and $r_5 = 0.8$.

(a) Find the average number of years that a newly hired instructor will spend working for this college (in any academic rank).

(b) Find the average number of years that a newly hired instructor will spend working for this college as an associate professor.

(c) Find the probability that a newly hired instructor will eventually be promoted to the rank of professor.

(d) Find the probability that an associate professor will eventually be promoted to the rank of professor, initially as a research professor.

(e) Assuming that a faculty member at this college begins her career as assistant professor, find the probability that she will be an associate professor after 2 years.

(f) In the long run, what proportion of associate professors will be applications professors?

(g) In the long run, what proportion of faculty members will be applications professors?

3.10 An investor buys a stock for $10. The monthly share price, rounded to the nearest $5, has been varying among $0, $5, $10, $15, and $20. She has been advised that the price will never exceed $20. She has been informed that the price of the stock can be modeled as a recurrent Markov chain in which the state, X_n, denotes the share price at the end of month n. The state space is $E = \{\$0, \$5, \$10, \$15, \$20\}$. Her financial advisor believes that the Markov chain will have the following transition probability matrix:

$$
P = \begin{array}{c|ccccc}
X_n \backslash X_{n+1} & 0 & 5 & 10 & 15 & 20 \\
\hline
0 & 0.34 & 0.12 & 0.26 & 0.10 & 0.18 \\
5 & 0.12 & 0.28 & 0.32 & 0.20 & 0.08 \\
10 & 0.22 & 0.24 & 0.30 & 0.14 & 0.10 \\
15 & 0.10 & 0.20 & 0.40 & 0.24 & 0.06 \\
20 & 0.08 & 0.18 & 0.38 & 0.22 & 0.14 \\
\end{array}
$$

The investor intends to sell the stock at the end of the first month in which the share price rises to $20 or falls to $0. Under this policy, the share price can be modeled as an absorbing multichain with two absorbing states, $0 and $20. An absorbing state is reached when the stock is sold. The remaining states, which are entered when the stock is held during the present month, are transient.

(a) Represent the transition probability matrix in canonical form.

(b) Find the probability that she will eventually sell the stock for $20.

(c) Find the probability that she will sell the stock after 3 months.

(d) Find the expected number of months until she sells the stock.

References

1. Bhat, N., *Elements of Applied Stochastic Processes*, 2nd ed., Wiley, New York, 1985.
2. Cinlar, E., *Introduction to Stochastic Processes*, Prentice-Hall, Englewood Cliffs, NJ, 1975.
3. Clarke, A. B. and Disney R. L., *Probability and Random Processes: A First Course with Applications*, 2nd ed., Wiley, New York, 1985.
4. Kemeny, J. G., Mirkil, H., Snell, J. L., and Thompson, G. L., *Finite Mathematical Structures*, Prentice-Hall, Englewood Cliffs, NJ, 1959.
5. Kemeny, J. G. and Snell, J. L., *Finite Markov Chains*, Van Nostrand, Princeton, NJ, 1960. Reprinted by Springer-Verlag, New York, 1976.

4

A Markov Chain with Rewards (MCR)

When income or costs are associated with the states of a Markov chain, the system is called a Markov chain with rewards, or MCR. This chapter, which treats an MCR, has two objectives. The first is to show how to calculate the economic value of an MCR. The second is to use an MCR to link a Markov chain to a Markov decision process (MDP), thereby unifying the treatment of both subjects. In Chapter 5, an MDP, is constructed by associating decision alternatives with a set of MCRs. Thus, an MDP can be viewed simply as a set of Markov chains with rewards plus decisions.

4.1 Rewards

This chapter develops procedures for constructing a reward vector and calculating the expected economic value of an MCR. As Section 1.2 indicates, all Markov chains with rewards treated in this book are assumed to have a finite number of states [2–4, 6, 7].

4.1.1 Planning Horizon

As Section 2.1 indicates, the set of consecutive, equally spaced time periods over which a Markov chain is analyzed is called a planning horizon. As Section 1.2 indicates, an epoch is a point in time that marks the end of a time period. That is, epoch n designates the end of period n, which is also the beginning of period $n + 1$. A planning horizon can be finite or infinite. If a planning horizon is finite, of length T periods, then it consists of T periods numbered consecutively 1, 2, … , T. The periods are separated by $T + 1$ epochs numbered consecutively 0, 1, 2, … , T. A planning horizon of length T periods is shown in Figure 4.1.

FIGURE 4.1
Planning horizon of length T periods.

Note that epoch 1 marks the end of period 1, which coincides with the beginning of period 2. The present time, denoted by epoch 0, marks the beginning of period 1. A future time is denoted by epoch n, where $n > 0$. Epoch T marks the end of period T, which is also the end of the planning horizon. Often, the following terms will be used interchangeably: epoch n, period n, time n, step n, and transition n.

4.1.2 Reward Vector

An MCR is a Markov chain that generates a sequence of rewards as it evolves over time from state to state, in accordance with the transition probabilities, p_{ij}, of the chain. Rewards are assumed to be generated at the epochs at which the Markov chain moves from state to state. Rewards are expressed as discrete, end-of-period cash flows, measured in dollars, which is the convention used in engineering economic analysis. (Dollar signs are often omitted.) A positive reward represents revenue, while a negative reward signifies cost. Each time the chain visits a state i at epoch n, a reward, denoted by q_i, is earned. Suppose that at epoch n a chain is observed to be in state i, so that $X_n = i$. Then at epoch n, the chain will receive a reward q_i. The chain will move with transition probability p_{ij} to state j at epoch $n + 1$, so that $X_{n+1} = j$. Both the rewards, q_i, and the transition probabilities, p_{ij}, are assumed to be stationary over time, that is, they do not depend on the epoch, n. For an MCR starting in state g, a sequence of states visited, the rewards earned in those states, and the state transition probabilities is shown in Figure 4.2.

In some cases the reward may depend on the state visited at the next epoch. In such cases, when the chain makes a transition from state i at epoch n to state j at epoch $n + 1$ with transition probability p_{ij}, it immediately earns an associated transition reward, denoted by r_{ij}. The transition rewards are assumed to be stationary over time. The reward received in state i, denoted by q_i, may now be called an expected immediate reward. For an N-state Markov chain, an expected immediate reward q_i received in state i is equal to the reward r_{ij} earned from a transition to state j weighted by the probability p_{ij} of the transition. That is,

$$q_i = \sum_{j=1}^{N} p_{ij} r_{ij}, \quad i = 1, 2, \ldots, N. \tag{4.1}$$

$X_0 = g$	→	$X_1 = h$	→	$X_2 = k$	$\cdots$	$X_n = i$	→	$X_{n+1} = j$	State
q_g		q_h		q_k	$\cdots$	q_i		q_j	Reward
$p_g^{(0)}$	p_{gh}		p_{hk}		$\cdots$		p_{ij}		Transition probability
0		1		2	$\cdots$	n		$n+1$	Epoch

FIGURE 4.2
Sequence of states, transitions, and rewards for an MCR.

In the most general case, the reward received in the present state is the sum of a constant term and an expected immediate reward earned from a transition to the next state. In this book, q_i will simply be termed a reward received in state i, regardless of how it is earned.

Figure 1.3 in Section 1.2 demonstrates that a small Markov chain can be represented by a transition probability graph. Similarly, a small MCR can be represented by a transition probability and reward graph. For example, consider a two-state generic MCR with the following transition probability matrix P and reward vector q:

$$P = \begin{array}{c|cc} \text{State} & 1 & 2 \\ \hline 1 & p_{11} & p_{12} \\ 2 & p_{21} & p_{22} \end{array}, \quad q = \begin{array}{c|c} \text{State} & \text{Reward} \\ \hline 1 & q_1 = p_{11}r_{11} + p_{12}r_{12} \\ 2 & q_2 = p_{21}r_{21} + p_{22}r_{22} \end{array}. \quad (4.2)$$

The corresponding transition probability and reward graph is shown in Figure 4.3.

The set of rewards for all states is collected in an N-component reward vector,

$$\mathbf{q} = [q_1 \quad q_2 \quad \cdots \quad q_N]^\mathsf{T}. \quad (4.3)$$

After each transition, a vector of expected rewards, which is a function of both P and q, is received. The vector of expected rewards received at epoch 0, after 0 steps, is equal to $\mathbf{q}$. The vector of expected rewards received at epoch 1, after one step, is dependent on the transitions made after one step. Those transitions are governed by the one-step transition probability matrix, $\mathbf{P}$. Therefore, the vector of expected rewards received at epoch 1 is $\mathbf{Pq}$. Similarly, the vector of expected rewards received at epoch 2, after two steps, is P^2q, where P^2 is the two-step transition probability matrix. Continuing to higher numbered epochs, the vector of expected rewards received at epoch n, after n steps, is given by Equation (4.4).

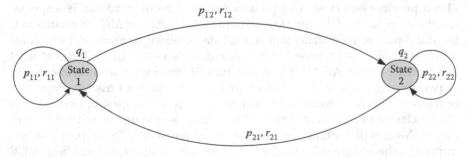

FIGURE 4.3
Transition probability and reward graph for a two-state Markov chain with rewards.

$$\text{Vector of expected rewards received after } n \text{ steps} = \mathbf{P}^n \mathbf{q}, \qquad (4.4)$$

where $\mathbf{P}^n$ is the n-step transition probability matrix.

If $\mathbf{p}^{(0)}$ is an initial probability state vector, then, as Equation (1.27) has indicated, $\mathbf{p}^{(n)} = p^{(0)}\mathbf{P}^n$ is the vector of state probabilities after n steps. The expected reward received after n steps is a scalar, given by

$$\text{Expected reward received after } n \text{ steps} = \mathbf{p}^{(0)}\mathbf{P}^n \mathbf{q} = \mathbf{p}^{(n)}\mathbf{q}. \qquad (4.5)$$

Vectors of expected rewards, $\mathbf{P}^n\mathbf{q}$, and expected reward scalars, $\mathbf{p}^{(n)}\mathbf{q}$, will be calculated for an example of an MCR model in Section 4.2.2.1.

4.2 Undiscounted Rewards

In Section 4.2, no interest is charged for the use of money. When no interest is charged, one dollar in the present has the same value as one dollar in the future. In other words, in this section, future cash flows over a planning horizon are not discounted. When cash flows are not discounted, Markov chain structure is relevant in calculating the expected economic value of an MCR. (When cash flows are discounted, as they are in Section 4.3, Markov chain structure is not relevant.)

4.2.1 MCR Chain Structure

The calculation of the expected value of an undiscounted cash flow generated by an MCR is dependent on chain structure [7]. The chain structure of an MCR is determined by the chain structure of the associated Markov chain, which is discussed in Sections 1.9 and 3.1. An MCR is unichain if the state space of the Markov chain consists of a single closed class of recurrent states plus a possibly empty set of transient states. A recurrent MCR is a special case of a unichain MCR with no transient states. Thus, an MCR is recurrent if the state space is irreducible, that is, if all states belong to a single closed communicating class of recurrent states. A reducible unichain MCR has at least one transient state. An MCR is multichain if the transition matrix consists of two or more closed classes of recurrent states plus a possibly empty set of transient states. A recurrent multichain MCR will consist of two or more closed classes of recurrent states with no transient states. A reducible multichain MCR will contain one or more transient states. To simplify the terminology, the adjective "reducible" will often be omitted in describing MCR chain structure. Thus, a reducible unichain MCR will simply be termed a unichain MCR, and a reducible multichain MCR will be called a multichain

MCR. To further simplify the terminology, a closed communicating class of recurrent states will often be termed a recurrent chain.

4.2.2 A Recurrent MCR over a Finite Planning Horizon

In Section 4.2.2, a recurrent MCR is analyzed over a finite planning horizon, while the planning horizon is assumed to be infinite in Section 4.2.3 [4, 7].

4.2.2.1 An MCR Model of Monthly Sales

Consider the following example of a recurrent MCR. (In Section 5.1.2.1 of Chapter 5 this model will be transformed into an MDP by introducing a set of MCRs plus decisions.) Suppose that a firm records its monthly sales at the end of every month. Monthly sales, which have fluctuated widely over the years, are ranked with respect to those of the firm's competitors. The rankings are expressed in quartiles. The fourth quartile is the highest rank, and indicates that a firm's monthly sales are greater than those of 75% of its competitors. At the beginning of every month, the firm follows a particular policy under which a unique action and reward are associated with each of the four quartile ranks. For example, when monthly sales are in the first quartile, the firm will always offer employee buyouts and earn a reward of –$20,000 (or incur a cost of $20,000).

By following the prescribed policy for many years, the firm has found that sales in any month can be used to forecast sales in the following month. Hence, the sequence of monthly sales is believed to form a Markov chain. The state X_{n-1} denotes the quartile rank of monthly sales at the beginning of month n. The chain has four states, which correspond to the four quartiles. Estimates of the transition probabilities are based on the firm's historical record of monthly sales. Table 4.1 identifies the four states accompanied by the associated actions and rewards prescribed by the firm's policy.

The firm constructs a model of a four-state MCR, which has the following transition probability matrix, P, and reward vector, q, with the entries of q expressed in thousands of dollars.

TABLE 4.1

States, Actions, and Rewards for Monthly Sales

Monthly Sales Quartile	State	Action	Reward
First (lowest)	1	Offer employee buyouts	–$20,000
Second	2	Reduce executive salaries	$5,000
Third	3	Invest in new technology	–$5,000
Fourth (highest)	4	Make strategic acquisitions	$25,000

$$P = \begin{matrix} 1 \\ 2 \\ 3 \\ 4 \end{matrix} \begin{bmatrix} 0.60 & 0.30 & 0.10 & 0 \\ 0.25 & 0.30 & 0.35 & 0.10 \\ 0.05 & 0.25 & 0.50 & 0.20 \\ 0 & 0.10 & 0.30 & 0.60 \end{bmatrix} \quad q = \begin{matrix} 1 \\ 2 \\ 3 \\ 4 \end{matrix} \begin{bmatrix} -20 \\ 5 \\ -5 \\ 25 \end{bmatrix}. \tag{4.6}$$

Suppose that the initial probability state vector is

$$\begin{aligned} \mathbf{p}^{(0)} &= [0.3 \quad 0.1 \quad 0.4 \quad 0.2] = \begin{bmatrix} p_1^{(0)} & p_2^{(0)} & p_3^{(0)} & p_4^{(0)} \end{bmatrix} \\ &= [P(X_0 = 1) \quad P(X_0 = 2) \quad P(X_0 = 3) \quad P(X_0 = 4)]. \end{aligned} \tag{4.7}$$

Using Equation (4.5), the expected reward received at the start of month one is

$$\mathbf{p}^{(0)}q = [0.3 \quad 0.1 \quad 0.4 \quad 0.2] \begin{bmatrix} -20 \\ 5 \\ -5 \\ 25 \end{bmatrix} = -2.5. \tag{4.8}$$

Using Equation (1.27), after 1 month the state probability vector is

$$\begin{aligned} \mathbf{p}^{(1)} = \mathbf{p}^{(0)}P &= [0.3 \quad 0.1 \quad 0.4 \quad 0.2] \begin{bmatrix} 0.60 & 0.30 & 0.10 & 0 \\ 0.25 & 0.30 & 0.35 & 0.10 \\ 0.05 & 0.25 & 0.50 & 0.20 \\ 0 & 0.10 & 0.30 & 0.60 \end{bmatrix} \\ &= [0.225 \quad 0.240 \quad 0.325 \quad 0.210]. \end{aligned} \tag{4.9}$$

Using Equation (4.5), the expected reward received after 1 month is equal to

$$\mathbf{p}^{(0)}Pq = \mathbf{p}^{(1)}q = [0.225 \quad 0.240 \quad 0.325 \quad 0.210] \begin{bmatrix} -20 \\ 5 \\ -5 \\ 25 \end{bmatrix} = 0.325. \tag{4.10}$$

Using Equation (4.4), the vector of expected rewards after 1 month, which is independent of the initial probability state vector, is equal to

$$\mathbf{Pq} = \begin{matrix} 1 \\ 2 \\ 3 \\ 4 \end{matrix} \begin{bmatrix} 0.60 & 0.30 & 0.10 & 0 \\ 0.25 & 0.30 & 0.35 & 0.10 \\ 0.05 & 0.25 & 0.50 & 0.20 \\ 0 & 0.10 & 0.30 & 0.60 \end{bmatrix} \begin{bmatrix} -20 \\ 5 \\ -5 \\ 25 \end{bmatrix} = \begin{matrix} 1 \\ 2 \\ 3 \\ 4 \end{matrix} \begin{bmatrix} -11 \\ -2.75 \\ 2.75 \\ 14 \end{bmatrix}. \tag{4.11}$$

Using Equation (1.28), after 2 months the state probability vector is

$$\mathbf{p}^{(2)} = \mathbf{p}^{(1)}\mathbf{P} = [0.225 \quad 0.240 \quad 0.325 \quad 0.210] \begin{bmatrix} 0.60 & 0.30 & 0.10 & 0 \\ 0.25 & 0.30 & 0.35 & 0.10 \\ 0.05 & 0.25 & 0.50 & 0.20 \\ 0 & 0.10 & 0.30 & 0.60 \end{bmatrix} \tag{4.12}$$

$$= [0.21125 \quad 0.24175 \quad 0.332 \quad 0.215].$$

Alternatively, using Equation (1.27), the state probability vector after 2 months is

$$\mathbf{p}^{(2)} = \mathbf{p}^{(0)}\mathbf{P}^2 = [0.3 \quad 0.1 \quad 0.4 \quad 0.2] \begin{bmatrix} 0.440 & 0.295 & 0.215 & 0.05 \\ 0.2425 & 0.2625 & 0.335 & 0.16 \\ 0.1175 & 0.235 & 0.4025 & 0.245 \\ 0.04 & 0.165 & 0.365 & 0.43 \end{bmatrix} \tag{4.13}$$

$$= [0.21125 \quad 0.24175 \quad 0.332 \quad 0.215],$$

where $\mathbf{P}^2$ is the two-step transition probability matrix.

Using Equation (4.5), the expected reward received after 2 months is equal to

$$\mathbf{p}^{(1)}\mathbf{Pq} = \mathbf{p}^{(2)}\mathbf{q} = [0.225 \quad 0.240 \quad 0.325 \quad 0.210] \begin{bmatrix} -20 \\ 5 \\ -5 \\ 25 \end{bmatrix} = 0.69875. \tag{4.14}$$

Using Equation (4.4), the vector of expected rewards received after 2 months, which is independent of the initial probability state vector, is equal to

$$\mathbf{P}^2\mathbf{q} = \begin{matrix} 1 \\ 2 \\ 3 \\ 4 \end{matrix} \begin{bmatrix} 0.440 & 0.295 & 0.215 & 0.05 \\ 0.2425 & 0.2625 & 0.335 & 0.16 \\ 0.1175 & 0.235 & 0.4025 & 0.245 \\ 0.04 & 0.165 & 0.365 & 0.43 \end{bmatrix} \begin{bmatrix} -20 \\ 5 \\ -5 \\ 25 \end{bmatrix} = \begin{matrix} 1 \\ 2 \\ 3 \\ 4 \end{matrix} \begin{bmatrix} -7.15 \\ -1.2125 \\ 2.9375 \\ 8.95 \end{bmatrix}. \tag{4.15}$$

4.2.2.2 Value Iteration over a Fixed Planning Horizon

Assume that a decision maker chooses a finite planning horizon of fixed length, equal to T time periods. Suppose that the expected total reward earned by an MCR from epoch n until epoch T, the end of the planning horizon, is of interest. Let $v_i(n)$ denote the expected total reward earned during the next $T - n$ periods, from epoch n to epoch T, if the system is in state i at epoch n. Since epoch 0 marks the beginning of period 1, the present time, the objective is to compute $v_i(0)$, which represents the expected total reward earned during the next $T - 0 = T$ periods, until the end of the planning horizon, if the system starts in state i. Since the planning horizon ends at epoch T, terminal values must be specified for the rewards earned in all states at epoch T. That is, terminal values must be specified for $v_j(T)$, for $j = 1, 2, \ldots, N$. The terminal values represent trade-in or salvage values that will be received at the end of the planning horizon. The set of expected total rewards for all states at epoch n is collected in an N-component, expected total reward vector [4, 7],

$$v(n) = [v_1(n) \quad v_2(n) \quad \ldots \quad v_N(n)]^{\mathsf{T}}. \tag{4.16}$$

4.2.2.2.1 Value Iteration Equation

Figure 4.4 shows a cash flow diagram involving a series of equal reward vectors, q, earned at the beginning of each period for T periods. At the end of period T, a vector of salvage values denoted by $v(T)$ is earned. Observe that vector q is earned at epochs 0 through $T - 1$, and that vector $v(T)$ is earned at epoch T.

The reward vector received at the present time, epoch 0, is q. As Equation (4.5) and Section 4.2.2.1 have demonstrated, the expected reward vector received one period from now, at epoch 1, is Pq, where P is the one-step transition probability matrix. The expected reward vector received two periods from now, at epoch 2, is P^2q, where P^2 is the two-step transition probability matrix. Similarly, the expected reward vector received $T - 1$ periods from now, at epoch $T-1$, is $P^{T-1}q$, where P^{T-1} is the $(T - 1)$-step transition probability matrix. Finally, at the end of the planning horizon, the expected salvage value is $P^Tv(T)$, where P^T is the T-step transition probability matrix, and $v(T)$ is the vector of salvage values received at the end of period T. Rewards are additive. Therefore, the expected total reward vector,

q	q	q	...	q	$v(T)$	Cash Flow
↑	↑	↑	...	↑	↑	
0	1	2	...	$T-1$	T	Epoch

FIGURE 4.4
Cash flow diagram for reward vectors over planning horizon of length T periods.

$v(0)$, is equal to the sum of the expected reward vectors, $P^k q$, earned at each epoch k, for $k = 0, 1, 2, \ldots, T - 1$, plus the expected salvage value, $P^T v(T)$, received at epoch T. That is,

$$
\left.
\begin{aligned}
v(0) &= P^0 q + P^1 q + P^2 q + P^3 q + \cdots + P^{T-1} q + P^T v(T) \\
&= I q + P q + P^2 q + P^3 q + \cdots + P^{T-1} q + P^T v(T) \\
&= q + P q + P^2 q + P^3 q + \cdots + P^{T-1} q + P^T v(T).
\end{aligned}
\right\}
\tag{4.17}
$$

A backward recursive solution procedure called value iteration will be developed to relate $v(n)$ to $v(n + 1)$. To simplify this development, consider a planning horizon of length $T = 4$ periods. The cash flow diagram is shown in Figure 4.5.

The backward recursive solution procedure begins by solving for $v(3)$ over a one-period horizon consisting of period 4 alone. Note that a salvage value, $v(4)$, is received one period after the end of period 3. The one-period cash flow diagram is shown in Figure 4.6.

Using Equation (4.17) with $T = 1$ and the origin moved to $n = 3$, the solution for $v(3)$ is

$$
v(3) = q + P v(4).
\tag{4.18}
$$

Next, $v(2)$ is calculated over a two-period horizon consisting of periods 3 and 4. The two-period cash flow diagram is shown in Figure 4.7.

Using Equation (4.17) with $T = 2$ and the origin moved to $n = 2$, the solution for $v(2)$ is given below:

$$
v(2) = q + P q + P^2 v(4).
\tag{4.19}
$$

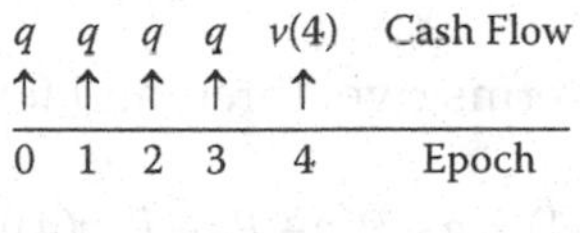

FIGURE 4.5
Cash flow diagram for reward vectors over four-period planning horizon.

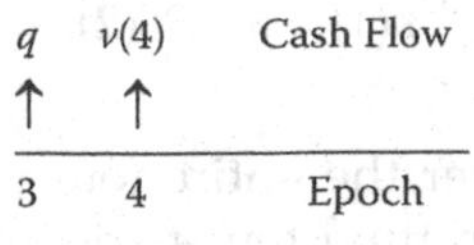

FIGURE 4.6
One-period cash flow diagram.

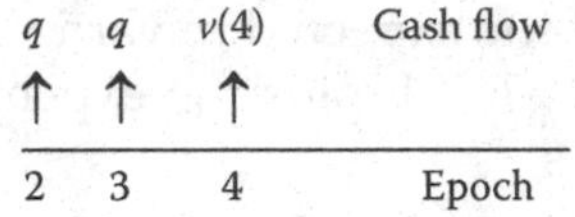

FIGURE 4.7
Two-period cash flow diagram.

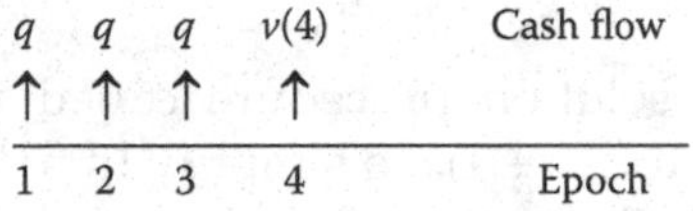

FIGURE 4.8
Three-period cash flow diagram.

When P is factored out of terms two and three

$$v(2) = q + P[q + Pv(4)]. \tag{4.20}$$

Substituting $v(3)$ from Equation (4.18),

$$v(2) = q + Pv(3). \tag{4.21}$$

Next, $v(1)$ is calculated over a three-period horizon consisting of periods 2, 3, and 4. The three-period cash flow diagram is shown in Figure 4.8.

Using Equation (4.17) with $T = 3$ and the origin moved to $n = 1$, the solution for $v(1)$ is given below:

$$v(1) = q + Pq + P^2q + P^3v(4). \tag{4.22}$$

When P is factored out of terms two, three, and four,

$$v(1) = q + P[q + Pq + P^2v(4)]. \tag{4.23}$$

Substituting $v(2)$ from Equation (4.19),

$$v(1) = q + Pv(2). \tag{4.24}$$

Finally, $v(0)$ is calculated over the entire four-period horizon consisting of periods 1, 2, 3, and 4. The original four-period cash flow diagram is shown in Figure 4.5.

Using Equation (4.17) with $T = 4$, the solution for $v(0)$ is given below:

$$v(0) = q + Pq + P^2q + P^3q + P^4v(4). \tag{4.25}$$

When P is factored out of terms two, three, four, and five,

$$v(0) = q + P[q + Pq + P^2q + P^3v(4)]. \tag{4.26}$$

Substituting $v(1)$ from Equation (4.22),

$$v(0) = q + Pv(1). \tag{4.27}$$

Observe that at the end of periods 0, 1, 2, and 3 in the four-period planning horizon, the following backward recursive equation in matrix form has been established:

$$v(n) = q + Pv(n+1), \quad \text{for } n = 0, 1, 2, \text{ and } 3, \tag{4.28}$$

where the salvage value $v(4)$ is specified.

By induction, the recursive equation (4.28) can be extended to any finite planning horizon. Thus, for a planning horizon of length T periods, $v(n)$ can be computed in terms of $v(n + 1)$ to produce Equation (4.29), which is called the value iteration equation in matrix form:

$$v(n) = q + Pv(n+1), \quad \text{for } n = 0, 1, 2, ..., T-1, \tag{4.29}$$

where the salvage value $v(T)$ is specified.

Consider a generic four-state MCR with transition probability matrix, P, and reward vector, q. When $N = 4$, the expanded matrix form of the recursive equation (4.28) is

$$\begin{bmatrix} v_1(n) \\ v_2(n) \\ v_3(n) \\ v_4(n) \end{bmatrix} = \begin{bmatrix} q_1 \\ q_2 \\ q_3 \\ q_4 \end{bmatrix} + \begin{bmatrix} p_{11} & p_{12} & p_{13} & p_{14} \\ p_{21} & p_{22} & p_{23} & p_{24} \\ p_{31} & p_{32} & p_{33} & p_{34} \\ p_{41} & p_{42} & p_{43} & p_{44} \end{bmatrix} \begin{bmatrix} v_1(n+1) \\ v_2(n+1) \\ v_3(n+1) \\ v_4(n+1) \end{bmatrix}. \tag{4.30}$$

In expanded algebraic form the four recursive equations (4.30) are

$$
\left.
\begin{aligned}
v_1(n) &= q_1 + p_{11}v_1(n+1) + p_{12}v_2(n+1) + p_{13}v_3(n+1) + p_{14}v_4(n+1) \\
v_2(n) &= q_2 + p_{21}v_1(n+1) + p_{22}v_2(n+1) + p_{23}v_3(n+1) + p_{24}v_4(n+1) \\
v_3(n) &= q_3 + p_{31}v_1(n+1) + p_{32}v_2(n+1) + p_{33}v_3(n+1) + p_{34}v_4(n+1) \\
v_4(n) &= q_4 + p_{41}v_1(n+1) + p_{42}v_2(n+1) + p_{43}v_3(n+1) + p_{44}v_4(n+1).
\end{aligned}
\right\} \quad (4.31)
$$

In compact algebraic form, using summation signs, the four recursive equations (4.31) are

$$
\left.
\begin{aligned}
v_1(n) &= q_1 + \sum_{j=1}^{4} p_{1j}v_j(n+1) \\
v_2(n) &= q_2 + \sum_{j=1}^{4} p_{2j}v_j(n+1) \\
v_3(n) &= q_3 + \sum_{j=1}^{4} p_{3j}v_j(n+1) \\
v_4(n) &= q_4 + \sum_{j=1}^{4} p_{4j}v_j(n+1).
\end{aligned}
\right\} \quad (4.32)
$$

The four algebraic equations (4.32) can be replaced by the single algebraic equation

$$
v_i(n) = q_i + \sum_{j=1}^{4} p_{ij}v_j(n+1), \tag{4.33}
$$

for $n = 0, 1, \ldots, T-1,$ and $i = 1, 2, 3,$ and 4.

This result can be generalized to apply to any N-state MCR. Therefore, $v_i(n)$ can be computed in terms of $v_j(n + 1)$ by using the following algebraic recursive equation, which is called the value iteration equation in algebraic form:

$$
v_i(n) = q_i + \sum_{j=1}^{N} p_{ij}v_j(n+1), \tag{4.34}
$$

for $n = 0, 1, \ldots, T-1,$ and $i = 1, 2, \ldots, N.$

Recall that $v_i(n)$ denotes the expected total reward earned until the end of the planning horizon if the system is in state i at epoch n. The value iteration equation (4.34) indicates that the total expected reward, $v_i(n)$, can be expressed as the sum of two terms. The first term is the reward, q_i, earned at epoch n. The second term is the expected total reward, $v_j(n + 1)$, that will be earned if the chain starts in state j at epoch $n + 1$, weighted by the probability, p_{ij}, that state j can be reached in one step from state i.

In summary, the value iteration equation expresses a backward recursive relationship because it starts with a known set of salvage values for vector $v(T)$ at epoch T, the end of the planning horizon. Next, vector $v(T - 1)$ is calculated in terms of $v(T)$. The value iteration procedure moves backward one epoch at a time by calculating $v(n)$ in terms of $v(n + 1)$. The vectors, $v(T - 2)$, $v(T - 3)$, ... , $v(1)$, and $v(0)$ are computed in succession. The backward recursive procedure, or backward recursion, stops at epoch 0 after $v(0)$ is calculated in terms of $v(1)$. The component $v_i(0)$ of $v(0)$ is the expected total reward earned until the end of the planning horizon if the system starts in state i at epoch 0. Figure 4.9 is a tree diagram of the value iteration equation in algebraic form for a two-state MCR.

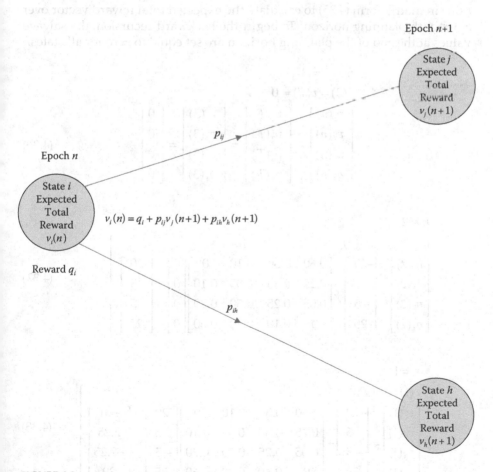

FIGURE 4.9
Tree diagram of value iteration for a two-state MCR.

4.2.2.2.2 *Value Iteration for MCR Model of Monthly Sales*

Consider the four-state MCR model of monthly sales introduced in Section 4.2.2.1 for which the transition probability matrix, P, and reward vector, q, with the entries of q expressed in thousands of dollars, are shown in Equation (4.6). Substituting the numerical values for P and q from Equation (4.6) into Equation (4.30), the matrix equation for the backward recursion is

$$\begin{bmatrix} v_1(n) \\ v_2(n) \\ v_3(n) \\ v_4(n) \end{bmatrix} = \begin{bmatrix} -20 \\ 5 \\ -5 \\ 25 \end{bmatrix} + \begin{bmatrix} 0.60 & 0.30 & 0.10 & 0 \\ 0.25 & 0.30 & 0.35 & 0.10 \\ 0.05 & 0.25 & 0.50 & 0.20 \\ 0 & 0.10 & 0.30 & 0.60 \end{bmatrix} \begin{bmatrix} v_1(n+1) \\ v_2(n+1) \\ v_3(n+1) \\ v_4(n+1) \end{bmatrix}. \tag{4.35}$$

Value iteration will be executed by solving the backward recursive equations in matrix form (4.35) to calculate the expected total reward vector over a 3-month planning horizon. To begin the backward recursion, the salvage values at the end of the planning horizon are set equal to zero for all states.

$$n = T = 3.$$
$$v(3) = v(T) = 0$$

$$\begin{bmatrix} v_1(n) \\ v_2(n) \\ v_3(n) \\ v_4(n) \end{bmatrix} = \begin{bmatrix} v_1(T) \\ v_2(T) \\ v_3(T) \\ v_4(T) \end{bmatrix} = \begin{bmatrix} v_1(3) \\ v_2(3) \\ v_3(3) \\ v_4(3) \end{bmatrix} = \begin{bmatrix} 0 \\ 0 \\ 0 \\ 0 \end{bmatrix} \tag{4.36}$$

$$n = 2$$
$$v(2) = q + Pv(3)$$
$$\begin{bmatrix} v_1(2) \\ v_2(2) \\ v_3(2) \\ v_4(2) \end{bmatrix} = \begin{bmatrix} -20 \\ 5 \\ -5 \\ 25 \end{bmatrix} + \begin{bmatrix} 0.60 & 0.30 & 0.10 & 0 \\ 0.25 & 0.30 & 0.35 & 0.10 \\ 0.05 & 0.25 & 0.50 & 0.20 \\ 0 & 0.10 & 0.30 & 0.60 \end{bmatrix} \begin{bmatrix} 0 \\ 0 \\ 0 \\ 0 \end{bmatrix} = \begin{bmatrix} -20 \\ 5 \\ -5 \\ 25 \end{bmatrix} \tag{4.37}$$

$$n = 1$$
$$v(1) = q + Pv(2)$$
$$\begin{bmatrix} v_1(1) \\ v_2(1) \\ v_3(1) \\ v_4(1) \end{bmatrix} = \begin{bmatrix} -20 \\ 5 \\ -5 \\ 25 \end{bmatrix} + \begin{bmatrix} 0.60 & 0.30 & 0.10 & 0 \\ 0.25 & 0.30 & 0.35 & 0.10 \\ 0.05 & 0.25 & 0.50 & 0.20 \\ 0 & 0.10 & 0.30 & 0.60 \end{bmatrix} \begin{bmatrix} -20 \\ 5 \\ -5 \\ 25 \end{bmatrix} = \begin{bmatrix} -31 \\ 2.25 \\ -2.25 \\ 39 \end{bmatrix}. \tag{4.38}$$

$n = 0$

$v(0) = q + Pv(1)$

$$\begin{bmatrix} v_1(0) \\ v_2(0) \\ v_3(0) \\ v_4(0) \end{bmatrix} = \begin{bmatrix} -20 \\ 5 \\ -5 \\ 25 \end{bmatrix} + \begin{bmatrix} 0.60 & 0.30 & 0.10 & 0 \\ 0.25 & 0.30 & 0.35 & 0.10 \\ 0.05 & 0.25 & 0.50 & 0.20 \\ 0 & 0.10 & 0.30 & 0.60 \end{bmatrix} \begin{bmatrix} -31 \\ 2.25 \\ -2.25 \\ 39 \end{bmatrix} = \begin{bmatrix} -38.15 \\ 1.0375 \\ 0.6875 \\ 47.95 \end{bmatrix}. \qquad (4.39)$$

The vector $v(0)$ indicates that if this MCR operates over a 3-month planning horizon, the expected total reward will be -38.15 if the system starts in state 1, 1.0375 if it starts in state 2, 0.6875 if it starts in state 3, and 47.95 if it starts in state 4. The calculations for the 3-month planning horizon are summarized in Table 4.2.

The solution by value iteration for $v(0)$ over a 3-month planning horizon has verified that, for $v(3) = 0$,

$$v(0) = q + Pq + P^2q + P^3v(3) \qquad (4.17)$$

$$v(0) = \begin{bmatrix} v_1(0) \\ v_2(0) \\ v_3(0) \\ v_4(0) \end{bmatrix} = \begin{bmatrix} -20 \\ 5 \\ -5 \\ 25 \end{bmatrix} + \begin{bmatrix} -11 \\ -2.75 \\ 2.75 \\ 14 \end{bmatrix} + \begin{bmatrix} -7.15 \\ -1.2125 \\ 2.9375 \\ 8.95 \end{bmatrix} + \begin{bmatrix} 0 \\ 0 \\ 0 \\ 0 \end{bmatrix} = \begin{bmatrix} -38.15 \\ 1.0375 \\ 0.6875 \\ 47.95 \end{bmatrix}, \qquad (4.40)$$

equal to the sum of the results obtained in Equations (4.11) and (4.15).

4.2.2.3 Lengthening a Finite Planning Horizon

The length of a finite planning horizon may be increased by adding additional periods during value iteration. Suppose the original planning horizon

TABLE 4.2

Expected Total Rewards for Monthly Sales Calculated by Value Iteration Over a 3-Month Planning Horizon

End of Month	n			
	0	1	2	3
$v_1(n)$	−38.15	−31	−20	0
$v_2(n)$	1.0375	2.25	5	0
$v_3(n)$	0.6875	−2.25	−5	0
$v_4(n)$	47.95	39	25	0

contains T periods. To continue the backward recursion of value iteration using the results already obtained for a horizon of length T periods, each additional period added is placed at the beginning of the horizon, and designated by a consecutively numbered negative epoch. For example, if one period is added, epoch (-1) is placed one period ahead of epoch 0, so that epoch (-1) becomes the present time or origin. If two periods are added, epoch (-2) is placed one period ahead of epoch (-1), or two periods ahead of epoch 0, so that epoch (-2) becomes the present time or origin. If Δ periods are added, epoch $(-\Delta)$ is placed one period ahead of epoch $(-\Delta + 1)$, or Δ periods ahead of epoch 0, so that epoch $(-\Delta)$ becomes the present time or origin. The lengthened planning horizon contains $T - (-\Delta) = T + \Delta$ periods. Value iteration is continued over the new horizon by starting at epoch 0 using a terminal salvage value equal to the vector $v(0)$ obtained for the original horizon of length T periods. The sequence of backward recursive value iteration equations is

$$v(-1) = q + Pv(0)$$
$$v(-2) = q + Pv(-1)$$
$$\vdots$$
$$v(-\Delta) = q + Pv(-\Delta + 1).$$

Note that if period 0 denotes the first period placed ahead of period 1 to lengthen the horizon, and the additional periods added are labeled with consecutively increasing negative integers, then epoch n denotes the end of period n, for $n = 0, -1, -2, \ldots, (-\Delta + 1)$, as well as for $n = 1, 2, \ldots, T$. For example, suppose that $T = 2$ and $\Delta = 3$. Then the lengthened planning horizon appears in Figure 4.10.

Lastly, all the epochs may be renumbered, so that the epochs of the lengthened planning horizon are numbered consecutively with nonnegative integers from epoch 0 at the origin to epoch $(T + \Delta)$ at the end, by adding Δ to the numerical index of each epoch, starting at epoch $(-\Delta)$ and ending at epoch T. If the epochs are renumbered, then

epoch $(-\Delta)$ becomes epoch$(-\Delta + \Delta)$ = epoch 0, the present time,

epoch $(-\Delta + 1)$ becomes epoch $[(-\Delta + 1) + \Delta]$ = epoch 1,

epoch 0 becomes epoch $(0 + \Delta)$ = epoch Δ,

epoch $(T - 1)$ becomes epoch $(T - 1 + \Delta)$, and

epoch T becomes epoch $(T + \Delta)$, the end of the lengthened planning horizon.

	Period − 2	Period − 1	Period 0	Period 1	Period 2	
−3		−2	−1	0	1	2 epoch

FIGURE 4.10
Two-period horizon lengthened by three periods.

For the example in which $T = 2$ and $\Delta = 3$, the lengthened planning horizon, with the epochs relabeled, appears in Figure 4.11.

For the example of monthly sales, suppose that four periods are added to the original three-period planning horizon to form a new 7-month planning horizon. Thus, $T = 3$ and $\Delta = 4$. As shown in Table 4.3, the four negative epochs, –1, –2, –3, and –4, are added sequentially at the beginning of the original 3-month horizon.

Value iteration is executed to compute the expected total rewards earned during the first 4 months of the 7-month horizon.

$$v(-1) = q + Pv(0)$$

$$\begin{bmatrix} v_1(-1) \\ v_2(-1) \\ v_3(-1) \\ v_4(-1) \end{bmatrix} = \begin{bmatrix} -20 \\ 5 \\ -5 \\ 25 \end{bmatrix} + \begin{bmatrix} 0.60 & 0.30 & 0.10 & 0 \\ 0.25 & 0.30 & 0.35 & 0.10 \\ 0.05 & 0.25 & 0.50 & 0.20 \\ 0 & 0.10 & 0.30 & 0.60 \end{bmatrix} \begin{bmatrix} -38.15 \\ 1.0375 \\ 0.6875 \\ 47.95 \end{bmatrix} = \begin{bmatrix} -42.51 \\ 0.8094 \\ 3.2856 \\ 54.08 \end{bmatrix} \quad (4.41)$$

$$\left. \begin{array}{l} v(-2) = q + Pv(-1) \\ v(-3) = q + Pv(-2). \\ v(-4) = q + Pv(-3) \end{array} \right\} \quad (4.42)$$

The calculations for the entire 7-month planning horizon are summarized in Table 4.4.

Finally, $\Delta = 4$ may be added to each end-of-month index, n, to renumber the epochs of the seven-period planning horizon consecutively from 0 to 7.

	Period 1	Period 2	Period 3	Period 4	Period 5	
0	1	2	3	4	5	epoch

FIGURE 4.11
Lengthened planning horizon with epochs renumbered.

TABLE 4.3

Three-Month Horizon for Monthly Sales Lengthened by 4 Months

					n			
End of Month	–4	–3	–2	–1	0	1	2	3
$v_1(n)$					–38.15	–31	–20	0
$v_2(n)$					1.0375	2.25	5	0
$v_3(n)$					0.6875	–2.25	–5	0
$v_4(n)$					47.95	39	25	0

TABLE 4.4

Expected Total Rewards for Monthly Sales Calculated by Value Iteration Over a 7-month Planning Horizon

	n							
End of Month	**−4**	**−3**	**−2**	**−1**	**0**	**1**	**2**	**3**
$v_1(n)$	−46.3092	−46.0552	−44.9346	−42.51	−38.15	−31	−20	0
$v_2(n)$	2.8782	1.9073	1.1733	0.8094	1.0375	2.25	5	0
$v_3(n)$	9.3101	7.5174	5.5357	3.2856	0.6875	−2.25	−5	0
$v_4(n)$	64.578	61.8868	58.5146	54.08	47.95	39	25	0

4.2.2.4 Numbering the Time Periods Forward

When value iteration is executed, many authors follow the convention of dynamic programming by numbering the time periods backward, starting with $n = 0$ at the end of the planning horizon, and ending with $n = T$ at the beginning. When time periods in the horizon are numbered backward, n denotes the number of periods remaining. However, numbering the time periods backward contradicts the convention used for Markov chains for which the time periods are numbered forward, starting with $n = 0$ at the beginning of the horizon. In this book, the time periods are always numbered forward for Markov chains, MCRs, and MDPs. When value iteration is executed over an infinite planning horizon, epochs are numbered as consecutive negative integers, starting with epoch 0 at the end of the horizon, and ending with epoch $(-T)$ at the beginning.

4.2.3 A Recurrent MCR over an Infinite Planning Horizon

Consider an N-state recurrent MCR with stationary transition probabilities and stationary rewards. As the planning horizon grows longer, the expected total return grows correspondingly larger. If the planning horizon becomes unbounded or infinite, then the expected total reward also becomes infinite. Two different measures of the earnings of an MCR over an infinite horizon are the expected average reward per period, called the gain, and the expected present value, which is also called the expected total discounted reward. Discounted cash flows are treated in Section 4.3.

Three approaches can be taken to calculate the gain, or average reward per period, of a recurrent MCR operating over an infinite planning horizon. The first approach is to compute the gain as the product of the steady-state probability vector and the reward vector. The second approach is to solve a system of linear equations called the value determination equations (VDEs) to find the gain and the relative values of the expected total rewards earned in every state. Since these two approaches both involve the solution of N simultaneous linear equations, they both require about the same computational effort. The third approach, which requires less computation, is to execute value iteration over a large number of periods. However, the sequence of expected total rewards produced by value iteration may not converge. When value iteration does converge, it gives only an approximate solution for the gain and the expected total rewards earned in every state.

4.2.3.1 Expected Average Reward or Gain

Consider an N-state MCR, irrespective of structure, in the steady state. Suppose that a vector called the gain vector is defined, equal to the limiting transition probability matrix multiplied by the reward vector. If the gain vector is denoted by $\mathbf{g} = [g_i]$, then

$$\mathbf{g} = \lim_{n \to \infty} \mathbf{P}^{(n)} \mathbf{q}. \tag{4.43}$$

Each component g_i of the gain vector represents the expected average reward per transition, or the gain, if the MCR starts in state i. Suppose now that the MCR is recurrent. Using Equation (2.7),

$$\mathbf{g} = \lim_{n \to \infty} \mathbf{P}^{(n)} \mathbf{q} = \Pi q = \begin{bmatrix} \pi_1 & \pi_2 & \cdots & \pi_N \\ \pi_1 & \pi_2 & \cdots & \pi_N \\ \vdots & \vdots & \vdots & \vdots \\ \pi_1 & \pi_2 & \cdots & \pi_N \end{bmatrix} \begin{bmatrix} q_1 \\ q_2 \\ \vdots \\ q_N \end{bmatrix} = \begin{bmatrix} \sum_{i=1}^{N} \pi_i q_i \\ \sum_{i=1}^{N} \pi_i q_i \\ \vdots \\ \sum_{i=1}^{N} \pi_i q_i \end{bmatrix} = \begin{bmatrix} \pi q \\ \pi q \\ \vdots \\ \pi q \end{bmatrix} = \begin{bmatrix} g_1 \\ g_2 \\ \vdots \\ g_N \end{bmatrix} = \begin{bmatrix} g \\ g \\ \vdots \\ g \end{bmatrix}.$$

$$\tag{4.44}$$

Hence, the gain for starting in every state of a recurrent MCR is the same. The gain in every state of a recurrent MCR is a scalar constant, denoted by g, where

$$g_1 = g_2 = \cdots = g_N = g. \tag{4.45}$$

The gain g in every state of a recurrent MCR is equal to the sum, over all states, of the reward q_i received in state i weighted by the steady-state probability, π_i. The algebraic form of the equation to calculate the gain g for a recurrent MCR is

$$g = \sum_{i=1}^{N} \pi_i q_i. \tag{4.46}$$

The matrix form of the equation to calculate the gain g for a recurrent MCR is

$$g = \pi q. \tag{4.47}$$

4.2.3.1.1 Gain of a Two-State Recurrent MCR

Consider a generic two-state recurrent MCR for which the transitional probability matrix and reward vector are given in Equation (4.2). The transition probability and reward graph is shown in Figure 4.3. The steady-state probability vector for the Markov chain is calculated in Equation (2.16). Using Equation (4.47), the gain of a two-state recurrent MCR is

$$\left.\begin{array}{l} g = \pi q = \begin{bmatrix} \pi_1 & \pi_2 \end{bmatrix} \begin{bmatrix} q_1 \\ q_2 \end{bmatrix} = \begin{bmatrix} \dfrac{p_{21}}{p_{12} + p_{21}} & \dfrac{p_{12}}{p_{12} + p_{21}} \end{bmatrix} \begin{bmatrix} q_1 \\ q_2 \end{bmatrix} \\[12pt] = \dfrac{p_{21}q_1 + p_{12}q_2}{p_{12} + p_{21}}. \end{array}\right\} \tag{4.48}$$

Alternatively, using Equation (4.44), the gain vector is

$$\begin{aligned} \mathbf{g} &= \begin{bmatrix} g_1 \\ g_2 \end{bmatrix} = \lim_{n \to \infty} \mathbf{P}^{(n)} \mathbf{q} = \Pi q = \begin{bmatrix} \pi_1 & \pi_2 \\ \pi_1 & \pi_2 \end{bmatrix} \begin{bmatrix} q_1 \\ q_2 \end{bmatrix} = \begin{bmatrix} \pi_1 q_1 + \pi_2 q_2 \\ \pi_1 q_1 + \pi_2 q_2 \end{bmatrix} \\[10pt] &= \begin{bmatrix} g \\ g \end{bmatrix} = \begin{bmatrix} \dfrac{p_{21}q_1 + p_{12}q_2}{p_{12} + p_{21}} \\[10pt] \dfrac{p_{21}q_1 + p_{12}q_2}{p_{12} + p_{21}} \end{bmatrix}. \end{aligned} \tag{4.49}$$

As the components of the gain vector have demonstrated, the gain in every state is equal to the scalar

$$g = g_1 = g_2 = \frac{p_{21}q_1 + p_{12}q_2}{p_{12} + p_{21}}. \tag{4.50}$$

For example, consider the following two-state recurrent MCR for which:

$$P = \begin{matrix} 1 \\ 2 \end{matrix}\begin{bmatrix} p_{11} & p_{12} \\ p_{21} & p_{22} \end{bmatrix} = \begin{matrix} 1 \\ 2 \end{matrix}\begin{bmatrix} 0.2 & 0.8 \\ 0.6 & 0.4 \end{bmatrix}, \quad q = \begin{matrix} 1 \\ 2 \end{matrix}\begin{bmatrix} q_1 \\ q_2 \end{bmatrix} = \begin{matrix} 1 \\ 2 \end{matrix}\begin{bmatrix} -300 \\ 200 \end{bmatrix} \tag{4.51}$$

Equation (2.18) calculates the vector of steady-state probabilities for the Markov chain.

The gain of the two-state recurrent MCR is

$$g = \pi \, q = \begin{bmatrix} \pi_1 & \pi_2 \end{bmatrix} \begin{bmatrix} q_1 \\ q_2 \end{bmatrix} = \begin{bmatrix} \dfrac{3}{7} & \dfrac{4}{7} \end{bmatrix} \begin{bmatrix} -300 \\ 200 \end{bmatrix} = \frac{p_{21} q_1 + p_{12} q_2}{p_{12} + p_{21}}$$

$$= \frac{(0.6)(-300) + (0.8)(200)}{0.8 + 0.6} = -\frac{100}{7} = -14.29. \tag{4.52}$$

4.2.3.1.2 *Gain of an MCR Model of Monthly Sales*

Consider the four-state MCR model of monthly sales for which the transition probability matrix, P, and reward vector, q, are shown in Equation (4.6). The steady-state vector, π, is calculated by solving equations (2.13). Using Equation (4.47), the gain is

$$g = \pi \, q = \begin{bmatrix} 0.1908 & 0.2368 & 0.3421 & 0.2303 \end{bmatrix} \begin{bmatrix} -20 \\ 5 \\ -5 \\ 25 \end{bmatrix} = 1.4143. \tag{4.53}$$

In the steady state, the process earns an average reward of \$1,414.30 per month.

4.2.3.2 *Value Determination Equations (VDEs)*

The second approach that can be taken to calculate the expected average reward of a recurrent MCR operating over an infinite planning horizon is to solve a system of equations called the VDEs. The solution of these equations will produce the gain and the relative values of the expected total rewards earned in every state.

4.2.3.2.1 *Optional Insight: Limiting Behavior of the Expected Total Reward*

The following informal derivation of the limiting behavior of the expected total reward is adapted from Bhat [1]. As Equation (4.17) indicates, at epoch 0, the beginning of a planning horizon of length T periods, the expected total reward vector is

$$v(0) = \sum_{k=0}^{T-1} P^k q + P^T v(T). \tag{4.17}$$

As T becomes large, the system enters the steady state. Then

$$
\begin{aligned}
v(0) &= \lim_{T \to \infty} \left[\sum_{k=0}^{T-1} P^k q + P^T v(T) \right] \\
&= \lim_{T \to \infty} \sum_{k=0}^{T-1} P^k q + \lim_{T \to \infty} P^T v(T).
\end{aligned} \tag{4.54}
$$

Interchanging the limit and the sum, $v(0)$ is expressed as the sum of the limiting transition probability matrices for P.

$$
v(0) = \sum_{k=0}^{T-1} \lim_{k \to \infty} P^k q + \lim_{T \to \infty} P^T v(T). \tag{4.55}
$$

Using Equation (2.7),

$$
\begin{aligned}
v(0) &= \sum_{k=0}^{T-1} \Pi q + \Pi v(T) \\
&= T \Pi q + \Pi v(T).
\end{aligned} \tag{4.56}
$$

The algebraic form of Equation (4.56) is

$$
v_i(0) = T \sum_{j=1}^{N} \pi_j q_j + \sum_{j=1}^{N} \pi_j v_j(T), \quad \text{for } i = 1, 2, \ldots, N. \tag{4.57}
$$

Recall from Section 4.2.3.1 that the average reward per period, or gain, for a recurrent MCR is

$$
g = \sum_{j=1}^{N} \pi_j q_j.
$$

Let

$$
v_i = \sum_{j=1}^{N} \pi_j v_j(T). \tag{4.58}
$$

Substituting quantities from Equations (4.46) and (4.58), into Equation (4.57), as T becomes large, $v_i(0)$ has the form

$$
v_i(0) \approx Tg + v_i. \tag{4.59}
$$

4.2.3.2.2 Informal Derivation of VDEs

Suppose the system starts in state i at epoch 0, and the planning horizon contains T periods. As T becomes large, the system enters the steady state. Then, as the optional insight in Section 4.2.3.2.1 has demonstrated, $v_i(0)$ becomes approximately equal to the sum of two terms. The first term, Tg, is independent of the starting state, i. This term can be interpreted as the number of periods, T, remaining in the planning horizon multiplied by the average reward per period or gain, g. The second term, v_i, can be interpreted as the expected total reward earned if the system starts in state i. Therefore [3, 4],

$$v_i(0) \approx Tg + v_i. \tag{4.59}$$

At epoch 1, $(T-1)$ periods remain. As T grows large, $(T-1)$ also grows large, so that

$$v_j(1) \approx (T-1)g + v_j. \tag{4.60}$$

Using the value iteration equation (4.34) with $n = 0$,

$$v_i(0) = q_i + \sum_{j=1}^{N} p_{ij} v_j(1), \quad \text{for } i = 1, 2, ..., N. \tag{4.61}$$

Substituting the linear approximations for $v_i(0)$ in Equation (4.59) and $v_j(1)$ in Equation (4.60), assumed to be valid for large T, into Equation (4.61) gives the result

$$Tg + v_i = q_i + \sum_{j=1}^{N} p_{ij}[(T-1)g + v_j], \quad \text{for } i = 1, 2, ..., N,$$

$$= q_i + Tg\sum_{j=1}^{N} p_{ij} - g\sum_{j=1}^{N} p_{ij} + \sum_{j=1}^{N} p_{ij} v_j.$$

Substituting $\sum_{j=1}^{N} p_{ij} = 1$,

$$Tg + v_i = q_i + Tg - g + \sum_{j=1}^{N} p_{ij} v_j$$

$$g + v_i = q_i + \sum_{j=1}^{N} p_{ij} v_j, \quad i = 1, 2, ..., N. \tag{4.62}$$

This system (4.62) of N linear equations in the $N + 1$ unknowns, $v_1, v_2, \ldots, v_N$, and g, is called the VDEs. Since the VDEs have one more unknown than equations, they will not have a unique solution. Therefore, the value of one variable may be chosen arbitrarily. By convention, the variable for the highest numbered state, v_N, is set equal to zero, and the remaining v_i are expressed in terms of v_N. Hence, the v_i are termed the relative values of the expected total rewards earned in every state or, more concisely, the relative values. If $v_N = 0$, then the relative value v_i represents the expected total reward earned by starting in state i relative to the expected total reward earned by starting in state N.

If the objective is simply to find the gain of the process, then it is not necessary to solve the VDEs (4.62). In that case, the first approach described in Section 4.2.3.1 is sufficient. However, by solving the VDEs, one can determine not only the gain but also the relative value in every state. The relative values are used in Chapter 5 to evaluate alternative policies, which are candidates to maximize the gain of an MDP. For an N-state recurrent MCR, both procedures for calculating the gain require the solution of N simultaneous linear equations in N unknowns.

4.2.3.2.3 VDEs for a Two-State Recurrent MCR

To verify that the gain of a recurrent MCR is obtained by solving the VDEs, consider once again the generic two-state MCR for which the gain was calculated in Equations (4.48) and (4.49). The VDEs are

$$\left. \begin{aligned} v_1 + g &= q_1 + p_{11}v_1 + p_{12}v_2 \\ v_2 + g &= q_2 + p_{21}v_1 + p_{22}v_2. \end{aligned} \right\} \tag{4.63}$$

Setting $v_2 = 0$, the VDEs become

$$\left. \begin{aligned} v_1 + g &= q_1 + p_{11}v_1 \\ g &= q_2 + p_{21}v_1 \end{aligned} \right\}. \tag{4.64}$$

The solution is

$$\left. \begin{aligned} v_1 &= \frac{q_1 - q_2}{1 - p_{11} + p_{21}} = \frac{q_1 - q_2}{p_{12} + p_{21}} \\ g &= q_2 + \frac{p_{21}(q_1 - q_2)}{1 - p_{11} + p_{21}} = \frac{q_2(p_{12} + p_{21}) + p_{21}(q_1 - q_2)}{p_{12} + p_{21}} = \frac{p_{21}q_1 + p_{12}q_2}{p_{12} + p_{21}}, \end{aligned} \right\} \tag{4.65}$$

which agrees with value of the gain obtained in Equations (4.48) and (4.49).

4.2.3.2.4 VDEs for an MCR Model of Monthly Sales

The VDEs for an N-state MCR can be expressed in matrix form. Let g be a column vector with all N entries equal to g. Let

$$v = [v_1 \quad v_2 \quad \cdots \quad v_N]^T, \tag{4.66}$$

and

$$q = [q_1 \quad q_2 \quad \cdots \quad q_N]^T. \tag{4..67}$$

Then

$$g + v = q + Pv. \tag{4.68}$$

When $N = 4$ the matrix form of the VDEs is

$$\begin{bmatrix} g \\ g \\ g \\ g \end{bmatrix} + \begin{bmatrix} v_1 \\ v_2 \\ v_3 \\ v_4 \end{bmatrix} = \begin{bmatrix} q_1 \\ q_2 \\ q_3 \\ q_4 \end{bmatrix} + \begin{bmatrix} p_{11} & p_{12} & p_{13} & p_{14} \\ p_{21} & p_{22} & p_{23} & p_{24} \\ p_{31} & p_{32} & p_{33} & p_{34} \\ p_{41} & p_{42} & p_{43} & p_{44} \end{bmatrix} \begin{bmatrix} v_1 \\ v_2 \\ v_3 \\ v_4 \end{bmatrix}. \tag{4.69}$$

Consider the four-state MCR model of monthly sales for which the transition probability matrix, P, and reward vector, q, are shown in Equation (4.6). The matrix equation for the VDE's, after setting $v_4 = 0$, is

$$\begin{bmatrix} g \\ g \\ g \\ g \end{bmatrix} + \begin{bmatrix} v_1 \\ v_2 \\ v_3 \\ 0 \end{bmatrix} = \begin{bmatrix} -20 \\ 5 \\ -5 \\ 25 \end{bmatrix} + \begin{bmatrix} 0.60 & 0.30 & 0.10 & 0 \\ 0.25 & 0.30 & 0.35 & 0.10 \\ 0.05 & 0.25 & 0.50 & 0.20 \\ 0 & 0.10 & 0.30 & 0.60 \end{bmatrix} \begin{bmatrix} v_1 \\ v_2 \\ v_3 \\ 0 \end{bmatrix}.$$

The solution for the gain and the relative values is

$$g = 1.4145, \quad v_1 = -116.5022, \quad v_2 = -64.9671, \quad v_3 = -56.9627, \text{ and } v_4 = 0. \tag{4.70}$$

Thus, if this MCR operates over an infinite planning horizon, the average return per period or the gain will be 1.4145. This agrees with the value of the gain calculated in Equation (4.53). (Discrepancies are due to roundoff error.) Suppose the system starts in state 4 for which $v_4 = 0$. Then it will earn 116.5022 more than it will earn if it starts in state 1. It will also earn 64.9671 more than it will earn if it starts in state 2, and it will earn 56.627 more than it will earn if it starts in state 3.

4.2.3.3 Value Iteration

The third approach that can be taken to calculate the gain of an MCR operating over an infinite planning horizon is to execute value iteration over a large number of periods. Value iteration requires less computational effort than the first two approaches. However, the sequence of expected total rewards produced by value iteration will not always converge. When value iteration does converge, it will yield upper and lower bounds on the gain as well as an approximate solution for the expected total rewards earned in every state [4, 7].

Value iteration is conducted over an infinite planning horizon by repeatedly adding one additional period to lengthen the horizon. Each period added is placed at the beginning of the horizon. To obtain convergence criteria, the epochs are numbered forward as consecutive negative integers, starting with epoch 0 at the end of the horizon, and ending with epoch $(-n)$ at the beginning.

4.2.3.3.1 Expected Relative Rewards

After a stopping condition has been satisfied, value iteration will produce an approximate solution for the expected total rewards earned in every state. An approximate solution for the relative values of the expected total rewards is obtained by subtracting the expected total reward earned in the highest numbered state from the expected total reward earned in every state. These differences are called the expected relative rewards. To see why the expected relative rewards represent approximate solutions for the relative values, suppose that n successive repetitions of value iteration have satisfied a stopping condition over a finite horizon consisting of n periods. The epochs in the planning horizon are numbered as consecutive negative integers from $-n$ to 0. That is, the epochs for a horizon of length n are numbered sequentially as $-n, -n + 1, -n + 2, \ldots , -2, -1, 0$. Epoch $(-n)$ denotes the beginning of the horizon, and epoch 0 denotes the end. The absolute value of each epoch represents the number of periods remaining in the horizon. For example, at epoch (-3), three periods remain until the end of the horizon at epoch 0. When a stopping condition is satisfied, let $T = |-n| = n$, so that T denotes the length of the planning horizon at which convergence is achieved. Thus, the epochs for the planning horizon of length T are numbered sequentially as $-T, -T + 1, -T + 2, \ldots , -2, -1, 0$.

Suppose that all the epochs are renumbered by adding T to the numerical index of each epoch, so that the epochs for the planning horizon of length T are numbered consecutively as $0, 1, 2, \ldots , T - 1, T$. If epoch 0 denotes the beginning of a finite horizon of length T periods, then for large T, using Equation (4.59),

$$v_i(0) = Tg + v_i, \tag{4.59}$$

and

$$v_N(0) = Tg + v_N. \tag{4.59b}$$

Hence, for large T, the difference

$$v_i(0) - v_N(0) \approx v_i - v_N. \tag{4.71}$$

The difference $v_i(0) - v_N(0)$ is termed the expected relative reward earned in state i because it is equal to the increase in the long run expected total reward earned if the system starts in state i rather than in state N. This difference is also the reason why v_i is called the relative value of the process when it starts in state i. If value iteration satisfies a stopping condition after a large number of iterations, then the expected relative rewards will converge to the relative values.

4.2.3.3.2 Bounds on the Gain

If epoch 0 denotes the beginning of a finite horizon of length T periods, then the quantity $v_i(0) - v_i(1)$ can be interpreted as the difference between the expected total rewards earned in state i over two planning horizons, which differ in length by one period. For long planning horizons of length T and $T - 1$, using Equations (4.59) and (4.60),

$$v_i(0) - v_i(1) = (Tg + v_i) - [(T-1)g + v_i] = (Tg + v_i) - (Tg - g + v_i) = g. \tag{4.72}$$

Hence, as T grows large, the difference in the expected total rewards earned in state i over two planning horizons, which differ in length by one period, approaches the gain, g. The convergence of $v_i(0) - v_i(1)$ to the gain can be used to obtain upper and lower bounds on the gain, and also to obtain an approximate solution for the gain. An upper bound on the gain over a planning horizon of length T is given by

$$g_U(T) = \max_{i=1,...,N} \{v_i(0) - v_i(1)\} \tag{4.73}$$

Similarly, a lower bound on the gain over a planning horizon of length T is given by

$$g_L(T) = \min_{i=1,...,N} \{v_i(0) - v_i(1)\} \tag{4.74}$$

Since upper and lower bounds on the gain have been obtained, the gain is approximately equal to the arithmetic average of its upper and lower bounds.

That is, an approximate solution for the gain is given by

$$g = \frac{g_U(T) + g_L(T)}{2}.$$

(4.75)

4.2.3.3.3 Stopping Rule for Value Iteration

The difference between the upper and lower bounds on the gain over planning horizons which differ in length by one period provides a stopping condition for value iteration. Value iteration can be stopped when the difference between the upper and lower bounds on the gain for the current planning horizon and a planning horizon one period shorter is less than a small positive number, ε, specified by the analyst. That is, value iteration can be stopped after a planning horizon of length T for which

$$g_U(T) - g_L(T) < \varepsilon,$$

(4.76)

or

$$\max_{i=1,\dots,N} [v_i(0) - v_i(1)] - \min_{i=1,\dots,N} [v_i(0) - v_i(1)] < \varepsilon.$$

(4.77)

Of course, the magnitude of ε is a matter of judgment, and the number of periods in T needed to achieve convergence cannot be predicted.

4.2.3.3.4 Value Iteration Algorithm

The foregoing discussion is the basis for the following value iteration algorithm in its simplest form (adapted from Puterman [7]), which can be applied to a recurrent MCR over an infinite planning horizon. Epochs are numbered as consecutive negative integers from $-n$ at the beginning of the horizon to 0 at the end.

Step 1. Select arbitrary salvage values for

$v_i(0)$, for $i = 1, 2, \dots, N$. For simplicity, set $v_i(0) = 0$.

Specify an $\varepsilon > 0$. Set $n = -1$.

Step 2. For each state i, use the value iteration equation to compute

$$v_i(n) = q_i + \sum_{j=1}^{N} p_{ij} v_j(n+1), \quad \text{for } i = 1, 2, \dots, N.$$

Step 3. If $\displaystyle\max_{i=1,2,\dots,N} [v_i(n) - v_i(n+1)] - \min_{i=1,2,\dots,N} [v_i(n) - v_i(n+1)] < \varepsilon$, go to step 4.
Otherwise, decrement n by 1 and return to step 2.

Step 4. Stop.

4.2.3.3.5 Solution by Value Iteration of MCR Model of Monthly Sales

Table 4.4 is repeated as Table 4.5 to show the expected total rewards earned from monthly sales calculated by value iteration over the last 7 months of an infinite planning horizon. (Recall that all the rewards are expressed in units of thousands of dollars.) The last eight epochs of the infinite planning horizon in Table 4.5 are numbered sequentially as −7, −6, −5, −4, −3, −2, −1, and 0. Epoch 0 denotes the end of the horizon. As Section 4.2.3.3.1 indicates, the absolute value of a negative epoch in Table 4.5 represents the number of months remaining in the infinite horizon.

Table 4.6 gives the differences between the expected total rewards earned over planning horizons, which differ in length by one period.

TABLE 4.5

Expected Total Rewards for Monthly Sales Calculated by Value Iteration During the Last 7 Months of an Infinite Planning Horizon

Epoch	−7	−6	−5	−4	−3	−2	−1	0
$v_1(n)$	−46.3092	−46.0552	−44.9346	−42.51	−38.15	−31	−20	0
$v_2(n)$	2.8782	1.9073	1.1733	0.8094	1.0375	2.25	5	0
$v_3(n)$	9.3101	7.5174	5.5357	3.2856	0.6875	−2.25	−5	0
$v_4(n)$	64.578	61.8868	58.5146	54.08	47.95	39	25	0

TABLE 4.6

Differences Between the Expected Total Rewards Earned Over Planning Horizons Which Differ in Length by One Period

Epoch	−7	−6	−5	−4	−3	−2	−1
i	$v_i(-7)$	$v_i(-6)$	$v_i(-5)$	$v_i(-4)$	$v_i(-3)$	$v_i(-2)$	$v_i(-1)$
	$-v_i(-6)$	$-v_i(-5)$	$-v_i(-4)$	$-v_i(-3)$	$-v_i(-2)$	$-v_i(-1)$	$-v_i(0)$
1	−0.254L	−1.1206L	−2.4246L	−4.36L	−7.15L	−11L	−20L
2	0.9709	0.734	0.3639	−0.2263	−1.2143	−2.75	5
3	1.7927	1.9817	2.2501	2.5981	2.9375	2.75	−5
4	2.6912U	3.3722U	4.4346U	6.13U	8.95U	14U	25U
Max $(v_i(n) - v_i(n+1))$ $= g_U(T)$	2.6912	3.3722	4.4346	6.13	8.95	14	25
Min $(v_i(n) - v_i(n+1))$ $= g_L(T)$	−0.254	−1.1206	−2.4246	−4.36	−7.15	−11	−20
$g_U(T) - g_L(T)$	2.9452	4.4928	6.8592	10.49	16.10	25	45

In Table 4.6, a suffix U identifies $g_U(T)$, the maximum difference for each epoch. The suffix L identifies $g_L(T)$, the minimum difference for each epoch. The differences, $g_U(T) - g_L(T)$, obtained for all the epochs are listed in the bottom row of Table 4.6. In Equations (4.53) and (4.70), the gain was found to be 1.4145. When seven periods remain in the planning horizon, Table 4.6 shows that the bounds on the gain obtained by value iteration are given by $-0.254 \le g \le 2.6912$. The gain is approximately equal to the arithmetic average of its upper and lower bounds, so that

$$g \approx (2.6912 + 0.254)/2 = 1.4726. \tag{4.78}$$

As the planning horizon is lengthened beyond seven periods, tighter bounds on the gain will be obtained. The bottom row of Table 4.6 shows that if an analyst chooses an $\varepsilon < 2.9452$, then more than seven iterations of value iteration will be needed before the value iteration algorithm can be assumed to have converged.

Table 4.7 gives the expected relative rewards, $v_i(n) - v_4(n)$, earned during the last 7 months of the planning horizon.

In Equation (4.70), which assumes that the planning horizon is infinite, the relative values obtained by solving the VDEs are

$$g = 1.4145, \ v_1 = -116.5022, \ v_2 = -64.9671, \ v_3 = -56.9627, \ \text{and} \ v_4 = 0. \quad (4.70)$$

Tables 4.7 demonstrates that as the horizon grows longer, the expected relative reward earned for starting in each state slowly approaches the corresponding relative value for that state when the relative value for the highest numbered state is set equal to zero.

4.2.3.4 Examples of Recurrent MCR Models

In this section, recurrent MCR models are constructed for an inventory system [3] and for component replacement [6].

4.2.3.4.1 Recurrent MCR Model of an Inventory System

In Section 1.10.1.1.3, a retailer who sells personal computers has modeled her inventory system as a Markov chain. She follows a (2, 3) inventory

TABLE 4.7

Expected Relative Rewards, $v_i(n) - v_4(n)$, Earned During the Last 7 Months of the Planning Horizon

Epoch				n				
	-7	-6	-5	-4	-3	-2	-1	0
$v_1(n) - v_4(n)$	-110.8872	-108.774	-103.4492	-96.59	-86.1	-70	-45	0
$v_2(n) - v_4(n)$	-61.6998	-59.9795	-57.3413	-53.2706	-46.9143	-36.75	-20	0
$v_3(n) - v_4(n)$	-55.2679	-54.3694	-52.9789	-50.7944	-47.2625	-41.25	-30	0

ordering policy, which has produced the transition probability matrix shown in Equation (1.48). The retailer will construct an expected reward vector to transform the Markov chain model into an MCR. She wishes to calculate her gain, that is, her expected average reward or profit per period, which is equal to the sum, over all states, of the expected reward or profit in each state multiplied by the steady-state probability for that state. The retailer has collected the following cost and revenue data. If she places an order for one or more computers, she must pay a $20 cost for placing the order, plus a cost of $120 per computer ordered. All orders are delivered immediately. The retailer sells the computers for $300 each. The holding cost for each computer not sold during the period is $50. A shortage cost of $40 is incurred for each computer that is not available to satisfy demand during the period.

The expected reward or profit equals the expected revenue minus the expected cost. The expected revenue equals the $300 selling price of a computer times the expected number of computers sold.

The expected cost equals the ordering cost plus the expected holding cost plus the expected shortage cost. Recall that c_{n-1} denotes the number of computers ordered at the beginning of period n. If one or more computers are ordered, then $c_{n-1} > 0$, and the ordering cost for the period is $20 + $120 c_{n-1}$. If no computers are ordered, then $c_{n-1} = 0$, and the ordering cost is zero dollars. When the beginning inventory plus the quantity ordered exceed demand during a period, the number of unsold computers left over at the end of the period, that is, the surplus of computers at the end of the period, is $X_{n-1} + c_{n-1} - d_n > 0$. The holding cost for computers not sold during a period is $50(X_{n-1} + c_{n-1} - d_n)$. When demand exceeds the beginning inventory plus the quantity ordered during a period, the shortage of computers at the end of the period is $d_n - X_{n-1} - c_{n-1} > 0$. The shortage cost for computers that are not available to satisfy demand during a period is $40(d_n - X_{n-1} - c_{n-1})$.

When $X_{n-1} = 0$, the retailer orders $c_{n-1} = 3$ computers. The next state is $X_n = 3 - d_n$, so that the expected number of computers sold equals the expected demand for computers. The expected demand is given by

$$E(d_n) = \sum_{d_n=0}^{3} d_n p(d_n) = \sum_{k=0}^{3} (d_n = k)P(d_n = k) = (d_n = 0)P(d_n = 0) + (d_n = 1)P(d_n = 1)$$
$$+ (d_n = 2)P(d_n = 2) + (d_n = 3)P(d_n = 3)$$
$$= (0)(0.3) + (1)(0.4) + (2)(0.1) + 3(0.2) = 1.2.$$

(4.79)

The expected revenue is $300(1.2) = $360. The ordering cost is $20 + $120 c_{n-1} = $20 + $120(3) = $380. The next state, which represents, the ending inventory, is given by $X_n = X_{n-1} + c_{n-1} - d_n = 0 + 3 - d_n = 3 - d_n$.

If the demand is less than three computers, the retailer will have $(3 - d_n)$ unsold computers remaining in stock at the end of a period. The expected number of computers not sold during a period is given by

$$
\left.
\begin{aligned}
E(3 - d_n) &= \sum_{d_n=0}^{3} (3 - d_n) p(d_n) \\
&= \sum_{k=0}^{3} (3 - k) P(d_n = k) \\
&= (3 - 0) P(d_n = 0) + (3 - 1) P(d_n = 1) \\
&\quad + (3 - 2) P(d_n = 2) + (3 - 3) P(d_n = 3) \\
&= 3(0.3) + 2(0.4) + 1(0.1) + 0(0.2) = 1.8.
\end{aligned}
\right\}
\tag{4.80}
$$

The expected holding cost is equal to the expected number of computers not sold multiplied by the holding cost per computer. Thus, the expected holding cost is equal to

$$
\$50 \sum_{k=0}^{3} (3 - k) P(d_n = k) = \$50[3(0.3) + 2(0.4) + 1(0.1) + 0(0.2)] = \$50(1.8) = \$90.
\tag{4.81}
$$

Since the ending inventory is given by $X_n = 3 - d_n$, and the demand will never exceed three computers, the retailer will never have a shortage of computers. Hence, she will never incur a shortage cost in state 0.

The expected reward or profit vector for the retailer's MCR model of her inventory system is denoted by $\mathbf{q} = [q_0 \quad q_1 \quad q_2 \quad q_3]^{\mathsf{T}}$. The expected reward or profit in state 0, denoted by q_0, equals the expected revenue minus the ordering cost minus the expected holding cost minus the expected shortage cost. The expected reward in state 0 is

$$
q_0 = \$360 - \$380 - \$90 - \$0 = -\$110.
\tag{4.82}
$$

When $X_{n-1} = 1$, the retailer orders $c_{n-1} = 2$ computers. In state 1, as in state 0, the expected revenue is $\$300(1.2) = \$360$.
    The ordering cost is

$$
\$20 + \$120 c_{n-1} = \$20 + \$120(2) = \$260.
$$

The next state is again equal to

$$
X_n = X_{n-1} + c_{n-1} - d_n = 1 + 2 - d_n = 3 - d_n.
$$

The expected number of computers not sold during a period is unchanged from its value when $X_{n-1} = 0$, and is given by

$$\sum_{k=0}^{3}(3-k)P(d_n = k) = 1.8. \qquad (4.83)$$

The expected holding cost is also equal to its value when $X_{n-1} = 0$, and is given by

$$\$50\sum_{k=0}^{3}(3-k)P(d_n = k) = \$90. \qquad (4.84)$$

In state 1, as in state 0, the retailer will never have a shortage of computers, and thus will never incur a shortage cost.

The expected reward or profit in state 1, denoted by q_1, equals the expected revenue minus the ordering cost minus the expected holding cost minus the expected shortage cost. The expected reward in state 1 is

$$q_1 = \$360 - \$260 - \$90 - \$0 = \$10. \qquad (4.85)$$

When $X_{n-1} = 2$, the order quantity is $c_{n-1} = 0$, so that no computers are ordered, and no ordering cost is incurred. In state 2, $X_n = \max(2-d_n, 0)$, so that at most two computers can be sold. The number of computers sold during the period is equal to $\min(2, d_n)$. The expected number of computers sold is given by

$$\left.\begin{array}{l}E[\min(2,d_n)] = \displaystyle\sum_{d_n=0}^{3}\min(2,d_n)p(d_n) = \sum_{k=0}^{3}\min(2,d_n = k)P(d_n = k) \\[2mm] = \min(2,d_n = 0)P(d_n = 0) + \min(2,d_n = 1)P(d_n = 1) \\[2mm] + \min(2,d_n = 2)P(d_n = 2) + \min(2,d_n = 3)P(d_n = 3) \\[2mm] = (0)(0.3) + (1)(0.4) + (2)(0.1) + 2(0.2) = 1.\end{array}\right\} \qquad (4.86)$$

In state 2, the expected revenue is $\$300(1) = \$300$.

The next state is

$$X_n = X_{n-1} + c_{n-1} - d_n = 2 + 0 - d_n = 2 - d_n \qquad (4.87)$$

provided that $2 - d_n \geq 0$. When the beginning inventory is represented by $X_{n-1} = 2$, the ending inventory is represented by $X_n = \max(2 - d_n, 0)$. If the demand is less than two computers, the retailer will have $(2 - d_n)$ unsold computers remaining in stock at the end of a period. The expected number of computers not sold during a period is given by

$$
\begin{aligned}
E[\max(2-d_n,0)] &= \sum_{d_n=0}^{2} (2-d_n)p(d_n) \\
&= \sum_{k=0}^{2} (2-k)P(d_n=k) = (2-0)P(d_n=0) + (2-1)P(d_n=1) \\
&\quad + (2-2)P(d_n=2) \\
&= 2(0.3) + 1(0.4) + 0(0.1) = 1.
\end{aligned}
\tag{4.88}
$$

Thus, the expected holding cost is equal to

$$
\$50 \sum_{k=0}^{2} (2-k)P(d_n=k) = \$50[2(0.3)+1(0.4)] = \$50(1) = \$50 \tag{4.89}
$$

When the demand exceeds two computers, the retailer will have a shortage of (d_n-2) computers, and thus will incur a shortage cost. Since the demand will never exceed three computers, the retailer can have a maximum shortage of one computer. The expected number of computers not available to satisfy the demand during a period is equal to the expected number of shortages. The expected number of shortages is given by

$$
\begin{aligned}
E[(\max(d_n-2,0)] &= \sum_{d_n=2}^{3} (d_n-2)p(d_n) = \sum_{k=2}^{3} (k-2)P(d_n=k) \\
&= (2-2)P(d_n=2)+(3-2)P(d_n=3) \\
&= (0)P(d_n=2)+(1)P(d_n=3) = 0(0.1)+1(0.2) = 0.2.
\end{aligned}
\tag{4.90}
$$

The expected shortage cost is equal to the expected number of shortages times the shortage cost per computer short. Thus, the expected shortage cost is equal to

$$
\$40 \sum_{k=2}^{3} (k-2)P(d_n=k) = \$40[1(0.2)] = \$40(0.2) = \$8. \tag{4.91}
$$

The expected reward in state 2, denoted by q_2, equals the expected revenue minus the ordering cost minus the expected holding cost minus the expected shortage cost. The expected reward in state 2 is

$$q_2 = \$300 - \$0 - \$50 - \$8 = \$242. \tag{4.92}$$

When $X_{n-1} = 3$, the order quantity is $c_{n-1} = 0$, so that no computers are ordered, and no ordering cost is incurred. In state 3, as in states 0 and 1, the expected revenue is $\$300(1.2) = \$360$.

The next state is

$$X_n = X_{n-1} + c_{n-1} - d_n = 3 + 0 - d_n = 3 - d_n. \tag{4.93}$$

The expected number of computers not sold during a period is unchanged from its value when $X_{n-1} = 0$ and $X_{n-1} = 1$, and is given by

$$\sum_{k=0}^{3} (3-k)P(d_n = k) = 1.8. \tag{4.94}$$

The expected holding cost is equal to its value when $X_{n-1} = 0$ and $X_{n-1} = 1$, and is given by

$$\$50 \sum_{k=0}^{3} (3-k)P(d_n = k) = \$50(1.8) = \$90. \tag{4.95}$$

In state 3, as in states 0 and 1, the retailer will never have a shortage of computers, and thus will never incur a shortage cost.

The expected reward in state 3, denoted by q_3, equals the expected revenue minus the ordering cost minus the expected holding cost minus the expected shortage cost. The expected reward in state 3 is

$$q_3 = \$360 - \$0 - \$90 - \$0 = \$270. \tag{4.96}$$

The retailer's inventory system, controlled by a (2, 3) ordering policy, has been modeled as a four-state MCR. The MCR model has the following transition probability matrix, P, and expected reward vector, q:

Beginning Inventory, X_{n-1}	Order, c_{n-1}	$X_{n-1} + c_{n-1}$	State	0	1	2	3		Reward
0	3	3	0	0.2	0.1	0.4	0.3		−110
1	2	3	1	0.2	0.1	0.4	0.3		10
2	0	2	2	0.3	0.4	0.3	0		242
3	0	3	3	0.2	0.1	0.4	0.3		270

with $P =$ and $q =$

$$\tag{4.97}$$

Her expected average reward per period, or gain, is equal to the sum, over all states, of the expected reward in each state multiplied by the steady-state probability for that state. The steady-state probability vector for the inventory system was calculated in Equation (2.27). Using Equation (4.47), the gain earned by following a (2, 3) inventory ordering policy is

$$g = \pi q = \begin{bmatrix} 0.2364 & 0.2091 & 0.3636 & 0.1909 \end{bmatrix} \begin{bmatrix} -110 \\ 10 \\ 242 \\ 270 \end{bmatrix} = 115.6212. \tag{4.98}$$

4.2.3.4.2 Recurrent MCR Model of Component Replacement

Consider the four-state recurrent Markov chain model of component replacement introduced in Section 1.10.1.1.5. The transition probability matrix is shown in Equation (1.50). The following replacement policy is implemented. All components are inspected every week. All components that have failed during the week are replaced with new components. No component is replaced later than age 4 weeks. By adding cost information, component replacement will be modeled as an MCR. Suppose that the cost to inspect a component is $2, and the cost to replace a component, which has failed, is $10. Therefore, the expected immediate cost incurred for a component of age i is $2 for inspecting a component plus $10 for replacing a component, which has failed times the probability p_{i0} that an i–week old component has failed. Letting q_i denote the cost incurred for a component of age i gives the result that

$$q_i = \$2 + \$10 p_{i0}. \tag{4.99}$$

The vector of costs for this replacement policy is

$$q = \begin{bmatrix} q_0 \\ q_1 \\ q_2 \\ q_3 \end{bmatrix} = \begin{bmatrix} \$2 + \$10 p_{00} \\ \$2 + \$10 p_{10} \\ \$2 + \$10 p_{20} \\ \$2 + \$10 p_{30} \end{bmatrix} = \begin{bmatrix} \$2 + \$10(0.2) \\ \$2 + \$10(0.375) \\ \$2 + \$10(0.8) \\ \$2 + \$10(1) \end{bmatrix} = \begin{bmatrix} \$4 \\ \$5.75 \\ \$10 \\ \$12 \end{bmatrix}. \tag{4.100}$$

The recurrent MCR model of component replacement has the following transition probability matrix, P, and cost vector, q:

State	0	1	2	3			Cost
0	0.2	0.8	0	0		0	$4
$P =$ 1	0.375	0	0.625	0	, $q =$	1	5.75
2	0.8	0	0	0.2		2	10
3	1	0	0	0		3	12

(4.101)

The expected average cost per week, or negative gain, for this component replacement policy is equal to the sum, over all states, of the cost in each state multiplied by the steady-state probability for that state. The steady-state probability vector for the component replacement model is calculated in Equation (2.29). Using Equation (4.47), the negative gain incurred by following this replacement policy is:

$$g = \pi q = \begin{bmatrix} 0.4167 & 0.3333 & 0.2083 & 0.0417 \end{bmatrix} \begin{bmatrix} \$4 \\ \$5.75 \\ \$10 \\ \$12 \end{bmatrix} = \$6.17. \quad (4.102)$$

4.2.4 A Unichain MCR

Enlarging Equation (3.1), the transition matrix P in canonical form and the reward vector q for a unichain MCR are (4.103)

$$P = \begin{bmatrix} S & 0 \\ D & Q \end{bmatrix}, \quad q = \begin{bmatrix} q_R \\ q_T \end{bmatrix}, \quad (4.103)$$

where the components of the reward vector q are the vectors q_R and q_T. The entries of vector q_R are the rewards received for visits to the recurrent states. The entries of vector q_T are the rewards received for visits to the transient states.

4.2.4.1 Expected Average Reward or Gain

The limiting transition probability matrix for a unichain is given by

$$\lim_{n \to \infty} P^n = \begin{bmatrix} \lim_{n \to \infty} S^n & 0 \\ \lim_{n \to \infty} D_n & 0 \end{bmatrix} = \begin{bmatrix} \Pi & 0 \\ \Pi & 0 \end{bmatrix}. \quad (3.148)$$

Using Equation (4.43), the gain vector g for a unichain MCR is

$$g = \lim_{n \to \infty} P^{(n)} q = \begin{bmatrix} \lim_{n \to \infty} S^n & 0 \\ \lim_{n \to \infty} R_n & 0 \end{bmatrix} \begin{bmatrix} q_R \\ q_T \end{bmatrix} = \begin{bmatrix} \Pi & 0 \\ \Pi & 0 \end{bmatrix} \begin{bmatrix} q_R \\ q_T \end{bmatrix} = \begin{bmatrix} \Pi q_R \\ \Pi q_R \end{bmatrix} = \begin{bmatrix} g_R \\ g_R \end{bmatrix}, \quad (4.104)$$

where each row of the matrix Π is the steady-state probability vector, $\pi = [\pi_i]$, for the recurrent chain S, and g_R is the gain vector for the recurrent states. Observe that $g_R = \Pi q_R$. By the rules of matrix multiplication, all the components of the gain vector for a unichain MCR are equal. All states, both recurrent and transient, have the same gain, which is equal to the gain of the closed class S of recurrent states. If g denotes the gain in every state of a unichain MCR, then the scalar gain g in every state of a unichain MCR with N recurrent states is

$$g = \sum_{i=1}^{N} \pi_i q_i. \tag{4.105}$$

In vector form,

$$g = \pi q_R. \tag{4.106}$$

Thus, the gain of a unichain MCR is equal to the steady-state probability vector for the recurrent chain multiplied by the reward vector for the recurrent chain. Since all states in a recurrent MCR have the same gain, and the gain is the same for every state in a reducible unichain MCR, the same three approaches given in Section 4.2.3 for calculating the gain of a recurrent MCR can be used to calculate the gain of a unichain MCR. Therefore, the gain of a unichain MCR can also be calculated by solving the VDEs or by executing value iteration over an infinite horizon.

4.2.4.1.1 Gain of a Four-State Unichain MCR

Consider the following generic four-state unichain MCR in which states 1 and 2 form a recurrent closed class, and states 3 and 4 are transient. The transition matrix and reward vector are given in Equation (4.107).

$$P = [p_{ij}] = \begin{matrix} 1 \\ 2 \\ 3 \\ 4 \end{matrix} \begin{bmatrix} p_{11} & p_{12} & 0 & 0 \\ p_{21} & p_{22} & 0 & 0 \\ p_{31} & p_{32} & p_{33} & p_{34} \\ p_{41} & p_{42} & p_{43} & p_{44} \end{bmatrix} = \begin{bmatrix} S & 0 \\ D & Q \end{bmatrix}, \quad q = \begin{bmatrix} q_1 \\ q_2 \\ q_3 \\ q_4 \end{bmatrix} = \begin{bmatrix} q_R \\ q_T \end{bmatrix},$$

$$q_R = \begin{bmatrix} q_1 \\ q_2 \end{bmatrix}, \quad q_T = \begin{bmatrix} q_3 \\ q_4 \end{bmatrix}. \tag{4.107}$$

The limiting transition probability matrix is calculated in Equation (3.149). The gain vector is

$$g = \lim_{n \to \infty} P^{(n)} q = \begin{matrix} 1 \\ 2 \\ 3 \\ 4 \end{matrix} \begin{bmatrix} \pi_1 & \pi_2 & 0 & 0 \\ \pi_1 & \pi_2 & 0 & 0 \\ \pi_1 & \pi_2 & 0 & 0 \\ \pi_1 & \pi_2 & 0 & 0 \end{bmatrix} \begin{bmatrix} q_1 \\ q_2 \\ q_3 \\ q_4 \end{bmatrix} = \begin{bmatrix} g_1 \\ g_2 \\ g_3 \\ g_4 \end{bmatrix} = \begin{bmatrix} g \\ g \\ g \\ g \end{bmatrix} = \begin{bmatrix} \pi_1 q_1 + \pi_2 q_2 \\ \pi_1 q_1 + \pi_2 q_2 \\ \pi_1 q_1 + \pi_2 q_2 \\ \pi_1 q_1 + \pi_2 q_2 \end{bmatrix}.$$

(4.108)

As Section 4.2.4.1 has indicated, the gain for the unichain MCR is the same in every state. Treating the two-state recurrent chain separately, the gain, calculated by using (4.47), is

$$g = \pi q = \pi q_R = \begin{bmatrix} \pi_1 & \pi_2 \end{bmatrix} \begin{bmatrix} q_1 \\ q_2 \end{bmatrix} = \pi_1 q_1 + \pi_2 q_2.$$

(4.109)

Since all rows of the limiting transition probability matrix for the four-state unichain are identical, the gain in every state of the unichain MCR can also be computed by applying Equation (4.47) to the unichain MCR. That is,

$$g = \pi q = \begin{bmatrix} \pi_1 & \pi_2 & 0 & 0 \end{bmatrix} \begin{bmatrix} q_1 \\ q_2 \\ q_3 \\ q_4 \end{bmatrix} = \begin{bmatrix} \pi_1 & \pi_2 \end{bmatrix} \begin{bmatrix} q_1 \\ q_2 \end{bmatrix} = \pi_1 q_1 + \pi_2 q_2. \quad (4.110)$$

The generic two-state recurrent chain has the transition probability matrix

$$S = \begin{matrix} 1 \\ 2 \end{matrix} \begin{bmatrix} p_{11} & p_{12} \\ p_{21} & p_{22} \end{bmatrix}.$$

(4.111)

The steady-state probability vector for S is calculated in Equation (2.16). Therefore, the gain in every state of the generic four-state unichain MCR is

$$g = \pi_1 q_1 + \pi_2 q_2 = \frac{p_{21} q_1 + p_{12} q_2}{p_{12} + p_{21}}.$$

(4.112)

4.2.4.1.2 Gain of an Absorbing Unichain MCR

In the special case of an absorbing unichain MCR, the recurrent chain consists of a single absorbing state. The canonical form of the transition matrix for an absorbing unichain MCR is given below:

$$P = \begin{bmatrix} 1 & 0 \\ D & Q \end{bmatrix}, \quad q = \begin{bmatrix} q_1 \\ q_T \end{bmatrix},$$

(4.113)

where the scalar q_1 is the reward received when state 1, the absorbing state, is visited. The steady-state probability for the absorbing state is $\pi_1 = 1$. Hence the gain in every state of an absorbing unichain MCR is equal to the reward received in that state. That is,

$$g = \pi_1 q_1 = 1 q_1 = q_1. \tag{4.114}$$

4.2.4.1.3 A Unichain MCR Model of Machine Maintenance

Consider the absorbing Markov chain model of machine deterioration introduced in Section 1.10.1.2.1.2. The engineer responsible for the machine observes the condition of the machine at the start of a day. The engineer always does nothing to respond to the deterioration of the machine. This model will be enlarged by adding revenue data to create an absorbing unichain MCR [3, 6]. Since all state transitions caused by deterioration occur at the end of the day, a full day's production is achieved by a working machine, which is left alone. If the machine is in state 4, working properly, revenue of $1,000 is earned for the day. If the machine is in state 3, working, with a minor defect, daily revenue of $500 is earned. If the machine is in state 2, working, with a major defect, the daily revenue is $200. Finally, if the machine is in state 1, not working, no daily revenue is received. The daily revenue earned in every state is summarized in Table 4.8.

Now consider the unichain model of machine maintenance introduced in Section 1.10.1.2.2.1. This unichain model will also be enlarged by adding cost data to the revenue data to create a unichain MCR. Assume that the engineer follows the original maintenance policy described in Section 1.10.1.2.2.1, under which a machine in state 3 is overhauled. All maintenance actions, which may include overhauling, repairing, or replacing the machine, take 1 day to complete. No revenue is earned on days during which maintenance is performed. For simplicity, the daily maintenance costs are assumed to be independent of the state of the machine. Suppose that the daily cost to overhaul the machine is $300, and the daily cost to

TABLE 4.8

Daily Revenue Earned in Every State

State	Condition	Daily Revenue
1	Not Working (NW)	$0
2	Working, with a Major Defect (WM)	$200
3	Working, with a Minor Defect (Wm)	$500
4	Working Properly (WP)	$1,000

TABLE 4.9

Daily Maintenance Costs

Decision	Action	Daily Cost
1	Do Nothing (DN)	$0
2	Overhaul (OV)	$300
3	Repair (RP)	$700
4	Replace (RL)	$1,200

repair it is $700. The daily cost to replace the machine with a new machine is $1,200. The daily maintenance costs in every state are summarized in Table 4.9.

The reward vector associated with the original maintenance policy of Section 1.10.1.2.2.1, under which the decision is to overhaul the machine in states 1 and 3, and do nothing in states 2 and 4, appears in Equation (4.115).

State, $X_{n-1} = i$	Decision, k	Reward, q_i	
1, Not Working	2, Overhaul	$-\$300 = q_1$	
2, Working, Major Defect	1, Do Nothing	$\$200 = q_2$	(4.115)
3, Working, Minor Defect	2, Overhaul	$\$300 = q_3$	
4, Working Properly	1, Do Nothing	$\$1,000 = q_4$	

The addition of a reward vector to the transition probability matrix associated with the original maintenance policy of Section 1.10.1.2.2.1 has transformed the Markov chain model into an MCR. The unichain MCR model is shown in below:

State, $X_{n-1} = i$	Decision, k		State	1	2	3	4	
1, Not Working	2, Overhaul		1	0.2	0.8	0	0	
2, Working, Major Defect	1, Do Nothing	$P =$	2	0.6	0.4	0	0	,
3, Working, Minor Defect	2, Overhaul		3	0	0	0.2	0.8	
4, Working Properly	1, Do Nothing		4	0.3	0.2	0.1	0.4	

$$q = \begin{vmatrix} \text{Reward} \\ -\$300 = q_1 \\ \$200 = q_2 \\ -\$300 = q_3 \\ \$1,000 = q_4 \end{vmatrix} \qquad (4.116)$$

Under the modified maintenance policy of Section 1.10.1.2.2.1, if the machine is in state 3, working, with a minor defect, the engineer will do nothing. In that case the revenue in state 3 will be $500, the daily income from working with a minor defect. The unichain MCR model under the modified maintenance policy is shown in below:

State, $X_{n-1} = i$	Decision, k		State	1	2	3	4
1, Not Working	2, Overhaul		1	0.2	0.8	0	0
2, Working, Major Defect	1, Do Nothing	$P =$	2	0.6	0.4	0	0
3, Working, Minor Defect	1, Do Nothing		3	0.2	0.3	0.5	0
4, Working Properly	1, Do Nothing		4	0.3	0.2	0.1	0.4

$$q = \begin{vmatrix} \text{Reward} \\ -\$300 = q_1 \\ \$200 = q_2 \\ \$500 = q_3 \\ \$1,000 = q_4 \end{vmatrix}. \qquad (4.117)$$

4.2.4.1.4 Gain of a Unichain MCR Model of Machine Maintenance

Using the steady-state probability vector calculated in Equation (3.150), the gain, under both maintenance policies, for the unichain MCR model of machine maintenance constructed in Section 4.2.4.1.3 is computed below by applying Equation (4.47) or Equation (4.110) for a four-state unichain MCR:

$$
g = \pi q = \begin{bmatrix} \pi_1 & \pi_2 & 0 & 0 \end{bmatrix} \begin{bmatrix} q_1 \\ q_2 \\ q_3 \\ q_4 \end{bmatrix} = \begin{bmatrix} \pi_1 & \pi_2 \end{bmatrix} \begin{bmatrix} q_1 \\ q_2 \end{bmatrix} = \pi_1 q_1 + \pi_2 q_2
$$

$$
= \begin{bmatrix} \frac{3}{7} & \frac{4}{7} \end{bmatrix} \begin{bmatrix} -300 \\ 200 \end{bmatrix} = \frac{-100}{7} = -14.29. \qquad (4.118)
$$

Note that the gain is the same under both maintenance policies.

4.2.4.2 Value Determination Equations

A second procedure for calculating the gain of a unichain MCR over an infinite planning horizon is to solve the VDEs. As the discussion for a recurrent MCR in Section 4.2.3.2.2 indicates, if the objective is simply to find the gain of the process, then it is not necessary to solve the VDEs. In that case, the approach taken in Section 4.2.4.1 to calculate the gain is sufficient. Of course, a solution of the VDEs will also produce the relative values. Since all states in a unichain MCR have the same gain, denoted by g, the VDEs for a unichain MCR are identical to those for a recurrent MCR. However, it is not necessary to simultaneously solve the full set of VDEs for a unichain MCR. It is sufficient to first solve the subset of the VDEs associated with the recurrent states to determine the gain and the relative values of the recurrent states. The relative values for the transient states may be obtained next by substituting the relative values of the recurrent states into the VDEs for the transient states. The relative values for all the states are used to execute policy improvement for an MDP in Section 5.1.2.3.3 of Chapter 5.

4.2.4.2.1 Solving the VDEs for a Four-State Unichain MCR

To illustrate the procedure briefly outlined in Section 4.2.4.2 for solving the VDEs for a unichain MCR, consider the generic four-state unichain MCR for which the transition matrix and reward vector are given in Equation (4.107). The gain was computed in Section 4.2.4.1.1. The system (4.62) of four VDEs is

$$g + v_i = q_i + \sum_{j=1}^{4} p_{ij} v_j, \ = q_i + p_{i1}v_1 + p_{i2}v_2 + p_{i3}v_3 + p_{i4}v_4, \quad \text{for } i = 1, 2, 3, 4. \quad (4.119)$$

The four individual value determination equation are

$$\left.\begin{array}{l} g + v_1 = q_1 + p_{11}v_1 + p_{12}v_2 + p_{13}v_3 + p_{14}v_4 \\ g + v_2 = q_2 + p_{21}v_1 + p_{22}v_2 + p_{23}v_3 + p_{24}v_4 \\ g + v_3 = q_3 + p_{31}v_1 + p_{32}v_2 + p_{33}v_3 + p_{34}v_4 \\ g + v_4 = q_4 + p_{41}v_1 + p_{42}v_2 + p_{43}v_3 + p_{44}v_4. \end{array}\right\} \quad (4.120)$$

Setting $p_{13} = p_{14} = 0$, and $p_{23} = p_{24} = 0$, for the unichain MCR, the four VDEs become

$$\left.\begin{array}{l} g + v_1 = q_1 + p_{11}v_1 + p_{12}v_2 \\ g + v_2 = q_2 + p_{21}v_1 + p_{22}v_2 \\ g + v_3 = q_3 + p_{31}v_1 + p_{32}v_2 + p_{33}v_3 + p_{34}v_4 \\ g + v_4 = q_4 + p_{41}v_1 + p_{42}v_2 + p_{43}v_3 + p_{44}v_4. \end{array}\right\} \quad (4.121)$$

The gain can be calculated by solving the subset of two VDEs for the recurrent chain separately. Since states 1 and 2 are the two recurrent states, the VDEs for the two-state recurrent chain within the four-state unichain are

$$\left.\begin{array}{l} g + v_1 = q_1 + p_{11}v_1 + p_{12}v_2 \\ g + v_2 = q_2 + p_{21}v_1 + p_{22}v_2. \end{array}\right\} \quad (4.122)$$

Setting $v_2 = 0$ for the highest numbered state in the recurrent chain, the VDEs for the recurrent chain become

$$\left.\begin{array}{r} g + v_1 = q_1 + p_{11}v_1 \\ g = q_2 + p_{21}v_1. \end{array}\right\} \tag{4.123}$$

The solutions for v_1 and g are

$$\left.\begin{array}{l} v_1 = \dfrac{q_1 - q_2}{1 - p_{11} + p_{21}} = \dfrac{q_1 - q_2}{p_{12} + p_{21}} \\[3mm] g = q_2 + \dfrac{p_{21}(q_1 - q_2)}{1 - p_{11} + p_{21}} = \dfrac{q_2(p_{12} + p_{21}) + p_{21}(q_1 - q_2)}{p_{12} + p_{21}} \\[3mm] = \dfrac{p_{21}q_1 + p_{12}q_2}{p_{12} + p_{21}} = \pi_1 q_1 + \pi_2 q_2, \end{array}\right\} \tag{4.124}$$

which agree with the gain and the relative values obtained for a generic two-state recurrent MCR in Equation (4.65). The solution for the gain also agrees with the solution obtained for the generic four-state MCR in Section 4.2.4.1.1.

By substituting the gain and the relative values of the recurrent states into the VDEs for the transient states, the latter two VDEs can be solved to find the relative values of the transient states. The two VDEs for the transient states are

$$\left.\begin{array}{l} g + v_3 = q_3 + p_{31}v_1 + p_{32}v_2 + p_{33}v_3 + p_{34}v_4 \\ g + v_4 = q_4 + p_{41}v_1 + p_{42}v_2 + p_{43}v_3 + p_{44}v_4. \end{array}\right\} \tag{4.125}$$

Rewriting the two VDEs with the two unknowns, the relative values v_3 and v_4 for the two transient states, on the left hand side,

$$\left.\begin{array}{l} (1 - p_{33})v_3 - p_{34}v_4 = -g + q_3 + p_{31}v_1 + p_{32}v_2 \\ -p_{43}v_3 + (1 - p_{44})v_4 = -g + q_4 + p_{41}v_1 + p_{42}v_2 \end{array}\right\} \tag{4.126}$$

The right-hand side constants are evaluated by setting $v_2 = 0$, and substituting from Equation (4.124)

$$\left.\begin{array}{l} v_1 = \dfrac{q_1 - q_2}{p_{12} + p_{21}} \\[3mm] g = \dfrac{p_{21}q_1 + p_{12}q_2}{p_{12} + p_{21}}. \end{array}\right\} \tag{4.124}$$

In matrix form, the two VDEs for the transient states become

$$\begin{bmatrix} (1-p_{33}) & -p_{34} \\ -p_{43} & (1-p_{44}) \end{bmatrix} \begin{bmatrix} v_3 \\ v_4 \end{bmatrix} = \begin{bmatrix} \dfrac{q_1(p_{31}-p_{21})-q_2(p_{12}+p_{31})+q_3(p_{12}+p_{21})}{p_{12}+p_{21}} \\ \dfrac{q_1(p_{41}-p_{21})-q_2(p_{12}+p_{41})+q_4(p_{12}+p_{21})}{p_{12}+p_{21}} \end{bmatrix}.$$

(4.127)

The solution, expressed as ratios of determinants, is

$$v_3 = \cfrac{\begin{vmatrix} \dfrac{q_1(p_{31}-p_{21})-q_2(p_{12}+p_{31})+q_3(p_{12}+p_{21})}{p_{12}+p_{21}} & -p_{34} \\ \dfrac{q_1(p_{41}-p_{21})-q_2(p_{12}+p_{41})+q_4(p_{12}+p_{21})}{p_{12}+p_{21}} & (1-p_{44}) \end{vmatrix}}{\begin{vmatrix} (1-p_{33}) & -p_{34} \\ -p_{43} & (1-p_{44}) \end{vmatrix}}$$

$$v_4 = \cfrac{\begin{vmatrix} (1-p_{33}) & \dfrac{q_1(p_{31}-p_{21})-q_2(p_{12}+p_{31})+q_3(p_{12}+p_{21})}{p_{12}+p_{21}} \\ -p_{43} & \dfrac{q_1(p_{41}-p_{21})-q_2(p_{12}+p_{41})+q_4(p_{12}+p_{21})}{p_{12}+p_{21}} \end{vmatrix}}{\begin{vmatrix} (1-p_{33}) & -p_{34} \\ -p_{43} & (1-p_{44}) \end{vmatrix}}.$$

(4.128)

4.2.4.2.2 *Procedure for Solving the VDEs for a Unichain MCR*

In summary, to find the gain and relative values for an N-state unichain MCR, it is not necessary to solve the N VDEs simultaneously. Instead, the gain and the relative values can be determined by executing the following three-step procedure:

Step 1. For the recurrent chain, solve the VDEs for the gain and the relative values of the recurrent states by setting equal to zero the relative value for the highest numbered recurrent state.

Step 2. Substitute, into the VDEs for the transient states, the gain and the relative values of the recurrent states obtained in step 1.

Step 3. Solve the VDEs for the transient states to obtain the relative values of the transient states.

4.2.4.2.3 *Solving the VDEs for a Unichain MCR Model of Machine Maintenance under Modified Policy of Doing Nothing in State 3*

The VDEs for the unichain MCR model of machine maintenance under the modified maintenance policy described in Section 4.2.4.1.3 will be solved to demonstrate the three-step procedure outlined in Section 4.2.4.2.2. The unichain MCR model under the modified maintenance policy is shown in Equation (4.117).

Step 1. The VDEs for the two recurrent states are

$$\left.\begin{aligned} g + v_1 &= -300 + 0.2v_1 + 0.8v_2 \\ g + v_2 &= 200 + 0.6v_1 + 0.4v_2 \end{aligned}\right\} \tag{4.129}$$

Setting $v_2 = 0$ for the highest numbered state in the recurrent chain, the VDEs for the recurrent chain become

$$\left.\begin{aligned} g + v_1 &= -300 + 0.2v_1 \\ g &= 200 + 0.6v_1. \end{aligned}\right\} \tag{4.130}$$

The solutions for v_1 and g are

$$\left.\begin{aligned} v_1 &= -357.14 \\ g &= -14.29. \end{aligned}\right\} \tag{4.131}$$

The same value for the gain was obtained in Equation (4.118).

Step 2. The VDEs for the two transient states are

$$\left.\begin{aligned} g + v_3 &= 500 + 0.2v_1 + 0.3v_2 + 0.5v_3 + 0v_4 \\ g + v_4 &= 1{,}000 + 0.3v_1 + 0.2v_2 + 0.1v_3 + 0.4v_4. \end{aligned}\right\} \tag{4.132}$$

Substituting the gain, $g = -14.29$, and the relative values, $v_1 = -357.14$, $v_2 = 0$, for the recurrent states obtained in step 1, the VDEs for the transient states are

$$\left.\begin{aligned} -14.29 + v_3 &= 500 + 0.2(-357.14) + 0.3(0) + 0.5v_3 + 0v_4 \\ -14.29 + v_4 &= 1{,}000 + 0.3(-357.14) + 0.2(0) + 0.1v_3 + 0.4v_4. \end{aligned}\right\} \tag{4.133}$$

Step 3. The solutions for v_3 and v_4, the relative values of the transient states, are

$$\left.\begin{aligned} v_3 &= 885.72 \\ v_4 &= 1659.53. \end{aligned}\right\} \tag{4.134}$$

4.2.4.3 Solution by Value Iteration of Unichain MCR Model of Machine Maintenance under Modified Policy of Doing Nothing in State 3

An approximate solution for the gain and the relative values of a unichain MCR can be obtained by executing value iteration over an infinite planning horizon. The value iteration algorithm for a unichain MCR is identical to the one given in Section 4.2.3.3.4 for a recurrent MCR. Hence, when value iteration is executed for a unichain MCR, bounds on the gain can be calculated by using the formula (4.75). In this section, seven iterations of value iteration will be executed to obtain an approximate solution for the gain and relative values of the unichain MCR model of machine maintenance under the modified maintenance policy described in Section 4.2.4.1.3. An exact solution for the relative values and the gain was obtained in Section 4.2.4.2.3 by solving the VDEs.

The four-state unichain MCR model under the modified maintenance policy is shown in Equation (4.117).

The value iteration equation in matrix form is

$$v(n) = q + Pv(n+1) \tag{4.29}$$

$$
\begin{bmatrix} v_1(n) \\ v_2(n) \\ v_3(n) \\ v_4(n) \end{bmatrix} = \begin{bmatrix} -300 \\ 200 \\ 500 \\ 1000 \end{bmatrix} + \begin{bmatrix} 0.2 & 0.8 & 0 & 0 \\ 0.6 & 0.4 & 0 & 0 \\ 0.2 & 0.3 & 0.5 & 0 \\ 0.3 & 0.2 & 0.1 & 0.4 \end{bmatrix} \begin{bmatrix} v_1(n+1) \\ v_2(n+1) \\ v_3(n+1) \\ v_4(n+1) \end{bmatrix}. \tag{4.135}
$$

To begin value iteration, the terminal values at the end of the planning horizon are set equal to zero for all states.

$$n = 0$$

$$
\begin{bmatrix} v_1(0) \\ v_2(0) \\ v_3(0) \\ v_4(0) \end{bmatrix} = \begin{bmatrix} 0 \\ 0 \\ 0 \\ 0 \end{bmatrix}.
$$

Value iteration is executed over the last three periods of an infinite planning horizon.

$n = -1$

$v(-1) = q + Pv(0)$

$$
\begin{bmatrix} v_1(-1) \\ v_2(-1) \\ v_3(-1) \\ v_4(-1) \end{bmatrix} = \begin{bmatrix} -300 \\ 200 \\ 500 \\ 1000 \end{bmatrix} + \begin{bmatrix} 0.2 & 0.8 & 0 & 0 \\ 0.6 & 0.4 & 0 & 0 \\ 0.2 & 0.3 & 0.5 & 0 \\ 0.3 & 0.2 & 0.1 & 0.4 \end{bmatrix} \begin{bmatrix} 0 \\ 0 \\ 0 \\ 0 \end{bmatrix} = \begin{bmatrix} -300 \\ 200 \\ 500 \\ 1000 \end{bmatrix}
$$ (4.136)

$n = -2$

$v(-2) = q + Pv(-1)$

$$
\begin{bmatrix} v_1(-2) \\ v_2(-2) \\ v_3(-2) \\ v_4(-2) \end{bmatrix} = \begin{bmatrix} -300 \\ 200 \\ 500 \\ 1000 \end{bmatrix} + \begin{bmatrix} 0.2 & 0.8 & 0 & 0 \\ 0.6 & 0.4 & 0 & 0 \\ 0.2 & 0.3 & 0.5 & 0 \\ 0.3 & 0.2 & 0.1 & 0.4 \end{bmatrix} \begin{bmatrix} -300 \\ 200 \\ 500 \\ 1000 \end{bmatrix} = \begin{bmatrix} -200 \\ 100 \\ 750 \\ 1400 \end{bmatrix}
$$ (4.137)

$n = -3$

$v(-3) = q + Pv(-2)$

$$
\begin{bmatrix} v_1(-3) \\ v_2(-3) \\ v_3(-3) \\ v_4(-3) \end{bmatrix} = \begin{bmatrix} -300 \\ 200 \\ 500 \\ 1000 \end{bmatrix} + \begin{bmatrix} 0.2 & 0.8 & 0 & 0 \\ 0.6 & 0.4 & 0 & 0 \\ 0.2 & 0.3 & 0.5 & 0 \\ 0.3 & 0.2 & 0.1 & 0.4 \end{bmatrix} \begin{bmatrix} -200 \\ 100 \\ 750 \\ 1400 \end{bmatrix} = \begin{bmatrix} -260 \\ 120 \\ 865 \\ 1595 \end{bmatrix}.
$$ (4.138)

After executing value iteration over four additional periods, the results are summarized in Table 4.10.

Table 4.11 gives the differences between the expected total rewards earned over planning horizons, which differ in length by one period.

TABLE 4.10

Expected Total Rewards for Machine Maintenance Under the Modified Maintenance Policy Calculated by Value Iteration During the Last Seven Periods of an Infinite Planning Horizon

Epoch	n							
	-7	-6	-5	-4	-3	-2	-1	0
$v_1(n)$	-304.416	-288.96	-277.6	-256	-260	-200	-300	0
$v_2(n)$	53.312	66.72	83.2	92	120	100	200	0
$v_3(n)$	930.6065	936.765	934.65	916.5	865	750	500	0
$v_4(n)$	1703.2945	1707.405	1701.45	1670.5	1595	1400	1000	0

TABLE 4.11

Differences between the Expected Total Rewards Earned Over Planning Horizons Which Differ in Length by One Period

Epoch		-7	-6	-5	-4	-3	-2	-1
i		$v_i(-7)$	$v_i(-6)$	$v_i(-5)$	$v_i(-4)$	$v_i(-3)$	$v_i(-2)$	$v_i(-1)$
		$-v_i(-6)$	$-v_i(-5)$	$-v_i(-4)$	$-v_i(-3)$	$-v_i(-2)$	$-v_i(-1)$	$-v_i(0)$
1		−15.456L	−11.36	−21.6L	4	−60L	100	−300L
2		−13.408	−16.48L	−8.8	−28L	20	−100L	200
3		−6.1585	2.115	18.15	51.5	115	250	500
4		−4.1105U	5.955U	30.95U	85.5U	185U	400U	1000U
$\mathrm{Max}\begin{pmatrix}v_i(n)-\\v_i(n+1)\end{pmatrix}$ $=g_U(T)$		−4.1105	5.955	30.95	85.5	185	400	1000
$\mathrm{Min}\begin{pmatrix}v_i(n)-\\v_i(n+1)\end{pmatrix}$ $=g_L(T)$		−15.456	−16.48	−21.6	−28	−60	−100	−300
$g_U(T)-g_L(T)$		11.3455	22.435	52.55	113.5	245	500	1300

In Table 4.11, a suffix U identifies $g_U(T)$, the maximum difference for each epoch. The suffix L identifies $g_L(T)$, the minimum difference for each epoch. The differences, $g_U(T) - g_L(T)$, obtained for all the epochs are listed in the bottom row of Table 4.11. In Equations (4.1.18) and (4.131), the gain was found to be −14.29 thousand dollars per period. When seven periods remain in the planning horizon, Table 4.11 shows that the bounds on the gain are given by $-15.456 \le g \le -4.1105$. The bounds are quite loose. After seven iterations, using Equation (4.75), the gain is approximately equal to the arithmetic average of its upper and lower bounds, so that

$$g \approx (-4.1105 - 15.456)/2 = -9.7833. \qquad (4.139)$$

Many more iterations will be needed to see if value iteration will generate a close approximation to the gain.

Subtracting $v_2(n)$, the expected total reward for the highest numbered state in the recurrent chain, from all of the other expected total rewards in Table 4.10 produces Table 4.12, which shows the expected relative rewards, $v_i(n) - v_2(n)$, during the last seven epochs of the planning horizon.

TABLE 4.12

Expected Relative Rewards, $v_i(n) - v_2(n)$, During the Last Seven Epochs of the Planning Horizon

	n							
Epoch	-7	-6	-5	-4	-3	-2	-1	0
$v_1(n) - v_2(n)$	-357.728	-355.68	-360.8	-348	-380	-300	-500	0
$v_3(n) - v_2(n)$	877.2945	870.045	851.45	824.5	745	650	300	0
$v_4(n) - v_2(n)$	1649.9825	1640.685	1618.25	1578.5	1465	1300	800	0

In Equations (4.131) and (4.134), the relative values obtained by solving the VDEs are, by comparison,

$$v_1 = -357.14, \quad v_2 = 0, \quad v_3 = 885.72, \quad \text{and} \quad v_4 = 1659.53. \quad (4.140)$$

4.2.4.4 Expected Total Reward before Passage to a Closed Set

In this section, the expected total reward earned by a unichain MCR before absorption or passage to a closed class of recurrent states, given that the chain started in a transient state, will be calculated [5]. As Section 4.2.5.4 will demonstrate, this calculation can also be made for a multichain MCR.

4.2.4.4.1 Four-State Unichain MCR

To see how to calculate the expected total reward received before passage to a closed class of recurrent states, given that the chain started in a transient state, consider the generic four-state unichain MCR, introduced in Section 4.2.4.1.1. The transition matrix and reward vector are given in Equation (4.107). The VDEs were solved in Section 4.2.4.2.1. By setting $v_2 = 0$ for the highest numbered state in the recurrent chain, the solutions for the other recurrent state v_1 and the gain g were calculated in Equation (4.124). The VDEs for the two transient states are shown in Equation (4.125). Referring to Equation (4.107), and letting

$$g_R = \begin{bmatrix} g \\ g \end{bmatrix}, \quad v_T = \begin{bmatrix} v_3 \\ v_4 \end{bmatrix}, \quad q_T = \begin{bmatrix} q_3 \\ q_4 \end{bmatrix}, \quad \text{and} \quad Q = \begin{bmatrix} p_{33} & p_{34} \\ p_{43} & p_{44} \end{bmatrix}, \quad (4.141)$$

in Equation (4.125), the matrix equation for the vector v_T of relative values for the transient states is

$$\left. \begin{array}{l} g_R + v_T = q_T + Qv_T \\ Iv_T - Qv_T = q_T - g_R \\ (I - Q)v_T = q_T - g_R \\ v_T = (I - Q)^{-1}(q_T - g_R). \end{array} \right\} \quad (4.142)$$

Substituting $U = (I - Q)^{-1}$ to represent the fundamental matrix, the vector of relative values for the transient states is

$$v_T = U(q_T - g_R) = Uq_T - Ug_R. \tag{4.143}$$

The first term, Uq_T, represents the vector of expected total rewards received before passage to the closed class of recurrent states, given that the chain started in a transient state. The second term, Ug_R, is the vector of expected total rewards received after passage to the recurrent closed class. Observe that when the components of vector q_R are set equal to zero, then the vector $g_R = 0$ because

$$g_R = \pi q_R = \begin{bmatrix} \pi_1 & \pi_2 \end{bmatrix} \begin{bmatrix} q_1 \\ q_2 \end{bmatrix} = \begin{bmatrix} \pi_1 & \pi_2 \end{bmatrix} \begin{bmatrix} 0 \\ 0 \end{bmatrix} = 0 \tag{4.144}$$

and

$$v_T = Uq_T - Ug_R = Uq_T - 0 = Uq_T. \tag{4.145}$$

Thus, if the rewards received in all the recurrent states are set equal to zero, then v_T, the vector of relative values for the transient states, is equal to Uq_T, the vector of expected total rewards received before passage to the closed class of recurrent states, given that the chain started in a transient state. In other words, by setting $q_R = 0$, $v_T = Uq_T$ can be found by solving the VDEs.

The following alternative approach to interpreting Uq_T as the vector of expected total rewards received before passage to a closed class of recurrent states, given that the chain started in a transient state, does not involve solving the VDEs for the relative values of the transient states. Recall from Section (3.2.2) that u_{ij}, the (i, j)th entry of the fundamental matrix, U, specifies the expected number of times that the chain is in transient state j before eventual passage to a recurrent state, given that the chain started in transient state i. The entry in row j of the vector q_T, denoted by $(q_T)_j$, is the reward received each time the chain is in transient state j. Therefore, the entry in row i of the vector Uq_T, denoted by $(Uq_T)_i$, is equal to the following sum of products:

$$\begin{aligned}
(Uq_T)_i &= \sum_{j \varepsilon T} [(\text{Expected number of times the chain is in transient state } j | \\
&\qquad \text{the chain started in transient state } i)(\text{Reward received each time} \\
&\qquad \text{the chain is in transient state } j)] \\
&= \sum_{j \varepsilon T} u_{ij}(q_T)_j.
\end{aligned} \tag{4.146}$$

Both of these approaches have demonstrated that the following result holds for any unichain MCR. Suppose that P is the transition probability matrix, U is the fundamental matrix, q_T is the vector of rewards received in the transient states, and T denotes the set of transient states. Then the ith component of the vector Uq_T represents the expected total reward earned before eventual passage to the recurrent closed class, given that the chain started in transient state i.

4.2.4.4.2 Unichain MCR Model of a Career Path with Lifetime Employment

Consider the company owned by its employees described in Section 1.10.1.2.2.2. The company offers both lifetime employment and career flexibility. After many years, the company has collected sufficient data to construct the transition probability matrix which is shown in canonical form in Equation (1.57). Monthly salary data has also been collected, and is summarized in the following reward vector:

$$q = \begin{array}{c} 1 \\ 2 \\ 3 \\ 4 \\ 5 \\ 6 \end{array}\begin{bmatrix} \$15,000 \\ 14,000 \\ 16,000 \\ 12,000 \\ 13,000 \\ 11,000 \end{bmatrix} = \begin{bmatrix} q_R \\ q_T \end{bmatrix}, \qquad (4.147)$$

where q_R is the vector of monthly salaries earned in recurrent states, which represent management positions, and q_T is the vector of monthly salaries earned in transient states, which represent engineering positions. By adding the reward vector q to the transition probability matrix P, the career path of an employee is modeled as a six-state MCR.

The fundamental matrix for this model is

$$U = (\mathbf{I} - \mathbf{Q})^{-1} = \begin{bmatrix} 0.70 & -0.16 & -0.24 \\ -0.26 & 0.60 & -0.18 \\ -0.14 & -0.32 & 0.72 \end{bmatrix}^{-1} = \begin{array}{c} \\ 4 \\ 5 \\ 6 \end{array}\begin{array}{ccc} 4 & 5 & 6 \\ \hline 1.9918 & 1.0215 & 0.9193 \\ 1.1300 & 2.5026 & 1.0023 \\ 0.8895 & 1.3109 & 2.0131 \end{array} . \quad (4.148)$$

The vector, Uq_T is computed below.

$$Uq_T = (\mathbf{I} - \mathbf{Q})^{-1}q_T = \begin{array}{c} 4 \\ 5 \\ 6 \end{array}\begin{bmatrix} 1.9918 & 1.0215 & 0.9193 \\ 1.1300 & 2.5026 & 1.0023 \\ 0.8895 & 1.3109 & 2.0131 \end{bmatrix}\begin{bmatrix} 12,000 \\ 13,000 \\ 11,000 \end{bmatrix} = \begin{array}{c} 4 \\ 5 \\ 6 \end{array}\begin{bmatrix} 47,293.40 \\ 57,119.10 \\ 49,859.80 \end{bmatrix} .$$

$$(4.149)$$

The entry in row i of the vector Uq_T represents the expected total salary earned by an engineer before she is promoted to a management position, given that she started in a transient state i. For example, an engineer who started in state 6, systems testing, can expect to earn \$49,859.80 prior to being promoted to management.

Of course, if the engineer is interested solely in calculating the expected total salary that she will earn before she is promoted to a management position, she can merge the three recurrent states, that is, states 1, 2, and 3, into a single absorbing state, denoted by 0, which will represent management. The result will be the following four-state absorbing unichain:

$$P = \begin{matrix} 0 \\ 4 \\ 5 \\ 6 \end{matrix} \begin{matrix} \text{Management} \\ \text{Engineering Product Design} \\ \text{Engineering Systems Integration} \\ \text{Engineering Systems Testing} \end{matrix} \left[\begin{array}{c|ccc} 1 & 0 & 0 & 0 \\ \hline 0.30 & 0.30 & 0.16 & 0.24 \\ 0.16 & 0.26 & 0.40 & 0.18 \\ 0.26 & 0.14 & 0.32 & 0.28 \end{array}\right] = \begin{bmatrix} 1 & 0 \\ D & Q \end{bmatrix}.$$

(4.150)

Since the matrix Q and the vector q_T are unchanged, the same entries will be obtained for the vector Uq_T.

Suppose that the fundamental matrix has not been calculated. If the components of q_R, the vector of monthly salaries earned in the recurrent states, are set equal to zero, then, as Section 4.2.4.4.1 indicates, solving the VDEs for the vector v_T of relative values for the transient states is an alternative way of calculating the expected total reward received before passage to the recurrent closed class, given that the chain started in a transient state. For example, suppose that the modified MCR is

$$P = \begin{matrix} 1 \\ 2 \\ 3 \\ 4 \\ 5 \\ 6 \end{matrix} \left[\begin{array}{ccc|ccc} 0.30 & 0.20 & 0.50 & 0 & 0 & 0 \\ 0.40 & 0.25 & 0.35 & 0 & 0 & 0 \\ 0.50 & 0.10 & 0.40 & 0 & 0 & 0 \\ \hline 0.05 & 0.15 & 0.10 & 0.30 & 0.16 & 0.24 \\ 0.04 & 0.07 & 0.05 & 0.26 & 0.40 & 0.18 \\ 0.08 & 0.06 & 0.12 & 0.14 & 0.32 & 0.28 \end{array}\right] = \begin{bmatrix} S & 0 \\ D & Q \end{bmatrix},$$

(4.151)

$$q = \begin{bmatrix} q_R \\ q_T \end{bmatrix} = \begin{matrix} 1 \\ 2 \\ 3 \\ 4 \\ 5 \\ 6 \end{matrix} \left[\begin{array}{c} 0 \\ 0 \\ 0 \\ \hline 12,000 \\ 13,000 \\ 11,000 \end{array}\right].$$

In vector form, the VDEs for the relative values of the transient states are

$$\left. \begin{array}{l} g_R + v_T = q_T + Q v_T \\ g_R = 0 \\ v_T = q_T + Q v_T. \end{array} \right\} \tag{4.152}$$

In algebraic form, the VDEs are

$$\left. \begin{array}{l} v_4 = 12,000 + 0.30 v_4 + 0.16 v_5 + 0.24 v_6 \\ v_5 = 13,000 + 0.26 v_4 + 0.4 v_5 + 0.18 v_6 \\ v_6 = 11,000 + 0.14 v_4 + 0.32 v_5 + 0.28 v_6. \end{array} \right\} \tag{4.153}$$

The solution is

$$v_T = \begin{matrix} 4 \\ 5 \\ 6 \end{matrix} \begin{bmatrix} v_4 \\ v_5 \\ v_6 \end{bmatrix} = \begin{matrix} 4 \\ 5 \\ 6 \end{matrix} \begin{bmatrix} 47,293.40 \\ 57,119.10 \\ 49,859.80 \end{bmatrix}.$$

These are the same values calculated in Equation (4.149) for $U q_T$, the vector of expected total salaries received before passage to the closed class of recurrent states, given that the employee started as an engineer.

4.2.4.5 Value Iteration over a Finite Planning Horizon

Value iteration can be executed to calculate the expected total reward received over a finite planning horizon by a reducible MCR, either unichain or multichain. In this section, value iteration will be executed over three periods for the following modified unichain MCR model of a career path treated in Section 4.2.4.4.2. Suppose that a significant decline in sales has induced the company to offer an engineer a buyout of $50,000 if she voluntarily leaves the company at the end of any month. The probabilities that an engineer will accept the buyout and leave are 0.30 if she is engaged in product design, 0.16 if she is engaged in systems integration, and 0.26 if she is engaged in systems testing. The probabilities that she will make monthly transitions among the three engineering positions remain unchanged. The three recurrent states that formerly represented management positions are merged to form a single absorbing state 0, which now represents an engineer's acceptance of the buyout. The resulting absorbing unichain MCR model is shown below:

$$P = \begin{matrix} 0 \\ 4 \\ 5 \\ 6 \end{matrix} \begin{matrix} \text{Accept Buyout} \\ \text{Engineering Product Design} \\ \text{Engineering Systems Integration} \\ \text{Engineering Systems Testing} \end{matrix} \left[\begin{array}{c|ccc} 1 & 0 & 0 & 0 \\ \hline 0.30 & 0.30 & 0.16 & 0.24 \\ 0.16 & 0.26 & 0.40 & 0.18 \\ 0.26 & 0.14 & 0.32 & 0.28 \end{array} \right] = \begin{bmatrix} 1 & 0 \\ D & Q \end{bmatrix},$$

$$q = \begin{matrix} 0 \\ 4 \\ 5 \\ 6 \end{matrix} \begin{matrix} \text{Accept Buyout} \\ \text{Engineering Product Design} \\ \text{Engineering Systems Integration} \\ \text{Engineering Systems Testing} \end{matrix} \begin{bmatrix} \$50,000 \\ \hline 12,000 \\ 13,000 \\ 11,000 \end{bmatrix} = \begin{bmatrix} q_A \\ q_T \end{bmatrix}$$

$$(4.154)$$

Value iteration using Equation (4.29) will be executed over a 3-month planning horizon to calculate the expected total cost to the company of offering a buyout to an engineer. To begin the backward recursion, the vector of expected terminal total costs received at the end of the 3-month planning horizon is set equal to zero for all states.

$$n = T = 3$$
$$v(3) = v(T) = 0$$

$$\begin{bmatrix} v_0(3) \\ v_4(3) \\ v_5(3) \\ v_6(3) \end{bmatrix} = \begin{bmatrix} 0 \\ 0 \\ 0 \\ 0 \end{bmatrix}$$

$n = 2$

$v(2) = q + Pv(3)$

$$\begin{bmatrix} v_0(2) \\ v_4(2) \\ v_5(2) \\ v_6(2) \end{bmatrix} = \begin{bmatrix} 50,000 \\ 12,000 \\ 13,000 \\ 11,000 \end{bmatrix} + \begin{bmatrix} 1 & 0 & 0 & 0 \\ 0.30 & 0.30 & 0.16 & 0.24 \\ 0.16 & 0.26 & 0.40 & 0.18 \\ 0.26 & 0.14 & 0.32 & 0.28 \end{bmatrix} \begin{bmatrix} 0 \\ 0 \\ 0 \\ 0 \end{bmatrix} = \begin{bmatrix} 50,000 \\ 12,000 \\ 13,000 \\ 11,000 \end{bmatrix} \qquad (4.155)$$

$n = 1$

$v(1) = q + Pv(2)$

$$\begin{bmatrix} v_0(1) \\ v_4(1) \\ v_5(1) \\ v_6(1) \end{bmatrix} = \begin{bmatrix} 50,000 \\ 12,000 \\ 13,000 \\ 11,000 \end{bmatrix} + \begin{bmatrix} 1 & 0 & 0 & 0 \\ 0.30 & 0.30 & 0.16 & 0.24 \\ 0.16 & 0.26 & 0.40 & 0.18 \\ 0.26 & 0.14 & 0.32 & 0.28 \end{bmatrix} \begin{bmatrix} 50,000 \\ 12,000 \\ 13,000 \\ 11,000 \end{bmatrix} = \begin{bmatrix} 100,000 \\ 35,320 \\ 31,300 \\ 32,920 \end{bmatrix}$$

$$(4.156)$$

$n = 0$

$v(0) = q + Pv(1)$

$$
\begin{bmatrix} v_1(0) \\ v_2(0) \\ v_3(0) \\ v_4(0) \end{bmatrix} = \begin{bmatrix} 50,000 \\ 12,000 \\ 13,000 \\ 11,000 \end{bmatrix} + \begin{bmatrix} 1 & 0 & 0 & 0 \\ 0.30 & 0.30 & 0.16 & 0.24 \\ 0.16 & 0.26 & 0.40 & 0.18 \\ 0.26 & 0.14 & 0.32 & 0.28 \end{bmatrix} \begin{bmatrix} 100,000 \\ 35,320 \\ 31,300 \\ 32,920 \end{bmatrix} = \begin{bmatrix} 150,000 \\ 65,504.80 \\ 56,628.80 \\ 61,178.40 \end{bmatrix}.
$$

(4.157)

Equation (4.157) indicates that after 3 months, the expected total cost to the company of offering a buyout to an engineer, given that she is currently engaged in product design, systems integration, or systems testing, will be $65,504.80, $56,628.80, or $61,178.40, respectively.

As an alternative to executing value iteration equation, Equation (4.17) with $T = 3$ can also be used to calculate $v(0)$ over a 3-month planning horizon.

$$
v(0) = q + Pq + P^2 q + P^3 v(3)
$$

(4.158)

$$
v(0) = \begin{bmatrix} v_0(0) \\ v_4(0) \\ v_5(0) \\ v_6(0) \end{bmatrix} = \begin{bmatrix} 50,000 \\ 12,000 \\ 13,000 \\ 11,000 \end{bmatrix} + \begin{bmatrix} 1 & 0 & 0 & 0 \\ 0.30 & 0.30 & 0.16 & 0.24 \\ 0.16 & 0.26 & 0.40 & 0.18 \\ 0.26 & 0.14 & 0.32 & 0.28 \end{bmatrix} \begin{bmatrix} 50,000 \\ 12,000 \\ 13,000 \\ 11,000 \end{bmatrix}
$$
$$
+ \begin{bmatrix} 1 & 0 & 0 & 0 \\ 0.4780 & 0.1652 & 0.1888 & 0.1680 \\ 0.3488 & 0.2072 & 0.2592 & 0.1848 \\ 0.4260 & 0.1644 & 0.2400 & 0.1696 \end{bmatrix} \begin{bmatrix} 50,000 \\ 12,000 \\ 13,000 \\ 11,000 \end{bmatrix} + \begin{bmatrix} 0 \\ 0 \\ 0 \\ 0 \end{bmatrix}
$$
$$
= \begin{bmatrix} 50,000 \\ 12,000 \\ 13,000 \\ 11,000 \end{bmatrix} + \begin{bmatrix} 50,000 \\ 23,320 \\ 18,300 \\ 21,920 \end{bmatrix} + \begin{bmatrix} 50,000 \\ 30,184.8 \\ 25,328.8 \\ 28,258.4 \end{bmatrix} = \begin{bmatrix} 150,000 \\ 65,504.80 \\ 56,628.80 \\ 61,178.40 \end{bmatrix}
$$

(4.159)

equal to the expected total cost vector calculated by value iteration in Equation (4.157).

4.2.5 A Multichain MCR

In this section, the expected average rewards, or gains, earned by the states in a multichain MCR over an infinite planning horizon are calculated. Value iteration will not be applied to a multichain MCR over an infinite horizon because no suitable stopping condition is available. The gain vector can be calculated either by multiplying the limiting transition probability matrix by the reward vector using Equation (4.43), or by solving two sets of equations, which together are called the reward evaluation equations (REEs) [4, 7].

Recall from Section 4.2.1 that a reducible multichain MCR, which is simply called a multichain MCR, consists of two or more closed classes of recurrent states plus one or more transient states. By enlarging the representation of the transition matrix for a multichain in Equation (3.4), the canonical form of the transition matrix for a multichain MCR with M recurrent chains is given below:

$$P = \begin{bmatrix} P_1 & 0 & \cdots & 0 & 0 \\ 0 & P_2 & \cdots & 0 & 0 \\ \vdots & \vdots & \ddots & \vdots & \vdots \\ 0 & 0 & \cdots & P_M & 0 \\ D_1 & D_2 & \cdots & D_M & Q \end{bmatrix}, \quad q = \begin{bmatrix} q_1 \\ q_2 \\ \vdots \\ q_M \\ q_T \end{bmatrix}. \tag{4.160}$$

The components of vectors $q_1, \dots, q_M$ are the rewards received by the recurrent states in the recurrent chains governed by the transition matrices $P_1, \dots, P_M$, respectively. The components of vector q_T are the rewards received in the transient states.

4.2.5.1 An Eight-State Multichain MCR Model of a Production Process

The eight-state multichain model of a three-stage production process introduced in Section 1.10.2.2 will be enlarged by introducing data for operation times and costs. (This model is adapted from one in Shamblin and Stevens [8].) The Markov chain model is revisited in Sections 3.1.3, 3.3.2.2.2, 3.3.3.2.4, 3.4.1.2, and 3.5.5.2. The objective of this section is to construct a cost vector (the negative of a reward vector) for a multichain MCR model, and calculate the expected total operation cost of an item sold.

Recall that the output of each manufacturing stage in the sequential production process is inspected. Output from stage 1 or 2 that is not defective is passed on to the next stage. Output with a minor defect is reworked at the current stage. Output with a major defect is scrapped. Nondefective but blemished output from stage 3 is sent to a training center where it is used to train engineers, technicians, and technical writers. Output from stage 3 that is neither defective nor blemished is sold. The transition probability matrix in canonical form is shown in Equation (3.7). Table 4.13 provides data for operation times and costs.

Each element q_i of the cost vector is simply the cost of an operation, which is equal to the operation time multiplied by the cost per hour. Thus, component q_i of the cost vector, which appears in the right hand column of

TABLE 4.13

Operation Times and Costs

State	Operation	Operation Time (h)	Cost Per Hour	Cost Per Operation
1	Scrap the output	2.6	$60 (disposal)	$156 = (2.6 h)($60)=q_1
2	Sell the output	1.4	$40	$56 = (1.4 h)($40)=q_2
3	Train engineers	3.2	$50	$160 = (3.2 h)($50)=q_3
4	Train technicians	4.1	$30	$123 = (4.1 h)($30)=q_4
5	Train tech. writers	5.3	$55	$291.50 = (5.3 h)($55)=q_5
6	Stage 3	10	$45	$450 = (10 h)($45)=q_6
7	Stage 2	16	$25	$400 = (16 h)($25)=q_7
8	Stage 1	12	$35	$420 = (12 h)($35)=q_8

Table 4.13, is equal to the product of the entries in columns three and four of row i of Table 4.13. The complete eight-state multichain MCR model of production is shown below:

$$P = \begin{array}{c} 1 \\ 2 \\ 3 \\ 4 \\ 5 \\ 6 \\ 7 \\ 8 \end{array} \begin{bmatrix} 1 & 0 & 0 & 0 & 0 & 0 & 0 & 0 \\ 0 & 1 & 0 & 0 & 0 & 0 & 0 & 0 \\ 0 & 0 & 0.50 & 0.30 & 0.20 & 0 & 0 & 0 \\ 0 & 0 & 0.30 & 0.45 & 0.25 & 0 & 0 & 0 \\ 0 & 0 & 0.10 & 0.35 & 0.55 & 0 & 0 & 0 \\ 0.20 & 0.16 & 0.04 & 0.03 & 0.02 & 0.55 & 0 & 0 \\ 0.15 & 0 & 0 & 0 & 0 & 0.20 & 0.65 & 0 \\ 0.10 & 0 & 0 & 0 & 0 & 0 & 0.15 & 0.75 \end{bmatrix}, \quad q = \begin{array}{c} 1 \\ 2 \\ 3 \\ 4 \\ 5 \\ 6 \\ 7 \\ 8 \end{array} \begin{bmatrix} 156 \\ 56 \\ 160 \\ 123 \\ 291.5 \\ 450 \\ 400 \\ 420 \end{bmatrix} \quad (4.161)$$

To calculate the expected operation cost of an item sold, it is first necessary to compute the expected cost that will be incurred by an entering item in each operation. These calculations are shown in Table 4.14. Recall that the three manufacturing stages correspond to transient states. Production stage i is represented by transient state $(9 - i)$. The fundamental matrix, U, is computed in Equation (3.102). The entries in the bottom row of U specify the expected number of visits that an entering item, starting at stage 1 in transient state 8, will make to each of the three manufacturing stages prior to being scrapped, sold, or sent to the training center. On average, an entering item will make $u_{88} = 4$ visits to stage 1, $u_{87} = 1.7143$ visits to stage 2, and $u_{86} = 0.7619$ visits to stage 3. The expected cost that will be incurred by an entering item in a manufacturing stage is equal to the operation cost for the stage multiplied by the expected number of visits to the stage. The expected cost that will be incurred by an entering item in each of the three manufacturing stages is calculated in rows 6, 7, and 8 in the right-hand column of Table 4.14.

TABLE 4.14

Expected Operation Costs Per Entering Item

State	Operation	Operation Cost	Expected Operation Cost Per Entering Item
1	Scrap the Output	$156	$q_1 f_{81} = (\$156)(0.8095) = \$126.28$
2	Sell the Output	$56	$q_2 f_{82} = (\$56)(0.1219) = \$6.83$
3	Train Engineers	$160	$q_3 \lim_{n\to\infty} p_{83}^{(n)} = (\$160)(0.0199) = \$3.18$
4	Train Technicians	$123	$q_4 \lim_{n\to\infty} p_{84}^{(n)} = (\$123)(0.0256) = \$3.15$
5	Train Technical Writers	$291.5	$q_5 \lim_{n\to\infty} p_{85}^{(n)} = (\$291.50)(0.0231) = \$6.73$
6	Stage 3	$450	$q_6 u_{86} = (\$450)(0.7619) = \$342.86$
7	Stage 2	$400	$q_7 u_{87} = (\$400)(1.7143) = \$685.72$
8	Stage 1	$420	$q_8 u_{88} = (\$420)(4) = \$1680$

As Equation (3.103) indicates, if the chain occupies a transient state, the matrix of absorption probabilities is

$$\mathbf{F} = U[D_1 \quad D_2] = \begin{array}{c|cc} & 1 & 2 \\ \hline 6 & 0.4444 & 0.3556 \\ 7 & 0.6825 & 0.2032 \\ 8 & 0.8095 & 0.1219 \end{array} = [F_1 \quad F_2]. \tag{3.103}$$

The operations "scrap the output," state 1, and "sell the output," state 2, are both represented by absorbing states. The expected cost that will be incurred by an entering item in an absorbing state j is equal to the operation cost for the absorbing state multiplied by the probability f_{8j} that an entering item will be absorbed in absorbing state j. The expected cost that will be incurred by an entering item in each of the two absorbing states is calculated in rows 1 and 2 in the right-hand column of Table 4.14.

As Section 3.5.5.2 indicates, if the chain occupies a transient state, the matrix of limiting probabilities of transitions from the set of transient states, $T = \{6, 7, 8\}$, to the three recurrent states, which belong to the recurrent chain, $R = \{3, 4, 5\}$, is given by Equation (3.163). As Equation (3.164) demonstrates, the expected cost that will be incurred by an entering item in a recurrent state j is equal to the operation cost for the recurrent state multiplied by

$$\lim_{n\to\infty} p_{8j}^{(n)} = (f_{83} + f_{84} + f_{85})\pi_j = f_{8R}\pi_j, \tag{4.162}$$

the limiting probability for recurrent state j. The expected cost that will be incurred by an entering item in each of the three recurrent states associated

224   *Markov Chains and Decision Processes for Engineers and Managers*

TABLE 4.15

Expected Operation Costs Per Item Sold

State	Operation	Expected Operation Cost per Item Sold = Expected Operation Cost per Entering Item)/f_{82}
1	Scrap the Output	($126.28)/$f_{82}$ = ($126.28)/(0.1219) = $1.035.95
2	Sell the Output	($6.83)/$f_{82}$ = ($6.83)/(0.1219) = $56
3	Train Engineers	($3.18)/$f_{82}$ = ($3.18)/(0.1219) = $26.09
4	Train Technicians	($3.15)/$f_{82}$ = ($3.15)/(0.1219) = $25.84
5	Train Technical Writers	($6.73)/$f_{82}$ = ($6.73)/(0.1219) = $55.21
6	Stage 3	($342.86)/$f_{82}$ = ($342.86)/(0.1219) = $2,812.63
7	Stage 2	($685.72)/$f_{82}$ = ($685.72)/(0.1219) = 5,625.27
8	Stage 1	($1,680)/$f_{82}$ = ($1,680)/(0.1219) = $13,781.79
		Sum = $23,418.78

with training is calculated in rows 3, 4, and 5 in the right-hand column of Table 4.14.

Recall from Equation (3.106) that the expected number of entering items required to enable 100 items to be sold is equal to $100/f_{82}$, where $f_{82}=0.1219$ is the probability that an entering item will be sold. Therefore, the expected operation cost per item sold is equal to the expected operation cost for an entering item, calculated in the right hand column of Table 4.14, divided by the probability that an entering item will be sold. The expected operation costs per item sold are calculated in the right-hand column of Table 4.15.

The expected total operation cost per item sold is $23,418.78, equal to the sum of the entries in the right-hand column in the bottom row of Table 4.15.

4.2.5.2 Expected Average Reward or Gain

As Section 4.2.3.1 indicates, every state i in a multichain MCR has its own gain g_i, which is the ith component of a gain vector **g**. Hence, the gain of a multichain MCR depends on the state in which it starts. The gain vector is calculated by Equation (4.43).

4.2.5.2.1 Gain of a Four-State Absorbing Multichain MCR

Consider a generic four-state absorbing multichain MCR with the following transition probability matrix P, which appears in Equation (3.85), and reward vector q. The Markov chain has two absorbing states and two transient states.

$$P = \begin{array}{c} 1 \\ 2 \\ 3 \\ 4 \end{array}\left[\begin{array}{cc|cc} 1 & 0 & 0 & 0 \\ 0 & 1 & 0 & 0 \\ \hline p_{31} & p_{32} & p_{33} & p_{34} \\ p_{41} & p_{42} & p_{43} & p_{44} \end{array}\right] = \left[\begin{array}{cc} I & 0 \\ D & Q \end{array}\right], \quad \mathbf{q} = \begin{array}{c} 1 \\ 2 \\ 3 \\ 4 \end{array}\left[\begin{array}{c} q_1 \\ q_2 \\ q_3 \\ q_4 \end{array}\right] = \left[\begin{array}{c} q_A \\ q_T \end{array}\right]. \quad (4.163)$$

Absorbing states 1 and 2, which belong to the set A, have gains denoted by g_1 and g_2, respectively. Transient states 3 and 4, which belong to the set T, have gains denoted by g_3 and g_4, respectively. The limiting transition probability matrix for the absorbing multichain was calculated in Equation (3.135). The gain vector for the four-state absorbing multichain MCR is computed using Equation (4.43).

$$
g = \lim_{n \to \infty} P^{(n)} q = \begin{matrix} 1 \\ 2 \\ 3 \\ 4 \end{matrix} \begin{bmatrix} 1 & 0 & 0 & 0 \\ 0 & 1 & 0 & 0 \\ f_{31} & f_{32} & 0 & 0 \\ f_{41} & f_{42} & 0 & 0 \end{bmatrix} \begin{bmatrix} q_1 \\ q_2 \\ q_3 \\ q_4 \end{bmatrix} = \begin{bmatrix} q_1 \\ q_2 \\ f_{31}q_1 + f_{32}q_2 \\ f_{41}q_1 + f_{42}q_2 \end{bmatrix} = \begin{bmatrix} g_1 \\ g_2 \\ g_3 \\ g_4 \end{bmatrix} = \begin{bmatrix} g_A \\ g_T \end{bmatrix}.
$$

(4.164)

The gain of the absorbing multichain MCR depends on the state in which it starts. If the system starts in an absorbing state i, for $i = 1$ or 2, the gain will be $g_i = q_i$. Thus, in a multichain MCR, the gain of an absorbing state is equal to the reward received in that state. If the system is initially in transient state i, for $i = 3$ or 4, the gain will be

$$
g_i = f_{i1}q_1 + f_{i2}q_2
$$

(4.165)

because the chain will eventually be absorbed in state 1 with probability f_{i1}, or in state 2 with probability f_{i2}.

4.2.5.2.2 Gain of a Five-State Multichain MCR

Consider a generic five-state multichain MCR with the transition probability matrix P, which appears in Equation (3.5), and reward vector q. The Markov chain, treated in Sections 1.9.2 and 3.1.2, has two recurrent closed sets, one of which is an absorbing state, plus two transient states.

$$
P = \begin{matrix} 1 \\ 2 \\ 3 \\ 4 \\ 5 \end{matrix} \begin{bmatrix} 1 & 0 & 0 & 0 & 0 \\ 0 & p_{22} & p_{23} & 0 & 0 \\ 0 & p_{32} & p_{33} & 0 & 0 \\ p_{41} & p_{42} & p_{43} & p_{44} & p_{45} \\ p_{51} & p_{52} & p_{53} & p_{54} & p_{55} \end{bmatrix}, \quad q = \begin{matrix} 1 \\ 2 \\ 3 \\ 4 \\ 5 \end{matrix} \begin{bmatrix} q_1 \\ q_2 \\ q_3 \\ q_4 \\ q_5 \end{bmatrix}.
$$

(4.166)

State 1, with a gain denoted by g_1, is an absorbing state, and constitutes the first recurrent closed set. States 2 and 3, with gains denoted by g_2 and g_3, respectively, are members of the second recurrent chain, denoted by $R_2 = \{2, 3\}$. Since both recurrent states belong to the same recurrent chain, they have the same gain, denoted by g_{R_2}, so that

$$
g_2 = g_3 = g_{R_2},
$$

(4.167)

where g_{R_2} denotes the gain of all states which belong to the recurrent chain R_2. States 4 and 5 are transient, with gains denoted by g_4 and g_5, respectively. The limiting transition probability matrix for the multichain was calculated in Equation (3.158). In the limiting transition probability matrix, π_2 and π_3 are the steady-state probabilities for states 2 and 3, respectively, within the recurrent chain R_2. The probabilities of absorption from transient states 4 and 5 are denoted by f_{41} and f_{51}, respectively. The probabilities of eventual passage from transient states 4 and 5 to the recurrent chain R_2 are denoted by f_{4R_2} and f_{5R_2}, respectively. All of these quantities are computed for the generic five-state multichain in Section 3.5.5.1. The gain vector for the five-state multichain MCR is computed using Equation (4.43).

$$
g = \lim_{n \to \infty} \mathbf{P}^{(n)} \mathbf{q} = \begin{matrix} 1 \\ 2 \\ 3 \\ 4 \\ 5 \end{matrix} \begin{bmatrix} 1 & 0 & 0 & 0 & 0 \\ 0 & \pi_2 & \pi_3 & 0 & 0 \\ 0 & \pi_2 & \pi_3 & 0 & 0 \\ f_{41} & f_{4R}\pi_2 & f_{4R}\pi_3 & 0 & 0 \\ f_{51} & f_{5R}\pi_2 & f_{5R}\pi_3 & 0 & 0 \end{bmatrix} \begin{bmatrix} q_1 \\ q_2 \\ q_3 \\ q_4 \\ q_5 \end{bmatrix} = \begin{bmatrix} g_1 \\ g_2 \\ g_3 \\ g_4 \\ g_5 \end{bmatrix} = \begin{bmatrix} g_1 \\ g_{R_2} \\ g_{R_2} \\ g_4 \\ g_5 \end{bmatrix}
$$

$$
= \begin{bmatrix} q_1 \\ \pi_2 q_2 + \pi_3 q_3 \\ \pi_2 q_2 + \pi_3 q_3 \\ f_{41} q_1 + f_{4R}(\pi_2 q_2 + \pi_3 q_3) \\ f_{51} q_1 + f_{5R}(\pi_2 q_2 + \pi_3 q_3) \end{bmatrix}. \tag{4.168}
$$

The gain of the multichain MCR depends on the state in which it starts. If the system starts in state 1, an absorbing state, the gain will be $g_1 = q_1$. Thus, in a multichain MCR, the gain of an absorbing state is equal to the reward received in that state, confirming the conclusion of Equation (4.114) in Section 4.2.4.1.2. If the system is initially in either of the two communicating recurrent states, states 2 or 3, the gain will be

$$
g_{R_2} = \pi_2 q_2 + \pi_3 q_3. \tag{4.169}
$$

If the chain starts in transient state 4, either it will eventually be absorbed with probability f_{41}, or it will eventually enter the recurrent chain R_2 with probability f_{4R_2}. Hence, if the chain starts in transient state 4, the gain will be

$$
g_4 = f_{41} q_1 + f_{4R_2} \pi_2 q_2 + f_{4R_2} \pi_3 q_3 = f_{41} q_1 + f_{4R_2}(\pi_2 q_2 + \pi_3 q_3) = f_{41} g_1 + f_{4R_2} g_{R_2}. \tag{4.170}
$$

Observe that the gain of transient state 4 has been expressed as the weighted average of the independent gains, g_1 and g_{R2}, of the two recurrent chains. The respective weights are f_{41} and f_{4R_2}. Similarly, if the chain starts in transient state 5, the gain will be

$$g_5 = f_{51}q_1 + f_{5R_2}(\pi_2 q_2 + \pi_3 q_3) = f_{51}g_1 + f_{5R_2}g_{R_2}. \tag{4.171}$$

The gain of transient state 5 has also been expressed as the weighted average of the independent gains, g_1 and g_{R2}, of the two recurrent chains. In this case the respective weights are f_{51} and f_{5R_2}.

Since each closed class of recurrent states in a multichain MCR can be treated as a separate recurrent chain, all states that belong to the same recurrent chain have the same gain. Hence, every recurrent chain has an independent gain. One may conclude that the gains of the recurrent chains can be found separately by finding the steady-state probability vector for each recurrent chain and multiplying it by the associated reward vector. The gain of every transient state can be expressed as a weighted average of the independent gains of the recurrent chains. The weights are the probabilities of eventual passage from the transient state to the recurrent chains.

4 2.5.2.3 Gain of un Eight-State Multichain MCR Model of a Production Process

Consider the eight-state multichain MCR model of production for which the transition matrix and reward vector are shown in Equation (4.161). The limiting transition probability matrix was calculated in Equation (3.165). Using Equation (4.43), the negative gain vector is

$$g = \lim_{n \to \infty} P^{(n)}q$$

	1	0	0	0	0	0	0	0		156		156
2	0	1	0	0	0	0	0	0		56		56
3	0	0	0.2909	0.3727	0.3364	0	0	0		160		190.45
4	0	0	0.2909	0.3727	0.3364	0	0	0		123		190.45
= 5	0	0	0.2909	0.3727	0.3364	0	0	0		291.5	=	190.45
6	0.4444	0.3556	0.0582	0.0745	0.0673	0	0	0		450		127.33
7	0.6825	0.2032	0.0332	0.0426	0.0385	0	0	0		400		139.62
8	0.8095	0.1219	0.0199	0.0256	0.0231	0	0	0		420		146.17

$$\tag{4.172}$$

4.2.5.3 Reward Evaluation Equations

As Sections 4.2.3.2.2 and 4.2.4.2 indicate, if the objective is simply to find the gain attained by starting in every state of a multichain MCR, then it is sufficient to find the gain vector, **g**, by using Equation (4.43). Alternatively, by solving two sets of equations, which together are called the (REEs) [4, 7], one can determine not only the gain in every state but also the relative value in every state. The advantage of obtaining the relative values is that they can be used to find an improved policy when decisions are introduced in Chapter 5 to transform an MCR into an MDP. The first set of REEs is called the gain state equations (GSEs), and the second set is called the VDEs. Thus, the REEs consist of the GSEs plus the VDEs.

4.2.5.3.1 *Informal Derivation of the REEs*

When an N-state multichain MCR starts in state i, and the length of the planning horizon, T, grows very large, the system enters the steady state. By modifying the argument given in Section 4.2.3.2.2 for a recurrent MCR to make it apply to a multichain MCR, Equation (4.59) for $v_i(0)$ becomes

$$v_i(0) \approx Tg_i + v_i \quad \text{for } i = 1, 2, ..., N, \tag{4.173}$$

where g_i is the gain in state i.

Similarly, Equation (4.60) for $v_j(1)$ becomes

$$v_j(1) \approx (T-1)g_j + v_j. \tag{4.174}$$

The recursive equation relating $v_i(0)$ to $v_j(1)$ is

$$v_i(0) = q_i + \sum_{j=1}^{N} p_{ij} v_j(1), \quad \text{for } i = 1, 2, ..., N$$

Substituting the linear approximations in Equations (4.173) and (4.174), assumed to be valid for large T, into Equation (4.61) gives the result

$$Tg_i + v_i = q_i + \sum_{j=1}^{N} p_{ij} [(T-1)g_j + v_j], \quad \text{for } i = 1, 2, ..., N,$$
$$= q_i + T\sum_{j=1}^{N} p_{ij} g_j - \sum_{j=1}^{N} p_{ij} g_j + \sum_{j=1}^{N} p_{ij} v_j. \tag{4.175}$$

In order for Equation (4.175) to be satisfied for any large T, let

$$g_i = \sum_{j=1}^{N} p_{ij} g_j, \quad \text{for } i = 1, 2, ..., N \tag{4.176}$$

Then

$$Tg_i + v_i = q_i + Tg_i - g_i + \sum_{j=1}^{N} p_{ij}v_j$$

Since

$$\sum_{j=1}^{N} p_{ij} = 1,$$

$$g_i + v_i = q_i + \sum_{j=1}^{N} p_{ij}v_j, \quad i = 1, 2, \dots, N.$$

(4.177)

Thus, the following two systems of RN linear equations each,

$$g_i = \sum_{j=1}^{N} p_{ij}g_j, \quad \text{for } i = 1, 2, \dots, N,$$

and

$$g_i + v_i = q_i + \sum_{j=1}^{N} p_{ij}v_j, \quad i = 1, 2, \dots, N,$$

(4.178)

have been obtained. These two systems may be solved for the N variables, g_i, and the N variables, v_i. The first system of N linear equations (4.176) is called the GSEs. The second system is the set of VDEs (4.177). These two systems of equations are together called the set of REEs (4.178).

4.2.5.3.2 *The REEs for a Five-State Multichain MCR*

Consider once again the generic five-state multichain MCR for which the transition matrix and reward vector are shown in Equation (4.166). The system of ten REEs (4.178) consists of a system of five VDEs (4.177) plus a system of five GSEs (4.176). The system of five VDEs is

$$g_i + v_i = q_i + \sum_{j=1}^{5} p_{ij}v_j = q_i + p_{i1}v_1 + p_{i2}v_2 + p_{i3}v_3 + p_{i4}v_4 + p_{i5}v_5, \quad \text{for } i = 1, 2, \dots, 5.$$

(4.179)

The individual value determination equations are

$$
\left.
\begin{aligned}
g_1 + v_1 &= q_1 + p_{11}v_1 + p_{12}v_2 + p_{13}v_3 + p_{14}v_4 + p_{15}v_5 \\
g_2 + v_2 &= q_2 + p_{21}v_1 + p_{22}v_2 + p_{23}v_3 + p_{24}v_4 + p_{25}v_5 \\
g_3 + v_3 &= q_3 + p_{31}v_1 + p_{32}v_2 + p_{33}v_3 + p_{34}v_4 + p_{35}v_5 \\
g_4 + v_4 &= q_4 + p_{41}v_1 + p_{42}v_2 + p_{43}v_3 + p_{44}v_4 + p_{45}v_5 \\
g_5 + v_5 &= q_5 + p_{51}v_1 + p_{52}v_2 + p_{53}v_3 + p_{54}v_4 + p_{55}v_5.
\end{aligned}
\right\}
$$

(4.180)

Setting $p_{12} = p_{13} = p_{14} = p_{15} = 0$, $p_{21} = p_{24} = p_{25} = 0$, and $p_{31} = p_{34} = p_{35} = 0$, and setting $g_2 = g_3 = g_{R_2}$, the five VDEs become

$$\left.\begin{aligned}
g_1 + v_1 &= q_1 + p_{11}v_1 \\
g_{R_2} + v_2 &= q_2 + p_{22}v_2 + p_{23}v_3 \\
g_{R_2} + v_3 &= q_3 + p_{32}v_2 + p_{33}v_3 \\
g_4 + v_4 &= q_4 + p_{41}v_1 + p_{42}v_2 + p_{43}v_3 + p_{44}v_4 + p_{45}v_5 \\
g_5 + v_5 &= q_5 + p_{51}v_1 + p_{52}v_2 + p_{53}v_3 + p_{54}v_4 + p_{55}v_5.
\end{aligned}\right\} \tag{4.181}$$

The system of five GSEs is given below:

$$g_i = \sum_{j=1}^{5} p_{ij}g_j = p_{i1}g_1 + p_{i2}g_2 + p_{i3}g_3 + p_{i4}g_4 + p_{i5}g_5, \quad \text{for } i = 1, 2, \ldots, 5. \tag{4.182}$$

Setting $p_{12} = p_{13} = p_{14} = p_{15} = 0$, $p_{21} = p_{24} = p_{25} = 0$, $p_{31} = p_{34} = p_{35} = 0$, and setting $g_2 = g_3 = g_{R_2}$, the five GSEs become

$$\left.\begin{aligned}
g_1 &= g_1 \\
g_2 &= p_{22}g_2 + p_{23}g_3 = p_{22}g_{R_2} + p_{23}g_{R_2} = (p_{22} + p_{23})g_{R_2} = g_{R_2} \\
g_3 &= p_{32}g_2 + p_{33}g_3 = p_{32}g_{R_2} + p_{33}g_{R_2} = (p_{32} + p_{33})g_{R_2} = g_{R_2} \\
g_4 &= p_{41}g_1 + p_{42}g_2 + p_{43}g_3 + p_{44}g_4 + p_{45}g_5 = p_{41}g_1 + (p_{42} + p_{43})g_{R_2} + p_{44}g_4 + p_{45}g_5 \\
g_5 &= p_{51}g_1 + p_{52}g_2 + p_{53}g_3 + p_{54}g_4 + p_{55}g_5 = p_{51}g_1 + (p_{52} + p_{53})g_{R_2} + p_{54}g_4 + p_{55}g_5.
\end{aligned}\right\} \tag{4.183}$$

4.2.5.3.3 Procedure for Solving the REEs for a Multichain MCR

As this section will indicate, both the independent gains and the relative values for the recurrent chains can be obtained by solving the VDEs for each recurrent chain separately. Next, the GSEs for the transient states can be solved to obtain the gains of the transient states. Finally, the VDEs for the transient states can be solved to obtain the relative values for the transient states [4].

Suppose that an N-state multichain MCR has L recurrent chains, where $L < N$. The MCR has L independent gains, each associated with a different recurrent chain. The unknown quantities are the L independent gains plus the N relative values, v_i. The number of unknowns can be reduced to N by equating to zero the relative value, v_i for the highest numbered state in each of the L recurrent chains. The remaining $(N - L)$ relative values for the MCR are not equated to zero. Hence, the total number of unknown quantities is

N, equal to L independent gains plus $(N - L)$ relative values. The L independent gains and the remaining $(N - L)$ relative values, v_i, can be determined by following the four-step procedure given below for solving the REEs for a multichain MCR:

Step 1. Assign an independent gain to each recurrent chain. For each recurrent chain, solve the VDEs for the independent gain and the relative values of the recurrent states by setting equal to zero the relative value for the highest numbered recurrent state.

Step 2. Solve the GSEs for the transient states to compute the gains of the transient states as weighted averages of the independent gains of the recurrent chains obtained in step 1.

Step 3. Substitute, into the VDEs for the transient states, the gains of the transient states obtained in step 2, and the relative values of the recurrent states obtained in step 1.

Step 4. Solve the VDEs for the transient states to obtain the relative values of the transient states.

4.2.5.3.4 Solving the REEs for a Five-State Multichain MCR

The four-step procedure outlined in Section 4.2.5.3.3 for solving the REEs will be demonstrated by applying it to solve the REEs obtained in Section 4.2.5.3.2 for the generic five-state multichain MCR. The MCR has $N = 5$ states, and $L = 2$ recurrent chains, with the relative value v_1 equated to zero in the first recurrent chain, and v_3 equated to zero in the second. The two independent gains, g_1 and g_{R_2}, and the remaining $(5 - 2)$ relative values, v_2, v_4, and v_5, are the five unknowns.

Step 1. Each recurrent chain in the five-state multichain MCR will be considered separately. The VDE for the first recurrent chain, consisting of the absorbing state 1, is

$$v_1 + g_1 = q_1 + p_{11}v_1 \tag{4.184}$$

Setting the relative value $v_1 = 0$ for the highest numbered state in the first recurrent chain, and setting $p_{11} = 1$ for the absorbing state 1, gives the solution $g_1 = q_1$, confirming the conclusion of Equation (4.114) in Section 4.2.4.1.2 that the gain of an absorbing state is equal to the reward received in that state. The VDEs for the second recurrent chain, consisting of states 2 and 3, and denoted by $R_2 = \{2, 3\}$, are

$$\left.\begin{array}{l} v_2 + g_2 = q_2 + p_{22}v_2 + p_{23}v_3 \\ v_3 + g_3 = q_3 + p_{32}v_2 + p_{33}v_3. \end{array}\right\} \tag{4.185}$$

Setting $v_3 = 0$ for the highest numbered state in the second recurrent chain, and setting $g_2 = g_3 = g_{R_2}$, the VDEs become

$$\left.\begin{aligned} v_2 + g_{R_2} &= q_2 + p_{22}v_2 \\ g_{R_2} &= q_3 + p_{32}v_2 \end{aligned}\right\}. \tag{4.186}$$

The solution is

$$\left.\begin{aligned} v_2 &= \frac{q_2 - q_3}{p_{23} + p_{32}} \\ g_{R_2} &= \frac{p_{32}q_2 + p_{23}q_3}{p_{23} + p_{32}} \end{aligned}\right\} \tag{4.187}$$

confirming the result obtained for the gain in Equation (4.169).

Step 2. The set of GSEs for the transient states that will be solved to express the gains of the two transient states as weighted averages of the independent gains of the recurrent chains is given below.

$$g_i = \sum_{j=1}^{5} p_{ij}g_j = p_{i1}g_1 + p_{i2}g_2 + p_{i3}g_3 + p_{i4}g_4 + p_{i5}g_5, \tag{4.188}$$

for transient states $i = 4, 5$.

For the transient states 4 and 5, Equation (4.185) becomes

$$\left.\begin{aligned} g_4 &= p_{41}g_1 + p_{42}g_2 + p_{43}g_3 + p_{44}g_4 + p_{45}g_5 = p_{41}g_1 + (p_{42} + p_{43})g_{R_2} + p_{44}g_4 + p_{45}g_5 \\ g_5 &= p_{51}g_1 + p_{52}g_2 + p_{53}g_3 + p_{54}g_4 + p_{55}g_5 = p_{51}g_1 + (p_{52} + p_{53})g_{R_2} + p_{54}g_4 + p_{55}g_5 \end{aligned}\right\}. \tag{4.189}$$

Rearranging the terms to place the two unknowns, g_4 and g_5, on the left-hand side,

$$\left.\begin{aligned} (1 - p_{44})g_4 - p_{45}g_5 &= p_{41}g_1 + (p_{42} + p_{43})g_{R_2} \\ -p_{54}g_4 + (1 - p_{55})g_5 &= p_{51}g_1 + (p_{52} + p_{53})g_{R_2} \end{aligned}\right\}. \tag{4.190}$$

The solution is

$$\left.\begin{aligned} g_4 &= \frac{p_{41}(1 - p_{55}) + p_{45}p_{51}}{(1 - p_{44})(1 - p_{55}) - p_{45}p_{54}}g_1 + \frac{(1 - p_{55})(p_{42} + p_{43}) + p_{45}(p_{52} + p_{53})}{(1 - p_{44})(1 - p_{55}) - p_{45}p_{54}}g_{R_2} \\ g_5 &= \frac{p_{51}(1 - p_{44}) + p_{54}p_{41}}{(1 - p_{44})(1 - p_{55}) - p_{45}p_{54}}g_1 + \frac{(1 - p_{44})(p_{52} + p_{53}) + p_{54}(p_{42} + p_{43})}{(1 - p_{44})(1 - p_{55}) - p_{45}p_{54}}g_{R_2} \end{aligned}\right\}. \tag{4.191}$$

These equations may be written more concisely to express the gains of the two transient states as weighted averages of the independent gains of the two recurrent chains.

$$
\left.
\begin{aligned}
g_4 &= f_{41} g_1 + f_{4R_2} g_{R_2} \\
g_5 &= f_{51} g_1 + f_{5R_2} g_{R_2}
\end{aligned}
\right\}, \tag{4.192}
$$

where the weights f_{41} and f_{51} are calculated in Equation (3.120), and the weights f_{4R_2} and f_{5R_2} are calculated in Equation (3.121). Thus, the gain, g_i, of each transient state, i, has been expressed as a weighted average of the independent gains of the recurrent chains. It is interesting to note that each weight, f_{iR}, can be interpreted as the probability of eventual passage from a transient state i to a recurrent chain R. This result confirms the conclusion reached earlier in Section 4.2.5.2.2 by solving Equation (4.43).

Step 3. The set of VDEs for the two transient states is given below:

$$
g_i + v_i = q_i + \sum_{j=1}^{5} p_{ij} v_j = q_i + p_{i1} v_1 + p_{i2} v_2 + p_{i3} v_3 + p_{i4} v_4 + p_{i5} v_5, \tag{4.193}
$$

for transient states $i = 4, 5$. The two VDEs for the transient states are

$$
\left.
\begin{aligned}
g_4 + v_4 &= q_4 + p_{41} v_1 + p_{42} v_2 + p_{43} v_3 + p_{44} v_4 + p_{45} v_5 \\
g_5 + v_5 &= q_5 + p_{51} v_1 + p_{52} v_2 + p_{53} v_3 + p_{54} v_4 + p_{55} v_5
\end{aligned}
\right\}. \tag{4.194}
$$

Rearranging the terms to place the two unknowns, v_4 and v_5, on the left hand side,

$$
\begin{aligned}
(1 - p_{44}) v_4 - p_{45} v_5 &= -g_4 + q_4 + p_{41} v_1 + p_{42} v_2 + p_{43} v_3 \\
- p_{54} v_4 + (1 - p_{55}) v_5 &= -g_5 + q_5 + p_{51} v_1 + p_{52} v_2 + p_{53} v_3
\end{aligned} \tag{4.195}
$$

Substituting the quantities

$$
g_1 = q_1, \tag{4.196}
$$

$$
g_2 = g_3 = g_{R_2} = \frac{p_{32} q_2 + p_{23} q_3}{p_{23} + p_{32}}, \tag{4.197}
$$

$$
g_4 = f_{41} g_1 + f_{4R_2} g_{R_2} = f_{41} q_1 + f_{4R_2} \frac{p_{32} q_2 + p_{23} q_3}{p_{23} + p_{32}}, \tag{4.198}
$$

$$g_5 = f_{51}g_1 + f_{5R_2}g_{K_2} = f_{51}q_1 + f_{5R_2}\frac{p_{32}q_2 + p_{23}q_3}{p_{23} + p_{32}}, \tag{4.199}$$

the VDEs for the two transient states appear as

$$(1-p_{44})v_4 - p_{45}v_5 = -f_{41}q_1 - f_{4R_2}\frac{p_{32}q_2 + p_{23}q_3}{p_{23} + p_{32}} + q_4 + p_{41}v_1 + p_{42}v_2 + p_{43}v_3$$

$$-p_{54}v_4 + (1-p_{55})v_5 = -f_{51}q_1 - f_{5R_2}\frac{p_{32}q_2 + p_{23}q_3}{p_{23} + p_{32}} + q_5 + p_{51}v_1 + p_{52}v_2 + p_{53}v_3.$$

$$\tag{4.200}$$

Setting the relative value $v_1 = 0$ for the highest numbered state in the first recurrent chain, and also setting $v_3 = 0$ for the highest numbered state in the second recurrent chain, the VDEs for the transient states are

$$(1-p_{44})v_4 - p_{45}v_5 = -f_{41}q_1 - f_{4R_2}\frac{p_{32}q_2 + p_{23}q_3}{p_{23} + p_{32}} + q_4 + p_{41}0 + p_{42}v_2 + p_{43}0$$

$$-p_{54}v_4 + (1-p_{55})v_5 = -f_{51}q_1 - f_{5R_2}\frac{p_{32}q_2 + p_{23}q_3}{p_{23} + p_{32}} + q_5 + p_{51}0 + p_{52}v_2 + p_{53}0.$$

$$\tag{4.201}$$

Substituting for v_2 from Equation (4.187), the VDEs for the transient states are

$$(1-p_{44})v_4 - p_{45}v_5 = -f_{41}q_1 - f_{4R_2}\frac{p_{32}q_2 + p_{23}q_3}{p_{23} + p_{32}} + q_4 + p_{42}\frac{q_2 - q_3}{p_{23} + p_{32}}$$

$$-p_{54}v_4 + (1-p_{55})v_5 = -f_{51}q_1 - f_{5R_2}\frac{p_{32}q_2 + p_{23}q_3}{p_{23} + p_{32}} + q_5 + p_{52}\frac{q_2 - q_3}{p_{23} + p_{32}}. \tag{4.202}$$

Step 4. The solution for the relative values of the transient states is

$$
v_4 = \frac{\left\{\begin{array}{l}\left[-f_{41}q_1 - f_{4R_2}\dfrac{p_{32}q_2 + p_{23}q_3}{p_{23} + p_{32}} + q_4 + p_{42}\dfrac{q_2 - q_3}{p_{23} + p_{32}}\right](1 - p_{55}) \\ + \left[-f_{51}q_1 - f_{5R_2}\dfrac{p_{32}q_2 + p_{23}q_3}{p_{23} + p_{32}} + q_5 + p_{52}\dfrac{q_2 - q_3}{p_{23} + p_{32}}\right]p_{45}\end{array}\right\}}{(1 - p_{44})(1 - p_{55}) - p_{45}p_{54}}
$$

$$
v_5 = \frac{\left\{\begin{array}{l}\left[-f_{51}q_1 - f_{5R_2}\dfrac{p_{32}q_2 + p_{23}q_3}{p_{23} + p_{32}} + q_5 + p_{52}\dfrac{q_2 - q_3}{p_{23} + p_{32}}\right](1 - p_{44}) \\ + \left[-f_{41}q_1 - f_{4R_2}\dfrac{p_{32}q_2 + p_{23}q_3}{p_{23} + p_{32}} + q_4 + p_{42}\dfrac{q_2 - q_3}{p_{23} + p_{32}}\right]p_{54}\end{array}\right\}}{(1 - p_{44})(1 - p_{55}) - p_{45}p_{54}}
$$

(4.203)

Note that in this generic five-state multichain example, $N = 5$ states, and $L = 2$ recurrent chains with the relative value v_1 equated to zero in the first recurrent chain, and v_3 equated to zero in the second. The two independent gains, g_1 and g_{R_2}, and the remaining $(5 - 2)$ relative values, v_2, v_4, and v_5, are the five unknowns.

4.2.5.3.5 Solving the REEs for an Eight-State Multichain MCR Model of a Production Process

The four-step procedure outlined in Section 4.2.5.3.3 for solving the REEs will be demonstrated on a numerical example by applying it to solve the REEs for the eight-state multichain MCR model of a production process for which the transition matrix and reward vector are shown in Equation (4.161). The system of 16 REEs consists of a system of eight VDEs plus a system of eight GSEs. The system of eight VDEs is

$$
g_i + v_i = q_i + p_{i1}v_1 + p_{i2}v_2 + p_{i3}v_3 + p_{i4}v_4 + p_{i5}v_5 + p_{i6}v_6 + p_{i7}v_7 + p_{i8}v_8,
$$
$$
\text{for } i = 1, 2, \dots, 8.
$$

(4.204)

The eight individual value determination equations are

$$
\left.
\begin{aligned}
g_1 + v_1 &= 156 + v_1 \\
g_2 + v_2 &= 56 + v_2 \\
g_3 + v_3 &= 160 + 0.5v_3 + 0.3v_4 + 0.2v_5 \\
g_4 + v_4 &= 123 + 0.3v_3 + 0.45v_4 + 0.25v_5 \\
g_5 + v_5 &= 291.5 + 0.1v_3 + 0.35v_4 + 0.55v_5 \\
g_6 + v_6 &= 450 + 0.2v_1 + 0.16v_2 + 0.04v_3 + 0.03v_4 + 0.02v_5 + 0.55v_6 \\
g_7 + v_7 &= 400 + 0.15v_1 + 0.2v_6 + 0.65v_7 \\
g_8 + v_8 &= 420 + 0.1v_1 + 0.15v_7 + 0.75v_8.
\end{aligned}
\right\}
\tag{4.205}
$$

The set of eight GSEs is

$$
g_i = p_{i1}g_1 + p_{i2}g_2 + p_{i3}g_3 + p_{i4}g_4 + p_{i5}g_5 + p_{i6}g_6 + p_{i7}g_7 + p_{i8}g_8,
\tag{4.206}
$$
for $i = 1, 2, \dots, 8$.

The eight individual GSEs are

$$
\left.
\begin{aligned}
g_1 &= g_1 \\
g_2 &= g_2 \\
g_3 &= 0.5g_3 + 0.3g_4 + 0.2g_5 \\
g_4 &= 0.3g_3 + 0.45g_4 + 0.25g_5 \\
g_5 &= 0.1g_3 + 0.35g_4 + 0.55g_5 \\
g_6 &= 0.2g_1 + 0.16g_2 + 0.04g_3 + 0.03g_4 + 0.02g_5 + 0.55g_6 \\
g_7 &= 0.15g_1 + 0.2g_6 + 0.65g_7 \\
g_8 &= 0.1g_1 + 0.15g_7 + 0.75g_8.
\end{aligned}
\right\}
\tag{4.207}
$$

Step 1. Letting $R_3 = \{3, 4, 5\}$ denote the third closed class of three recurrent states that has an independent gain denoted by g_{R_3}, and substituting

$$
g_3 = g_4 = g_5 = g_{R_3},
$$

the eight VDEs appear as

$$
\left.\begin{aligned}
g_1 + v_1 &= 156 + v_1 \\
g_2 + v_2 &= 56 + v_2 \\
g_R + v_3 &= 160 + 0.5v_3 + 0.3v_4 + 0.2v_5 \\
g_R + v_4 &= 123 + 0.3v_3 + 0.45v_4 + 0.25v_5 \\
g_R + v_5 &= 291.5 + 0.1v_3 + 0.35v_4 + 0.55v_5 \\
g_6 + v_6 &= 450 + 0.2v_1 + 0.16v_2 + 0.04v_3 + 0.03v_4 + 0.02v_5 + 0.55v_6 \\
g_7 + v_7 &= 400 + 0.15v_1 + 0.2v_6 + 0.65v_7 \\
g_8 + v_8 &= 420 + 0.1v_1 + 0.15v_7 + 0.75v_8.
\end{aligned}\right\} \quad (4.208)
$$

Setting $v_1 = 0$ for the absorbing state in the first recurrent chain, the first VDE yields the gain $g_1 = 156$ for absorbing state 1. Similarly, setting $v_2 = 0$ for the absorbing state in the second recurrent chain, the second VDE produces the gain $g_2 = 56$ for absorbing state 2. Setting $v_5 = 0$ for the highest numbered state in the third recurrent closed class, the three VDEs for the third recurrent chain appear as

$$
\begin{aligned}
g_{R_3} + v_3 &= 160 + 0.5v_3 + 0.3v_4 \\
g_{R_3} + v_4 &= 123 + 0.3v_3 + 0.45v_4 \\
g_{R_3} &= 291.5 + 0.1v_3 + 0.35v_4.
\end{aligned} \quad (4.209)
$$

These three VDEs are solved simultaneously for g_{R_3}, v_3, and v_4 to obtain the quantities $g_{R_3} = 190.44$, $v_3 = -199.86$, and $v_4 = -231.64$ for the third recurrent chain.

Step 2. Substituting

$$
g_3 = g_4 = g_5 = g_{R_3}, \quad (4.210)
$$

the three GSEs for the transient states are

$$
\left.\begin{aligned}
g_6 &= 0.2g_1 + 0.16g_2 + (0.04 + 0.03 + 0.02)g_{R_3} + 0.55g_6 \\
g_7 &= 0.15g_1 + 0.2g_6 + 0.65g_7 \\
g_8 &= 0.1g_1 + 0.15g_7 + 0.75g_8.
\end{aligned}\right\} \quad (4.211)
$$

After algebraic simplification, the following result is obtained for the gains of the three transient states expressed as weighted averages of the independent gains of the two absorbing states and of the recurrent closed class.

$$\left.\begin{array}{l} g_6 = 0.4444g_1 + 0.3556g_2 + 0.2g_{R_3} \\ g_7 = 0.6825g_1 + 0.2032g_2 + 0.1143g_{R_3} \\ g_8 = 0.8095g_1 + 0.1219g_2 + 0.0686g_{R_3}. \end{array}\right\} \qquad (4.212)$$

Substituting the independent gains, $g_1 = 156$, $g_2 = 56$, and $g_{R_3} = 190.44$, computed for the three recurrent chains in step 1, the gains of the three transient states are

$$\left.\begin{array}{l} g_6 = 0.4444(156) + 0.3556(56) + 0.2(190.44) = 127.33 \\ g_7 = 0.6825(156) + 0.2032(56) + 0.1143(190.44) = 139.62 \\ g_8 = 0.8095(156) + 0.1219(56) + 0.0686(190.44) = 146.17. \end{array}\right\} \qquad (4.213)$$

Step 3. The three VDEs for the three transient states are

$$\left.\begin{array}{l} g_6 + v_6 = 450 + 0.2v_1 + 0.16v_2 + 0.04v_3 + 0.03v_4 + 0.02v_5 + 0.55v_6 \\ g_7 + v_7 = 400 + 0.15v_1 + 0.2v_6 + 0.65v_7 \\ g_8 + v_8 = 420 + 0.1v_1 + 0.15v_7 + 0.75v_8. \end{array}\right\} \qquad (4.214)$$

Substituting the gains, $g_6 = 127.33$, $g_7 = 139.62$, and $g_8 = 146.17$, of the three transient states computed in step 2, and substituting the relative values, $v_3 = -199.86$, $v_4 = -231.64$, and $v_1 = v_2 = v_5 = 0$, determined in step 1, the three VDEs for the transient states are

$$\left.\begin{array}{l} 127.33 + v_6 = 450 + 0.04(-199.86) + 0.03(-231.64) + 0.55v_6 \\ 139.62 + v_7 = 400 + 0.2v_6 + 0.65v_7 \\ 146.17 + v_8 = 420 + 0.15v_7 + 0.75v_8. \end{array}\right\} \qquad (4.215)$$

Step 4. The solution of the three VDEs for the transient states produces the relative values

$$v_6 = 683.84, \ v_7 = 1{,}134.71, \quad \text{and} \quad v_8 = 1{,}776.14.$$

Note that in this multichain model, $N = 8$ states, and $L = 3$ recurrent chains. The unknown quantities are the $L = 3$ independent gains and the $N = 8$ relative values, v_i. In step 1, the relative value v_1 is equated to zero in the first recurrent chain, v_2 is equated to zero in the second, and v_5 is equated to zero in the third. Also in step 1, the three independent gains, g_1, g_2, and g_{R_3}, are determined by solving the VDEs for each recurrent chain. The remaining $(N - L)$ or $(8 - 3)$ relative values are v_3, v_4, v_6, v_7, and v_8. Step 1 has further reduced the number of independent unknowns from 5 to 3 because the relative values v_3 and v_4 for the third recurrent chain are obtained by solving the VDEs for the third recurrent chain. In step 2, the GSEs for the transient states are solved to calculate the dependent gains of the transient states, g_6, g_7, and g_8, as weighted averages of the independent gains, g_1, g_2, and g_{R_3}, of the recurrent chains. Finally, in step 4, the VDEs for the transient states are solved to obtain the relative values of the transient states, v_6, v_7, and v_8.

The complete solutions for the gain vector and the vector of relative values within each class of states are given below:

$$\mathbf{g} = \begin{array}{c} 1 \\ 2 \\ 3 \\ 4 \\ 5 \\ 6 \\ 7 \\ 8 \end{array} \begin{bmatrix} 156 \\ \hline 56 \\ \hline 190.44 \\ 190.44 \\ 190.44 \\ \hline 127.33 \\ 139.62 \\ 146.17 \end{bmatrix}, \quad \mathbf{v} = \begin{array}{c} 1 \\ 2 \\ 3 \\ 4 \\ 5 \\ 6 \\ 7 \\ 8 \end{array} \begin{bmatrix} 0 \\ \hline 0 \\ \hline -199.86 \\ -231.64 \\ 0 \\ \hline 683.84 \\ 1,134.71 \\ 1,776.14 \end{bmatrix}. \tag{4.216}$$

Thus, if the process starts in any of the three recurrent states associated with the training center, the expected cost per item sold will be \$190.44. If the process starts in recurrent state 3, the expected total cost will be \$199.86 lower than if it starts in recurrent state 5. Similarly, if the process starts in recurrent state 4, the expected total cost will be \$231.64 lower than if it starts in recurrent state 5.

4.2.5.4 Expected Total Reward before Passage to a Closed Set

In Section 4.2.4.4, the expected total reward earned by a unichain MCR model before passage to a closed class of recurrent states, given that the chain started in a transient state, was calculated. This section will demonstrate that this calculation can also be made for a multichain MCR model [5].

4.2.5.4.1 Multichain Eight-State MCR Model of a Production Process

Consider the eight-state multichain MCR model of a production process for which the transition matrix and reward vector are shown in Equation (4.161). Recall that $T = \{6, 7, 8\}$ denotes the class of transient states associated with the three production stages. Transient state i represents production stage $(9 - i)$. Suppose that the operation costs for the transient states, calculated in the right-hand column of Table 4.13, are treated as the components of a production stage operation cost vector denoted by q_T. The production operation cost vector for the three transient states is

$$\mathbf{q}_T = [\$450 \quad \$400 \quad \$420]^T. \tag{4.217}$$

The expected total cost vector $U q_T$ is computed below.

$$U q_T = (\mathbf{I} - \mathbf{Q})^{-1} q_T = \begin{array}{c} 6 \\ 7 \\ 8 \end{array} \begin{bmatrix} 2.2222 & 0 & 0 \\ 1.2698 & 2.8571 & 0 \\ 0.7619 & 1.7143 & 4 \end{bmatrix} \begin{bmatrix} \$450 \\ \$400 \\ \$420 \end{bmatrix} = \begin{array}{c} 6 \\ 7 \\ 8 \end{array} \begin{bmatrix} \$10,000 \\ \$1,714.25 \\ \$2,708.58 \end{bmatrix} \tag{4.218}$$

Since q_T is the vector of operation costs for the transient states, the ith component of the vector $U q_T$ represents the expected total operation cost for an item before its eventual passage to an absorbing state or to the recurrent closed class, given that the item started in transient state i. An item enters the production process at production stage 1, which is transient state 8. With reference to the last three rows in the right-hand column of Table 4.14, note that:

Expected operation cost that will be incurred by an entering item in state 6

+ Expected operation cost that will be incurred by an entering item in state 7

+ Expected operation cost that will be incurred by an entering item in state 8

$= q_6 u_{86} + q_7 u_{87} + q_8 u_{88}$

$= (\$450)(0.7619) + (\$400)(1.7143) + (\$420)(4)$

$= \$342.86$ in state 6 $+ \$685.72$ in state 7 $+ \$\$1,680$ in state 8

$= \$2,708.58,$ \hfill (4.219)

which is the expected total operation cost for an item before its eventual passage to an absorbing state or to the recurrent closed class, given that the item started in transient state 8. The expected total operation cost of $2,708.58 for an item entering the production process at stage 1 is the last entry in the vector $U q_T$ calculated in Equation (4.218), and is also the sum of the entries in the last three rows of the right-hand column of Table 4.14.

This example has demonstrated that the following result holds for any multichain MCR. Suppose that P is the transition matrix for a multichain MCR, U is the fundamental matrix, and q_T is the vector of rewards received in the transient states. Then the ith component of the vector $U q_T$ represents the expected total reward earned before eventual passage to any recurrent closed class, given that the chain started in transient state i.

4.2.5.4.2 Absorbing Multichain MCR Model of Selling a Stock with Two Target Prices

This section gives a second example of the calculation of expected total rewards earned by a multichain MCR model before passage to a recurrent chain, given that the chain started in a transient state [5]. Suppose that on March 31, at the end of a 3-month quarter, a woman buys one share of a certain stock for $40. The share price, rounded to the nearest $5, has been varying among $30, $35, $40, $45, and $50 from quarter to quarter. She plans to sell her stock at the end of the first quarter in which the share price rises to $50 or falls to $30. She believes that the price of the stock can be modeled as a Markov chain in which the state, X_n, denotes the share price at the end of quarter n. The state space is $E = \{\$30, \$35, \$40, \$45, \$50\}$. The two states $X_n = \$30$ and $X_n = \$50$ are absorbing states, reached when the stock is sold. The three remaining states, which are entered when the stock is held, are transient. (A model for selling a stock with one target price was constructed in Section 1.10.1.2.1.) The quarterly dividend is $2 per share. No dividend is received when the stock is sold. She believes that her investment can be represented as a multichain MCR with the following transition probability matrix, expressed in canonical form, and the associated reward vector.

$$P = \begin{array}{c|ccccc} \text{State} & 30 & 50 & 35 & 40 & 45 \\ \hline 30 & 1 & 0 & 0 & 0 & 0 \\ 50 & 0 & 1 & 0 & 0 & 0 \\ 35 & 0.20 & 0.10 & 0.40 & 0.20 & 0.10 \\ 40 & 0.10 & 0.06 & 0.30 & 0.28 & 0.26 \\ 45 & 0.14 & 0.12 & 0.24 & 0.30 & 0.20 \end{array} = \begin{bmatrix} 1 & 0 & 0 \\ 0 & 1 & 0 \\ D_1 & D_2 & Q \end{bmatrix}, \quad (4.220a)$$

$$q = \begin{array}{c|c} \text{State} & \text{Reward} \\ \hline 30 & 30 \\ 50 & 50 \\ 35 & 2 \\ 40 & 2 \\ 45 & 2 \end{array} = \begin{bmatrix} q_{A1} \\ q_{A2} \\ q_T \end{bmatrix}, \quad (4.220b)$$

where

$$D_1 = \begin{array}{c|c} \text{State} & 30 \\ \hline 35 & 0.20 \\ 40 & 0.10 \\ 45 & 0.14 \end{array}, \quad D_2 = \begin{array}{c|c} \text{State} & 50 \\ \hline 35 & 0.10 \\ 40 & 0.06 \\ 45 & 0.12 \end{array}, \quad Q = \begin{array}{c|ccc} \text{State} & 35 & 40 & 45 \\ \hline 35 & 0.40 & 0.20 & 0.10 \\ 40 & 0.30 & 0.28 & 0.26 \\ 45 & 0.24 & 0.30 & 0.20 \end{array},$$

$$q_T = \begin{array}{c|c} \text{State} & \text{Reward} \\ \hline 35 & 2 \\ 40 & 2 \\ 45 & 2 \end{array},$$

$$q_A = \begin{array}{c|c} \text{State} & \text{Reward} \\ \hline 30 & q_{A1} \\ 50 & q_{A2} \end{array} = \begin{array}{c|c} \text{State} & \text{Reward} \\ \hline 30 & 30 \\ 50 & 50 \end{array}. \tag{4.220c}$$

Note that q_A is the vector of selling prices received in the two absorbing states, and q_T is the vector of dividends received in the three transient states. The fundamental matrix for the submatrix Q is

$$U = (I-Q)^{-1} = \begin{bmatrix} 0.60 & -0.20 & -0.10 \\ -0.30 & 0.72 & -0.26 \\ -0.24 & -0.30 & 0.80 \end{bmatrix}^{-1} = \begin{array}{c|ccc} & 35 & 40 & 45 \\ \hline 35 & 2.3486 & 0.8961 & 0.5848 \\ 40 & 1.4261 & 2.1505 & 0.8772 \\ 45 & 1.2394 & 1.0753 & 1.7544 \end{array}. \tag{4.221}$$

The vector, Uq_T, is computed below:

$$Uq_T = (I-Q)^{-1}q_T = \begin{array}{c} 35 \\ 40 \\ 45 \end{array}\begin{bmatrix} 2.3486 & 0.8961 & 0.5848 \\ 1.4261 & 2.1505 & 0.8772 \\ 1.2394 & 1.0753 & 1.7544 \end{bmatrix}\begin{bmatrix} 2 \\ 2 \\ 2 \end{bmatrix} = \begin{array}{c} 35 \\ 40 \\ 45 \end{array}\begin{bmatrix} 7.66 \\ 8.91 \\ 8.14 \end{bmatrix}. \tag{4.222}$$

As Sections 4.2.4.4 and 4.2.5.4.1 indicate, the ith component of the vector Uq_T represents the expected total dividend earned before eventual passage to an absorbing state when the stock is sold, given that the chain started in transient state i. For example, if she buys the stock for \$40, she will earn an expected total dividend of $(Uq_T)_{40} = \$8.91$ before the stock is sold.

When the stock is sold, the chain will be absorbed in state \$30 or in state \$50. Using Equation (3.98), the matrix of absorption probabilities for an absorbing multichain, given that the process starts in a transient state, is

$$F = (I-Q)^{-1}D = UD = U[D_1 \quad D_2] = \begin{array}{c|cc} & 30 & 50 \\ \hline 35 & f_{35,30} & f_{35,50} \\ 40 & f_{40,30} & f_{40,50} \\ 45 & f_{45,30} & f_{45,50} \end{array}$$

$$= \begin{array}{c|ccc} & 35 & 40 & 45 \\ \hline 35 & 2.3486 & 0.8961 & 0.5848 \\ 40 & 1.4261 & 2.1505 & 0.8772 \\ 45 & 1.2394 & 1.0753 & 1.7544 \end{array}\begin{array}{|cc} 30 & 50 \\ \hline 0.20 & 0.10 \\ 0.10 & 0.06 \\ 0.14 & 0.12 \end{array} = \begin{array}{c|cc} & 30 & 50 \\ \hline 35 & 0.6412 & 0.3588 \\ 40 & 0.6231 & 0.3769 \\ 45 & 0.6010 & 0.3990 \end{array}. \tag{4.223}$$

If the investor buys the stock for \$40, she will eventually sell it for either \$30 with probability $f_{40,30} = 0.6231$ or \$50 with probability $f_{40,50} = 0.3769$. Hence, when the stock is sold, she will receive an expected selling price of

$$\$30 f_{40,30} + \$50 f_{40,50} = \$30(0.6231) + \$50(0.3769) = \$37.54. \tag{4.224}$$

To extend this result, suppose that a reward is received when an absorbing multichain MCR enters an absorbing state. In this example the reward is the selling price. The vector, Fq_A, of expected rewards received when the process is absorbed, given that the MCR started in a transient state, is computed below:

$$Fq_A = \begin{array}{c|cc} & 30 & 50 \\ \hline 35 & f_{35,30} & f_{35,50} \\ 40 & f_{40,30} & f_{40,50} \\ 45 & f_{45,30} & f_{45,50} \end{array} \begin{bmatrix} q_{A1} \\ q_{A2} \end{bmatrix} = \begin{array}{c} 35 \\ 40 \\ 45 \end{array} \begin{bmatrix} f_{35,30}q_{A1} + f_{35,50}q_{A2} \\ f_{40,30}q_{A1} + f_{40,50}q_{A2} \\ f_{45,30}q_{A1} + f_{45,50}q_{A2} \end{bmatrix}$$

$$= \begin{array}{c|cc} & 30 & 50 \\ \hline 35 & 0.6412 & 0.3588 \\ 40 & 0.6231 & 0.3769 \\ 45 & 0.6010 & 0.3990 \end{array} \begin{bmatrix} 30 \\ 50 \end{bmatrix} = \begin{array}{c} 35 \\ 40 \\ 45 \end{array} \begin{bmatrix} \$37.18 \\ \$37.54 \\ \$37.98 \end{bmatrix}. \tag{4.225}$$

Thus, the ith component of the vector Fq_A represents the expected selling price received when the stock is sold, given that the chain started in transient state i. For example, if the investor buys the stock for \$40, she will receive an expected selling price of

$$(Fq_A)_{40} = f_{40,30}q_{A1} + f_{40,50}q_{A2} = \$37.54 \tag{4.226}$$

when the stock is sold, confirming the earlier result. The investor's expected total reward, given that she bought the stock for \$40, is the expected dividends received before selling plus the expected selling price, or

$$(Uq_T)_{40} + (Fq_A)_{40} = (Uq_T)_{40} + (\$30 f_{40,30} + \$50 f_{40,50}) = \$8.91 + \$37.54 = \$46.45. \tag{4.227}$$

4.2.5.5 Value Iteration over a Finite Planning Horizon

As Section 4.2.4.5 indicates, value iteration can be executed to calculate the expected total reward received by a reducible MCR over a finite planning horizon. To demonstrate this calculation for a multichain MCR, consider the absorbing multichain MCR model of selling a stock with two target prices described in Section 4.2.5.4.2. The transition matrix P and reward vector q are shown in Equation (4.220). Value iteration will be executed to calculate the expected total income that will be earned by the investor's one share of stock after three quarters, given that she paid $40 for the stock. She may not keep her share for three quarters because she will sell her share at the end of the first quarter in which the share price rises to $50 or falls to $30.

The matrix form of the value iteration equation (4.29) is

$$v(n) = q + Pv(n+1) \quad \text{for } n = 0,1,2. \tag{4.228}$$

To begin the backward recursion, the vector of expected total income received at the end of the three quarter planning horizon is set equal to zero for all states.

$$n = T = 3 \tag{4.229a}$$

$$v(3) = v(T) = 0 \tag{4.229b}$$

$$\begin{bmatrix} v_{30}(3) \\ v_{50}(3) \\ v_{35}(3) \\ v_{40}(3) \\ v_{45}(3) \end{bmatrix} = \begin{bmatrix} 0 \\ 0 \\ 0 \\ 0 \\ 0 \end{bmatrix} \tag{4.229c}$$

$n = 2$

$v(2) = q + Pv(3)$

$$\begin{bmatrix} v_{30}(2) \\ v_{50}(2) \\ v_{35}(2) \\ v_{40}(2) \\ v_{45}(2) \end{bmatrix} = \begin{bmatrix} 30 \\ 50 \\ 2 \\ 2 \\ 2 \end{bmatrix} + \begin{bmatrix} 1 & 0 & 0 & 0 & 0 \\ 0 & 1 & 0 & 0 & 0 \\ 0.20 & 0.10 & 0.40 & 0.20 & 0.10 \\ 0.10 & 0.06 & 0.30 & 0.28 & 0.26 \\ 0.14 & 0.12 & 0.24 & 0.30 & 0.20 \end{bmatrix} \begin{bmatrix} 0 \\ 0 \\ 0 \\ 0 \\ 0 \end{bmatrix} = \begin{bmatrix} 30 \\ 50 \\ 2 \\ 2 \\ 2 \end{bmatrix} \Bigg\} \tag{4.230}$$

$n = 1$

$v(1) = q + Pv(2)$

$$\begin{bmatrix} v_{30}(1) \\ v_{50}(1) \\ v_{35}(1) \\ v_{40}(1) \\ v_{45}(1) \end{bmatrix} = \begin{bmatrix} 30 \\ 50 \\ 2 \\ 2 \\ 2 \end{bmatrix} + \begin{bmatrix} 1 & 0 & 0 & 0 & 0 \\ 0 & 1 & 0 & 0 & 0 \\ 0.20 & 0.10 & 0.40 & 0.20 & 0.10 \\ 0.10 & 0.06 & 0.30 & 0.28 & 0.26 \\ 0.14 & 0.12 & 0.24 & 0.30 & 0.20 \end{bmatrix} \begin{bmatrix} 30 \\ 50 \\ 2 \\ 2 \\ 2 \end{bmatrix} = \begin{bmatrix} 60 \\ 100 \\ 14.4 \\ 9.68 \\ 13.68 \end{bmatrix} \Bigg\} \tag{4.231}$$

$n = 0$

$v(0) = q + Pv(1)$

$$\begin{bmatrix} v_{30}(0) \\ v_{50}(0) \\ v_{35}(0) \\ v_{40}(0) \\ v_{45}(0) \end{bmatrix} = \begin{bmatrix} 30 \\ 50 \\ 2 \\ 2 \\ 2 \end{bmatrix} + \begin{bmatrix} 1 & 0 & 0 & 0 & 0 \\ 0 & 1 & 0 & 0 & 0 \\ 0.20 & 0.10 & 0.40 & 0.20 & 0.10 \\ 0.10 & 0.06 & 0.30 & 0.28 & 0.26 \\ 0.14 & 0.12 & 0.24 & 0.30 & 0.20 \end{bmatrix} \begin{bmatrix} 60 \\ 100 \\ 14.4 \\ 9.68 \\ 13.68 \end{bmatrix} = \begin{bmatrix} 90 \\ 150 \\ 33.064 \\ 24.5872 \\ 31.496 \end{bmatrix}.$$

(4.232)

If the woman paid \$40 for the stock, her expected total income after three quarters will be $v_{40}(0) = \$24.59$.

4.3 Discounted Rewards

In Section 4.2, the earnings of an MCR have been measured by the expected total reward earned over a finite planning horizon, or by the expected average reward or gain earned over an infinite horizon. An alternative measure is one that reflects the time value of money. The measure of an MCR's earnings used in this section is based on the time value of money, and is called either the present value of the expected total reward, or the expected present value, or the expected total discounted reward. In this book, all three terms are used interchangeably. When the earnings of an MCR are measured by the expected present value, the process is called an MCR with discounted rewards, or a discounted MCR. Markov chain structure is not relevant when a discounted MCR is analyzed [3, 4, 7].

4.3.1 Time Value of Money

The time value of money is a consequence of the fact that a dollar received now is worth more than a dollar received in the future because money received now can be invested to earn interest. Conversely, one dollar in the future is worth less than one dollar in the present. Suppose that money earns compound interest at a positive interest rate of $i\%$ per period. A discount factor, denoted by α, is defined as

$$\alpha = \frac{1}{1+i}, \tag{4.233}$$

where $0 < \alpha < 1$.

Clearly, α is a fraction between zero and one. Thus, q dollars received one period in the future is equivalent to αq dollars received now, a smaller quantity. The present value, at epoch 0, of q dollars received n periods in the future, at epoch n, is $\alpha^n q$ dollars. Rewards of q dollars received at epochs 0, 1, 2, ... , n have the respective present values of q, αq, $\alpha^2 q$, ... , $\alpha^n q$ dollars. Note that α^n is the single-payment present-worth factor used in engineering economic analysis to compute the present worth of a single payment received n periods in the future.

4.3.2 Value Iteration over a Finite Planning Horizon

When the planning horizon is finite, of length T periods, the expected total discounted reward received in every state can be found by using value iteration with a discount factor α [2–4, 7].

4.3.2.1 Value Iteration Equation

Suppose that a Markov chain with discounted rewards, which is called a discounted MCR, has N states. Let $v_i(n)$ denote the value, at epoch n, of the expected total discounted reward earned during the $T - n$ periods from epoch n to epoch T, the end of the planning horizon, if the system is in state i at epoch n. The value, at epoch 0, of the expected total discounted reward earned during the T periods from epoch 0 to epoch T, if the system starts in state i at epoch 0, is denoted by $v_i(0)$. For brevity, $v_i(0)$ will be termed either an expected total discounted reward or an expected present value if the system starts in state i. By following a procedure similar to the one used in Section 4.2.2.2.1 for a Markov chain with undiscounted rewards, a backward recursive value iteration equation relating $v_i(n)$ to $v_j(n + 1)$ will be developed. The set of expected total discounted rewards for all states received at epoch n is collected in an N-component expected total discounted reward vector,

$$v(n) = [v_1(n) \quad v_2(n) \quad \cdots \quad v_N(n)]^{\mathrm{T}}. \tag{4.234}$$

Figure 4.12 is a discounted cash flow diagram, which shows the present values, with a discount factor, α, of a series of equal reward vectors, q, received at epochs 0 through $T - 1$. The present value of a reward vector, q,

$$
\begin{array}{cccccc}
q & \alpha q & \alpha^2 q & \cdots & \alpha^{T-1}q & \alpha^T v(T) & \text{Present value of reward} \\
\uparrow & \uparrow & \uparrow & \cdots & \uparrow & \uparrow & \\
\hline
0 & 1 & 2 & \cdots & T-1 & T & \text{Epoch}
\end{array}
$$

FIGURE 4.12
Present values of a discounted cash flow diagram for reward vectors over a planning horizon of length T periods.

received at epoch n, is $\alpha^n q$. At epoch T, a vector of salvage values denoted by $v(T)$ is received.

The expected total discounted reward vector received at epoch 0 is denoted by $v(0)$. Note that the reward vector received now, at epoch 0, is q. The expected present value of the reward vector received one period from now, at epoch 1, is $P(\alpha q) = (\alpha P)q$. The expected present value of the reward vector received two periods from now, at epoch 2, is $P^2(\alpha^2 q) = (\alpha P)^2 q$. Similarly, the expected present value of the reward vector received $T - 1$ periods from now, at epoch $T - 1$, is $(\alpha P)^{T-1} q$. Finally, the expected present value of the vector of salvage values is $(\alpha P)^T v(T)$, where $v(T)$ is the vector of salvage values received at epoch T. Expected reward vectors are additive. That is, the expected value of a sum of expected reward vectors is the sum of their expected values. Therefore,

$$v(0) = q + \alpha Pq + (\alpha P)^2 q + (\alpha P)^3 q + \cdots + (\alpha P)^{T-2} q + (\alpha P)^{T-1} q + (\alpha P)^T v(T)$$
$$= q + \alpha P[q + \alpha Pq + (\alpha P)^2 q + (\alpha P)^3 q + \cdots + (\alpha P)^{T-3} q + (\alpha P)^{T-2} q + (\alpha P)^{T-1} v(T)].$$

$$(4.235)$$

Observe that the term in brackets, obtained by factoring out αP, is equal to $v(1)$, which represents the vector, at epoch 1, of the expected total discounted income received from epoch 1 until the end of the planning horizon. That is,

$$v(1) = q + \alpha Pq + (\alpha P)^2 q + (\alpha P)^3 q + \cdots + (\alpha P)^{T-2} q + (\alpha P)^{T-1} v(T). \quad (4.236)$$

Thus, the recursive relationship between the vectors $v(0)$ and $v(1)$ is

$$v(0) = q + \alpha Pv(1). \quad (4.237)$$

By following an argument analogous to the one used in Section 4.2.2.2.1 for a Markov chain with undiscounted rewards, this result can be generalized to produce the following recursive value iteration equation in matrix form for a discounted MCR relating the vector $v(n)$ at epoch n to the vector $v(n + 1)$ at epoch $n + 1$.

$$v(n) = q + \alpha Pv(n+1), \quad \text{for } n = 0, 1, ..., T-1, \text{ where } \mathbf{v(T)} \text{ is specified.} \quad (4.238)$$

When rewards are discounted, chain structure is not relevant because all row sums in a matrix, αP, are less than one. Thus, the entries of αP are not probabilities. For this reason, the value iteration procedure is the same for all discounted MCRs, irrespective of whether the associated Markov chains are recurrent, unichain, or multichain.

Consider a generic four-state discounted MCR with transition probability matrix, P, discount factor, α, and reward vector, q. When $N = 4$ the expanded matrix form of the recursive value iteration equation (4.238) is

$$
\begin{bmatrix} v_1(n) \\ v_2(n) \\ v_3(n) \\ v_4(n) \end{bmatrix} = \begin{bmatrix} q_1 \\ q_2 \\ q_3 \\ q_4 \end{bmatrix} + \begin{bmatrix} \alpha p_{11} & \alpha p_{12} & \alpha p_{13} & \alpha p_{14} \\ \alpha p_{21} & \alpha p_{22} & \alpha p_{23} & \alpha p_{24} \\ \alpha p_{31} & \alpha p_{32} & \alpha p_{33} & \alpha p_{34} \\ \alpha p_{41} & \alpha p_{42} & \alpha p_{43} & \alpha p_{44} \end{bmatrix} \begin{bmatrix} v_1(n+1) \\ v_2(n+1) \\ v_3(n+1) \\ v_4(n+1) \end{bmatrix}. \tag{4.239}
$$

In expanded algebraic form the four recursive value iteration equations are

$$
\left.
\begin{aligned}
v_1(n) &= q_1 + \alpha p_{11} v_1(n+1) + \alpha p_{12} v_2(n+1) + \alpha p_{13} v_3(n+1) + \alpha p_{14} v_4(n+1) \\
v_2(n) &= q_2 + \alpha p_{21} v_1(n+1) + \alpha p_{22} v_2(n+1) + \alpha p_{23} v_3(n+1) + \alpha p_{24} v_4(n+1) \\
v_3(n) &= q_3 + \alpha p_{31} v_1(n+1) + \alpha p_{32} v_2(n+1) + \alpha p_{33} v_3(n+1) + \alpha p_{34} v_4(n+1) \\
v_4(n) &= q_4 + \alpha p_{41} v_1(n+1) + \alpha p_{42} v_2(n+1) + \alpha p_{43} v_3(n+1) + \alpha p_{44} v_4(n+1)
\end{aligned}
\right\}
$$

$$\tag{4.240}$$

In compact algebraic form the four recursive value iteration equations are

$$
\left.
\begin{aligned}
v_1(n) &= q_1 + \alpha \sum_{j=1}^{4} p_{1j} v_j(n+1) \\[6pt]
v_2(n) &= q_2 + \alpha \sum_{j=1}^{4} p_{2j} v_j(n+1) \\[6pt]
v_3(n) &= q_3 + \alpha \sum_{j=1}^{4} p_{3j} v_j(n+1) \\[6pt]
v_4(n) &= q_4 + \alpha \sum_{j=1}^{4} p_{4j} v_j(n+1).
\end{aligned}
\right\}
$$

$$\tag{4.241}$$

The four algebraic equations can be represented by the single equation

$$
v_i(n) = q_i + \alpha \sum_{j=1}^{4} p_{ij} v_j(n+1), \tag{4.242}
$$

for $n = 0, 1, \ldots, T-1$, and $i = 1, 2, 3,$ and 4.

This result can be generalized to apply to any N-state Markov chain with discounted rewards. Therefore, $v_i(n)$ can be computed by using the following recursive value iteration equation in algebraic form which relates $v_i(n)$ to $v_j(n + 1)$.

$$v_i(n) = q_i + \alpha \sum_{j=1}^{N} p_{ij} v_j(n+1),$$

for $n = 0, 1, ..., T-1$, $i = 1, 2, ..., N$, where $v_i(T)$ is specified for all states i.

(4.243)

In summary, the value iteration equation expresses a backward recursive relationship because it starts with a known set of salvage values for vector $v(T)$ at the end of the planning horizon. Next, vector $v(T-1)$ is calculated in terms of $v(T)$. The value iteration procedure moves backward one epoch at a time by calculating $v(n)$ in terms of $v(n+1)$. The backward recursive procedure stops at epoch 0 after $v(0)$ is calculated in terms of $v(1)$. The component $v_i(0)$ of $v(0)$ is the expected total discounted reward received until the end of the planning horizon if the system starts in state i at epoch 0.

4.3.2.2 Value Iteration for Discounted MCR Model of Monthly Sales

Consider a discounted MCR model of monthly sales. This model was introduced without discounting in Section 4.2.2.1 and solved by value iteration over a finite horizon in Section 4.2.2.2.2. The four-state discounted MCR model has the following transition probability matrix P, discount factor $\alpha = 0.9$, matrix αP, and reward vector q:

$$P = \begin{matrix} 1 \\ 2 \\ 3 \\ 4 \end{matrix} \begin{bmatrix} 0.60 & 0.30 & 0.10 & 0 \\ 0.25 & 0.30 & 0.35 & 0.10 \\ 0.05 & 0.25 & 0.50 & 0.20 \\ 0 & 0.10 & 0.30 & 0.60 \end{bmatrix}, \; \alpha P = \begin{matrix} 1 \\ 2 \\ 3 \\ 4 \end{matrix} \begin{bmatrix} 0.540 & 0.270 & 0.090 & 0 \\ 0.225 & 0.270 & 0.315 & 0.090 \\ 0.045 & 0.225 & 0.450 & 0.180 \\ 0 & 0.090 & 0.270 & 0.540 \end{bmatrix}, \; q = \begin{bmatrix} -20 \\ 5 \\ -5 \\ 25 \end{bmatrix}.$$

(4.244)

(Rewards are in thousands of dollars.) In matrix form, the value iteration equation (4.239) for the discounted MCR model of monthly sales is

$$\begin{bmatrix} v_1(n) \\ v_2(n) \\ v_3(n) \\ v_4(n) \end{bmatrix} = \begin{bmatrix} -20 \\ 5 \\ -5 \\ 25 \end{bmatrix} + \begin{bmatrix} 0.540 & 0.270 & 0.090 & 0 \\ 0.225 & 0.270 & 0.315 & 0.090 \\ 0.045 & 0.225 & 0.450 & 0.180 \\ 0 & 0.090 & 0.270 & 0.540 \end{bmatrix} \begin{bmatrix} v_1(n+1) \\ v_2(n+1) \\ v_3(n+1) \\ v_4(n+1) \end{bmatrix}.$$

(4.245)

Value iteration will be executed by solving the backward recursive equations in matrix form to compute the expected total discounted reward vector over a planning horizon consisting of three periods. To begin value iteration, in order to simplify computations, the salvage values at the end of the planning horizon are set equal to zero for all states.

$$n = T = 3.$$

$$\begin{bmatrix} v_1(n) \\ v_2(n) \\ v_3(n) \\ v_4(n) \end{bmatrix} = \begin{bmatrix} v_1(T) \\ v_2(T) \\ v_3(T) \\ v_4(T) \end{bmatrix} = \begin{bmatrix} v_1(3) \\ v_2(3) \\ v_3(3) \\ v_4(3) \end{bmatrix} = \begin{bmatrix} 0 \\ 0 \\ 0 \\ 0 \end{bmatrix} \tag{4.246}$$

$n = 2$

$$v(2) = q + \alpha P v(3) = q + (0.9P)v(3) = q + (0.9P)(0) = q$$

$$\begin{bmatrix} v_1(2) \\ v_2(2) \\ v_3(2) \\ v_4(2) \end{bmatrix} = \begin{bmatrix} -20 \\ 5 \\ -5 \\ 25 \end{bmatrix} + \begin{bmatrix} 0.540 & 0.270 & 0.090 & 0 \\ 0.225 & 0.270 & 0.315 & 0.09 \\ 0.045 & 0.225 & 0.450 & 0.180 \\ 0 & 0.090 & 0.270 & 0.540 \end{bmatrix} \begin{bmatrix} 0 \\ 0 \\ 0 \\ 0 \end{bmatrix} = \begin{bmatrix} -20 \\ 5 \\ -5 \\ 25 \end{bmatrix} \tag{4.247}$$

$n = 1$

$$v(1) = q + \alpha P v(2) = q + (0.9P)v(2)$$

$$\begin{bmatrix} v_1(1) \\ v_2(1) \\ v_3(1) \\ v_4(1) \end{bmatrix} = \begin{bmatrix} -20 \\ 5 \\ -5 \\ 25 \end{bmatrix} + \begin{bmatrix} 0.540 & 0.270 & 0.090 & 0 \\ 0.225 & 0.270 & 0.315 & 0.09 \\ 0.045 & 0.225 & 0.450 & 0.180 \\ 0 & 0.090 & 0.270 & 0.540 \end{bmatrix} \begin{bmatrix} -20 \\ 5 \\ -5 \\ 25 \end{bmatrix} = \begin{bmatrix} -29.9 \\ 2.525 \\ -2.525 \\ 37.6 \end{bmatrix} \tag{4.248}$$

$n = 0$

$$v(0) = q + \alpha P v(1) = q + (0.9P)v(1)$$

$$\begin{bmatrix} v_1(0) \\ v_2(0) \\ v_3(0) \\ v_4(0) \end{bmatrix} = \begin{bmatrix} -20 \\ 5 \\ -5 \\ 25 \end{bmatrix} + \begin{bmatrix} 0.540 & 0.270 & 0.090 & 0 \\ 0.225 & 0.270 & 0.315 & 0.09 \\ 0.045 & 0.225 & 0.450 & 0.180 \\ 0 & 0.090 & 0.270 & 0.540 \end{bmatrix} \begin{bmatrix} -29.9 \\ 2.525 \\ -2.525 \\ 37.6 \end{bmatrix} = \begin{bmatrix} -35.6915 \\ 1.5429 \\ -0.1456 \\ 44.8495 \end{bmatrix} \tag{4.249}$$

The vector $v(0)$ indicates that if this discounted MCR operates over a three-period planning horizon with a discount factor of 0.9, the expected total discounted reward will be −35.6915 if the system starts in state 1, 1.5429 if it starts in state 2, −0.1456 if it starts in state 3, and 44.8495 if it starts in state 4.

Using Equation (4.235) with $T = 3$, the solution by value iteration for $v(0)$ in Equation (4.249) can be verified by calculating

$$v(0) = q + \alpha Pq + (\alpha P)^2 q$$

$$
\begin{bmatrix} v_1(0) \\ v_2(0) \\ v_3(0) \\ v_4(0) \end{bmatrix} = \begin{bmatrix} -20 \\ 5 \\ -5 \\ 25 \end{bmatrix} + \begin{bmatrix} 0.540 & 0.270 & 0.090 & 0 \\ 0.225 & 0.270 & 0.315 & 0.09 \\ 0.045 & 0.225 & 0.450 & 0.180 \\ 0 & 0.090 & 0.270 & 0.540 \end{bmatrix} \begin{bmatrix} -20 \\ 5 \\ -5 \\ 25 \end{bmatrix}
$$

$$
+ \begin{bmatrix} 0.540 & 0.270 & 0.090 & 0 \\ 0.225 & 0.270 & 0.315 & 0.09 \\ 0.045 & 0.225 & 0.450 & 0.180 \\ 0 & 0.090 & 0.270 & 0.540 \end{bmatrix}^2 \begin{bmatrix} -20 \\ 5 \\ -5 \\ 25 \end{bmatrix} = \begin{bmatrix} -35.6915 \\ 1.5429 \\ -0.1456 \\ 44.8495 \end{bmatrix}
$$

$$\text{(4.250)}$$

Suppose that one period is added to the three-period planning horizon to create a four-period horizon. The expected total discounted reward vector, $v(-1)$, for the four-period horizon is calculated below, using the procedure described in Section 4.2.2.3:

$$v(-1) = q + \alpha Pv(0) = q + (0.9P)v(0)$$

$$
\begin{bmatrix} v_1(-1) \\ v_2(-1) \\ v_3(-1) \\ v_4(-1) \end{bmatrix} = \begin{bmatrix} -20 \\ 5 \\ -5 \\ 25 \end{bmatrix} + \begin{bmatrix} 0.540 & 0.270 & 0.090 & 0 \\ 0.225 & 0.270 & 0.315 & 0.090 \\ 0.045 & 0.225 & 0.450 & 0.180 \\ 0 & 0.090 & 0.270 & 0.540 \end{bmatrix} \begin{bmatrix} -35.6915 \\ 1.5429 \\ -0.1456 \\ 44.8495 \end{bmatrix} = \begin{bmatrix} -38.8699 \\ 1.3766 \\ 1.7484 \\ 49.3183 \end{bmatrix}.
$$

$$\text{(4.251)}$$

4.3.3 An Infinite Planning Horizon

Two approaches can be followed to calculate the expected total discounted rewards received by an N-state MCR operating over an infinite planning horizon. The first approach is to solve a set of VDEs for a discounted MCR to find the expected total discounted rewards earned in all states. This approach involves the solution of N simultaneous linear equations in N unknowns. The second approach, which requires less computation, is to execute value iteration over a large number of periods. However, value iteration gives only an approximate solution.

4.3.3.1 VDEs for Expected Total Discounted Rewards

Recall Equation (4.235) for calculating the vector $v(0)$ over a finite planning horizon of length T periods,

$$v(0) = [I + \alpha P + (\alpha P)^2 + (\alpha P)^3 + \cdots + (\alpha P)^{T-2} + (\alpha P)^{T-1}]q + (\alpha P)^T v(T). \quad \text{(4.235)}$$

As T approaches infinity, the finite-state Markov chain with discounted rewards enters the steady state. Since αP is a substochastic matrix for which all row sums are less than one, $\lim_{T \to \infty}(\alpha P)^T = 0$. Using formula (3.73) for the sum of an infinite series of substochastic matrices,

$$\lim_{T \to \infty}(I + (\alpha P) + (\alpha P)^2 + (\alpha P)^3 + \cdots + (\alpha P)^{T-1}) = (I - \alpha P)^{-1}. \tag{4.252}$$

Using this result,

$$\lim_{T \to \infty} v(0) = v = (I - \alpha P)^{-1} q. \tag{4.253}$$

Thus, the expected total discounted reward vector, **v**, of income received in each state over an infinite planning horizon is finite, and is equal to the inverse of the matrix $(I - \alpha P)$ times the reward vector, q. When both sides of the matrix equation (4.253) are premultiplied by the matrix $(I - \alpha P)$, the result is

$$v = (I - \alpha P)^{-1} q \tag{4.253}$$

$$(I - \alpha P)v = q \tag{4.254}$$

$$\begin{aligned} Iv - \alpha Pv &= q \\ v - \alpha Pv &= q \\ v &= q + \alpha Pv. \end{aligned} \tag{4.255}$$

For an N-state MCR, the matrix equation (4.253) relates the expected total discounted reward vector, $v = [\ (v_1, v_2, \ldots, v_N)]^T$, to the reward vector, $q = [(q_1, q_2, \ldots, q_N)]^T$. The matrix equation (4.255) represents a system of N linear equations in N unknowns. The system of equations (4.255) is called the matrix form of VDEs, for a discounted MCR. The N unknowns, $v_1, v_2, \ldots, v_N$, represent the expected total discounted rewards received in every state. An alternate form of the VDEs is the matrix equation (4.253). However, Gaussian reduction is a more efficient procedure than matrix inversion for solving a system of linear equations.

When $N = 4$ the expanded matrix form of the VDEs (4.255) is

$$\begin{bmatrix} v_1 \\ v_2 \\ v_3 \\ v_4 \end{bmatrix} = \begin{bmatrix} q_1 \\ q_2 \\ q_3 \\ q_4 \end{bmatrix} + \begin{bmatrix} \alpha p_{11} & \alpha p_{12} & \alpha p_{13} & \alpha p_{14} \\ \alpha p_{21} & \alpha p_{22} & \alpha p_{23} & \alpha p_{24} \\ \alpha p_{31} & \alpha p_{32} & \alpha p_{33} & \alpha p_{34} \\ \alpha p_{41} & \alpha p_{42} & \alpha p_{43} & \alpha p_{44} \end{bmatrix} \begin{bmatrix} v_1 \\ v_2 \\ v_3 \\ v_4 \end{bmatrix}. \tag{4.256}$$

In expanded algebraic form the four VDEs are

$$
\left.\begin{aligned}
v_1 &= q_1 + \alpha p_{11} v_1 + \alpha p_{12} v_2 + \alpha p_{13} v_3 + \alpha p_{14} v_4 \\
v_2 &= q_2 + \alpha p_{21} v_1 + \alpha p_{22} v_2 + \alpha p_{23} v_3 + \alpha p_{24} v_4 \\
v_3 &= q_3 + \alpha p_{31} v_1 + \alpha p_{32} v_2 + \alpha p_{33} v_3 + \alpha p_{34} v_4 \\
v_4 &= q_4 + \alpha p_{41} v_1 + \alpha p_{42} v_2 + \alpha p_{43} v_3 + \alpha p_{44} v_4 .
\end{aligned}\right\}
\tag{4.257}
$$

In compact algebraic form the four VDEs are

$$
\left.\begin{aligned}
v_1 &= q_1 + \alpha \sum_{j=1}^{4} p_{1j} v_j \\
v_2 &= q_2 + \alpha \sum_{j=1}^{4} p_{2j} v_j \\
v_3 &= q_3 + \alpha \sum_{j=1}^{4} p_{3j} v_j \\
v_4 &= q_4 + \alpha \sum_{j=1}^{4} p_{4j} v_j .
\end{aligned}\right\}
\tag{4.258}
$$

A more compact algebraic form of the four VDEs is

$$
v_i = q_i + \alpha \sum_{j=1}^{4} p_{ij} v_j, \quad \text{for } i = 1, 2, 3, \text{ and } 4.
\tag{4.259}
$$

This result can be generalized to apply to any N-state Markov chain with discounted rewards. Therefore, the compact algebraic form of the VDEs is

$$
v_i = q_i + \alpha \sum_{j=1}^{N} p_{ij} v_j, \quad \text{for } i = 1, 2, \ldots, N
\tag{4.260}
$$

Consider the matrix equation (4.253).

$$
v = \lim_{T \to \infty} v(0) = (I - \alpha P)^{-1} q.
\tag{4.253}
$$

Since v_i is the ith element of the vector v, and $v_i(0)$ is the ith element of the vector $v(0)$, it follows that, for large T,

$$
v_i = \lim_{T \to \infty} v_i(0).
\tag{4.261}
$$

Similarly, $v(1)$ is the expected total discounted reward vector earned at epoch 1 over a planning horizon of length T periods. As T grows large, $(T - 1)$ also grows large, so that

$$v_i = \lim_{T \to \infty} v_i(1). \tag{4.262}$$

4.3.3.1.1 VDEs for a Two-State Discounted MCR

Consider a generic two-state discounted MCR with a discount factor, α, for which the transition probability matrix and reward vector are given in Equation (4.2). As Equation (4.254) indicates, the VDEs can be expressed in matrix form as

$$(I - \alpha P)v = q \tag{4.254}$$

$$\left(\begin{bmatrix} 1 & 0 \\ 0 & 1 \end{bmatrix} - \begin{bmatrix} \alpha p_{11} & \alpha p_{12} \\ \alpha p_{21} & \alpha p_{22} \end{bmatrix} \right) \begin{bmatrix} v_1 \\ v_2 \end{bmatrix} = \begin{bmatrix} q_1 \\ q_2 \end{bmatrix}$$

$$\begin{bmatrix} 1 - \alpha p_{11} & -\alpha p_{12} \\ -\alpha p_{21} & 1 - \alpha p_{22} \end{bmatrix} \begin{bmatrix} v_1 \\ v_2 \end{bmatrix} = \begin{bmatrix} q_1 \\ q_2 \end{bmatrix} . \tag{4.263}$$

The solution for the expected total discounted rewards is

$$\begin{aligned} v_1 &= \frac{q_1(1 - \alpha p_{22}) + q_2 \alpha p_{12}}{(1 - \alpha p_{11})(1 - \alpha p_{22}) - \alpha^2 p_{12} p_{21}} \\ v_2 &= \frac{q_2(1 - \alpha p_{11}) + q_1 \alpha p_{21}}{(1 - \alpha p_{11})(1 - \alpha p_{22}) - \alpha^2 p_{12} p_{21}} \end{aligned} . \tag{4.264}$$

For example, consider the following two-state recurrent MCR for which the gain was calculated in Equation (4.51).

$$P = \begin{matrix} 1 \\ 2 \end{matrix} \begin{bmatrix} p_{11} & p_{12} \\ p_{21} & p_{22} \end{bmatrix} = \begin{matrix} 1 \\ 2 \end{matrix} \begin{bmatrix} 0.2 & 0.8 \\ 0.6 & 0.4 \end{bmatrix}, \quad q = \begin{matrix} 1 \\ 2 \end{matrix} \begin{bmatrix} q_1 \\ q_2 \end{bmatrix} = \begin{matrix} 1 \\ 2 \end{matrix} \begin{bmatrix} -300 \\ 200 \end{bmatrix}. \tag{4.67}$$

When a discount factor of $\alpha = 0.9$ is used for the MCR, the expected total discounted rewards are

$$
\left.
\begin{aligned}
v_1 &= \frac{q_1(1-\alpha p_{22})+q_2\alpha p_{12}}{(1-\alpha p_{11})(1-\alpha p_{22})-\alpha^2 p_{12}p_{21}} = \frac{(-300)[(1-0.9(0.4)]+200(0.9)0.8}{[(1-0.9(0.2)][(1-0.9(0.4)]-(0.9)^2(0.8)(0.6)} \\
&= -352.9412 \\[4pt]
v_2 &= \frac{q_2(1-\alpha p_{11})+q_1\alpha p_{21}}{(1-\alpha p_{11})(1-\alpha p_{22})-\alpha^2 p_{12}p_{21}} = \frac{200[(1-0.9(0.2)]-300(0.9)0.6}{[(1-0.9(0.2)][(1-0.9(0.4)]-(0.9)^2(0.8)(0.6).} \\
&= 14.706
\end{aligned}
\right\}
\tag{4.265}
$$

4.3.3.1.2 Optional Insight: Limiting Relationship Between Expected Total Discounted Rewards and the Gain

Consider a recurrent Markov chain with discounted rewards. If $\alpha = 1$ so that rewards are not discounted, the gain of the recurrent chain is g. Now suppose that rewards are discounted. Suppose that the discount factor, α, approaches the limiting value of one. When a discount factor $\alpha \to 1$ for an N-state MCR, which has a recurrent chain with a gain of g, then

$$
\lim_{\alpha \to 1}(1-\alpha)v_i = g,
\tag{4.266}
$$

where v_i is the expected total discounted reward received in state i [2, 7]. This limiting relationship will be demonstrated with respect to the discounted two-state MCR for which v_1 and v_2 were calculated in Equation (4.265).

Recall from Equation (4.48) or (4.50) that the gain of an undiscounted recurrent two-state MCR is

$$
g = \frac{p_{21}q_1 + p_{12}q_2}{p_{12} + p_{21}}.
\tag{4.50}
$$

The expected total discounted rewards, v_1 and v_2, calculated in Equation (4.264) for the discounted two-state MCR will be expressed in terms of the transition probabilities p_{12} and p_{21} by making the substitutions $p_{11} = 1 - p_{12}$ and $p_{22} = 1 - p_{21}$.

$$
v_1 = \frac{q_1[1-\alpha(1-p_{21})]+q_2\alpha p_{12}}{[1-\alpha(1-p_{12})][1-\alpha(1-p_{21})]-\alpha^2 p_{12}p_{21}}.
\tag{4.267}
$$

After simplification of the denominator,

$$
v_1 = \frac{q_1[1-\alpha(1-p_{21})]+q_2\alpha p_{12}}{[1-2\alpha+\alpha p_{12}+\alpha p_{21}+\alpha^2 -\alpha^2 p_{12}-\alpha^2 p_{21}] + \alpha^2 p_{12}p_{21}-\alpha^2 p_{12}p_{21}}.
\tag{4.268}
$$

Since the last two terms in the denominator sum to zero, they can be dropped, and

$$
\begin{aligned}
v_1 &= \frac{q_1[1-\alpha(1-p_{21})] + q_2\alpha p_{12}}{(1-2\alpha+\alpha^2) + \alpha p_{12}(1-\alpha) + \alpha p_{21}(1-\alpha)} \\
&= \frac{q_1[1-\alpha(1-p_{21})] + q_2\alpha p_{12}}{(1-\alpha)^2 + \alpha p_{12}(1-\alpha) + \alpha p_{21}(1-\alpha)}.
\end{aligned}
\tag{4.269}
$$

Observe that when $\alpha = 1$,

$$
v_1 = (q_1 p_{21} + q_2 p_{12})/0 = \infty
\tag{4.270}
$$

because over an infinite planning horizon the expected total reward without discounting is infinite. Hence α can approach one but can never equal one. Note that when both sides of Equation (4.269) are multiplied by $(1 - \alpha)$,

$$
(1-\alpha)v_1 = \frac{(1-\alpha)\{q_1[1-\alpha(1-p_{21})] + q_2\alpha p_{12}\}}{(1-\alpha)^2 + \alpha p_{12}(1-\alpha) + \alpha p_{21}(1-\alpha)} = \frac{q_1[1-\alpha(1-p_{21})] + q_2\alpha p_{12}}{(1-\alpha) + \alpha p_{12} + \alpha p_{21}}
\tag{4.271}
$$

$$
\begin{aligned}
\lim_{\alpha\to1}(1-\alpha)v_1 &= \lim_{\alpha\to1}\frac{q_1[1-\alpha(1-p_{21})] + q_2\alpha p_{12}}{(1-\alpha) + \alpha p_{12} + \alpha p_{21}} = \frac{q_1[1-1(1-p_{21})] + q_2(1)p_{12}}{(1-1) + (1)p_{12} + (1)p_{21}} \\
&= \frac{q_1 p_{21} + q_2 p_{12}}{p_{12} + p_{21}} = g.
\end{aligned}
\tag{4.272}
$$

A similar argument will show that

$$
\lim_{\alpha\to1}(1-\alpha)v_2 = \frac{q_1 p_{21} + q_2 p_{12}}{p_{12} + p_{21}} = g.
\tag{4.273}
$$

When the discount factor for the two-state MCR represented in Equation (4.51) of Section 4.3.3.1.1 is increased to $\alpha = 0.999$, the expected total discounted rewards are

$$\left.\begin{aligned}
v_1 &= \frac{q_1(1-\alpha p_{22}) + q_2 \alpha p_{12}}{(1-\alpha p_{11})(1-\alpha p_{22}) - \alpha^2 p_{12}p_{21}} \\
&= \frac{(-300)[(1-0.999(0.4)] + 200(0.999)0.8}{[(1-0.999(0.2)][(1-0.999(0.4)] - (0.999)^2(0.8)(0.6)} = -14,489.854 \\
v_2 &= \frac{q_2(1-\alpha p_{11}) + q_1 \alpha p_{21}}{(1-\alpha p_{11})(1-\alpha p_{22}) - \alpha^2 p_{12}p_{21}} \\
&= \frac{200[(1-0.999(0.2)] - 300(0.999)0.6}{[(1-0.999(0.2)][(1 - 0.999(0.4)] - (0.999)^2(0.8)(0.6)} = -14,132.609.
\end{aligned}\right\} \quad (4.274)$$

As Equation (4.52) indicates, the gain without discounting is $g = -14.29$. Observe that

$$\left.\begin{aligned}
(1-\alpha)v_1 &= (1-0.999)(-14,489.854) = -14.49 \rightarrow -14.29 = g \\
(1-\alpha)v_2 &= (1-0.999)(-14,132.609) = -14.13 \rightarrow -14.29 = g.
\end{aligned}\right\} \quad (4.275)$$

Although these results have been demonstrated to hold for a two-state MCR, they can be extended to show that Equation (4.266) is true for any N-state MCR, which has a recurrent chain with a gain g.

4.3.3.1.3 VDEs for a Discounted MCR Model of Monthly Sales

Consider the four-state discounted MCR model of monthly sales treated in Section 4.3.2.2 over a finite planning horizon. The transition probability matrix P, discount factor $\alpha = 0.9$, matrix αP, and reward vector q are shown in Equation (4.244).

The expanded matrix form of the VDEs (4.256) is

$$\begin{bmatrix} v_1 \\ v_2 \\ v_3 \\ v_4 \end{bmatrix} = \begin{bmatrix} -20 \\ 5 \\ -5 \\ 25 \end{bmatrix} + \begin{bmatrix} 0.540 & 0.270 & 0.090 & 0 \\ 0.225 & 0.270 & 0.315 & 0.090 \\ 0.045 & 0.225 & 0.450 & 0.180 \\ 0 & 0.090 & 0.270 & 0.540 \end{bmatrix} \begin{bmatrix} v_1 \\ v_2 \\ v_3 \\ v_4 \end{bmatrix}. \quad (4.276)$$

The solution for the expected total discounted rewards is

$$v_1 = -35.7733, \quad v_2 = 8.9917, \quad v_3 = 12.4060, \text{ and } v_4 = 63.3889.$$

Thus, if this Markov chain with discounted rewards operates over an infinite planning horizon, the expected total discounted reward will be −$35,773.30, $8,991.70, $12,406.00, or $63,388.90 if it starts in state 1, 2, 3, or 4, respectively.

4.3.3.1.4 Optional Insight: Probabilistic Interpretation of Discount Factor

A discount factor α, with $0 < \alpha < 1$, has an interesting probabilistic interpretation with respect to an undiscounted MCR of uncertain duration [4, 5, 7]. To reveal this probabilistic interpretation, consider the four-state MCR model of monthly sales introduced without discounting in Section 4.2.2.1. The transition probability matrix P and reward vector q for the undiscounted MCR model are shown in Equation (4.6).

$$P = \begin{matrix} 1 \\ 2 \\ 3 \\ 4 \end{matrix} \begin{bmatrix} 0.60 & 0.30 & 0.10 & 0 \\ 0.25 & 0.30 & 0.35 & 0.10 \\ 0.05 & 0.25 & 0.50 & 0.20 \\ 0 & 0.10 & 0.30 & 0.60 \end{bmatrix}, \quad q = \begin{bmatrix} -20 \\ 5 \\ -5 \\ 25 \end{bmatrix}. \tag{4.6}$$

Assume that a discount factor of α is specified. Suppose that the matrix αP is embedded in the following transition matrix denoted by Y:

State	0	1	2	3	4
0	1	0	0	0	0
1	$1-\alpha$	$\alpha(0.60)$	$\alpha(0.30)$	$\alpha(0.10)$	$\alpha(0)$
2	$1-\alpha$	$\alpha(0.25)$	$\alpha(0.30)$	$\alpha(0.35)$	$\alpha(0.10)$
3	$1-\alpha$	$\alpha(0.05)$	$\alpha(0.25)$	$\alpha(0.50)$	$\alpha(0.20)$
4	$1-\alpha$	$\alpha(0)$	$\alpha(0.10)$	$\alpha(0.30)$	$\alpha(0.60)$

$$Y = \begin{bmatrix} 1 & 0 \\ 1-\alpha & \alpha P \end{bmatrix} = \begin{bmatrix} 1 & 0 \\ D & Q \end{bmatrix}. \tag{4.277}$$

Since the entries in every row of P sum to 1, the entries in all rows of αP sum to α. Hence, Y is the transition probability matrix for an absorbing unichain. The state space $E=(1, 2, 3, 4\}$ of the undiscounted process has been augmented by an absorbing state 0. States 1 through 4 are transient. There is a probability $1 - \alpha$ that the undiscounted process will reach the absorbing state on the next transition, and stop. There is also a probability α that the process will not reach the absorbing state on the next transition, and will therefore continue. One may conclude that if the duration of an undiscounted MCR is indefinite, one minus a discount factor can be interpreted as the probability that the process will stop after the next transition. Therefore, a discount factor can be interpreted as the probability that the process will continue after the next transition.

Matrix αP governs transitions among transient states before the process is absorbed. The matrix $(I - \alpha P)^{-1}$ is the fundamental matrix of the absorbing unichain. Hence, the alternate matrix form of the VDEs (4.253) for a discounted MCR,

$$v = (I - \alpha P)^{-1} q, \tag{4.253}$$

indicates that the vector of expected total discounted rewards can be obtained by multiplying the fundamental matrix by the reward vector.

4.3.3.1.5 VDEs for an Eight-State Discounted Multichain MCR Model of a Production Process

To illustrate the solution of the VDEs for a multichain MCR model with discounting, consider the eight-state sequential production process for which the transition matrix and reward vector without discounting are shown in Equation (4.161). Assume in this case the discount factor is $\alpha = 0.9$. The matrix αP and the cost vector q are shown below:

$$\alpha P = \begin{array}{c} 1 \\ 2 \\ 3 \\ 4 \\ 5 \\ 6 \\ 7 \\ 8 \end{array} \begin{bmatrix} 0.9 & 0 & 0 & 0 & 0 & 0 & 0 & 0 \\ 0 & 0.9 & 0 & 0 & 0 & 0 & 0 & 0 \\ 0 & 0 & 0.45 & 0.27 & 0.18 & 0 & 0 & 0 \\ 0 & 0 & 0..27 & 0.405 & 0.225 & 0 & 0 & 0 \\ 0 & 0 & 0.09 & 0.315 & 0.495 & 0 & 0 & 0 \\ 0.18 & 0.144 & 0.036 & 0.027 & 0.018 & 0.495 & 0 & 0 \\ 0.135 & 0 & 0 & 0 & 0 & 0.18 & 0.585 & 0 \\ 0.09 & 0 & 0 & 0 & 0 & 0 & 0.135 & 0.675 \end{bmatrix}, \quad (4.278)$$

$$q = \begin{array}{c} 1 \\ 2 \\ 3 \\ 4 \\ 5 \\ 6 \\ 7 \\ 8 \end{array} \begin{bmatrix} 156 \\ 56 \\ 160 \\ 123 \\ 291.5 \\ 450 \\ 400 \\ 420 \end{bmatrix}.$$

In matrix form the VDEs (4.255) are

$$\begin{bmatrix} v_1 \\ v_2 \\ v_3 \\ v_4 \\ v_5 \\ v_6 \\ v_7 \\ v_8 \end{bmatrix} = \begin{bmatrix} 156 \\ 56 \\ 160 \\ 123 \\ 291.5 \\ 450 \\ 400 \\ 420 \end{bmatrix} + \begin{bmatrix} 0.900 & 0 & 0 & 0 & 0 & 0 & 0 & 0 \\ 0 & 0.900 & 0 & 0 & 0 & 0 & 0 & 0 \\ 0 & 0 & 0.450 & 0.270 & 0.180 & 0 & 0 & 0 \\ 0 & 0 & 0.270 & 0.405 & 0.225 & 0 & 0 & 0 \\ 0 & 0 & 0.090 & 0.315 & 0.495 & 0 & 0 & 0 \\ 0.180 & 0.144 & 0.036 & 0.027 & 0.018 & 0.495 & 0 & 0 \\ 0.135 & 0 & 0 & 0 & 0 & 0.180 & 0.585 & 0 \\ 0.090 & 0 & 0 & 0 & 0 & 0 & 0.135 & 0.675 \end{bmatrix} \begin{bmatrix} v_1 \\ v_2 \\ v_3 \\ v_4 \\ v_5 \\ v_6 \\ v_7 \\ v_8 \end{bmatrix}.$$

$$(4.279)$$

The solution for the vector of expected total discounted costs is

$$[v_1 \ v_2 \ v_3 \ v_4 \ v_5 \ v_6 \ v_7 \ v_8]^T = [1,560 \quad 560 \quad 1,852.83 \quad 1,819.93$$
$$2,042.64 \quad 1,909 \quad 2,299.33 \quad 2,679.41]^T$$

(4.280)

The solution vector shows, for example, that if this discounted multichain MCR model of a serial production process operates over an infinite planning horizon, the expected total discounted cost will be $1,560 if it starts in state 1, and $2,679.41 if it starts in state 8.

4.3.3.2 Value Iteration for Expected Total Discounted Rewards

The second approach that can be taken to calculate the expected total discounted rewards received by an MCR over an infinite planning horizon is to execute value iteration over a large number of periods. Value iteration requires less computational effort than solving a system of VDEs, but gives only an approximate solution [7].

Value iteration is conducted over an infinite planning horizon by repeatedly adding one additional period at a time to lengthen the horizon, following the procedure described in Section 4.2.2.3. The value iteration equation is solved over each most recently lengthened horizon until a stopping condition has been satisfied. The beginning of each new period, which is added at the beginning of the horizon, is designated by a consecutively numbered negative epoch. In Section 4.2.3.3, this procedure was applied to develop a value iteration algorithm for an undiscounted recurrent MCR.

4.3.3.2.1 Stopping Rule for Value Iteration

As Section 4.2.3.3.1 indicates, when a value iteration algorithm stops, the planning horizon contains n periods. The epochs are numbered sequentially as $-n, -n+1, -n+2, \ldots, -2, -1, 0$. Epoch $(-n)$ denotes the beginning of the horizon. Epoch 0 denotes the end. As Section 4.2.3.3.1 indicates, the absolute value of each epoch represents the number of periods remaining. When a stopping condition is satisfied, let $T = |-n| = n$, so that T denotes the length of the horizon at which convergence of a value iteration algorithm is achieved.

Suppose that all the epochs are renumbered by adding T to the numerical index of each epoch, so that the planning horizon begins at epoch 0 and ends at epoch T. As T grows large,

$$\lim_{T \to \infty}[v_i(0) - v_i(1)] = \lim_{T \to \infty} v_i(0) - \lim_{T \to \infty} v_i(1) = v_i - v_i = 0, \qquad (4.281)$$

using Equations (4.261) and (4.262). The eventual convergence of the expected total discounted rewards received over a finite horizon to the expected total discounted rewards received over an infinite horizon suggests the following rule for ending value iteration. Value iteration will stop when the absolute value of the maximum difference, over all states, between the expected total discounted rewards for the current planning horizon and a planning horizon one period shorter is less than a small positive number, ε, specified by the analyst. That is, value iteration can be stopped after a planning horizon of length T for which

$$\max_{i=1,\dots,N} |v_i(0) - v_i(1)| < \varepsilon. \tag{4.282}$$

Of course, the magnitude of ε is a matter of judgment, and the number of periods in T cannot be predicted.

4.3.3.2.2 Value Iteration Algorithm

The foregoing discussion is the basis for the following value iteration algorithm in its simplest form (adapted from Puterman [7], p. 161) that can be applied to a discounted MCR over an infinite horizon. Epochs are numbered as consecutive negative integers from $-n$ at the beginning of the horizon to 0 at the end.

Step 1. Select arbitrary salvage values for

$v_i(0)$, for $i = 1, 2, \dots, N$. For simplicity, set $v_i(0) = 0$. Specify $\varepsilon > 0$. Set $n = -1$.

Step 2. For each state i, use the value iteration equation to compute

$$v_i(n) = q_i + \alpha \sum_{j=1}^{N} p_{ij} v_j(n+1), \quad \text{for } i = 1, 2, \dots, N.$$

Step 3. If $\max_{i=1,2,\dots,N} |v_i(n) - v_i(n+1)| < \varepsilon$, go to step 4. Otherwise, decrement n by 1 and return to step 2.

Step 4. Stop.

4.3.3.2.3 Solution by Value Iteration of Discounted MCR Model of Monthly Sales

In Section 4.3.2.2, value iteration was applied to a four-state discounted MCR model of monthly sales over a three-period planning horizon, which was lengthened to four periods. By applying value iteration to a planning horizon lengthened to seven periods, Table 4.15 is constructed to show the expected total discounted rewards for monthly sales during the last 7 months of an infinite horizon. The last eight epochs of an infinite horizon are numbered sequentially in Table 4.15 as −7, −6, −5, −4, −3, −2, −1, and 0. Epoch 0 denotes the end of the horizon.

TABLE 4.15

Expected Total Discounted Rewards for Monthly Sales Calculated by Value Iteration During the Last 7 Months of an Infinite Planning Horizon

Epoch	n							
	-7	-6	-5	-4	-3	-2	-1	0
$v_1(n)$	-41.2574	-41.1224	-40.4607	-38.8699	-35.6915	-29.9	-20	0
$v_2(n)$	2.5647	2.0488	1.6153	1.3766	1.5429	2.525	5	0
$v_3(n)$	5.3476	4.3948	3.2247	1.7484	-0.1456	-2.525	-5	0
$v_4(n)$	55.6493	54.2191	52.2278	49.3183	44.8495	37.6	25	0

TABLE 4.16

Absolute Values of the Differences Between the Expected Total Discounted Rewards Earned Over Planning Horizons, Which Differ in Length by One Period

Epoch	n						
	-7	-6	-5	-4	-3	-2	-1
	$\begin{vmatrix} v_i(-7) \\ -v_i(-6) \end{vmatrix}$	$\begin{vmatrix} v_i(-6) \\ -v_i(-5) \end{vmatrix}$	$\begin{vmatrix} v_i(-5) \\ -v_i(-4) \end{vmatrix}$	$\begin{vmatrix} v_i(-4) \\ -v_i(-3) \end{vmatrix}$	$\begin{vmatrix} v_i(-3) \\ -v_i(-2) \end{vmatrix}$	$\begin{vmatrix} v_i(-2) \\ -v_i(-1) \end{vmatrix}$	$\begin{vmatrix} v_i(-1) \\ -v_i(0) \end{vmatrix}$
i							
1	-0.135	-0.6617	-1.5908	-3.1784	-5.7915	-9.9	-20
2	0.5159	0.4335	0.2387	-0.1663	-0.9821	-2.475	5
3	0.9528	1.1701	1.4763	1.894	2.3794	2.475	-5
4	1.4302U	1.9913U	2.9095U	4.4688U	7.2495U	12.6U	25U
$\text{Max}\left(\begin{vmatrix} v_i(n) \\ -v_i(n+1) \end{vmatrix}\right)$	1.4302	1.9913	2.9095	4.4688	7.2495	12.6	25

Table 4.16 gives the absolute values of the differences between the expected total discounted rewards earned over planning horizons, which differ in length by one period. Note that the maximum absolute differences become progressively smaller for each epoch added to the planning horizon.

In Table 4.16, a suffix U identifies the maximum absolute difference for each epoch. The maximum absolute differences obtained for all seven epochs are listed in the bottom row of Table 4.16. The bottom row of Table 4.16 shows that if an analyst chooses an $\varepsilon < 1.4302$, then more than seven repetitions of value iteration will be needed before the algorithm can be assumed to have converged.

Table 4.15 shows that the approximate expected total discounted rewards,

$$v_1 = -41.2574, \quad v_2 = 2.5647, \quad v_3 = 5.3476, \quad v_4 = 55.6493,$$

obtained after seven repetitions of value iteration are significantly different from the actual expected total discounted rewards,

$$v_1 = -35.7733, \quad v_2 = 8.9917, \quad v_3 = 12.4060, \text{ and } \quad v_4 = 63.3889,$$

obtained by solving the VDEs in Section 4.3.3.1.3. Thus, value iteration for the discounted MCR model of monthly sales has not converged after seven repetitions to the actual expected total discounted rewards.

PROBLEMS

4.1 This problem adds daily revenue to the Markov chain model of Problems 1.1 and 2.1. Since a repair takes 1 day to complete, no revenue is earned on the day during which maintenance is performed. The daily revenue earned in every state is shown below:

State	Condition	Daily Revenue
1	Not Working, under repair (NW)	$0
2	Working, with a Major Defect (MD)	$200
3	Working, with a Minor Defect (mD)	$400
4	Working Properly (WP)	$600

The transition probability matrix and the daily reward vector for the recurrent MCR are given below:

State	Maintenance Action		State	1	2	3	4		Reward
1, NW	Under Repair		1	0.1	0.2	0.2	0.5		$0 = q_1
2, MD	Do Nothing	$P =$	2	0.3	0.7	0	0	$, q =$	$200 = q_2
3, mD	Do Nothing		3	0.2	0.5	0.3	0		$400 = q_3
4, WP	Do Nothing		4	0.1	0.2	0.3	0.4		$600 = q_4

(a) Given that the machine begins day one with an equal probability of starting in any state, what is the expected total revenue that will be earned after 3 days?

(b) Find the expected average reward, or gain.

(c) Find the expected total discounted reward vector using a discount factor of $\alpha = 0.9$.

4.2 A married couple owns a consulting firm. The firm's weekly income is a random variable, which may equal $10,000, $15,000, or $20,000. The income earned next week depends only on the income earned this week. In any week, the married couple may sell the firm to either their adult son or their adult daughter. If the income this week is $10,000, the couple may sell the firm next week to their son with probability 0.05 or to their daughter with probability 0.10. If the income this week is $10,000 and they do not sell, the income next week will be either $10,000 with probability 0.40, $15,000 with probability 0.25, or $20,000 with probability 0.20. If the income this week is $15,000, then next week there is a 0.15 probability that the couple will sell the firm to their son, and a 0.20 probability that they will sell the firm to their daughter. If the income this week is $15,000 and they do not sell, the income next week will be either $10,000 with probability 0.20, $15,000 with probability 0.35, or $20,000 with probability 0.10. Finally, if the income this week is $20,000, the couple may sell the firm next week to their son with probability 0.15 or to their daughter with probability 0.10. If the income this week is $20,000 and they do not sell, the income next week will be either $10,000 with probability 0.25, $15,000 with probability 0.20, or $20,000 with probability 0.30.

(a) Formulate this problem as an absorbing multichain MCR. Represent the transition probability matrix in canonical form. Construct the reward vector.

(b) Given that the firm's income in the first week is $10,000, what is the probability that the firm will eventually be sold to the son?

(c) Given that the firm's income in the first week is $20,000, what is the probability that the firm will eventually be sold to the daughter?

(d) Given that the firm's income in the first week is $15,000, what is the expected number of weeks before the firm is sold to the daughter?

(e) Given that the firm's income in the first week is $15,000, what is the expected total income that will be earned before the firm is sold?

4.3   This problem adds cost data to the Markov chain model of Problems 1.9 and 2.5. Suppose that the cost to inspect and test an IC varies inversely with its age as shown below:

Age i of an IC in years, i	0	1	2	3
Cost to inspect and test an IC	$57	$34.29	$23.33	$10

The cost to replace an IC, which has failed, is $10. The cost in dollars of buying and operating an IC of age i is denoted by q_i.

(a) Construct a cost vector for the IC replacement problem.

(b) Model the IC replacement problem as a recurrent MCR by adding the cost vector to the transition probability matrix, which was constructed for Problem: 2.5.

(c) Calculate the average cost per year of operation and replacement.

(d) If operation starts with a 1-year-old component, calculate the expected total cost of operation and replacement relative to the expected total cost of starting with a 3-year-old component.

(e) If operation starts with a 2-year-old component, calculate the expected total cost of operation before the component fails for the first time.

(f) If operation starts with a 1-year-old component, calculate the expected total discounted cost of operation and replacement, for a discount factor of $\alpha = 0.9$.

(g) If operation starts with a 1-year-old component, calculate the expected total cost of operation after 3 years. Do not use value iteration.

(h) If operation starts with a 1-year-old component, calculate the expected total cost of operation after 3 years. Use value iteration.

4.4 This problem refers to Problem 3.3.

(a) Given that the woman starts with $2,000, and will bet $1,000 each time, find the expected amount of money that she will have after two bets. Do not use value iteration.

(b) Given that the woman starts with $2,000, and will bet $1,000 each time, use value iteration to find the expected amount of money that she will have after two bets.

(c) Given that the woman starts with $2,000, and will bet $1,000 each time, find the expected amount of money that she will have when the game ends.

4.5 This problem refers to Problem 3.6. The following revenue and costs are specified for the states of the production process. A cost of $60 is incurred to scrap a defective item. A cost of $20 is incurred to sell an acceptable item. An acceptable item is sold for $1,000. The cost per item of input material to stage 1 is $100. Each time an item is processed and inspected at stage 1, the cost is $300. The corresponding costs per item processed and inspected at stages 2 and 3 are $400 and $500, respectively.

(a) Model this production process as a five-state absorbing multichain MCR. Construct the transition probability matrix and the reward vector.

(b) Find the expected total cost of an item sold.

(c) Find the expected total cost of an item before it is scrapped or sold, given that an entering item starts in the first manufacturing stage.

(d) Find the expected total cost of an item after two inspections, given that an entering item starts in the first manufacturing stage.

4.6 In Problem 3.7, the following revenue and cost data are specified. If the mission is aborted, a cost of $6,000,000 is incurred for the loss of the satellite payload. When the rocket is launched, the cost of fuel needed to reach HEO is $2,000,000. The revenue earned by delivering the satellite to HEO is $4,000,000. If the satellite is delivered instead to LEO, the revenue will be reduced to $1,000,000. The cost of fuel for each minor course correction is $200,000, and is $700,000 for each major course correction.

(a) Model this rocket launch as a five-state absorbing multichain MCR. Construct the transition probability matrix and the reward vector.

(b) If, upon launch, the rocket is observed to start in state 3, find the expected total revenue earned by launching the rocket.

(c) If, upon launch, the rocket is observed to start in state 3, find the expected total revenue earned by launching the rocket into LEO.

(d) If, upon launch, the rocket is observed to start in state 3, find the expected total revenue earned by the rocket launch after two course corrections.

4.7   In Problem 3.10, suppose that the investor receives a monthly dividend of d dollars per share. No dividend is received when the stock is sold. The investor wants to compare the following two alternative policies for deciding when to sell the stock. Policy 1 is to sell the stock at the end of the first month in which the share price rises to $20 or falls to $0. Policy 2 is to sell the stock at the end of the first month in which the share price rises to $15 or falls to $5. If the investor bought the stock for $10, find the monthly dividend d for which both policies will produce equal expected total rewards when the investor sells the stock.

4.8   A dam is used for generating electric power and for irrigation. The dam has a capacity of 4 units of water. Assume that the volume of water stored in the dam is always an integer. At the beginning of every week, a volume of water denoted by W_n flows into the dam. The stationary probability distribution of W_n, which is an independent, identically distributed, and integer random variable, is given below.

Volume W_n of water flowing into dam at start of week n, $W_n = k$	0	1	2	3
Probability, $P(W_n = k)$	0.2	0.3	0.4	0.1

The dam has the following policy for releasing water at the beginning of every week. If the volume of water stored in the dam plus the volume flowing into the dam at the beginning of the week exceeds 2 units, then 2 units of water are released. The first unit of water released is used to generate electricity, which is sold for $5. The second unit released is used for irrigation, which earns $4. If the volume of water stored in the dam plus the volume flowing into the dam at the beginning of the week equals 2 units, then only 1 unit of water is released. The 1 unit of water released is used to generate electricity, which is sold for $5. No water is released for irrigation. The volume of water stored in the dam is normally never allowed to drop below 1 unit to provide a reserve in the event of a natural disaster. Hence, if the volume of water stored in the dam plus the volume flowing into the dam at the beginning of the week is less than 2 units, no water is released. If the volume of water stored in the dam plus the volume flowing into the dam at the beginning of the week exceeds 6 units, then 2 units of water are released to generate electricity and for irrigation. In addition, surplus water is released through the spillway and lost, causing flood damage at a cost of $3 per unit.

The volume of water in the dam can be modeled as a Markov chain. Let the state X_n be the volume of water in the dam at the beginning of week n.

(a) Assuming that this release policy is followed, model the volume of water in the dam and the associated rewards as a four-state recurrent MCR. Construct the transition probability matrix and the reward vector.

(b) Execute value iteration to find the vector of expected total rewards that will be earned in every state after 3 weeks. Assume zero terminal rewards at the end of week 3.

(c) Find the expected average reward, or gain.

(d) Find the vector of expected total discounted rewards using a discount factor of $\alpha = 0.9$.

4.9 At 9:00 AM every day an independent financial advisor will see a client who is concerned about retirement. A client has one of the following three types of concerns: creating a new retirement plan, managing investments under an existing retirement plan, or generating income during retirement. The retirement concerns of the clients form a Markov chain. The states are the clients' three retirement concerns. Let the state X_n denote the retirement concern of the client who is seen on day n. The financial advisor's historical records indicate that if today's client is interested in creating a new retirement plan, the probabilities that tomorrow's client will be interested in managing investments under an existing retirement plan, or generating income during retirement, are 0.2 and 0.5, respectively. If a client today is interested in managing investments under an existing retirement plan, the probabilities that a client tomorrow will be interested in creating a new retirement plan, or generating income during retirement, are 0.4 and 0.1, respectively. Finally, if today's client is interested in generating income during retirement, the probabilities that tomorrow's client will be interested in creating a new retirement plan, or managing investments under an existing retirement plan, are 0.3 and 0.6, respectively. The financial advisor charges the following fixed fees each time that a client visits. The fees for creating a new retirement plan, managing investments under an existing retirement plan, and generating income during retirement are $500, $300, and $400, respectively.

(a) Model the financial advisor's daily 9:00 AM meetings with clients as a three-state recurrent MCR. Construct the transition probability matrix and the reward vector.

(b) Execute value iteration to find the vector of expected total rewards that the financial advisor will earn from her retirement plan clients after 3 days. Assume zero terminal rewards at the end of day 3.

(c) Find the financial advisor's expected average reward, or gain.

(d) Find the financial advisor's vector of expected total discounted rewards using a discount factor of $\alpha = 0.9$.

4.10 Suppose that in a certain state, every licensed professional engineer (P.E.) is required to pass an annual proficiency test in the practice of engineering in order to renew her license. A P.E. who passes the test has her license renewed for 1 year. A P.E. who fails the test has her license suspended. If a person whose license is suspended passes the test after the first or second year of suspension, her license is restored. However, if a person whose license is suspended fails the test after the second year of suspension, her license is revoked.

Assume that the status of an engineer's license can be modeled as a four-state absorbing unichain. Let the state X_n be the status of an engineer's license after the nth annual proficiency test. The four states are identified below:

State	Status of an engineer's license
0	Renewed
1	In first year of suspension
2	In second year of suspension
3	Revoked

Suppose that a licensed P.E. has a 0.85 probability of passing the test, and earns $200,000 per year as a licensed engineer. An engineer with a suspended P.E. license is employed as an associate engineer at an annual salary of $120.000. An engineer whose license is suspended has a 0.65 probability of passing the test on her first try, and a 0.45 probability of passing the test on her second try. An engineer without a P.E. license is employed as an assistant engineer at an annual salary of $80.000.

(a) Model the licensure status of an engineer as a four-state absorbing unichain MCR. Construct the transition probability matrix and the reward vector.

(b) Find the expected total reward that a licensed P.E. will earn after 3 years.

(c) Find the expected total reward that a licensed P.E. will earn until her license is revoked.

Suppose that an engineer can increase her likelihood of passing the annual proficiency test by enrolling in courses of continuing professional development (CPD). The tuition for such courses is $1,000 per hour. The recommended annual course loads are 5 h for a licensee, 10 h for an engineer beginning her first year of suspension, and 15 h for an engineer beginning her second year of suspension. By following the recommended annual course loads, an engineer can add 0.10 to each of her probabilities of passing the annual proficiency test.

(d) If an engineer always follows the recommended annual course loads, model the licensure status of an engineer as a four-state absorbing unichain MCR.

(e) Find the expected total reward that a licensed P.E. who always follows the recommended annual course loads will earn until her license is revoked.

4.11 Every December 15, a certain company offers its employees a choice of one of the following three health care plans for the new year beginning on January 1:

Health Plan	Description
P.P.O.	Preferred provider organization
H.M.O.	Health maintenance organization
H.D.P.	High-deductible plan

An employee may keep her current plan or switch to a different plan. During the past several years the Human Resources Department (HRD) has compiled the following data on the numbers of employees who have switched health care plans:

From Plan\To Plan	P.P.O.	H.M.O.	H.D.P.	Total
P.P.O.	170	24	6	200
H.M.O.	45	225	30	300
H.D.P.	4	16	80	100
Total	219	265	116	600

The HRD believes that the choice of a health care plan by an employee can be modeled as a three-state recurrent Markov chain. They let the state X_n be the health care plan selected at

State	Health care plan selected
0	P.P.O.
1	H.M.O.
2	H.D.P.

the end of year n. The three states are identified below:
The HRD has estimated that the company incurs the following administrative costs for each employee who switches her health care plan.

From Plan\To Plan	P.P.O.	H.M.O.	H.D.P.
P.P.O.	$0	$30	$25
H.M.O.	20	0	24
H.D.P.	32	26	0

(a) Model the choice of a health care plan by an employee as a three-state recurrent MCR. Construct the transition probability matrix and the cost vector.

(b) Execute value iteration to find the vector of expected total costs that the company will incur for every health care plan after 3 years. Assume zero terminal rewards at the end of year 3.

(c) Find the company's expected average cost, or negative gain.

(d) Find the company's vector of expected total discounted costs using a discount factor of $\alpha = 0.9$.

Suppose that the HRD has proposed that the company try to reduce its administrative costs by offering all employees a cash incentive of $D to stay with their current health care plan rather than switch plans. The goal of the HRD proposal is to increase by 0.06 the probability that an employee will stay with her current plan, and to decrease by 0.03 the probability that she will switch to either of the other two plans.

(e) Assume that the desired modified probabilities of switching health care plans can be achieved by offering cash incentives of $D to each employee. Find the value of $D for which the expected total costs of the original policy and the policy proposed by the HRD are the same.

References

1. Bhat, N., *Elements of Applied Stochastic Processes*, 2nd ed., Wiley, New York, 1985.
2. Feldman, R. M. and Valdez-Flores, C., *Applied Probability & Stochastic Processes*, PWS, Boston, MA, 1996.
3. Hillier, F. S. and Lieberman G. J., *Introduction to Operations Research*, 8th ed., McGraw-Hill, New York, 2005.
4. Howard, R. A., *Dynamic Programming and Markov Processes*, M.I.T. Press, Cambridge, MA, 1960.
5. Kemeny, J. G., Schleifer, Jr., A., Snell, J. L., and Thompson, G. L., *Finite Mathematics with Business Applications*, 2nd ed., Prentice-Hall, Englewood Cliffs, NJ, 1972.
6. Jensen, P. A. and Bard, J. F., *Operations Research: Models and Methods*, Wiley, New York, 2003.
7. Puterman, M. L., *Markov Decision Processes: Discrete Stochastic Dynamic Programming*, Wiley, New York, 1994.
8. Shamblin, J. E. and Stevens, G. T., *Operations Research: A Fundamental Approach*, McGraw-Hill, New York, 1974.

5

A Markov Decision Process (MDP)

A Markov decision process (MDP) is a sequential decision process for which the decisions produce a sequence of Markov chains with rewards or MCRs. As the introduction to Chapter 4 indicates, when decisions are added to a set of MCRs, the augmented system is called an MDP. An MDP generates a sequence of states and an associated sequence of rewards as it evolves over time from state to state, governed by both its transition probabilities and the series of decisions made. A rule that prescribes a set of decisions for all states is called a policy. When the planning horizon is infinite, a policy is assumed to be independent of time, or stationary. When the planning horizon is infinite, the objective of an MDP model is to determine a stationary policy that is optimal in the sense that it will either maximize the gain, or expected reward per period, or maximize the expected total discounted reward received in every state. As in Chapter 4, both transition probabilities and rewards are assumed to be stationary.

An MCR can be viewed as a building block for an MDP. Therefore, many of the definitions, equations, and algorithms developed in Chapter 4 for an MCR will be modified in Chapter 5 to enable them to be applied, in an extended form, to an MDP. All MDPs are assumed to have a finite number of states.

5.1 An Undiscounted MDP

When cash flows for an MDP are not discounted, a Markov decision process is called an undiscounted MDP.

5.1.1 MDP Chain Structure

In Section 4.2.1, an MCR is classified as unichain, or multichain. Similarly, an MDP can also be termed unichain, or multichain. An MDP is unichain if the transition matrix associated with every stationary policy consists of a recurrent chain (which is a single closed communicating class of recurrent states) plus a possibly empty set of transient states. A recurrent MDP is a special case of a unichain MDP with no transient states. Thus, an MDP is recurrent if the transition matrix associated with every stationary policy is irreducible or recurrent. An MDP is multichain if at least one transition matrix associated with a stationary policy consists of two or more recurrent chains plus a possibly empty set of transient states.

In this chapter, all unichain MDPs and all multichain MDPs are assumed to have one or more transient states, so that they are reducible, although the latter term is omitted. Hence, an MDP will be classified as either recurrent, unichain, or multichain. Chain structure affects the algorithm used to find an optimal policy when rewards are not discounted [2, 3].

5.1.2 A Recurrent MDP

A recurrent MDP will be analyzed first over a finite planning horizon, and next over an infinite horizon.

5.1.2.1 A Recurrent MDP Model of Monthly Sales

A recurrent MCR model of monthly sales was introduced in Table 4.1 of Section 4.2.2.1. To transform this model into an MDP model, the MCR model will be augmented by adding decisions in every state. Each decision in every state will be associated with a set of transition probabilities and a reward. Recall that a firm tracks its monthly sales, which have fluctuated widely over many months. Monthly sales are ranked with respect to those of the firm's competitors. The rankings are expressed in quartiles. The fourth quartile is the highest ranking, while the first quartile is the lowest. The sequence of monthly sales is believed to form a Markov chain. The state X_{n-1} denotes the quartile rank of monthly sales at the beginning of month n. The chain has four states, which correspond to the four quartiles. Estimates of the transition probabilities are based on a historical record of monthly sales. At the beginning of each month, in contrast to the MCR model of Section 4.2.2.1, the firm must now select one of several alternatives actions in every state. The alternative selected is called the decision for that state. Table 5.1 identifies the four states accompanied by their decision alternatives and the associated rewards.

Note, for example, that when monthly sales are in the third quartile, state 3, the firm will either make decision 1 to design more appealing products for a reward of −$10,000 (which is a cost of $10,000), or make decision 2 to invest in new technology for a reward of −$5,000 (which is a cost of $5,000). Observe that the firm can consider three decision alternatives in state 1, and can choose between two decision alternatives in each of the other three states.

Over a finite horizon, epoch n denotes the end of month n. A policy prescribes the decision to be made in every state at every epoch. The selection of a policy identifies which transition probabilities will govern the behavior of the system as it evolves over time from state to state. For an N-state

TABLE 5.1

States, Decisions, and Rewards for Monthly Sales

Monthly Sales Quartile	State	Decision	Action	Reward
First (Lowest)	1	1	Sell noncore assets	-$30,000
		2	Take firm private	-25,000
		3	Offer employee buyouts	-20,000
Second	2	1	Reduce executive salaries	$5,000
		2	Reduce employee benefits	10,000
Third	3	1	Design more appealing products	-$10,000
		2	Invest in new technology	-5,000
Fourth (Highest)	4	1	Invest in new projects	$35,000
		2	Make strategic acquisitions	25,000

process, a policy at epoch n can be specified by the following $N \times 1$ column vector:

$$d(n) = \begin{matrix} 1 \\ 2 \\ \vdots \\ N \end{matrix} \begin{bmatrix} d_1(n) \\ d_2(n) \\ \vdots \\ d_N(n) \end{bmatrix} = [d_1(n) \ d_2(n) \ \cdots \ d_N(n)]^T, \qquad (5.1, 5.2)$$

called a decision vector. The elements of vector $d(n)$ indicate which decision is made in every state at epoch n. Thus, the element $d_i(n) = k$ indicates that in state i at epoch n, decision k is made.

Each decision made in a state has an associated reward and probability distribution for transitions out of that state. A superscript k is used to designate the decision in a state. Thus, p_j^k indicates that the transition probability from state i to state j is determined by the decision k in state i at epoch n. That is,

$$p_{ij}^k = P(X_{n+1} = j \mid X_n = i \cap d_i(n) = k). \qquad (5.3)$$

Similarly, q_i^k indicates that the reward earned by a transition in state i is determined by decision k. Suppose that at epoch n an MDP is in state i, so that $X_n = i$. At epoch n, decision $d_i(n) = k$ is made, and a reward q_i^k is received. The MDP will move with transition probability p_{ij}^k to state j at epoch $n + 1$, so that $X_{n+1} = j$. For an MDP starting in state g, a sequence of states visited, the decisions made, the rewards earned, and the state transition probabilities are shown in Figure 5.1.

$X_0 = g$	$\rightarrow$	$X_1 = h$	$\rightarrow$	$X_2 = u$	$\cdots$	$X_n = i$	$\rightarrow$	$X_{n+1} = j$	State
q_g^a		q_h^b		q_u^c	$\cdots$	q_i^k		q_j^f	Reward
$d_g(0) = a$		$d_h(1) = b$		$d_u(2) = c$	$\cdots$	$d_i(n) = k$		$d_j(n+1) = f$	Decision
$p_g^{(0)}$	p_{gh}^a		p_{hu}^b		$\cdots$		p_{ij}^k		Transition probability
0		1		2		n		$n+1$	Epoch

FIGURE 5.1
Sequence of states, decisions, transitions, and rewards for an MDP.

TABLE 5.2

Two-State Generic MDP with Two Decisions in Each State

		Transition Probability		
State i	Decision k	p_{i1}^k	p_{i2}^k	Reward q_i^k
1	1	p_{11}^1	p_{12}^1	q_1^1
	2	p_{11}^2	p_{12}^2	q_1^1
2	1	p_{21}^1	p_{22}^1	q_2^1
	2	p_{21}^2	p_{12}^2	q_2^2

To see how an MDP can be represented in a tabular format, Table 5.2 specifies a 2-state generic MDP with two decisions in each state.

By matching the actions in Table 4.1 with those in Table 5.1, it is apparent that the MCR model of monthly sales constructed in equation (4.6) is generated by the decision vector $d(n) = [3 \quad 1 \quad 2 \quad 2]^T$. For example, $d_1(n) = 3$ in Table 5.1 means that in state 1 at epoch n, when monthly sales are in the first quartile, the firm will make decision 3 to offer employee buyouts. The decision vector $d(n)$ produces the same MCR that was introduced without decision alternatives in Section 4.2.2.1. Hence the MCR model in Section 4.2.2.1 corresponds to the following transition probability matrix P, reward vector q with entries expressed in thousands of dollars, and decision vector $d(n)$.

$$P = \begin{matrix} 1 \\ 2 \\ 3 \\ 4 \end{matrix}\begin{bmatrix} 0.60 & 0.30 & 0.10 & 0 \\ 0.25 & 0.30 & 0.35 & 0.10 \\ 0.05 & 0.25 & 0.50 & 0.20 \\ 0 & 0.10 & 0.30 & 0.60 \end{bmatrix}, \quad q = \begin{matrix} 1 \\ 2 \\ 3 \\ 4 \end{matrix}\begin{bmatrix} -20 \\ 5 \\ -5 \\ 25 \end{bmatrix}, \quad d(n) = \begin{matrix} 1 \\ 2 \\ 3 \\ 4 \end{matrix}\begin{bmatrix} 3 \\ 1 \\ 2 \\ 2 \end{bmatrix}. \tag{5.4}$$

In state 3 decision 2 is made. Therefore, $d_3(n) = 2$, and

$$p_{3j}^2 = [0.05 \quad 0.25 \quad 0.50 \quad 0.20] \quad \text{and} \quad q_3^2 = -5. \tag{5.5}$$

TABLE 5.3

Data for Recurrent MDP Model of Monthly Sales

State i	Decision k	Transition Probabilities				Reward
		p_{i1}^k	p_{i2}^k	p_{i3}^k	p_{i4}^k	q_i^k
1	1 Sell noncore assets	0.15	0.40	0.35	0.10	−30
	2 Take firm private	0.45	0.05	0.20	0.30	−25
	3 Offer employee buyouts	0.60	0.30	0.10	0	−20
2	1 Reduce management salaries	0.25	0.30	0.35	0.10	5
	2 Reduce employee benefits	0.30	0.40	0.25	0.05	10
3	1 Design more appealing products	0.05	0.65	0.25	0.05	−10
	2 Invest in new technology	0.05	0.25	0.50	0.20	−5
4	1 Invest in new projects	0.05	0.20	0.40	0.35	35
	2 Make strategic acquisitions	0	0.10	0.30	0.60	25

The data collected by the firm on the rewards and the transition probabilities associated with the decision made in every state of Table 5.1 are summarized in Table 5.3. Table 5.3 represents a recurrent MDP model of monthly sales in which the state denotes the monthly sales quartile.

5.1.2.2 Value Iteration over a Finite Planning Horizon

In this section value iteration is executed over a finite planning horizon containing T periods [2, 3].

5.1.2.2.1 Value Iteration Equation for Expected Total Reward

An optimal policy over a finite planning horizon is defined as one which maximizes the vector of expected total rewards received until the end of the horizon. When the planning horizon is finite, an optimal policy can be found by using value iteration which was applied to an MCR in Section 4.2.2.2. When value iteration is applied to an MDP, the definition of $v_i(n)$ is changed so that $v_i(n)$ now denotes the expected total reward earned if the system is in state i at the end of period n and an optimal policy is followed until the end of the planning horizon. By extending the informal derivation of the value iteration equation (4.34) for an MCR, the algebraic form of the value iteration equation for an MDP is

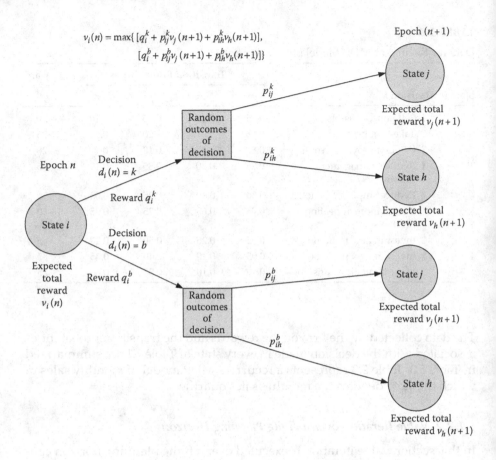

$$v_i(n) = \max\{[q_i^k + p_{ij}^k v_j(n+1) + p_{ih}^k v_h(n+1)],$$
$$[q_i^b + p_{ij}^b v_j(n+1) + p_{ih}^b v_h(n+1)]\}$$

FIGURE 5.2
Tree diagram of the value iteration equation for an MDP

$$\left.v_i(n) = \max_k \left[q_i^k + \sum_{j=1}^{N} p_{ij}^k v_j(n+1) \right], \right\}$$
$$\left. \text{for } n = 0, 1, \ldots, T-1, \text{ and } i = 1, 2, \ldots, N \right] \quad\quad (5.6)$$

The salvage values $v_i(T)$ at the end of the planning horizon must be specified for all states $i = 1, 2, \ldots, N$. Figure 5.2 is a tree diagram of the value iteration equation in algebraic form for an MDP with two decisions in the current state.

5.1.2.2.2 Value Iteration for MDP Model of Monthly Sales

Value iteration will be applied to the MDP model of monthly sales constructed in Section 5.1.2.1 to determine which decision to make in every state in each month of a seven month planning horizon so that the vector of expected total rewards is maximized. The data is given in Table 5.3.

To begin, the following salvage values are specified for all states at the end of month 7:

$$v_1(7) = v_2(7) = v_3(7) = v_4(7) = 0. \tag{5.7}$$

For the four-state MDP the value iteration equations are

$$v_i(7) = 0, \quad \text{for } i = 1, 2, 3, 4 \tag{5.8}$$
$$v_i(7) = 0, \quad \text{for } i = 1, 2, 3, 4$$

$$
\left.
\begin{aligned}
v_i(n) &= \max_k \left[q_i^k + \sum_{j=1}^{4} p_{ij}^k v_j(n+1) \right], \quad \text{for } n = 0, 1, \dots, 6, \text{ and } i = 1, 2, 3, 4 \\
&= \max_k [q_i^k + p_{i1}^k v_1(n+1) + p_{i2}^k v_2(n+1) + p_{i3}^k v_3(n+1) + p_{i4}^k v_4(n+1)], \\
&\quad \text{for } n = 0, 1, \dots, 6.
\end{aligned}
\right\} \tag{5.9}
$$

Since value iteration with decision alternatives is a form of dynamic programming, the detailed calculations for each epoch will be expressed in a tabular format.

The calculations for month 6, denoted by $n=6$, are indicated in Table 5.4.

$$
\left.
\begin{aligned}
v_i(6) &= \max_k [q_i^k + p_{i1}^k v_1(7) + p_{i2}^k v_2(7) + p_{i3}^k v_3(7) + p_{i4}^k v_4(7)], \quad \text{for } i = 1,2,3,4 \\
&= \max_k [q_i^k + p_{i1}^k(0) + p_{i2}^k(0) + p_{i3}^k(0) + p_{i4}^k(0)], \quad \text{for } i = 1,2,3,4 \\
&= \max_k [q_i^k], \quad \text{for } i = 1,2,3,4.
\end{aligned}
\right\}
$$

$$(5.10, 5.11)$$

At the end of month 6, the optimal decision is to select the alternative which will maximize the reward in every state. Thus, the decision vector at the end of month 6 is $\mathbf{d}(6) = [3 \quad 2 \quad 2 \quad 1]^{\mathrm{T}}$.

The calculations for month 5, denoted by $n=5$, are indicated in Table 5.5. The $\leftarrow$ symbol in column 3 of Tables 5.5 through 5.10 identifies the maximum

TABLE 5.4

Value Iteration for $n = 6$

State	q_i^1	q_i^2	q_i^3	Expected Total Reward	Decision
i	$k = 1$	$k = 2$	$k = 3$	$v_i(6) = \max_k[q_i^1, q_i^2, q_i^3]$, for $i = 1, 2, 3, 4$	k
1	−30	−25	−20	$v_1(6) = \max[-30, -25, -20] = -20$	3
2	5	10	—	$v_2(6) = \max[5, 10] = 10$	2
3	−10	−5	—	$v_3(6) = \max[-10, -5] = -5$	2
4	35	25	—	$v_4(6) = \max[35, 25] = 35$	1

TABLE 5.5

Value Iteration for $n = 5$

i	k	$q_i^k + p_{i1}^k v_1(6) + p_{i2}^k v_2(6) + p_{i3}^k v_3(6) + p_{i4}^k v_4(6)$ $= q_i^k + p_{i1}^k(-20) + p_{i2}^k(10) + p_{i3}^k(-5) + p_{i4}^k(35)$	$v_i(5) =$ Expected Total Reward	Decision $d_i(5) = k$
1	1	$-30 + 0.15(-20) + 0.40(10) + 0.35(-5) + 0.10(35)$ $= -27.25$		
1	2	$-25 + 0.45(-20) + 0.05(10) + 0.20(-5) + 0.30(35)$ $= -24 \leftarrow$	$\max[-27.25, -24, -29.5]$ $= -24 = v_1(5)$	$d_1(5) = 2$
1	3	$-20 + 0.60(-20) + 0.30(10) + 0.10(-5) + 0(35)$ $= -29.5$		
2	1	$5 + 0.25(-20) + 0.30(10) + 0.35(-5) + 0.10(35)$ $= 4.75$		
2	2	$10 + 0.30(-20) + 0.40(10) + 0.25(-5) + 0.05(35)$ $= 8.5 \leftarrow$	$\max[4.75, 8.5]$ $= 8.5 = v_2(5)$	$d_2(5) = 2$
3	1	$-10 + 0.05(-20) + 0.65(10) + 0.25(-5) + 0.05(35)$ $= -4$		
3	2	$-5 + 0.05(-20) + 0.25(10) + 0.50(-5) + 0.20(35)$ $= 1 \leftarrow$	$\max[-4, 1]$ $= 1 = v_3(5)$	$d_3(5) = 2$
4	1	$35 + 0.05(-20) + 0.20(10) + 0.40(-5) + 0.35(35)$ $= 46.25 \leftarrow$	$\max[46.25, 45.5]$ $= 46.25 = v_4(5)$	$d_4(5) = 1$
4	2	$25 + 0(-20) + 0.10(10) + 0.30(-5) + 0.60(35)$ $= 45.5$		

expected total reward. Both the maximum expected total reward and the associated decision are indicated in columns 4 and 5, respectively, of the row flagged by the $\leftarrow$ symbol.

$$
\left.\begin{aligned}
v_i(5) &= \max_k[q_i^k + p_{i1}^k v_1(6) + p_{i2}^k v_2(6) + p_{i3}^k v_3(6) + p_{i4}^k v_4(6)], \quad \text{for } i = 1,2,3,4 \\
&= \max_k[q_i^k + p_{i1}^k(-20) + p_{i2}^k(10) + p_{i3}^k(-5) + p_{i4}^k(35)], \quad \text{for } i = 1,2,3,4
\end{aligned}\right\} \cdot \quad (5.12)
$$

At the end of month 5, the optimal decision is to select the second alternative in states 1, 2, and 3, and the first alternative in state 4. Thus, the decision vector at the end of month 5 is $\mathbf{d}(5) = [2 \quad 2 \quad 2 \quad 1]^T$.

The calculations for month 4, denoted by $n = 4$, are indicated in Table 5.6.

TABLE 5.6

Value Iteration for $n = 4$

i	k	$q_i^k + p_{i1}^k v_1(5) + p_{i2}^k v_2(5) + p_{i3}^k v_3(5) + p_{i4}^k v_4(5)$ $= q_i^k + p_{i1}^k(-24) + p_{i2}^k(8.5) + p_{i3}^k(1) + p_{i4}^k(46.25)$	$v_i(4)$ = Expected Total Reward	Decision $d_i(4) = k$
1	1	$-30 + 0.15(-24) + 0.40(8.5) + 0.35(1) + 0.10(46.25)$ $= -25.225$		
1	2	$-25 + 0.45(-24) + 0.05(8.5) + 0.20(1) + 0.30(46.25)$ $= -21.3 \leftarrow$	max$[-25.225, -21.3,$ $-31.75] = -21.3 = v_1(4)$	$d_1(4) = 2$
1	3	$-20 + 0.60(-24) + 0.30(8.5) + 0.10(1) + 0(46.25)$ $= -31.75$		
2	1	$5 + 0.25(-24) + 0.30(8.5) + 0.35(1) + 0.10(46.25)$ $= 6.525$		
2	2	$10 + 0.30(-24) + 0.40(8.5) + 0.25(1) + 0.05(46.25)$ $= 8.7625 \leftarrow$	max $[6.525, 8.7625$ $= 8.7625 = v_2(4)$	$d_2(4) = 2$
3	1	$-10 + 0.05(-24) + 0.65(8.5) + 0.25(1) + 0.05(46.25)$ $= -3.1125$		
3	2	$-5 + 0.05(-24) + 0.25(8.5) + 0.50(1) + 0.20(46.25)$ $= 5.675 \leftarrow$	max $[-3.1125, 5.675$ $= 5.675 = v_3(4)$	$d_3(4) = 2$
4	1	$35 + 0.05(-24) + 0.20(8.5) + 0.40(1) + 0.35(46.25)$ $= 52.0875$		
4	2	$25 + 0(-24) + 0.10(8.5) + 0.30(1) + 0.60(46.25)$ $= 53.9 \leftarrow$	max$[52.0875, 53.9]$ $= 53.9 = v_4(4)$	$d_4(4) = 2$

TABLE 5.7

Value Iteration for $n = 3$

i	k	$q_i^k + p_{i1}^k v_1(4) + p_{i2}^k v_2(4) + p_{i3}^k v_3(4) + p_{i4}^k v_4(4)$ $= q_i^k + p_{i1}^k(-21.3) + p_{i2}^k(8.7625)$ $+ p_{i3}^k(5.675) + p_{i4}^k(53.9)$	$v_i(3)$ = Expected Total Reward	Decision $d_i(3) = k$
1	1	$-30 + 0.15(-21.3) + 0.40(8.7625) + 0.35(5.675)$ $+ 0.10(53.9) = -22.3138$		
1	2	$-25 + 0.45(-21.3) + 0.05(8.7625) + 0.20(5.675)$ $+ 0.30(53.9) = -16.8419 \leftarrow$	max$[-22.3138, -16.8419,$ $-29.5838] = -16.8419$ $= v_1(3)$	$d_1(3) = 2$
1	3	$-20 + 0.60(-21.3) + 0.30(8.7625) + 0.10(5.675)$ $+ 0(53.9) = -29.58375$		
2	1	$5 + 0.25(-21.3) + 0.30(8.7625) + 0.35(5.675)$ $+ 0.10(53.9) = 9.68$		

continued

TABLE 5.7 (continued)

Value Iteration for $n = 3$

i	k	$q_i^k + p_{i1}^k v_1(4) + p_{i2}^k v_2(4) + p_{i3}^k v_3(4) + p_{i4}^k v_4(4)$ $= q_i^k + p_{i1}^k(-21.3) + p_{i2}^k(8.7625)$ $\quad + p_{i3}^k(5.675) + p_{i4}^k(53.9)$	$v_i(3)$ = Expected Total Reward	Decision $d_i(3) = k$
2	2	$10 + 0.30(-21.3) + 0.40(8.7625) + 0.25(5.675)$ $\quad + 0.05(53.9) = 11.22875 \leftarrow$	$\max[9.68, 11.2288]$ $= 11.2288 = v_2(3)$	$d_2(3) = 2$
3	1	$-10 + 0.05(-21.3) + 0.65(8.7625) + 0.25(5.675)$ $\quad + 0.05(53.9) = -1.255625$		
3	2	$-5 + 0.05(-21.3) + 0.25(8.7625) + 0.50(5.675)$ $\quad + 0.20(53.9) = 9.743125 \leftarrow$	$\max[-1.2556, 9.7431]$ $= 9.7431 = v_3(3)$	$d_3(3) = 2$
4	1	$35 + 0.05(-21.3) + 0.20(8.7625) + 0.40(5.675)$ $\quad + 0.35(53.9) = 56.8225$		
4	2	$25 + 0(-21.3) + 0.10(8.7625) + 0.30(5.675)$ $\quad + 0.60(53.9) = 59.91875 \leftarrow$	$\max[56.8225, 59.9188]$ $= 59.9188 = v_4(3)$	$d_4(3) = 2$

$$\left.\begin{aligned}
v_i(4) &= \max_k [q_i^k + p_{i1}^k v_1(5) + p_{i2}^k v_2(5) + p_{i3}^k v_3(5) + p_{i4}^k v_4(5)], \text{ for } i = 1,2,3,4 \\
&= \max_k [q_i^k + p_{i1}^k(-24) + p_{i2}^k(8.5) + p_{i3}^k(1) + p_{i4}^k(46.25)], \text{ for } i = 1,2,3,4
\end{aligned}\right\} \quad (5.13)$$

At the end of month 4, the optimal decision is to select the second alternative in every state. Thus, the decision vector at the end of month 4 is $\mathbf{d}(4) = [2 \quad 2 \quad 2 \quad 2]^T$.

The calculations for month 3, denoted by $n = 3$, are indicated in Table 5.7.

$$\left.\begin{aligned}
v_i(3) &= \max_k [q_i^k + p_{i1}^k v_1(4) + p_{i2}^k v_2(4) + p_{i3}^k v_3(4) + p_{i4}^k v_4(4)], \text{ for } i = 1,2,3,4 \\
&= \max_k [q_i^k + p_{i1}^k(-21.3) + p_{i2}^k(8.7625) + p_{i3}^k(5.675) + p_{i4}^k(53.9)], \\
&\quad \text{for } i = 1,2,3,4.
\end{aligned}\right\} \quad (5.14)$$

At the end of month 3, the optimal decision is to select the second alternative in every state. Thus, the decision vector at the end of month 3 is $\mathbf{d}(3) = [2 \quad 2 \quad 2 \quad 2]^T$.

The calculations for month 2, denoted by $n = 2$, are indicated in Table 5.8.

$$\left.\begin{aligned}
v_i(2) &= \max_k [q_i^k + p_{i1}^k v_1(3) + p_{i2}^k v_2(3) + p_{i3}^k v_3(3) + p_{i4}^k v_4(3)], \text{ for } i = 1,2,3,4 \\
&= \max_k [q_i^k + p_{i1}^k(-16.8419) + p_{i2}^k(11.2288) + p_{i3}^k(9.7431) + p_{i4}^k(59.9188)], \\
&\quad \text{for } i = 1,2,3,4.
\end{aligned}\right\} \quad (5.15)$$

TABLE 5.8

Value Iteration for $n = 2$

i	k	$q_i^k + p_{i1}^k v_1(3) + p_{i2}^k v_2(3) + p_{i3}^k v_3(3) + p_{i4}^k v_4(3)$ $= q_i^k + p_{i1}^k(-16.8419) + p_{i2}^k(11.2288)$ $+ p_{i3}^k(9.7431) + p_{i4}^k(59.9188)$	$v_i(2)$ = Expected Total Reward	Decision $d_i(2) = k$
1	1	$-30 + 0.15(-16.8419) + 0.40(11.2288)$ $+ 0.35(9.7431) + 0.10(59.9188) = -18.6328$		
1	2	$-25 + 0.45(-16.8419) + 0.05(11.2288)$ $+ 0.20(9.7431) + 0.30(59.9188) = -12.0932 \leftarrow$	$\max[-18.6328, -12.0932,$ $-25.7622] = -12.0932 = v_1$	$d_1(2) = 2$
1	3	$-20 + 0.60(-16.8419) + 0.30(11.2288)$ $+ 0.10(9.7431) + 0(59.9188) = -25.7622$		
2	1	$5 + 0.25(-16.8419) + 0.30(11.2288)$ $+ 0.35(9.7431) + 0.10(59.9188) = 13.5601$		
2	2	$10 + 0.30(-16.8419) + 0.40(11.2288)$ $+ 0.25(9.7431) + 0.05(59.9188) = 14.8707 \leftarrow$	$\max[13.5601, 14.8707]$ $= 14.8707 = v_2(2)$	$d_2(2) = 2$
3	1	$-10 + 0.05(-16.8419) + 0.65(11.2288)$ $+ 0.25(9.7431) + 0.05(59.9188) = 1.8883$		
3	2	$-5 + 0.05(-16.8419) + 0.25(11.2288)$ $+ 0.50(9.7431) + 0.20(59.9188) = 13.8204 \leftarrow$	$\max[1.8883, 13.8204]$ $= 13.8204 = v_3(2)$	$d_3(2) = 2$
4	1	$35 + 0.05(-16.8419) + 0.20(11.2288)$ $+ 0.40(9.7431) + 0.35(59.9188) = 61.2725$		
4	2	$25 + 0(-16.8419) + 0.10(11.2288)$ $+ 0.30(9.7431) + 0.60(59.9188) = 64.9971 \leftarrow$	$\max[61.2725, 64.9971]$ $= 64.9971 = v_4(2)$	$d_4(2) = 2$

TABLE 5.9

Value Iteration for $n = 1$

i	k	$q_i^k + p_{i1}^k v_1(2) + p_{i2}^k v_2(2) + p_{i3}^k v_3(2) + p_{i4}^k v_4(2)$ $= q_i^k + p_{i1}^k(-12.0932) + p_{i2}^k(14.8707)$ $+ p_{i3}^k(13.8204) + p_{i4}^k(64.9971)$	$v_i(1)$ = Expected Total Reward	Decision $d_i(1) = k$
1	1	$-30 + 0.15(-12.0932) + 0.40(14.8707)$ $+ 0.35(13.8204) + 0.10(64.9971) = -14.5289$		
1	2	$-25 + 0.45(-12.0932) + 0.05(14.8707)$ $+ 0.20(13.8204) + 0.30(64.9971) = -7.4352 \leftarrow$	$\max[-14.5289, -7.4352,$ $-21.4127]$ $= -7.4352 = v_1(1)$	$d_1(1) = 2$

continued

282   *Markov Chains and Decision Processes for Engineers and Managers*

TABLE 5.9 (continued)

Value Iteration for $n = 1$

i	k	$q_i^k + p_{i1}^k v_1(2) + p_{i2}^k v_2(2) + p_{i3}^k v_3(2) + p_{i4}^k v_4(2)$ $= q_i^k + p_{i1}^k(-12.0932) + p_{i2}^k(14.8707)$ $+ p_{i3}^k(13.8204) + p_{i4}^k(64.9971)$	$v_i(1)$ = Expected Total Reward	Decision $d_i(1) = k$
1	3	$-20 + 0.60(-12.0932) + 0.30(14.8707)$ $+ 0.10(13.8204) + 0(64.9971) = -21.4127$		
2	1	$5 + 0.25(-12.0932) + 0.30(14.8707)$ $+ 0.35(13.8204) + 0.10(64.9971) = 17.7748$		
2	2	$10 + 0.30(-12.0932) + 0.40(14.8707)$ $+ 0.25(13.8204) + 0.05(64.9971) = 19.0253 \leftarrow$	max[17.7748, 19.0253] $= 19.0253 = v_2(1)$	$d_2(1) = 2$
3	1	$-10 + 0.05(-12.0932) + 0.65(14.8707)$ $+ 0.25(13.8204) + 0.05(64.9971) = 5.7663$		
3	2	$-5 + 0.05(-12.0932) + 0.25(14.8707)$ $+ 0.50(13.8204) + 0.20(64.9971) = 18.0226 \leftarrow$	max[5.7663, 18.0226] $= 18.0226 = v_3(1)$	$d_3(1) = 2$
4	1	$35 + 0.05(-12.0932) + 0.20(14.8707)$ $+ 0.40(13.8204) + 0.35(64.9971) = 65.6466$		
4	2	$25 + 0(-12.0932) + 0.10(14.8707)$ $+ 0.30(13.8204) + 0.60(64.9971) = 69.6315 \leftarrow$	max [65.6466, 69.6315] $= 69.6315 = v_4(1)$	$d_4(1) = 2$

TABLE 5.10

Value Iteration for $n = 0$

i	k	$q_i^k + p_{i1}^k v_1(1) + p_{i2}^k v_2(1) + p_{i3}^k v_3(1) + p_{i4}^k v_4(1)$ $= q_i^k + p_{i1}^k(-7.4352) + p_{i2}^k(19.0253)$ $+ p_{i3}^k(18.0226) + p_{i4}^k(69.6315)$	$v_i(0)$ = Expected Total Reward	Decision $d_i(0) = k$
1	1	$-30 + 0.15(-7.4352) + 0.40(19.0253)$ $+ 0.35(18.0226) + 0.10(69.6315) = -10.2341$		
1	2	$-25 + 0.45(-7.4352) + 0.05(19.0253)$ $+ 0.20(18.0226) + 0.30(69.6315) = -2.9006 \leftarrow$	max [$-10.2341, -2.9006,$ -16.9513] $= -2.9006 = v_1(0)$	$d_1(0) = 2$
1	3	$-20 + 0.60(-7.4352) + 0.30(19.0253)$ $+ 0.10(18.0226) + 0(69.6315) = -16.9513$		
2	1	$5 + 0.25(-7.4352) + 0.30(19.0253)$ $+ 0.35(18.0226) + 0.10(69.6315) = 22.1199$		

TABLE 5.10 (continued)

Value Iteration for $n = 0$

i	k	$q_i^k + p_{i1}^k v_1(1) + p_{i2}^k v_2(1) + p_{i3}^k v_3(1) + p_{i4}^k v_4(1)$ $= q_i^k + p_{i1}^k(-7.4352) + p_{i2}^k(19.0253)$ $+ p_{i3}^k(18.0226) + p_{i4}^k(69.6315)$	$v_i(0) =$ Expected Total Reward	Decision $d_i(0) = k$
2	2	$10 + 0.30(-7.4352) + 0.40(19.0253)$ $+ 0.25(18.0226) + 0.05(69.6315) = 23.3668 \leftarrow$	max[22.1199, 23.3668] $= 23.3668 = v_2(0)$	$d_2(0) = 2$
3	1	$-10 + 0.05(-7.4352) + 0.65(19.0253)$ $+ 0.25(18.0226) + 0.05(69.6315) = 9.9819$		
3	2	$-5 + 0.05(-7.4352) + 0.25(19.0253)$ $+ 0.50(18.0226) + 0.20(69.6315) = 22.3222 \leftarrow$	max[9.9819, 22.3222] $= 22.3222 = v_3(0)$	$d_3(0) = 2$
4	1	$35 + 0.05(-7.4352) + 0.20(19.0253)$ $+ 0.40(18.0226) + 0.35(69.6315) = 70.0134$		
4	2	$25 + 0(-7.4352) + 0.10(19.0253)$ $+ 0.30(18.0226) + 0.60(69.6315) = 74.0882 \leftarrow$	max[70.0134, 74.0882] $= 74.0882 = v_4(0)$	$d_4(0) = 2$

At the end of month 2, the optimal decision is to select the second alternative in every state. Thus, the decision vector at the end of month 2 is $\mathbf{d}(2) = [2\ 2\ 2\ 2]^T$.

The calculations for month 1, denoted by $n = 1$, are indicated in Table 5.9.

$$\left. \begin{aligned} v_i(1) &= \max_k [q_i^k + p_{i1}^k v_1(2) + p_{i2}^k v_2(2) + p_{i3}^k v_3(2) + p_{i4}^k v_4(2)], \quad \text{for } i = 1,2,3,4 \\ &= \max_k [q_i^k + p_{i1}^k(-12.0932) + p_{i2}^k(14.8707) + p_{i3}^k(13.8204) + p_{i4}^k(64.9971)], \\ &\quad \text{for } i = 1,2,3,4. \end{aligned} \right\} \quad (5.16)$$

At the end of month 1, the optimal decision is to select the second alternative in every state. Thus, the decision vector at the end of month 1 is $\mathbf{d}(2) = [2\ 2\ 2\ 2]^T$.

Finally, the calculations for month 0, denoted by $n = 0$, are indicated in Table 5.10.

$$\left. \begin{aligned} v_i(0) &= \max_k [q_i^k + p_{i1}^k v_1(1) + p_{i2}^k v_2(1) + p_{i3}^k v_3(1) + p_{i4}^k v_4(1)], \quad \text{for } i = 1,2,3,4 \\ &= \max_k [q_i^k + p_{i1}^k(-7.4352) + p_{i2}^k(19.0253) + p_{i3}^k(18.0226) + p_{i4}^k(69.6315)], \\ &\quad \text{for } i = 1,2,3,4. \end{aligned} \right\}$$

$$(5.17)$$

At the end of month 0, which is the beginning of month 1, the optimal decision is to select the second alternative in every state. Thus, the decision vector at the beginning of month 1 is $\mathbf{d}(0) = [2\ \ 2\ \ 2\ \ 2]^T$.

TABLE 5.11

Expected Total Rewards and Optimal Decisions for a Planning
Horizon of 7 Months

End of Month	n							
	0	1	2	3	4	5	6	7
$v_1(n)$	−2.9006	−7.4352	−12.0932	−16.8419	−21.3	−24	−20	0
$v_2(n)$	23.3668	19.0253	14.8707	11.2288	8.7625	8.5	10	0
$v_3(n)$	22.3222	18.0226	13.8204	9.7431	5.675	1	−5	0
$v_4(n)$	74.0882	69.6315	64.9971	59.9188	53.9	46.25	35	0
$d_1(n)$	2	2	2	2	2	2	3	−
$d_2(n)$	2	2	2	2	2	2	2	−
$d_3(n)$	2	2	2	2	2	2	2	−

The results of these calculations for the expected total rewards and the optimal decisions at the end of each month of the 7-month planning horizon are summarized in Table 5.11.

If the process starts in state 4, the expected total reward is 74.0882, the highest for any state. On the other hand, the lowest expected total reward is −2.9006, which is an expected total cost of 2.9006, if the process starts in state 1.

5.1.2.3 An Infinite Planning Horizon

To find an optimal policy for a recurrent MDP over an infinite planning horizon, the following four computational procedures can, in principle, be applied: exhaustive enumeration, value iteration, policy iteration (PI), and linear programming (LP). However, exhaustive enumeration, which involves calculating the steady-state probability distribution and the associated gain for every possible policy, is computationally prohibitive unless the problem is extremely small. Nevertheless, exhaustive enumeration will be applied to the MDP model of monthly sales to display the scope of the enumeration process, and to show why a formal optimization procedure is needed. Value iteration requires fewer arithmetic operations than alternative computational procedures, but may never satisfy a stopping condition. If value iteration does satisfy a stopping condition, it yields only approximate solutions for the relative values and the gain of an optimal policy.

To illustrate the four methods for finding an optimal policy for a recurrent MDP over an infinite planning horizon, consider the MDP model of monthly sales specified in Table 5.3. The expected total rewards and the associated decisions were obtained by value iteration executed over a finite horizon of length seven periods and summarized in Table 5.11. Suppose that the firm plans to track its monthly sales over a large enough number of months to justify using an infinite planning horizon. Over an infinite horizon, the system enters a steady state. In the steady state, a decision is no longer a function of the epoch n. Thus, over an infinite horizon, $d_i = k$ indicates that in state i the same decision k will be always made, irrespective of the epoch, n.

The set of decisions for all states is called a policy. For an N-state process, a policy over an infinite horizon can be specified by a decision vector,

$$d = [d_1 \quad d_2 \quad \cdots \quad d_N]^T. \tag{5.18}$$

The elements of d indicate which decision is to be made in every state. Over an infinite planning horizon, the policy specified by the vector d is called a stationary policy because the policy will always specify the same decision in a given state, independent of time. For the model of monthly sales, if the horizon is infinite and the policy is chosen, then in state 1, when monthly sales are in the first quartile, $d_1 = 3$ given by the decision vector in Equation (5.19). Therefore, every month in which monthly sales are in the first quartile, the firm will always make decision 3, which is to offer employee buyouts.

$$d = \begin{array}{c} 1 \\ 2 \\ 3 \\ 4 \end{array} \begin{bmatrix} 3 \\ 1 \\ 2 \\ 2 \end{bmatrix}. \tag{5.19}$$

Over an infinite horizon, an optimal policy is defined as one which maximizes the gain, or average reward per period. In Equation (4.53), the gain g for the policy $d = [3 \quad 1 \quad 2 \quad 2]^T$ was calculated by using Equation (4.47).

Under this policy the firm earns an average return of 1.4143 per period. Suppose that the firm wishes to determine whether this policy is optimal. The firm may proceed by applying exhaustive enumeration to this four-state MDP model.

5.1.2.3.1 Exhaustive Enumeration

The simplest but least efficient way to find an optimal stationary policy is to use exhaustive enumeration [1, 2]. This procedure involves listing every possible policy, computing the steady-state probability vector for each policy, and then using Equation (4.47) to compute the gain for each policy. The policy with the highest gain is optimal. Since the decisions in every state can be made independently, the number of possible policies is equal to the product, over all states, of the number of decisions permitted in every state. For example, for the MDP model of monthly sales, three decisions are allowed in state 1 while two decisions are permitted in each of the other three states. Hence, there are $(3)(2)(2)(2) = 24$ alternative policies which must be compared. Letting $^r d$ denote the rth policy, for $r = 1, 2, \ldots, 24$, the 24 possible policies are enumerated in Table 5.12.

TABLE 5.12

Exhaustive Enumeration of Decision Vectors for the 24 Possible Policies for the MDP Model of Monthly Sales

$$^1d = \begin{bmatrix} 1 \\ 2 \\ 3 \\ 4 \end{bmatrix}\begin{bmatrix} 1 \\ 1 \\ 1 \\ 1 \end{bmatrix}, \quad ^2d = \begin{bmatrix} 1 \\ 2 \\ 3 \\ 4 \end{bmatrix}\begin{bmatrix} 1 \\ 1 \\ 1 \\ 2 \end{bmatrix}, \quad ^3d = \begin{bmatrix} 1 \\ 2 \\ 3 \\ 4 \end{bmatrix}\begin{bmatrix} 1 \\ 1 \\ 2 \\ 1 \end{bmatrix}, \quad ^4d = \begin{bmatrix} 1 \\ 2 \\ 3 \\ 4 \end{bmatrix}\begin{bmatrix} 1 \\ 1 \\ 2 \\ 2 \end{bmatrix}, \quad ^5d = \begin{bmatrix} 1 \\ 2 \\ 3 \\ 4 \end{bmatrix}\begin{bmatrix} 1 \\ 2 \\ 1 \\ 1 \end{bmatrix}, \quad ^6d = \begin{bmatrix} 1 \\ 2 \\ 3 \\ 4 \end{bmatrix}\begin{bmatrix} 1 \\ 2 \\ 1 \\ 2 \end{bmatrix}, \quad ^7d = \begin{bmatrix} 1 \\ 2 \\ 3 \\ 4 \end{bmatrix}\begin{bmatrix} 1 \\ 2 \\ 2 \\ 1 \end{bmatrix}, \quad ^8d = \begin{bmatrix} 1 \\ 2 \\ 3 \\ 4 \end{bmatrix}\begin{bmatrix} 1 \\ 2 \\ 2 \\ 2 \end{bmatrix},$$

$$^9d = \begin{bmatrix} 1 \\ 2 \\ 3 \\ 4 \end{bmatrix}\begin{bmatrix} 2 \\ 1 \\ 1 \\ 1 \end{bmatrix}, \quad ^{10}d = \begin{bmatrix} 1 \\ 2 \\ 3 \\ 4 \end{bmatrix}\begin{bmatrix} 2 \\ 1 \\ 1 \\ 2 \end{bmatrix}, \quad ^{11}d = \begin{bmatrix} 1 \\ 2 \\ 3 \\ 4 \end{bmatrix}\begin{bmatrix} 2 \\ 1 \\ 2 \\ 1 \end{bmatrix}, \quad ^{12}d = \begin{bmatrix} 1 \\ 2 \\ 3 \\ 4 \end{bmatrix}\begin{bmatrix} 2 \\ 1 \\ 2 \\ 2 \end{bmatrix}, \quad ^{13}d = \begin{bmatrix} 1 \\ 2 \\ 3 \\ 4 \end{bmatrix}\begin{bmatrix} 2 \\ 2 \\ 1 \\ 1 \end{bmatrix}, \quad ^{14}d = \begin{bmatrix} 1 \\ 2 \\ 3 \\ 4 \end{bmatrix}\begin{bmatrix} 2 \\ 2 \\ 1 \\ 2 \end{bmatrix}, \quad ^{15}d = \begin{bmatrix} 1 \\ 2 \\ 3 \\ 4 \end{bmatrix}\begin{bmatrix} 2 \\ 2 \\ 2 \\ 1 \end{bmatrix}, \quad ^{16}d = \begin{bmatrix} 1 \\ 2 \\ 3 \\ 4 \end{bmatrix}\begin{bmatrix} 2 \\ 2 \\ 2 \\ 2 \end{bmatrix},$$

$$^{17}d = \begin{bmatrix} 1 \\ 2 \\ 3 \\ 4 \end{bmatrix}\begin{bmatrix} 3 \\ 1 \\ 1 \\ 1 \end{bmatrix}, \quad ^{18}d = \begin{bmatrix} 1 \\ 2 \\ 3 \\ 4 \end{bmatrix}\begin{bmatrix} 3 \\ 1 \\ 1 \\ 2 \end{bmatrix}, \quad ^{19}d = \begin{bmatrix} 1 \\ 2 \\ 3 \\ 4 \end{bmatrix}\begin{bmatrix} 3 \\ 1 \\ 2 \\ 1 \end{bmatrix}, \quad ^{20}d = \begin{bmatrix} 1 \\ 2 \\ 3 \\ 4 \end{bmatrix}\begin{bmatrix} 3 \\ 1 \\ 2 \\ 2 \end{bmatrix}, \quad ^{21}d = \begin{bmatrix} 1 \\ 2 \\ 3 \\ 4 \end{bmatrix}\begin{bmatrix} 3 \\ 2 \\ 1 \\ 1 \end{bmatrix}, \quad ^{22}d = \begin{bmatrix} 1 \\ 2 \\ 3 \\ 4 \end{bmatrix}\begin{bmatrix} 3 \\ 2 \\ 1 \\ 2 \end{bmatrix}, \quad ^{23}d = \begin{bmatrix} 1 \\ 2 \\ 3 \\ 4 \end{bmatrix}\begin{bmatrix} 3 \\ 2 \\ 2 \\ 1 \end{bmatrix}, \quad ^{24}d = \begin{bmatrix} 1 \\ 2 \\ 3 \\ 4 \end{bmatrix}\begin{bmatrix} 3 \\ 2 \\ 2 \\ 2 \end{bmatrix}.$$

Each policy corresponds to a different MCR. Let ^rP, ^rq, $^r\pi$, and ^rg denote the transition probability matrix, the reward vector, the steady-state probability vector, and the gain, respectively, associated with policy rd. For example, for policy $^3d = [1 \quad 1 \quad 2 \quad 1]^T$, the associated transition probability matrix, the reward vector, the steady-state probability vector, and the gain are indicated below.

$$^3d = \begin{bmatrix} 1 \\ 1 \\ 2 \\ 1 \end{bmatrix}, \quad ^3P = \begin{array}{c} 1 \\ 2 \\ 3 \\ 4 \end{array}\begin{bmatrix} 0.15 & 0.40 & 0.35 & 0.10 \\ 0.25 & 0.30 & 0.35 & 0.10 \\ 0.05 & 0.25 & 0.50 & 0.20 \\ 0.05 & 0.20 & 0.40 & 0.35 \end{bmatrix}, \quad ^3q = \begin{array}{c} 1 \\ 2 \\ 3 \\ 4 \end{array}\begin{bmatrix} -30 \\ 5 \\ -5 \\ 35 \end{bmatrix},$$

$$^3\pi = [0.1159 \quad 0.2715 \quad 0.4229 \quad 0.1897], \text{ and } ^3g = 2.4055.$$

Observe that policy $^{20}d = [3 \quad 1 \quad 2 \quad 2]^T$, the policy used for the MCR model of monthly sales, with a gain of 1.4143 calculated in Equation (4.53), is inferior to policy $^3d = [1 \quad 1 \quad 2 \quad 1]^T$ which has a higher gain of 2.4055. All 24 policies and their associated transition probability matrices, reward vectors, steady-state probability vectors, and gains are enumerated in Table 5.13a.

Exhaustive enumeration has identified $^{16}d = [2 \quad 2 \quad 2 \quad 2]^T$ as the optimal policy with a gain of 4.39. However, as this example has demonstrated, exhaustive enumeration is not feasible for larger problems because of the huge computational burden it imposes.

TABLE 5.13a

Exhaustive Enumeration of 24 Policies and their Parameters for MDP Model of Monthly Sales

$$^1d = \begin{bmatrix} 1 \\ 1 \\ 1 \\ 1 \end{bmatrix}, \quad ^1P = \begin{matrix} 1 \\ 2 \\ 3 \\ 4 \end{matrix}\begin{bmatrix} 0.15 & 0.40 & 0.35 & 0.10 \\ 0.25 & 0.30 & 0.35 & 0.10 \\ 0.05 & 0.65 & 0.25 & 0.05 \\ 0.05 & 0.20 & 0.40 & 0.35 \end{bmatrix}, \quad ^1q = \begin{matrix} 1 \\ 2 \\ 3 \\ 4 \end{matrix}\begin{bmatrix} -30 \\ 5 \\ -10 \\ 35 \end{bmatrix}'$$

$^1\pi = [0.1482 \quad 0.4168 \quad 0.3233 \quad 0.1118]$, and $^1g = -1.682$

$$^2d = \begin{bmatrix} 1 \\ 1 \\ 1 \\ 2 \end{bmatrix}, \quad ^2P = \begin{matrix} 1 \\ 2 \\ 3 \\ 4 \end{matrix}\begin{bmatrix} 0.15 & 0.40 & 0.35 & 0.10 \\ 0.25 & 0.30 & 0.35 & 0.10 \\ 0.05 & 0.65 & 0.25 & 0.05 \\ 0 & 0.10 & 0.30 & 0.60 \end{bmatrix}, \quad ^2q = \begin{matrix} 1 \\ 2 \\ 3 \\ 4 \end{matrix}\begin{bmatrix} -30 \\ 5 \\ -10 \\ 25 \end{bmatrix}'$$

$^2\pi = [0.1324 \quad 0.3881 \quad 0.3105 \quad 0.1689]$, and $^2g = -0.914$

$$^3d = \begin{bmatrix} 1 \\ 1 \\ 2 \\ 1 \end{bmatrix}, \quad ^3P = \begin{matrix} 1 \\ 2 \\ 3 \\ 4 \end{matrix}\begin{bmatrix} 0.15 & 0.40 & 0.35 & 0.10 \\ 0.25 & 0.30 & 0.35 & 0.10 \\ 0.05 & 0.25 & 0.50 & 0.20 \\ 0.05 & 0.20 & 0.40 & 0.35 \end{bmatrix}, \quad ^3q = \begin{matrix} 1 \\ 2 \\ 3 \\ 4 \end{matrix}\begin{bmatrix} -30 \\ 5 \\ -5 \\ 35 \end{bmatrix}'$$

$^3\pi = [0.1159 \quad 0.2715 \quad 0.4229 \quad 0.1897]$, and $^3g = 2.4055$

$$^4d = \begin{bmatrix} 1 \\ 1 \\ 2 \\ 2 \end{bmatrix}, \quad ^4P = \begin{matrix} 1 \\ 2 \\ 3 \\ 4 \end{matrix}\begin{bmatrix} 0.15 & 0.40 & 0.35 & 0.10 \\ 0.25 & 0.30 & 0.35 & 0.10 \\ 0.05 & 0.25 & 0.50 & 0.20 \\ 0 & 0.10 & 0.30 & 0.60 \end{bmatrix}, \quad ^4q = \begin{matrix} 1 \\ 2 \\ 3 \\ 4 \end{matrix}\begin{bmatrix} -30 \\ 5 \\ -5 \\ 25 \end{bmatrix}'$$

$^4\pi = [0.0920 \quad 0.2336 \quad 0.3953 \quad 0.2791]$, and $^4g = 3.409$

$$^5d = \begin{bmatrix} 1 \\ 2 \\ 1 \\ 1 \end{bmatrix}, \quad ^5P = \begin{matrix} 1 \\ 2 \\ 3 \\ 4 \end{matrix}\begin{bmatrix} 0.15 & 0.40 & 0.35 & 0.10 \\ 0.30 & 0.40 & 0.25 & 0.05 \\ 0.05 & 0.65 & 0.25 & 0.05 \\ 0.05 & 0.20 & 0.40 & 0.35 \end{bmatrix}, \quad ^5q = \begin{matrix} 1 \\ 2 \\ 3 \\ 4 \end{matrix}\begin{bmatrix} -30 \\ 10 \\ -10 \\ 35 \end{bmatrix}'$$

$^5\pi = [0.1815 \quad 0.4533 \quad 0.2808 \quad 0.0844]$, and $^5g = -0.766$

$$^6d = \begin{bmatrix} 1 \\ 2 \\ 1 \\ 2 \end{bmatrix}, \quad ^6P = \begin{matrix} 1 \\ 2 \\ 3 \\ 4 \end{matrix}\begin{bmatrix} 0.15 & 0.40 & 0.35 & 0.10 \\ 0.30 & 0.40 & 0.25 & 0.05 \\ 0.05 & 0.65 & 0.25 & 0.05 \\ 0 & 0.10 & 0.30 & 0.60 \end{bmatrix}, \quad ^6q = \begin{matrix} 1 \\ 2 \\ 3 \\ 4 \end{matrix}\begin{bmatrix} -30 \\ 10 \\ -10 \\ 25 \end{bmatrix}'$$

$^6\pi = [0.1676 \quad 0.4294 \quad 0.2732 \quad 0.1297]$, and $^6g = -0.2235$

continued

TABLE 5.13a (continued)

Exhaustive Enumeration of 24 Policies and their Parameters for MDP Model of Monthly Sales

$$
{}^{7}d = \begin{bmatrix} 1 \\ 2 \\ 2 \\ 1 \end{bmatrix}, \quad
{}^{7}P = \begin{matrix} 1 \\ 2 \\ 3 \\ 4 \end{matrix}\begin{bmatrix} 0.15 & 0.40 & 0.35 & 0.10 \\ 0.30 & 0.40 & 0.25 & 0.05 \\ 0.05 & 0.25 & 0.50 & 0.20 \\ 0.05 & 0.20 & 0.40 & 0.35 \end{bmatrix}, \quad
{}^{7}q = \begin{matrix} 1 \\ 2 \\ 3 \\ 4 \end{matrix}\begin{bmatrix} -30 \\ 10 \\ -5 \\ 35 \end{bmatrix},
$$

${}^{7}\pi = [0.1415 \quad 0.3094 \quad 0.3850 \quad 0.1640]$, and ${}^{7}g = 2.664$

$$
{}^{8}d = \begin{bmatrix} 1 \\ 2 \\ 2 \\ 2 \end{bmatrix}, \quad
{}^{8}P = \begin{matrix} 1 \\ 2 \\ 3 \\ 4 \end{matrix}\begin{bmatrix} 0.15 & 0.40 & 0.35 & 0.10 \\ 0.30 & 0.40 & 0.25 & 0.05 \\ 0.05 & 0.25 & 0.50 & 0.20 \\ 0 & 0.10 & 0.30 & 0.60 \end{bmatrix}, \quad
{}^{8}q = \begin{matrix} 1 \\ 2 \\ 3 \\ 4 \end{matrix}\begin{bmatrix} -30 \\ 10 \\ -5 \\ 25 \end{bmatrix},
$$

${}^{8}\pi = [0.1173 \quad 0.2714 \quad 0.3654 \quad 0.2459]$, and ${}^{8}g = 3.5155$

$$
{}^{9}d = \begin{bmatrix} 2 \\ 1 \\ 1 \\ 1 \end{bmatrix}, \quad
{}^{9}P = \begin{matrix} 1 \\ 2 \\ 3 \\ 4 \end{matrix}\begin{bmatrix} 0.45 & 0.05 & 0.20 & 0.30 \\ 0.25 & 0.30 & 0.35 & 0.10 \\ 0.05 & 0.65 & 0.25 & 0.05 \\ 0.05 & 0.20 & 0.40 & 0.35 \end{bmatrix}, \quad
{}^{9}q = \begin{matrix} 1 \\ 2 \\ 3 \\ 4 \end{matrix}\begin{bmatrix} -25 \\ 5 \\ -10 \\ 35 \end{bmatrix},
$$

${}^{9}\pi = [0.1963 \quad 0.3390 \quad 0.2989 \quad 0.1658]$, and ${}^{9}g = -0.4006$

$$
{}^{10}d = \begin{bmatrix} 2 \\ 1 \\ 1 \\ 2 \end{bmatrix}, \quad
{}^{10}P = \begin{matrix} 1 \\ 2 \\ 3 \\ 4 \end{matrix}\begin{bmatrix} 0.45 & 0.05 & 0.20 & 0.30 \\ 0.25 & 0.30 & 0.35 & 0.10 \\ 0.05 & 0.65 & 0.25 & 0.05 \\ 0 & 0.10 & 0.30 & 0.60 \end{bmatrix}, \quad
{}^{10}q = \begin{matrix} 1 \\ 2 \\ 3 \\ 4 \end{matrix}\begin{bmatrix} -25 \\ 5 \\ -10 \\ 25 \end{bmatrix},
$$

${}^{10}\pi = [0.1669 \quad 0.3102 \quad 0.2846 \quad 0.2383]$, and ${}^{10}g = 0.49$

$$
{}^{11}d = \begin{bmatrix} 2 \\ 1 \\ 2 \\ 1 \end{bmatrix}, \quad
{}^{11}P = \begin{matrix} 1 \\ 2 \\ 3 \\ 4 \end{matrix}\begin{bmatrix} 0.45 & 0.05 & 0.20 & 0.30 \\ 0.25 & 0.30 & 0.35 & 0.10 \\ 0.05 & 0.25 & 0.50 & 0.20 \\ 0.05 & 0.20 & 0.40 & 0.35 \end{bmatrix}, \quad
{}^{11}q = \begin{matrix} 1 \\ 2 \\ 3 \\ 4 \end{matrix}\begin{bmatrix} -25 \\ 5 \\ -5 \\ 35 \end{bmatrix},
$$

${}^{11}\pi = [0.1561 \quad 0.2183 \quad 0.3976 \quad 0.2280]$, and ${}^{11}g = -6.9415$

$$
{}^{12}d = \begin{bmatrix} 2 \\ 1 \\ 2 \\ 2 \end{bmatrix}, \quad
{}^{12}P = \begin{matrix} 1 \\ 2 \\ 3 \\ 4 \end{matrix}\begin{bmatrix} 0.45 & 0.05 & 0.20 & 0.30 \\ 0.25 & 0.30 & 0.35 & 0.10 \\ 0.05 & 0.25 & 0.50 & 0.20 \\ 0 & 0.10 & 0.30 & 0.60 \end{bmatrix}, \quad
{}^{12}q = \begin{matrix} 1 \\ 2 \\ 3 \\ 4 \end{matrix}\begin{bmatrix} -25 \\ 5 \\ -5 \\ 25 \end{bmatrix},
$$

TABLE 5.13a (continued)

Exhaustive Enumeration of 24 Policies and their Parameters for
MDP Model of Monthly Sales

$^{12}\boldsymbol{\pi} = [0.1189 \quad 0.1873 \quad 0.3718 \quad 0.3219]$, and $^{12}g = 4.1524$

$$^{13}\mathbf{d} = \begin{bmatrix} 2 \\ 2 \\ 1 \\ 1 \end{bmatrix}, \quad ^{13}\mathbf{P} = \begin{matrix} 1 \\ 2 \\ 3 \\ 4 \end{matrix}\begin{bmatrix} 0.45 & 0.05 & 0.20 & 0.30 \\ 0.30 & 0.40 & 0.25 & 0.05 \\ 0.05 & 0.65 & 0.25 & 0.05 \\ 0.05 & 0.20 & 0.40 & 0.35 \end{bmatrix}, \quad ^{13}\mathbf{q} = \begin{matrix} 1 \\ 2 \\ 3 \\ 4 \end{matrix}\begin{bmatrix} -25 \\ 10 \\ -10 \\ 35 \end{bmatrix}$$

$^{13}\boldsymbol{\pi} = [0.2308 \quad 0.3538 \quad 0.2615 \quad 0.1538]$, and $^{13}g = 0.536$

$$^{14}\mathbf{d} = \begin{bmatrix} 2 \\ 2 \\ 1 \\ 2 \end{bmatrix}, \quad ^{14}\mathbf{P} = \begin{matrix} 1 \\ 2 \\ 3 \\ 4 \end{matrix}\begin{bmatrix} 0.45 & 0.05 & 0.20 & 0.30 \\ 0.30 & 0.40 & 0.25 & 0.05 \\ 0.05 & 0.65 & 0.25 & 0.05 \\ 0 & 0.10 & 0.30 & 0.60 \end{bmatrix}, \quad ^{14}\mathbf{q} = \begin{matrix} 1 \\ 2 \\ 3 \\ 4 \end{matrix}\begin{bmatrix} -25 \\ 10 \\ -10 \\ 25 \end{bmatrix}$$

$^{14}\boldsymbol{\pi} = [0.2006 \quad 0.3258 \quad 0.2511 \quad 0.2225]$, and $^{14}g = 1.2945$

$$^{15}\mathbf{d} = \begin{bmatrix} 2 \\ 2 \\ 2 \\ 1 \end{bmatrix}, \quad ^{15}\mathbf{P} = \begin{matrix} 1 \\ 2 \\ 3 \\ 4 \end{matrix}\begin{bmatrix} 0.45 & 0.05 & 0.20 & 0.30 \\ 0.30 & 0.40 & 0.25 & 0.05 \\ 0.05 & 0.25 & 0.50 & 0.20 \\ 0.05 & 0.20 & 0.40 & 0.35 \end{bmatrix}, \quad ^{15}\mathbf{q} = \begin{matrix} 1 \\ 2 \\ 3 \\ 4 \end{matrix}\begin{bmatrix} -25 \\ 10 \\ -5 \\ 35 \end{bmatrix}$$

$^{15}\boldsymbol{\pi} = [0.1827 \quad 0.2385 \quad 0.3641 \quad 0.2147]$, and $^{15}g = 3.5115$

$$^{16}\mathbf{d} = \begin{bmatrix} 2 \\ 2 \\ 2 \\ 2 \end{bmatrix}, \quad ^{16}\mathbf{P} = \begin{matrix} 1 \\ 2 \\ 3 \\ 4 \end{matrix}\begin{bmatrix} 0.45 & 0.05 & 0.20 & 0.30 \\ 0.30 & 0.40 & 0.25 & 0.05 \\ 0.05 & 0.25 & 0.50 & 0.20 \\ 0 & 0.10 & 0.30 & 0.60 \end{bmatrix}, \quad ^{16}\mathbf{q} = \begin{matrix} 1 \\ 2 \\ 3 \\ 4 \end{matrix}\begin{bmatrix} -25 \\ 10 \\ -5 \\ 25 \end{bmatrix}$$

$^{16}\boldsymbol{\pi} = [0.1438 \quad 0.2063 \quad 0.3441 \quad 0.3057]$, and $^{16}g = 4.39$

$$^{17}\mathbf{d} = \begin{bmatrix} 3 \\ 1 \\ 1 \\ 1 \end{bmatrix}, \quad ^{17}\mathbf{P} = \begin{matrix} 1 \\ 2 \\ 3 \\ 4 \end{matrix}\begin{bmatrix} 0.60 & 0.30 & 0.10 & 0 \\ 0.25 & 0.30 & 0.35 & 0.10 \\ 0.05 & 0.65 & 0.25 & 0.05 \\ 0.05 & 0.20 & 0.40 & 0.35 \end{bmatrix}, \quad ^{17}\mathbf{q} = \begin{matrix} 1 \\ 2 \\ 3 \\ 4 \end{matrix}\begin{bmatrix} -20 \\ 5 \\ -10 \\ 35 \end{bmatrix}$$

$^{17}\boldsymbol{\pi} = [0.2811 \quad 0.3824 \quad 0.2579 \quad 0.0787]$, and $^{17}g = -3.5345$

continued

TABLE 5.13a (continued)

Exhaustive Enumeration of 24 Policies and their Parameters for
MDP Model of Monthly Sales

$$
^{18}\mathbf{d} = \begin{bmatrix} 3 \\ 1 \\ 1 \\ 2 \end{bmatrix}, \quad
^{18}\mathbf{P} = \begin{array}{r} 1 \\ 2 \\ 3 \\ 4 \end{array}\!\!\begin{bmatrix} 0.60 & 0.30 & 0.10 & 0 \\ 0.25 & 0.30 & 0.35 & 0.10 \\ 0.05 & 0.65 & 0.25 & 0.05 \\ 0 & 0.10 & 0.30 & 0.60 \end{bmatrix}, \quad
^{18}\mathbf{q} = \begin{array}{r} 1 \\ 2 \\ 3 \\ 4 \end{array}\!\!\begin{bmatrix} -20 \\ 5 \\ -10 \\ 25 \end{bmatrix},
$$

$^{18}\boldsymbol{\pi} = [0.2594 \quad 0.3642 \quad 0.2537 \quad 0.1228]$, and $^{18}g = -2.834$

$$
^{19}\mathbf{d} = \begin{bmatrix} 3 \\ 1 \\ 2 \\ 1 \end{bmatrix}, \quad
^{19}\mathbf{P} = \begin{array}{r} 1 \\ 2 \\ 3 \\ 4 \end{array}\!\!\begin{bmatrix} 0.60 & 0.30 & 0.10 & 0 \\ 0.25 & 0.30 & 0.35 & 0.10 \\ 0.05 & 0.25 & 0.50 & 0.20 \\ 0.05 & 0.20 & 0.40 & 0.35 \end{bmatrix}, \quad
^{19}\mathbf{q} = \begin{array}{r} 1 \\ 2 \\ 3 \\ 4 \end{array}\!\!\begin{bmatrix} -20 \\ 5 \\ -5 \\ 35 \end{bmatrix},
$$

$^{19}\boldsymbol{\pi} = [0.2299 \quad 0.2674 \quad 0.3529 \quad 0.1497]$, and $^{19}g = 0.214$

$$
^{20}\mathbf{d} = \begin{bmatrix} 3 \\ 1 \\ 2 \\ 2 \end{bmatrix}, \quad
^{20}\mathbf{P} = \begin{array}{r} 1 \\ 2 \\ 3 \\ 4 \end{array}\!\!\begin{bmatrix} 0.60 & 0.30 & 0.10 & 0 \\ 0.25 & 0.30 & 0.35 & 0.10 \\ 0.05 & 0.25 & 0.50 & 0.20 \\ 0 & 0.10 & 0.30 & 0.60 \end{bmatrix}, \quad
^{20}\mathbf{q} = \begin{array}{r} 1 \\ 2 \\ 3 \\ 4 \end{array}\!\!\begin{bmatrix} -20 \\ 5 \\ -5 \\ 25 \end{bmatrix},
$$

$^{20}\boldsymbol{\pi} = [0.1908 \quad 0.2368 \quad 0.3421 \quad 0.2303]$, and $^{20}g = 1.4143$

$$
^{21}\mathbf{d} = \begin{bmatrix} 3 \\ 2 \\ 1 \\ 1 \end{bmatrix}, \quad
^{21}\mathbf{P} = \begin{array}{r} 1 \\ 2 \\ 3 \\ 4 \end{array}\!\!\begin{bmatrix} 0.60 & 0.30 & 0.10 & 0 \\ 0.30 & 0.40 & 0.25 & 0.05 \\ 0.05 & 0.65 & 0.25 & 0.05 \\ 0.05 & 0.20 & 0.40 & 0.35 \end{bmatrix}, \quad
^{21}\mathbf{q} = \begin{array}{r} 1 \\ 2 \\ 3 \\ 4 \end{array}\!\!\begin{bmatrix} -20 \\ 10 \\ -10 \\ 35 \end{bmatrix},
$$

$^{21}\boldsymbol{\pi} = [0.3380 \quad 0.4083 \quad 0.2064 \quad 0.0473]$, and $^{21}g = -3.0855$

$$
^{22}\mathbf{d} = \begin{bmatrix} 3 \\ 2 \\ 1 \\ 2 \end{bmatrix}, \quad
^{22}\mathbf{P} = \begin{array}{r} 1 \\ 2 \\ 3 \\ 4 \end{array}\!\!\begin{bmatrix} 0.60 & 0.30 & 0.10 & 0 \\ 0.30 & 0.40 & 0.25 & 0.05 \\ 0.05 & 0.65 & 0.25 & 0.05 \\ 0 & 0.10 & 0.30 & 0.60 \end{bmatrix}, \quad
^{22}\mathbf{q} = \begin{array}{r} 1 \\ 2 \\ 3 \\ 4 \end{array}\!\!\begin{bmatrix} -20 \\ 10 \\ -10 \\ 25 \end{bmatrix},
$$

$^{22}\boldsymbol{\pi} = [0.3230 \quad 0.3965 \quad 0.2053 \quad 0.0752]$, and $^{22}g = -2.668$

$$
^{23}\mathbf{d} = \begin{bmatrix} 3 \\ 2 \\ 2 \\ 1 \end{bmatrix}, \quad
^{23}\mathbf{P} = \begin{array}{r} 1 \\ 2 \\ 3 \\ 4 \end{array}\!\!\begin{bmatrix} 0.60 & 0.30 & 0.10 & 0 \\ 0.30 & 0.40 & 0.25 & 0.05 \\ 0.05 & 0.25 & 0.50 & 0.20 \\ 0.05 & 0.20 & 0.40 & 0.35 \end{bmatrix}, \quad
^{23}\mathbf{q} = \begin{array}{r} 1 \\ 2 \\ 3 \\ 4 \end{array}\!\!\begin{bmatrix} -20 \\ 10 \\ -5 \\ 35 \end{bmatrix},
$$

$^{23}\boldsymbol{\pi} = [0.2799 \quad 0.3038 \quad 0.3005 \quad 0.1158]$, and $^{23}g = -0.01$

TABLE 5.13a (continued)

Exhaustive Enumeration of 24 Policies and their Parameters for
MDP Model of Monthly Sales

$$
{}^{24}\mathbf{d} = \begin{bmatrix} 3 \\ 2 \\ 2 \\ 2 \end{bmatrix}, \quad
{}^{24}\mathbf{P} = \begin{matrix} 1 \\ 2 \\ 3 \\ 4 \end{matrix}\begin{bmatrix} 0.60 & 0.30 & 0.10 & 0 \\ 0.30 & 0.40 & 0.25 & 0.05 \\ 0.05 & 0.25 & 0.50 & 0.20 \\ 0 & 0.10 & 0.30 & 0.60 \end{bmatrix}, \quad
{}^{24}\mathbf{q} = \begin{matrix} 1 \\ 2 \\ 3 \\ 4 \end{matrix}\begin{bmatrix} -20 \\ 10 \\ -5 \\ 25 \end{bmatrix},
$$

$${}^{24}\boldsymbol{\pi} = [0.2443 \quad 0.2762 \quad 0.2967 \quad 0.1829], \text{ and } {}^{24}g = 0.9655$$

5.1.2.3.2 Value Iteration

An optimal policy for a recurrent MDP over an infinite planning horizon
can be found with a relatively small computational effort by executing value
iteration over a large number of periods. Value iteration over an infinite horizon for a recurrent MDP is similar to value iteration for a recurrent MCR,
which is treated in Section 4.2.3.3. As Section 4.2.3.3 indicates, the sequence
of expected total rewards produced by value iteration will not always converge. When value iteration does converge, it will identify an optimal policy
for an MDP, and will also produce approximate solutions for the gain and
the expected relative rewards earned in every state.

5.1.2.3.2.1 Value Iteration Algorithm   By extending the value iteration algorithm for an MCR in Section 4.2.3.3.4, the following value iteration algorithm in its most basic form (adapted from Puterman [3]) can be applied to
a recurrent MDP over an infinite horizon. This value iteration algorithm
assumes that esspochs are numbered as consecutive negative integers from
−n at the beginning of the horizon to 0 at the end.

Step 1. Select arbitrary salvage values for

$v_i(0)$, for $i = 1, 2, \ldots , N$. For simplicity, set $v_i(0) = 0$. Specify $\varepsilon > 0$. Set $n = -1$.

Step 2. For each state i, use the value iteration equation to compute

$$v_i(n) = \max_{k}\left[q_i^k + \sum_{j=1}^{N} p_{ij}^k v_j(n+1) \right], \text{ for } i = 1, 2, \ldots , N.$$

Step 3. If $\max\limits_{i=1,2,\ldots,N} [v_i(n) - v_i(n+1)] - \min\limits_{i=1,2,\ldots,N} [v_i(n) - v_i(n+1)] < \varepsilon$, go to step 4.
Otherwise, decrement n by 1 and return to step 2.

Step 4. For each state i, choose the decision $d_i = k$ which maximizes the value
of $v_i(n)$, and stop.

TABLE 5.13b

Expected Total Rewards and Optimal Decisions during the Last
7 Months of an Infinite Planning Horizon

Epoch	n							
	-7	-6	-5	-4	-3	-2	-1	0
$v_1(n)$	-2.9006	-7.4352	-12.0932	-16.8419	-21.3	-24	-20	0
$v_2(n)$	23.3668	19.0253	14.8707	11.2288	8.7625	8.5	10	0
$v_3(n)$	22.3222	18.0226	13.8204	9.7431	5.675	1	-5	0
$v_4(n)$	74.0882	69.6315	64.9971	59.9188	53.9	46.25	35	0
$d_1(n)$	2	2	2	2	2	2	3	$-$
$d_2(n)$	2	2	2	2	2	2	2	$-$
$d_3(n)$	2	2	2	2	2	2	2	$-$
$d_4(n)$	2	2	2	2	2	1	1	$-$

5.1.2.3.2.2 Solution by Value Iteration of MDP Model of Monthly Sales In
Section 5.1.2.2.2, value iteration was executed to find an optimal policy for
a four-state MDP model of monthly sales over a 7-month planning horizon.
The data are given in Table 5.3. The solution is summarized in Table 5.11 of
Section 5.1.2.2.2, and is repeated in Table 5.13b to show the expected total
rewards received and the optimal decisions made during the last 7 months
of an infinite planning horizon. In accordance with the treatment of an MCR
model of monthly sales in Section 4.2.3.3.5, the last eight epochs of an infinite
horizon are numbered sequentially in Table 5.13b as -7, -6, -5, -4, -3, -2,
-1, and 0. Epoch 0 denotes the end of the horizon. As Sections 4.2.3.3.1 and
4.2.3.3.5 indicate, the absolute value of a negative epoch in Table 5.13b represents the number of months remaining in the horizon.

Table 5.13b shows that when 1 month remains in an infinite planning horizon, the optimal policy is to maximize the expected reward in every state.
When 2 months remain, the expected total rewards are maximized by making decision 2 in states 1, 2, and 3, and decision 1 in state 4. When 3 or more
months remain, the optimal policy is to make decision 2 in every state. Thus,
as value iteration proceeds backward from the end of the horizon, convergence to an optimal policy given by the decision vector $d(n) = [2 \quad 2 \quad 2 \quad 2]^T$
appears to have occurred at $n = -3$. This optimal policy was identified by
exhaustive enumeration in Section 5.1.2.3.1. It can be shown that as n becomes
very large, value iteration will converge to an optimal policy.

Table 5.14 gives the differences between the expected total rewards earned
over planning horizons which differ in length by one period.

In Table 5.14, a suffix U identifies $g_U(T)$, the maximum difference for each
epoch. The suffix L identifies $g_L(T)$, the minimum difference for each epoch.
The differences, $g_U(T) - g_L(T)$, obtained for all the epochs are listed in the bottom
row of Table 5.14. In Section 5.1.2.3.1, the optimal policy, $^{16}d = [2 \quad 2 \quad 2 \quad 2]^T$, found
by exhaustive enumeration, has a gain of 4.39. When seven periods remain in

TABLE 5.14

Differences between the Expected Total Rewards Earned Over Planning Horizons Which Differ in Length by One Period

Epoch	n						
	-7	-6	-5	-4	-3	-2	-1
	$v_i(-7)$	$v_i(-6)$	$v_i(-5)$	$v_i(-4)$	$v_i(-3)$	$v_i(-2)$	$v_i(-1)$
	$-v_i(-6)$	$-v_i(-5)$	$-v_i(-4)$	$-v_i(-3)$	$-v_i(-2)$	$-v_i(-1)$	$-v_i(0)$
i							
1	4.5346U	4.658U	4.7487	4.4581	2.7	-4L	-20L
2	4.3415	4.1546L	3.6419L	2.4663L	0.2625L	-1.5	10
3	4.2996L	4.2022	4.0773	4.0681	4.675	6	-5
4	4.4567	4.6344	5.0783U	6.0188U	7.65U	11.25U	35U
$\mathrm{Max}\begin{pmatrix} V_i(n)- \\ v_i(n+1) \end{pmatrix}$ $=g_U(T)$	4.5346	4.6580	5.0783	6.0188	7.65	11.25	35
$\mathrm{Min}\begin{pmatrix} V_i(n)- \\ v_i(n+1) \end{pmatrix}$ $=g_L(T)$	4.2996	4.1564	3.6419	2.4663	0.2625	-4	-20
$g_U(T)-g_L(T)$	0.235	0.5016	1.4364	3.5525	7.3875	15.25	55

the planning horizon, Table 5.14 shows that the bounds on the gain obtained by value iteration are given by $4.2996 \le g \le 4.5346$. The gain is approximately equal to the arithmetic average of its upper and lower bounds, so that

$$g \approx (4.5346 + 4.2996)/2 = 4.4171.$$

As the planning horizon is lengthened beyond seven periods, tighter bounds on the gain will be obtained. The bottom row of Table 5.14 shows that if an analyst chooses an $\varepsilon < 0.235$, then more than seven iterations of value iteration will be needed before the value iteration algorithm can be assumed to have converged to an optimal policy.

Table 5.15 gives the expected relative rewards, $v_i(n) - v_4(n)$, received during the last seven epochs of the planning horizon.

Table 5.16 compares the expected relative rewards, $v_i(-7) - v_4(-7)$, received after seven repetitions of value iteration, with the relative values, v_i, for $i = 1, 2$, and 3. The relative values are obtained by policy iteration (PI) in Section 5.1.2.3.3.3, and by linear programming (LP) as dual variables in Section 5.1.2.3.4.3.

Tables 5.15 and 5.16 have demonstrated that as the horizon grows longer and $n \to \infty$, the expected relative reward $v_i(n) - v_4(n)$ received for starting in each state i approaches the relative value v_i when the relative value for the highest numbered state, v_4, is set equal to 0.

TABLE 5.15

Expected Relative Rewards, $v_i(n) - v_4(n)$, Received during the Last Seven Epochs of the Planning Horizon

				n				
Epoch	−7	−6	−5	−4	−3	−2	−1	0
$v_1(n) - v_4(n)$	−76.9888	−77.0667	−77.0903	−76.7607	−75.2	−70.25	−55	0
$v_2(n) - v_4(n)$	−50.7214	−50.6062	−50.1264	−48.69	−45.1375	−37.75	−25	0
$v_3(n) - v_4(n)$	−51.766	−51.6089	−51.1767	−50.1757	−48.225	−45.25	−40	0

TABLE 5.16

Expected Relative Rewards, $v_i(-7) - v_4(-7)$, Received after Seven Repetitions of Value Iteration, Compared with the Relative Values

Seven Period Horizon (Value Iteration)	Infinite Horizon (Policy Iteration, Linear Programming)
$v_1(-7) - v_4(-7) = -76.9888$	$v_1 = -76.8825$
$v_2(-7) - v_4(-7) = -50.7214$	$v_2 = -50.6777$
$v_3(-7) - v_4(-7) = -51.766$	$v_3 = -51.8072$

5.1.2.3.3 Policy Iteration (PI)

In 1960, Ronald A. Howard [2] published an efficient algorithm, which he called PI, for finding an optimal stationary policy for a unichain MDP over an infinite planning horizon. He extended PI to a multichain MDP, which is treated in Section 5.1.4.2, and to a discounted MDP, which is treated in Section 5.2.2.2. In this section, PI is applied to a recurrent MDP. Recall that an MDP is termed recurrent when the transition probability matrix corresponding to every stationary policy is irreducible. A recurrent MDP is a special case of a unichain MDP without transient states. As Section 5.1.3 indicates, Howard's PI algorithm for a recurrent MDP also applies to a unichain MDP. He proved that the algorithm for a recurrent MDP will converge to an optimal policy which will maximize the gain. The PI algorithm has two main steps: the value determination (VD) operation and the policy improvement (IM) routine. The algorithm begins by arbitrarily choosing an initial policy. During the VD operation, the VD equations (VDEs) (4.62) corresponding to the current policy are solved for the gain and the relative values. (Recall that the VDEs for a recurrent MCR were informally derived in Section 4.2.3.2.2.) The IM routine attempts to find a better policy. If a better policy is found, the VD operation is repeated using the new policy to identify the appropriate transition probabilities, rewards, and VDEs. The algorithm stops when two successive iterations lead to identical policies.

5.1.2.3.3.1 Test Quantity for Policy Improvement (IM)   The IM routine is based on the value iteration Equation (5.4) for an MDP. Equation (5.4) indicates that if an optimal policy is known over a planning horizon starting at epoch $n + 1$ and ending at epoch T, then the best decision in state i at epoch n can be found by maximizing a test quantity,

$$q_i^k + \sum_{j=1}^{N} p_{ij}^k v_j(n+1), \tag{5.20}$$

over all decisions in state i. Therefore, if an optimal policy is known over a planning horizon starting at epoch 1 and ending at epoch T, then at epoch 0, the beginning of the planning horizon, the best decision in state i can be found by maximizing a test quantity,

$$q_i^k + \sum_{j=1}^{N} p_{ij}^k v_j(1), \tag{5.21}$$

over all decisions in state i. Recall from Section 4.2.3.2.2 that when T, the length of the planning horizon, is very large, $(T - 1)$ is also very large, so that

$$v_j(1) \approx (T-1)g + v_j. \tag{4.60}$$

Substituting this expression for $v_j(1)$ in the test quantity produces the result

$$\left. \begin{aligned} q_i^k + \sum_{j=1}^{N} p_{ij}^k v_j(1) &= q_i^k + \sum_{j=1}^{N} p_{ij}^k[(T-1)g + v_j] \\ &= q_i^k + (T-1)g\sum_{j=1}^{N} p_{ij}^k + \sum_{j=1}^{N} p_{ij}^k v_j \\ &= q_i^k + (T-1)g + \sum_{j=1}^{N} p_{ij}^k v_j \end{aligned} \right\}, \tag{5.22}$$

as the test quantity to be maximized with respect to all alternatives in every state. Since the term $(T - 1)g$ is independent of k, the test quantity to be maximized when making decisions in state i is

$$q_i^k + \sum_{j=1}^{N} p_{ij}^k v_j, \tag{5.23}$$

for $i = 1, 2, \ldots, N$. The relative values produced by solving the VDEs associated with the most recent policy can be used.

5.1.2.3.3.2 Policy Iteration Algorithm   The detailed steps of Howard's PI algorithm are given below:

Step 1. Initial policy
Arbitrarily choose an initial policy by selecting for each state i a decision $d_i = k$.

Step 2. VD operation
Use p_{ij} and q_i for a given policy to solve the VDEs (4.62),

$$g + v_i = q_i + \sum_{j=1}^{N} p_{ij} v_{j}, \quad i = 1, 2, ..., N$$

for all relative values v_i and the gain g by setting $v_N = 0$.

Step 3. IM routine
For each state i, find the decision $k*$ that maximizes the test quantity,

$$q_i^k + \sum_{j=1}^{N} p_{ij}^k v_j$$

using the relative values v_i of the previous policy. Then $k*$ becomes the new decision in state i, so that $d_i = k*$, q_i^{k*} becomes q_i, and p_{ij}^{k*} becomes p_{ij}.

Step 4. Stopping rule
When the policies on two successive iterations are identical, the algorithm stops because an optimal policy has been found. Leave the old d_i unchanged if the test quantity for the old d_i is equal to the test quantity for any other alternative in the new policy determination. If the new policy is different from the previous policy in at least one state, go to step 2.

   Howard proved that the gain of each policy will be greater than or equal to that of its predecessor. The algorithm will terminate after a finite number of iterations.

5.1.2.3.3.3 Solution by PI of MDP Model of Monthly Sales   Policy iteration will be executed to find an optimal policy over an infinite horizon for the MDP model of monthly sales specified in Table 5.3. An optimal policy was obtained by exhaustive enumeration in Section 5.1.2.3.1, and by value iteration in Section 5.1.2.3.2.2.

First iteration

Step 1. Initial policy
   Arbitrarily choose the initial policy $^{23}\mathbf{d} = [3 \quad 2 \quad 2 \quad 1]^T$ by making decision 3 in state 1, decision 2 in states 2 and 3, and decision 1 in state 4. Thus, $d_1 = 3$, $d_2 = d_3 = 2$, and $d_4 = 1$. The initial decision vector $^{23}\mathbf{d}$, along with the associated transition probability matrix $^{23}\mathbf{P}$ and the reward vector $^{23}\mathbf{q}$, are shown below:

$$^{23}\mathbf{d} = \begin{bmatrix} 3 \\ 2 \\ 2 \\ 1 \end{bmatrix}, \quad {}^{23}\mathbf{P} = \begin{matrix} 1 \\ 2 \\ 3 \\ 4 \end{matrix}\begin{bmatrix} 0.60 & 0.30 & 0.10 & 0 \\ 0.30 & 0.40 & 0.25 & 0.05 \\ 0.05 & 0.25 & 0.50 & 0.20 \\ 0.05 & 0.20 & 0.40 & 0.35 \end{bmatrix}, \quad {}^{23}\mathbf{q} = \begin{matrix} 1 \\ 2 \\ 3 \\ 4 \end{matrix}\begin{bmatrix} -20 \\ 10 \\ -5 \\ 35 \end{bmatrix}$$

Step 2. VD operation

Use p_{ij} and q_i for the initial policy, $^{23}\mathbf{d} = [3 \ 2 \ 2 \ 1]^T$, to solve the VDEs (4.62)

$$g + v_1 = q_1 + p_{11}v_1 + p_{12}v_2 + p_{13}v_3 + p_{14}v_4$$
$$g + v_2 = q_2 + p_{21}v_1 + p_{22}v_2 + p_{23}v_3 + p_{24}v_4$$
$$g + v_3 = q_3 + p_{31}v_1 + p_{32}v_2 + p_{33}v_3 + p_{34}v_4$$
$$g + v_4 = q_4 + p_{41}v_1 + p_{42}v_2 + p_{43}v_3 + p_{44}v_4$$

for all relative values v_i and the gain g.

$$g + v_1 = -20 + 0.60v_1 + 0.30v_2 + 0.10v_3 + 0v_4$$
$$g + v_2 = 10 + 0.30v_1 + 0.40v_2 + 0.25v_3 + 0.05v_4$$
$$g + v_3 = -5 + 0.05v_1 + 0.25v_2 + 0.50v_3 + 0.20v_4$$
$$g + v_4 = 35 + 0.05v_1 + 0.20v_2 + 0.40v_3 + 0.35v_4.$$

Setting $v_4 = 0$, the solution of the VDEs is

$$g = -0.01, \quad v_1 = -102.6731, \quad v_2 = -54.4350, \quad v_3 = -47.4686, \quad v_4 = 0.$$

Step 3. IM routine

For each state i, find the decision k^* that maximizes the test quantity

$$q_i^k + \sum_{j=1}^{4} p_{ij}^k v_j$$

using the relative values v_i of the initial policy. Then k^* becomes the new decision in state i, so that $d_i = k^*$, $q_i^{k^*}$ becomes q_i, and $p_{ij}^{k^*}$ becomes p_{ij}. The first policy improvement routine is s executed in Table 5.17.

Step 4. Stopping rule

The new policy is $^{12}\mathbf{d} = [2 \ 1 \ 2 \ 2]^T$, which is different from the initial policy. Therefore, go to step 2. The new decision vector $^{12}\mathbf{d}$, along with the associated transition probability matrix $^{12}\mathbf{P}$ and the reward vector $^{12}\mathbf{q}$, are shown below:

$$^{12}\mathbf{d} = \begin{bmatrix} 2 \\ 1 \\ 2 \\ 2 \end{bmatrix}, \quad ^{12}\mathbf{P} = \begin{matrix} 1 \\ 2 \\ 3 \\ 4 \end{matrix}\begin{bmatrix} 0.45 & 0.05 & 0.20 & 0.30 \\ 0.25 & 0.30 & 0.35 & 0.10 \\ 0.05 & 0.25 & 0.50 & 0.20 \\ 0 & 0.10 & 0.30 & 0.60 \end{bmatrix}, \quad ^{12}\mathbf{q} = \begin{matrix} 1 \\ 2 \\ 3 \\ 4 \end{matrix}\begin{bmatrix} -25 \\ 5 \\ -5 \\ 25 \end{bmatrix}.$$

TABLE 5.17

First IM for Monthly Sales Example

State i	Decision Alternative k	Test Quantity $q_i^k + p_{i1}^k v_1 + p_{i2}^k v_2 + p_{i3}^k v_3 + p_{i4}^k v_4$ $= q_i^k + p_{i1}^k(-102.6731) + p_{i2}^k(-54.4350)$ $+ p_{i3}^k(-47.4686) + p_{i4}^k(0)$	Maximum Value of Test Quantity	Decision $d_i = k^*$
1	1	$-30 + 0.15(-102.6731) + 0.40(-54.4350)$ $+ 0.35(-47.4686) + 0.10(0) = -83.7890$		
1	2	$-25 + 0.45(-102.6731) + 0.05(-54.4350)$ $+ 0.20(-47.4686) + 0.30(0) = -83.4184 \leftarrow$	max $[-83.7890,$ $-83.4184,$ $-102.6812]$ $= -83.4184$	$d_1 = 2$
1	3	$-20 + 0.60(-102.6731) + 0.30(-54.4350)$ $+ 0.10(-47.4686) + 0(0) = -102.6812$		
2	1	$5 + 0.25(-102.6731) + 0.30(-54.4350)$ $+ 0.35(-47.4686) + 0.10(0) = -53.6128 \leftarrow$	max $[-53.6128,$ $-54.4431]$ $= -53.6128$	$d_2 = 1$
2	2	$10 + 0.30(-102.6731) + 0.40(-54.4350)$ $+ 0.25(-47.4686) + 0.05(0) = -54.4431$		
3	1	$-10 + 0.05(-102.6731) + 0.65(-54.4350)$ $+ 0.25(-47.4686) + 0.05(0) = -62.3836$		
3	2	$-5 + 0.05(-102.6731) + 0.25(-54.4350)$ $+ 0.50(-47.4686) + 0.20(0) = -47.4767 \leftarrow$	max $[-62.3836,$ $-47.4767]$ $= -47.4767$	$d_3 = 2$
4	1	$35 + 0.05(-102.6731) + 0.20(-54.4350)$ $+ 0.40(-47.4686) + 0.35(0) = -0.0081$		
4	2	$25 + 0(-102.6731) + 0.10(-54.4350)$ $+ 0.30(-47.4686) + 0.60(0) = 5.3159 \leftarrow$	max $[-0.0081,$ $5.3159]$ $= 5.3159$	$d_4 = 2$

Step 2. VD operation

Use p_{ij} and q_i for the first new policy, $^{12}\mathbf{d} = [2 \quad 1 \quad 2 \quad 2]^T$, to solve the VDEs (4.62) for all relative values v_i and the gain g.

$$g + v_1 = -25 + 0.45v_1 + 0.05v_2 + 0.20v_3 + 0.30v_4$$
$$g + v_2 = 5 + 0.25v_1 + 0.30v_2 + 0.35v_3 + 0.10v_4$$
$$g + v_3 = -5 + 0.05v_1 + 0.25v_2 + 0.50v_3 + 0.20v_4$$
$$g + v_4 = 25 + 0v_1 + 0.10v_2 + 0.30v_3 + 0.60v_4.$$

Setting $v_4 = 0$, the solution of the VDEs is

$$g = 4.15, \quad v_1 = -76.6917, \quad v_2 = -52.2215, \quad v_3 = -52.0848, \quad v_4 = 0.$$

Step 3. IM routine
For each state i, find the decision k^* that maximizes the test quantity

$$q_i^k + \sum_{j=1}^{4} p_{ij}^k v_j$$

using the relative values v_i of the previous policy. Then k^* becomes the new decision in state i, so that $d_i = k^*$, $q_i^{k^*}$ becomes q_i, and $p_{ij}^{k^*}$ becomes p_{ij}. The second policy improvement routine is executed in Table 5.18.

Step 4. Stopping rule
The new policy is $^{16}\mathbf{d} = [2 \ \ 2 \ \ 2 \ \ 2]^T$, which is different from the previous policy. Therefore, go to step 2. The new decision vector $^{16}\mathbf{d}$, along with the associated transition probability matrix $^{16}\mathbf{P}$, and the reward vector $^{16}\mathbf{q}$, are shown below:

$$^{16}\mathbf{d} = \begin{bmatrix} 2 \\ 2 \\ 2 \\ 2 \end{bmatrix}, \quad ^{16}\mathbf{P} = \begin{matrix} 1 \\ 2 \\ 3 \\ 4 \end{matrix} \begin{bmatrix} 0.45 & 0.05 & 0.20 & 0.30 \\ 0.30 & 0.40 & 0.25 & 0.05 \\ 0.05 & 0.25 & 0.50 & 0.20 \\ 0 & 0.10 & 0.30 & 0.60 \end{bmatrix}, \quad ^{16}\mathbf{q} = \begin{matrix} 1 \\ 2 \\ 3 \\ 4 \end{matrix} \begin{bmatrix} -25 \\ 10 \\ -5 \\ 25 \end{bmatrix} \quad (5.24)$$

Step 2. VD operation
Use p_{ij} and q_i for the second new policy, $^{16}\mathbf{d} = [2 \ \ 2 \ \ 2 \ \ 2]^T$, to solve the VDEs (4.62) for all relative values v_i and the gain g.

$$g + v_1 = -25 + 0.45v_1 + 0.05v_2 + 0.20v_3 + 0.30v_4$$
$$g + v_2 = 10 + 0.30v_1 + 0.40v_2 + 0.25v_3 + 0.05v_4$$
$$g + v_3 = -5 + 0.05v_1 + 0.25v_2 + 0.50v_3 + 0.20v_4$$
$$g + v_4 = 25 + 0v_1 + 0.10v_2 + 0.30v_3 + 0.60v_4.$$

Setting $v_4 = 0$, the solution of the VDE's is

$$g = 4.39, \quad v_1 = -76.8825, \quad v_2 = -50.6777, \quad v_3 = -51.8072, \quad v_4 = 0. \quad (5.25)$$

TABLE 5.18

Second IM for Monthly Sales Example

State i	Decision Alternative k	Test Quantity $q_i^k + p_{i1}^k v_1 + p_{i2}^k v_2 + p_{i3}^k v_3 + p_{i4}^k v_4$ $= q_i^k + p_{i1}^k(-76.6917) + p_{i2}^k(-52.221$ $+ p_{i3}^k(-52.0848) + p_{i4}^k(0)$	Maximum Value of Test Quantity	Decision $d_i = k^*$
1	1	$-30 + 0.15(-76.6917) + 0.40(-52.2215)$ $+ 0.35(-52.0848) + 0.10(0) = -80.6220$		
1	2	$-25 + 0.45(-76.6917) + 0.05(-52.2215)$ $+ 0.20(-52.0848) + 0.30(0) = -72.5393 \leftarrow$	max $[-80.6220,$ $-72.5393,$ $-86.8900]$ $= -72.5393$	$d_1 = 2$
1	3	$-20 + 0.60(-76.6917) + 0.30(-52.2215)$ $+ 0.10(-52.0848) + 0(0) = -86.8900$		
2	1	$5 + 0.25(-76.6917) + 0.30(-52.2215)$ $+ 0.35(-52.0848) + 0.10(0) = -48.0691$		
2	2	$10 + 0.30(-76.6917) + 0.40(-52.2215)$ $+ 0.25(-52.0848) + 0.05(0) = -46.9173 \leftarrow$	max $[-48.0691,$ $-46.9173]$ $= -46.9173$	$d_2 = 2$
3	1	$-10 + 0.05(-76.6917) + 0.65(-52.2215)$ $+ 0.25(-52.0848) + 0.05(0) = -60.7998$		
3	2	$-5 + 0.05(-76.6917) + 0.25(-52.2215)$ $+ 0.50(-52.0848) + 0.20(0) = -47.9324 \leftarrow$	max $[-60.7998,$ $-47.9324]$ $= -47.9324$	$d_3 = 2$
4	1	$35 + 0.05(-76.6917) + 0.20(-52.2215)$ $+ 0.40(-52.0848) + 0.35(0) = -0.1128$		
4	2	$25 + 0(-76.6917) + 0.10(-52.2215)$ $+ 0.30(-52.0848) + 0.60(0) = 4.1524 \leftarrow$	max $[-0.1128,$ $4.1524]$ $= 4.1524$	$d_4 = 2$

Step 3. IM routine
   For each state i, find the decision k^* that maximizes the test quantity

$$q_i^k + \sum_{j=1}^{4} p_{ij}^k v_j$$

TABLE 5.19

Third IM for Monthly Sales Example

State i	Decision Alternative k	Test Quantity $q_i^k + p_{i1}^k v_1 + p_{i2}^k v_2 + p_{i3}^k v_3 + p_{i4}^k v_4$ $= q_i^k + p_{i1}^k(-76.8825) + p_{i2}^k(-50.6777)$ $+ p_{i3}^k(-51.8072) + p_{i4}^k(0)$	Maximum Value of Test Quantity	Decision $d_i = k^*$
1	1	$-30 + 0.15(-76.8825) + 0.40(-50.6777)$ $+ 0.35(-51.8072) + 0.10(0) = -79.9360$		
1	2	$-25 + 0.45(-76.8825) + 0.05(-50.6777)$ $+ 0.20(-51.8072) + 0.30(0) = -72.4925 \leftarrow$	max $[-79.9360,$ $-72.4925,$ $-86.5135]$ $= -72.4925$	$d_1 = 2$
1	3	$-20 + 0.60(-76.8825) + 0.30(-50.6777)$ $+ 0.10(-51.8072) + 0(0) = -86.5135$		
2	1	$5 + 0.25(-76.8825) + 0.30(-50.6777)$ $+ 0.35(-51.8072) + 0.10(0) = -47.5565$		
2	2	$10 + 0.30(-76.8825) + 0.40(-50.6777)$ $+ 0.25(-51.8072) + 0.05(0) = -46.2876 \leftarrow$	max $[-47.5565,$ $-46.2876]$ $= -46.2876$	$d_2 = 2$
3	1	$-10 + 0.05(-76.8825) + 0.65(-50.6777)$ $+ 0.25(-51.8072) + 0.05(0) = -59.7364$		
3	2	$-5 + 0.05(-76.8825) + 0.25(-50.6777)$ $+ 0.50(-51.8072) + 0.20(0) = -47.4172 \leftarrow$	max $[-59.7364,$ $-47.4172]$ $= -47.4172$	$d_3 = 2$
4	1	$35 + 0.05(-76.8825) + 0.20(-50.6777)$ $+ 0.40(-51.8072) + 0.35(0) = 0.2975$		
4	2	$25 + 0(-76.8825) + 0.10(-50.6777)$ $+ 0.30(-51.8072) + 0.60(0) = 4.3901 \leftarrow$	max $[0.2975,$ $4.3901]$ $= 4.3901$	$d_4 = 2$

using the relative values v_i of the previous policy. Then k^* becomes the new decision in state i, so that $d_i = k^*$, $q_i^{k^*}$ becomes q_i, and $p_{ij}^{k^*}$ becomes p_{ij}. The third policy improvement routine is executed in Table 5.19.

Step 4. Stopping rule

Stop because the new policy, given by the vector $^{16}\mathbf{d} = [2 \quad 2 \quad 2 \quad 2]^T$, is identical to the previous policy. Therefore, this policy is optimal. The transition probability matrix $^{16}\mathbf{P}$ and the reward vector $^{16}\mathbf{q}$ for the optimal policy are shown in Equation (5.20). The relative values and the gain are given in Equation (5.21). Equation (5.21) shows that the relative values obtained by PI are identical to the corresponding dual variables obtained by LP in Table 5.23 of Section 5.1.2.3.4.3.

5.1.2.3.3.4 Optional Insight: IM for a Two-State MDP Step 3 of the PI algorithm in Section 5.1.2.3.3.2 is the IM routine. To gain insight (adapted from Howard [2]) into why the IM routine works, it is instructive to examine the special case of a generic two-state recurrent MDP with two decisions per state, which is shown in Table 5.2. To simplify the notation, all symbols for transition probabilities and rewards in Table 5.2 are replaced by letters without subscripts or superscripts in Table 5.20.

Since the process has two states with two decisions per state, there are $2^2 = 4$ possible policies, identified by the letters A, B, C, and D. The decision vectors for the four policies are

$$d^A = \frac{1}{2}\begin{bmatrix}1\\1\end{bmatrix}, \quad d^B = \frac{1}{2}\begin{bmatrix}1\\2\end{bmatrix}, \quad d^C = \frac{1}{2}\begin{bmatrix}2\\1\end{bmatrix}, \quad d^D = \frac{1}{2}\begin{bmatrix}2\\2\end{bmatrix}$$

Suppose that policy A is selected. Policy A is characterized by the following decision vector, transition probability matrix, reward vector, and relative value vector:

$$\mathbf{d}^A = \begin{bmatrix}1\\1\end{bmatrix}, \quad \mathbf{P}^A = \frac{1}{2}\begin{bmatrix}1-a & a\\ c & 1-c\end{bmatrix}, \quad \mathbf{q}^A = \frac{1}{2}\begin{bmatrix}e\\s\end{bmatrix}, \quad \mathbf{v}^A = \frac{1}{2}\begin{bmatrix}v_1^A\\v_2^A\end{bmatrix} = \frac{1}{2}\begin{bmatrix}v_1^A\\0\end{bmatrix},$$

after setting $v_2^A = 0$.

Under policy A, using Equation (4.65), the gain is

$$g^A = \frac{ce + as}{a+c}$$

and the relative value for state 1 is

$$v_1^A = \frac{e-s}{a+c}.$$

TABLE 5.20

Two-State Generic MDP with Two Decisions Per State

State i	Decision k	Transition Probability		Reward q_i^k
		p_{i1}^k	p_{i2}^k	
1	1	$1-a$	a	e
	2	$1-b$	b	f
2	1	c	$1-c$	s
	2	d	$1-d$	h

Suppose that the evaluation of policy A by the IM routine has produced a new policy, policy D. Policy D is characterized by the following decision vector, transition probability matrix, and reward vector:

$$\mathbf{d}^D = \begin{bmatrix} 2 \\ 2 \end{bmatrix}, \quad \mathbf{P}^D = \frac{1}{2}\begin{bmatrix} 1-b & b \\ d & 1-d \end{bmatrix}, \quad \mathbf{q}^D = \frac{1}{2}\begin{bmatrix} f \\ h \end{bmatrix}.$$

Using Equation (4.65), the gain under policy D is

$$g^D = \frac{df + bh}{b + d}.$$

The objective of this insight is to show that $g^D > g^A$. Since the IM routine has chosen policy D over policy A, the test quantity for policy D must be greater than or equal to the test quantity for policy A in every state. Therefore,

For $i = 1$,

$$q_1^D + p_{11}^D v_1^A + p_{12}^D v_2^A \geq q_1^A + p_{11}^A v_1^A + p_{12}^A v_2^A$$
$$f + (1-b)v_1^A + b(0) \geq e + (1-a)v_1^A + a(0)$$
$$f + (1-b)v_1^A \geq e + (1-a)v_1^A.$$

Let γ_1 denote the improvement in the test quantity achieved in state 1.

$$\gamma_1 = f + (1-b)v_1^A - e - (1-a)v_1^A \geq 0$$
$$= f - e + (a - b)v_1^A \geq 0.$$

For $i = 2$,

$$q_2^D + p_{21}^D v_1^A + p_{22}^D v_2^A \geq q_2^A + p_{21}^A v_1^A + p_{22}^A v_2^A$$
$$h + dv_1^A + (1-d)(0) \geq s + cv_1^A + (1-c)(0)$$
$$h + dv_1^A \geq s + cv_1^A.$$

Let γ_2 denote the improvement in the test quantity achieved in state 2.

$$\gamma_2 = h + dv_1^A - s - cv_1^A \geq 0$$
$$= h - s + (d - c)v_1^A \geq 0.$$

To show that $g^D > g^A$, calculate

$$g^D - g^A = \frac{df + bh}{b+d} - \frac{ce + as}{a+c}$$

$$= \frac{(a+c)(df+bh) - (b+d)(ce+as)}{(b+d)(a+c)}$$

$$= \frac{d}{b+d}\left[\frac{(a-b)(e-s)+(a+c)(f-e)}{a+c}\right] + \frac{b}{b+d}\left[\frac{(d-c)(e-s)+(a+c)(h-s)}{a+c}\right]$$

$$= \frac{d}{b+d}\left[(a-b)v_1^A + f - e\right] + \frac{b}{b+d}\left[(d-c)v_1^A + h - s\right]$$

$$= \frac{d}{b+d}\gamma_1 + \frac{b}{b+d}\gamma_2$$

Equation (2.16) indicates that the steady-state probability vector under policy D is

$$\pi^D = \left[\frac{d}{b+d} \quad \frac{b}{b+d}\right].$$

Hence,

$$g^D - g^A = \pi_1^D\gamma_1 + \pi_2^D\gamma_2.$$

Since $\pi_1^D > 0$, $\pi_2^D > 0$, $\gamma_1 \geq 0$, and $\gamma_2 \geq 0$, it follows that $g^D - g^A \geq 0$ for the two-state process. Howard proves that for an N-state MDP, g^D will be greater than g^A if an improvement in the test quantity can be made in any state that will be recurrent under policy D.

5.1.2.3.4 Linear Programming

An MDP can be formulated as a linear program (LP) when the planning horizon is infinite. (In this section, and in Sections 5.1.3.2, 5.1.4.3, and 5.2.2.3, a basic knowledge of LP is assumed.) In this section, MDPs will be formulated as linear programs which will be solved by using a computer software package. The advantage of an LP formulation is the availability of various computer software packages for solving LPs. For example, the Excel spreadsheet has an add-in called Solver which will solve both linear and nonlinear programs. The disadvantage of an LP formulation is that it produces a much larger model of an MDP than the corresponding model solved by PI. (In this book, both "linear programming" and "linear program" are abbreviated as "LP.")

5.1.2.3.4.1 Formulation of a LP Model

The LP formulation for a recurrent N-state MDP assumes that the Markov chain associated with every transition matrix is regular [1,3]. As Section 2.1 indicates, the entries of the steady-state probability vector for a regular Markov chain are strictly positive and sum to one. To formulate an LP, one must identify decision variables, an objective function, and a set of constraints.

Decision Variables
The decision variable in the LP formulation is the joint probability of being in state i and making decision k. This decision variable is denoted by

$$y_i^k = P(\text{state} = i \cap \text{decision} = k).$$

The marginal probability of being in state i is obtained by summing the joint probabilities over all values of k. Letting K_i denote the number of possible decisions in state i,

$$P(\text{state} = i) = \sum_{k=1}^{K_i} P(\text{state} = i \cap \text{decision} = k) = \sum_{k=1}^{K_i} y_i^k. \tag{5.26}$$

In the steady state,

$$\pi_i = P(\text{state} = i) \sum_{k=1}^{K_i} y_i^k. \tag{5.27}$$

By the multiplication rule of probability,

$$P(\text{state} = i \cap \text{decision} = k) = P(\text{state} = i)\, P(\text{decision} = k \mid \text{state} = i), \text{ or}$$

$$y_i^k = \pi_i\, P(\text{decision} = k \mid \text{state} = i). \tag{5.28}$$

Dividing both sides by $\pi_i = P(\text{state} = i)$,

$$P(\text{decision} = k \mid \text{state} = i) = y_i^k / \pi_i. \tag{5.29}$$

Substituting $\pi_i = \sum_{k=1}^{K_i} y_i^k$

$$P(\text{decision} = k \mid \text{state} = i) = \frac{y_i^k}{\sum_{k=1}^{K_i} y_i^k} \tag{5.30}$$

To interpret these probabilities, consider the recurrent MDP model of monthly sales specified in Table 5.3. Consider the decision vector [20] $\mathbf{d} = [3\ 1\ 2\ 2]^{\mathsf{T}}$

in Table 5.12. Expressed in terms of the conditional probabilities, the decision vector is

$$^{20}\mathbf{d} = \begin{bmatrix} d_1 \\ d_2 \\ d_3 \\ d_N \end{bmatrix} = \begin{array}{c} 1 \\ 2 \\ 3 \\ 4 \end{array} \begin{bmatrix} 3 \\ 1 \\ 2 \\ 2 \end{bmatrix} = \begin{bmatrix} P(\text{decision} = 3 \mid \text{state} = 1) \\ P(\text{decision} = 1 \mid \text{state} = 2) \\ P(\text{decision} = 2 \mid \text{state} = 3) \\ P(\text{decision} = 2 \mid \text{state} = 4) \end{bmatrix}. \tag{5.31}$$

The conditional probabilities in a decision vector must assume values of zero or one. For each state, only one conditional probability has a value of one; the others have values of zero. The reason is that in each state i exactly one decision alternative is selected, and the remaining alternatives are rejected. The decision alternative which is selected in state i has a conditional probability of one. The remaining alternatives which are not selected in state i have conditional probabilities of zero. Thus, if decision alternative h is selected in state i, $P(\text{decision} = h \mid \text{state} = i) = 1$, and $P(\text{decision} \neq h \mid \text{state} = i) = 0$, for all other decision alternatives in state i. For example, the model of monthly sales has three decision alternatives in state 1. For the decision vector $^{20}\mathbf{d} = [3\ 1\ 2\ 2]^T$, alternative 3 is selected in state 1. Therefore, the conditional probabilities in state 1 are

$$P(\text{decision} = 1 \mid \text{state} = 1) = 0, \ P(\text{decision} = 2 \mid \text{state} = 1) = 0,$$

$$P(\text{decision} = 3 \mid \text{state} = 1) = 1.$$

For each state, only one joint probability has a positive value; the others have values of zero. The joint probability $y_i^k = P(\text{state} = i \cap \text{decision} = k) > 0$ if the associated conditional probability, $P(\text{decision} = k \mid \text{state} = i) = 1$. The reason is that when $P(\text{decision} = k \mid \text{state} = i) = 1$, then $P(\text{state} = i \cap \text{decision} = k) = P(\text{state} = i) P(\text{decision} = k \mid \text{state} = i)$

$$y_i^k = \pi_i \, P(\text{decision} = k \mid \text{state} = i) = \pi_i(1) = \pi_i > 0.$$

Similarly, the joint probability $y_i^k = P(\text{state} = i \cap \text{decision} = k) = 0$ if the associated conditional probability, $P(\text{decision} = k \mid \text{state} = i) = 0$ because when $P(\text{decision} = k \mid \text{state} = i) = 0$ then,

$$y_i^k = \pi_i \, P(\text{decision} = k \mid \text{state} = i) = \pi_i(0) = 0.$$

For the decision vector $^{20}\mathbf{d} = [3\ 1\ 2\ 2]^T$, the joint probabilities in state 1 are $y_i^1 = \pi_1 P(\text{decision} = 1 \mid \text{state} = 1) = \pi_1(0) = 0$, $y_i^2 = \pi_1 P(\text{decision} = 2 \mid \text{state} = 1) = \pi_1(0) = 0$, $y_i^3 = \pi_1 P(\text{decision} = 3 \mid \text{state} = 1) = \pi_1(1) = \pi_1 = 0.1908 > 0$, from Table 5.12.

Objective Function

The objective of the LP formulation for a recurrent MDP is to find a policy, which will maximize the gain. The reward received in state i of an MCR is denoted by q_i. The reward received in state i of an MDP when decision k is made is denoted by q_i^k. The reward received in state i of an MDP is equal to the reward received for each decision in state i weighted by the conditional probability of making that decision, $P(\text{decision} = k \mid \text{state} = i)$, so that

$$q_i = \sum_{k=1}^{K_i} P(\text{decision} = k \mid \text{state} = i) \, q_i^k. \tag{5.32}$$

Using Equation (4.46), the equation for the gain of a recurrent MDP is

$$g = \pi q = \sum_{i=1}^{N} \pi_i q_i = \sum_{i=1}^{N} \pi_i \sum_{k=1}^{K_i} P(\text{decision} = k \mid \text{state} = i) \, q_i^k.$$

Bringing π_i inside the second summation,

$$g = \sum_{i=1}^{N} \sum_{k=1}^{K_i} \pi_i \, P(\text{decision} = k \mid \text{state} = i) \, q_i^k$$

Using Equation (5.28),

$$g = \sum_{i=1}^{N} \sum_{k=1}^{K_i} y_i^k \, q_i^k$$

Therefore, the objective function is to find the decision variables y_i^k that will

$$\text{Maximize } g = \sum_{i=1}^{N} \sum_{k=1}^{K_i} q_i^k \, y_i^k. \tag{5.33}$$

Constraints

The LP formulation has the following three sets of constraints. They represent Equation (2.12), which must be solved to find the steady-state probability distribution for a regular Markov chain.

1. $\pi_j = \sum_{i=1}^{N} \pi_i p_{ij}$, for $j = 1, 2, \dots, N$ $\qquad$ (5.34)

2. $\sum_{i=1}^{N} \pi_i = 1$ $\qquad$ (5.35)

3. $\pi_i > 0$, for $i = 1, 2, \dots, N$. $\qquad$ (5.36)

The three sets of constraints will be expressed in terms of the decision variables, y_i^k. The left-hand side of the set of constraints (5.34) is

$$\pi_j = \sum_{k=1}^{K_j} y_j^k, \quad \text{for } j = 1, 2, \dots, N. \tag{5.27}$$

The right-hand side of the set of constraints (5.34) is $\sum_{i=1}^{N} \pi_i p_{ij}$.

The probability of a transition from state i to state j is denoted by p_{ij}. The probability of a transition from state i to state j when decision k is made is denoted by p_{ij}^k. The probability of a transition from state i to state j is equal to the probability of a transition from state i to state j for each decision k in state i weighted by $P(\text{decision} = k \mid \text{state} = i)$, the conditional probability of making such a decision, so that

$$p_{ij} = \sum_{k=1}^{K_i} P(\text{decision} = k \mid \text{state} = i) \, p_{ij}^k. \tag{5.37}$$

Therefore, the right-hand side of the set of constraints (5.34) is

$$\sum_{i=1}^{N} \pi_i p_{ij} = \sum_{i=1}^{N} \pi_i \sum_{k=1}^{K_i} P(\text{decision} = k \mid \text{state} = i) \, p_{ij}^k$$

Bringing π_i inside the second summation,

$$\sum_{i=1}^{N} \pi_i p_{ij} = \sum_{i=1}^{N} \sum_{k=1}^{K_i} \pi_i \, P(\text{decision} = k \mid \text{state} = i) \, p_{ij}^k$$

Substituting $y_i^k = \pi_i \, P(\text{decision} = k \mid \text{state} = i)$, the right-hand side of the set of constraints (5.34) is

$$\sum_{i=1}^{N} \pi_i p_{ij} = \sum_{i=1}^{N} \sum_{k=1}^{K_i} y_i^k \, p_{ij}^k$$

Hence, the set of constraints (5.34) expressed in terms of y_i^k is

$$\sum_{k=1}^{K_j} y_j^k = \sum_{i=1}^{N} \sum_{k=1}^{K_i} y_i^k \, p_{ij}^k, \quad \text{for } j = 1, 2, \dots, N.$$

The set of constraints (5.35) is the normalizing equation,

$$\sum_{i=1}^{N} \pi_i = 1.$$

Substituting $\pi_i = \sum_{k=1}^{K_i} y_i^k$, the normalizing equation expressed in terms of y_i^k is

$$\sum_{i=1}^{N} \sum_{k=1}^{K_i} y_i^k = 1.$$

The set of constraints (5.36) is the nonnegativity condition imposed by LP on the decision variables, y_i^k,

$$y_i^k \geq 0, \quad \text{for } i = 1, 2, \ldots, N, \quad \text{and} \quad k = 1, 2, \ldots, K_i.$$

The complete LP formulation is

$$\text{Maximize } g = \sum_{i=1}^{N} \sum_{k=1}^{K_i} q_i^k \, y_i^k \qquad (5.38)$$

subject to

$$\sum_{k=1}^{K_i} y_j^k - \sum_{i=1}^{N} \sum_{k=1}^{K_i} y_i^k \, p_{ij}^k = 0, \quad \text{for } j = 1, 2, \ldots, N \qquad (5.39)$$

$$\sum_{i=1}^{N} \sum_{k=1}^{K_i} y_i^k = 1 \qquad (5.40)$$

$$y_i^k \geq 0, \quad \text{for } i = 1, 2, \ldots, N, \text{ and } k = 1, 2, \ldots, K_i. \qquad (5.41)$$

Although this LP formulation has $N + 1$ constraints, only N of them are independent. One of the first N constraints (5.39) is redundant. The reason is that these constraints are based on the steady-state equations (5.34), one of which is redundant. The Nth constraint in this set will be deleted. That is, the constraint associated with a transition to state $j = N$,

$$\sum_{k=1}^{K_i} y_N^k - \sum_{i=1}^{N} \sum_{k=1}^{K_i} y_i^k \, p_{iN}^k = 0, \qquad (5.42)$$

will be deleted. The complete LP formulation, with the redundant constraint for $j = N$ omitted, is

$$\text{Maximize } g = \sum_{i=1}^{N} \sum_{k=1}^{K_i} q_i^k \, y_i^k \qquad (5.43a)$$

subject to

$$\left.\begin{array}{c} \displaystyle\sum_{k=1}^{K_i} y_j^k - \sum_{i=1}^{N} \sum_{k=1}^{K_i} y_i^k \, p_{ij}^k = 0, \quad \text{for } j = 1, 2, \dots, N{-}1 \\[4mm] \displaystyle\sum_{i=1}^{N} \sum_{k=1}^{K_i} y_i^k = 1 \\[6mm] y_i^k \geq 0, \quad \text{for } i = 1, 2, \dots, N, \quad \text{and} \quad k = 1, 2, \dots, K_i. \end{array}\right\} \qquad (5.43\text{b})$$

The LP formulation for a regular N-state MDP with K_i decisions per state has $\sum_{i=1}^{N} K_i$ decision variables and N constraints, all of which are equations. In contrast, the N VDEs, which must be solved for each repetition of PI, have only N unknowns consisting of the $N - 1$ relative values, $v_1, v_2, \dots, v_{N-1}$, and the gain, g.

When an optimal solution for the LP decision variables, y_i^k, has been found, the conditional probabilities,

$$P(\text{decision} = k| \text{ state} = i) = \frac{y_i^k}{\displaystyle\sum_{k=1}^{K_i} y_i^k},$$

can be calculated to determine the optimal decision in every state, thereby specifying an optimal policy. In the final tableau of the LP solution, with the redundant constraint (5.42) deleted so that the first $N - 1$ constraints remain, the dual variable associated with constraint i is equal to the relative value, v_i, earned for starting in state i, for $i = 1, 2, \dots, N{-}1$. The gain, which is the maximum value of the objective function, is also equal to the dual variable associated with the constraint for the normalizing equation, constraint (5.40). As the constraints are equations, the dual variables, and hence the relative values, are unrestricted in sign.

5.1.2.3.4.2 Formulation of MDP Model of Monthly Sales as an LP The recurrent MDP model of monthly sales introduced in Section 5.1.2.1 will be formulated in this section as an LP, and solved in Section 5.1.2.3.4.3 to find an optimal policy. Data for the recurrent MDP model of monthly sales appear in Table 5.21 which is the same as Table 5.3, augmented by a right-hand side column of LP decision variables.

TABLE 5.21

Data for Recurrent MDP Model of Monthly Sales

State i	Decision k	Transition Probability				Reward q_i^k	LP Variable
		p_{i1}^k	p_{i2}^k	p_{i3}^k	p_{i4}^k		
1	1 Sell noncore assets	0.15	0.40	0.35	0.10	−30	y_1^1
	2 Take firm private	0.45	0.05	0.20	0.30	−25	y_1^2
	3 Offer employee buyouts	0.60	0.30	0.10	0	−20	y_1^3
2	1 Reduce management salaries	0.25	0.30	0.35	0.10	5	y_2^1
	2 Reduce employee benefits	0.30	0.40	025	0.05	10	y_2^2
3	1 Design more appealing products	0.05	0.65	0.25	0.05	−10	y_3^1
	2 Invest in new technology	0.05	0.25	0.50	0.20	−5	y_3^2
4	1 Invest in new projects	0.05	0.20	0.40	0.35	35	y_4^1
	2 Make strategic acquisitions	0	0.10	0.30	0.60	25	y_4^2

In this example $N = 4$ states, $K_1 = 3$ decisions in state 1, and $K_2 = K_3 = K_4 = 2$ decisions in states 2, 3, and 4.

Objective Function

The objective function for the LP is

$$\text{Maximize } g = \sum_{i=1}^{4}\sum_{k=1}^{K_i} q_i^k\, y_i^k = \sum_{k=1}^{3} q_1^k\, y_1^k + \sum_{k=1}^{2} q_2^k\, y_2^k + \sum_{k=1}^{2} q_3^k\, y_3^k + \sum_{k=1}^{2} q_4^k\, y_4^k$$

$$= (q_1^1 y_1^1 + q_1^2 y_1^2 + q_1^3 y_1^3) + (q_2^1 y_2^1 + q_2^2 y_2^2) + (q_3^1 y_3^1 + q_3^2 y_3^2) + (q_4^1 y_4^1 + q_4^2 y_4^2)$$

$$= (-30 y_1^1 - 25 y_1^2 - 20 y_1^3) + (5 y_2^1 + 10 y_2^2) + (-10 y_3^1 - 5 y_3^2) + (35 y_4^1 + 25 y_4^2)$$

Constraints

State 1 has $K_1 = 3$ possible decisions. The constraint associated with a transition to state $j = 1$ is

$$\sum_{k=1}^{3} y_1^k - \sum_{i=1}^{4} \sum_{k=1}^{K_i} y_i^k \ p_{i1}^k = 0$$

$$(y_1^1 + y_1^2 + y_1^3) - (\sum_{k=1}^{3} y_1^k p_{11}^k + \sum_{k=1}^{2} y_2^k p_{21}^k + \sum_{k=1}^{2} y_3^k p_{31}^k + \sum_{k=1}^{2} y_4^k p_{41}^k) = 0$$

$$(y_1^1 + y_1^2 + y_1^3) - (y_1^1 p_{11}^1 + y_1^2 p_{11}^2 + y_1^3 p_{11}^3) - (y_2^1 p_{21}^1 + y_2^2 p_{21}^2) - (y_3^1 p_{31}^1 + y_3^2 p_{31}^2)$$
$$- (y_4^1 p_{41}^1 + y_4^2 p_{41}^2) = 0$$

$$(y_1^1 + y_1^2 + y_1^3) - (0.15 y_1^1 + 0.45 y_1^2 + 0.60 y_1^3) - (0.25 y_2^1 + 0.30 y_2^2) - (0.05 y_3^1 + 0.05 y_3^2)$$
$$- (0.05 y_4^1 + 0 y_4^2) = 0.$$

State 2 has $K_2 = 2$ possible decisions. The constraint associated with a transition to state $j = 2$ is

$$\sum_{k=1}^{2} y_2^k - \sum_{i=1}^{4} \sum_{k=1}^{K_i} y_i^k \ p_{i2}^k = 0$$

$$(y_2^1 + y_2^2) - (\sum_{k=1}^{3} y_1^k p_{12}^k + \sum_{k=1}^{2} y_2^k p_{22}^k + \sum_{k=1}^{2} y_3^k p_{32}^k + \sum_{k=1}^{2} y_4^k p_{42}^k) = 0$$

$$(y_2^1 + y_2^2) - (y_1^1 p_{12}^1 + y_1^2 p_{12}^2 + y_1^3 p_{12}^3) - (y_2^1 p_{22}^1 + y_2^2 p_{22}^2) - (y_3^1 p_{32}^1 + y_3^2 p_{32}^2)$$
$$- (y_4^1 p_{42}^1 + y_4^2 p_{42}^2) = 0$$

$$(y_2^1 + y_2^2) - (0.40 y_1^1 + 0.05 y_1^2 + 0.30 y_1^3) - (0.30 y_2^1 + 0.40 y_2^2) - (0.65 y_3^1 + 0.25 y_3^2)$$
$$- (0.20 y_4^1 + 0.10 y_4^2) = 0.$$

State 3 has $K_3 = 2$ possible decisions. The constraint associated with a transition to state $j = 3$ is

$$\sum_{k=1}^{2} y_3^k - \sum_{i=1}^{4} \sum_{k=1}^{K_i} y_i^k \ p_{i3}^k = 0.$$

$$(y_3^1 + y_3^2) - (\sum_{k=1}^{3} y_1^k p_{13}^k + \sum_{k=1}^{2} y_2^k p_{23}^k + \sum_{k=1}^{2} y_3^k p_{33}^k + \sum_{k=1}^{2} y_4^k p_{43}^k) = 0$$

$$(y_3^1 + y_3^2) - (y_1^1 p_{13}^1 + y_1^2 p_{13}^2 + y_1^3 p_{13}^3) - (y_2^1 p_{23}^1 + y_2^2 p_{23}^2) - (y_3^1 p_{33}^1 + y_3^2 p_{33}^2)$$
$$-(y_4^1 p_{43}^1 + y_4^2 p_{43}^2) = 0$$

$$(y_3^1 + y_3^2) - (0.35y_1^1 + 0.20y_1^2 + 0.10y_1^3) - (0.35y_2^1 + 0.25y_2^2) - (0.25y_3^1 + 0.50y_3^2)$$
$$-(0.40y_4^1 + 0.30y_4^2) = 0.$$

The redundant constraint associated with the state $j = 4$ will be omitted. The normalizing equation for the four-state MDP is

$$\sum_{i=1}^{4} \sum_{k=1}^{K_i} y_i^k = (y_1^1 + y_1^2 + y_1^3) + (y_2^1 + y_2^2) + (y_3^1 + y_3^2) + (y_4^1 + y_4^2) = 1.$$

The complete LP formulation for the recurrent MDP model of monthly sales, with the redundant constraint associated with the state $j = 4$ omitted, and the four remaining constraints numbered consecutively from 1 to 4, is

Maximize

$$g = (-30y_1^1 - 25y_1^2 - 20y_1^3) + (5y_2^1 + 10y_2^2) + (-10y_3^1 - 5y_3^2) + (35y_4^1 + 25y_4^2)$$

subject to

(1) $(0.85y_1^1 + 0.55y_1^2 + 0.40y_1^3) - (0.25y_2^1 + 0.30y_2^2) - (0.05y_3^1 + 0.05y_3^2)$
$-(0.05y_4^1 + 0y_4^2) = 0$

(2) $-(0.40y_1^1 + 0.05y_1^2 + 0.30y_1^3) + (0.70y_2^1 + 0.60y_2^2) - (0.65y_3^1 + 0.25y_3^2)$
$-(0.20y_4^1 + 0.10y_4^2) = 0$

(3) $-(0.35y_1^1 + 0.20y_1^2 + 0.10y_1^3) - (0.35y_2^1 + 0.25y_2^2) + (0.75y_3^1 + 0.50y_3^2)$
$-(0.40y_4^1 + 0.30y_4^2) = 0$

(4) $(y_1^1 + y_1^2 + y_1^3) + (y_2^1 + y_2^2) + (y_3^1 + y_3^2) + (y_4^1 + y_4^2) = 1$

$y_1^1 \geq 0,\ y_1^2 \geq 0,\ y_1^3 \geq 0,\ y_2^1 \geq 0,\ y_2^2 \geq 0,\ y_3^1 \geq 0,\ y_3^2 \geq 0,\ y_4^1 \geq 0,\ y_4^2 \geq 0.$

5.1.2.3.4.3 Solution by LP of MDP Model of Monthly Sales The LP formulation of the recurrent MDP model of monthly sales is solved on a personal computer using LP software to find an optimal policy. The output of the LP software is summarized in Table 5.22.

TABLE 5.22

LP Solution of MDP Model of Monthly Sales

Objective i	Function Value k	$G = 4.3901$ y_i^k		
1	1	0		
	2	0.1438	Row	Dual Variable
	3	0	Constraint 1	−76.8825
2	1	0	Constraint 2	−50.6777
	2	0.2063	Constraint 3	−51.8072
3	1	0	Constraint 4	4.3901
	2	0.3441		
4	1	0		
	2	0.3057		

Table 5.22 shows that the value of the objective function is 4.3901. All the y_i^k equal zero, except for $y_1^2 = 0.1438$, $y_2^2 = 0.2063$, $y_3^2 = 0.3441$, and $y_4^2 = 0.3057$. These values are the steady-state probabilities for the optimal policy given by $^{16}\mathbf{d} = [2\ 2\ 2\ 2\ ]^T$, which was previously found by exhaustive enumeration in Section 5.1.2.3.1, by value iteration in Section 5.1.2.3.2.2, and by PI in Section 5.1.2.3.3.3. The conditional probabilities,

$$P(\text{decision} = k | \text{state} = i) = \frac{y_i^k}{\displaystyle\sum_{k=1}^{K_i} y_i^k},$$

are calculated below:

$P(\text{decision} = 2 \mid \text{state} = 1) = P(\text{decision} = 2 \mid \text{state} = 2) = P(\text{decision} = 2 \mid \text{state} = 3)$
$= P(\text{decision} = 2 \mid \text{state} = 4) = 1.$

All the remaining conditional probabilities are zero.

Table 5.23 shows that the dual variables associated with constraints 1, 2, and 3 are identical to the relative values, v_1, v_2, and v_3, respectively, obtained by PI in Equation (5.25) of Section 5.1.2.3.3.3. The dual variable associated with constraint 4, the normalizing constraint, is the gain, $g=4.3901$, which is also the value of the objective function. Table 5.23 also indicates how closely the expected relative rewards, $v_i(-7) - v_4(-7)$, obtained by value iteration over a seven-period planning horizon in Section 5.1.2.3.2.2, approach the relative values, v_i, for $i = 1, 2,$ and 3, confirming the results shown in Table 5.16.

TABLE 5.23

Comparison of Dual Variables Obtained by LP, Relative Values Obtained by PI, and Expected Relative Rewards, $v_i(-7) - v_4(-7)$, Obtained by Value Iteration Over a Seven-Period Planning Horizon

LP Row	LP Dual Variable	Relative Value	Expected Relative Reward
Constraint 1	−76.8825	$v_1 = -76.8825$	$v_1(-7) - v_4(-7) = -76.9888$
Constraint 2	−50.6777	$v_2 = -50.6777$	$v_2(-7) - v_4(-7) = -50.7214$
Constraint 3	−51.8072	$v_3 = -51.8072$	$v_3(-7) - v_4(-7) = -51.766$
Constraint 4	4.3901	$g = 4.3901$	

5.1.3 A Unichain MDP

Recall that an MDP is unichain if the transition matrix associated with every stationary policy contains one closed communicating class of recurrent states plus a possibly empty set of transient states. Section 5.1.2 treats a recurrent MDP, which is a unichain MDP without any transient states. Section 5.1.3 will treat a unichain MDP which has one or more transient states. An optimal policy for a unichain MDP over an infinite planning horizon will be found by PI in Section 5.1.3.1, and by LP in Section 5.1.3.2. Value iteration for a unichain MDP is identical to value iteration for a recurrent MDP. Section 5.1.3.3.3 will formulate a problem of when to stop a sequence of independent trials over a finite planning horizon, and solve this optimal stopping problem by executing value iteration.

5.1.3.1 Policy Iteration (PI)

Recall from Section 4.2.4.1 that all states in a unichain MCR, both recurrent and transient, have the same gain, which is equal to the gain of the closed set of recurrent states. Since a policy for an MDP specifies a particular MCR, this property holds also for a unichain MDP. Hence, in a unichain MDP, the gain of every state is equal to g, the independent gain of the closed class of recurrent states. Since all states in a unichain MDP have the same gain, the PI algorithm for a unichain MDP with transient states is unchanged from the algorithm given in Section 5.1.2.3.3.2 for a recurrent MDP, which is a unichain MDP without transient states. The gain and the relative values for an MCR associated with a particular policy for a unichain MDP can be determined by executing the same three-step procedure given in Section 4.2.4.2.2 for a unichain MCR [2, 3].

5.1.3.1.1 A Unichain Four-State MDP Model of an Experimental Production Process

To see how PI can be applied to find an optimal policy for a unichain MDP in which a state may be recurrent under one stationary policy and transient under another, the following simplified unichain model of an experimental production line will be constructed. (An enlarged multichain MDP model of a flexible production system is constructed in Section 5.1.4.1.) Suppose that an experimental production line consists of three manufacturing stages in series. The three-stage line can be used to manufacture two types of products: industrial (I) or consumer (C). Each stage can be programmed to make either type of product independently of the other stages. Because this sequential production process is experimental, no output will be sold or scrapped. The output of each manufacturing stage is inspected. Output with a defect is reworked at the current stage. Output from stage 1 or stage 2 that is not defective is passed on to the next stage. Output from stage 3 that is not defective is sent to a training center. An industrial product sent from stage 3 to the training center will remain there. A consumer product sent from stage 3 to the training center will be disassembled, and returned as input to stage one in the following period.

By assigning a state to represent each operation, and adding a reward vector, the production process is modeled as a four-state unichain MDP. The four states are indexed in Table 5.24.

Observe that production stage i is represented by transient state $(5 - i)$.

In every state, two decisions are possible. The decision is the type of customer, either industrial (I) or consumer (C), for whom the product is designed. The probability of defective or nondefective output from a stage is affected by whether the product will be industrial or commercial. When the training center decides that a product is industrial, the product will never leave the training center, so that state 1 is absorbing and the remaining states are transient. When the training center decides that a product is designed for consumers, all four states are members of one closed class of recurrent states. With four states and two decisions per state, the number of possible policies is equal to $2^4 = 16$. Table 5.25 represents the four-state unichain MDP model of the experimental production process with numerical data furnished for the transition probabilities and the rewards.

TABLE 5.24

States for Unichain Model of an Experimental Production Process

State	Operation
1	Training center
2	Stage 3
3	Stage 2
4	Stage 1

The passage of an item through the experimental production process is shown in Figure 5.3.

5.1.3.1.2 Solution by PI of a Unichain Four-State MDP Model of an Experimental Production Process

Policy iteration will be executed to find an optimal policy for the unichain MDP model of an experimental production process.

First iteration

TABLE 5.25

Data for Unichain Model of an Experimental Production Process

State i	Operation	Decision k	p_{i1}^k	p_{i2}^k	p_{i3}^k	p_{i4}^k	q_i^k
					Transition Probability		Reward
1	Training	$1 = I$	1	0	0	0	56
		$2 = C$	0	0	0	1	62
2	Stage 3	$1 = I$	0.45	0.55	0	0	423.54
		$2 = C$	0.55	0.45	0	0	468.62
3	Stage 2	$1 = I$	0	0.35	0.65	0	847.09
		$2 = C$	0	0.50	0.50	0	796.21
4	Stage 1	$1 = I$	0	0	0.25	0.75	1935.36
		$2 = C$	0	0	0.20	0.80	2042.15

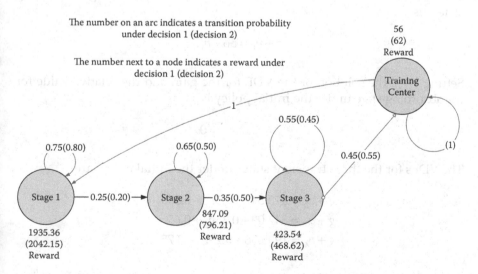

FIGURE 5.3

Passage of an item through a three-stage experimental production process.

Step 1. Initial policy

Arbitrarily choose an initial policy which consists of making decision 1 in every state. Thus, at every manufacturing stage and in the training center, the product is designed for an industrial customer. The initial decision vector $^1\mathbf{d}$, along with the associated transition probability matrix $^1\mathbf{P}$ and the reward vector, $^1\mathbf{q}$, are shown below:

$$^1\mathbf{d} = \begin{bmatrix} 1 \\ 1 \\ 1 \\ 1 \end{bmatrix}, \quad ^1\mathbf{P} = \begin{array}{c} 1 \\ 2 \\ 3 \\ 4 \end{array}\begin{bmatrix} 1 & 0 & 0 & 0 \\ 0.45 & 0.55 & 0 & 0 \\ 0 & 0.35 & 0.65 & 0 \\ 0 & 0 & 0.25 & 0.75 \end{bmatrix}, \quad ^1\mathbf{q} = \begin{bmatrix} 56 \\ 423.54 \\ 847.09 \\ 1935.36 \end{bmatrix}.$$

Note that state 1 is absorbing and states 2, 3, and 4 are transient.

Step 2. VD operation

Use p_{ij} and q_i for the initial policy, $^1\mathbf{d}=[1\ 1\ 1\ 1]^T$, to solve the VDEs (4.62) for the relative values v_i and the gain g. The VDEs (4.62) for the initial policy are

$$g+v_1 = 56+v_1$$
$$g+v_2 = 423.54+0.45v_1+0.55v_2$$
$$g+v_3 = 847.09+0.35v_2+0.65v_3$$
$$g+v_4 = 1935.36+0.25v_3+0.75v_4.$$

The VDE for the recurrent closed set, which consists of the single absorbing state 1, is

$$g+v_1 = 56+v_1.$$

Setting $v_1=0$, the solution of the VDE for the gain and the relative value for the absorbing state under the initial policy is

$$g = 56, \quad v_1 = 0.$$

The VDEs for the three transient states under the initial policy are

$$g+v_2 = 423.54+0.45v_1+0.55v_2$$
$$g+v_3 = 847.09+0.35v_2+0.65v_3$$
$$g+v_4 = 1935.36+0.25v_3+0.75v_4.$$

Substituting $g = 56$, $v_1 = 0$, the solution for the relative values of the transient states under the initial policy is

$$v_2 = 816.7556, \quad v_3 = 3077.0127, \quad v_4 = 10594.453.$$

Step 3. IM routine

For each state i, find the decision $k*$ that maximizes the test quantity

$$q_i^k + \sum_{j=1}^{4} p_{ij}^k v_j$$

using the relative values v_i of the initial policy. Then $k*$ becomes the new decision in state i, so that $d_i=k*$, q_i^{k*} becomes q_i, and p_{ij}^{k*} becomes p_{ij}. The first policy improvement routine is shown in Table 5.26.

TABLE 5.26

First IM for a Uichain Model of an Experimental Production Process

State i	Decision Alternative k	Test Quantity $q_i^k + p_{i_1}^k v_1 + p_{i_2}^k v_2 + p_{i_3}^k v_3 + p_{i_4}^k v_4$ $= q_i^k + p_{i_1}^k (0) + p_{i_2}^k (816.7556)$ $+ p_{i_3}^k (3077.0127) + p_{i_4}^k (10594.453)$	Maximum Value of Test Quantity	Decision $d_i = k*$
1	1	$56 + 1(0) = 56$		
1	2	$62 + 1(10594.453) = 10656.453 \leftarrow$	10656.453	$d_1 = 2$
2	1	$423.54 + 0.45(0) + 0.55(816.7556)$ $= 872.7556 \leftarrow$	872.7556	$d_2 = 1$
2	2	$468.62 + 0.55(0) + 0.45(816.7556)$ $= 836.1600$		
3	1	$847.09 + 0.35(816.7556)$ $+0.65(3077.0127) = 3133.0127 \leftarrow$	3133.0127	$d_3 = 1$
3	2	$796.21 + 0.50(816.7556)$ $+0.50(3077.0127) = 2743.0942$		
4	1	$1935.36 + 0.25(3077.0127)$ $+0.75(10594.453) = 10650.453$		
4	2	$2042.15 + 0.20(3077.0127)$ $+0.80(10594.453) = 11133.115 \leftarrow$	11133.115	$d_4 = 2$

Step 4. Stopping rule
The new policy is $^2\mathbf{d} = [2\ 1\ 1\ 2]^T$, which is different from the initial policy. Therefore, go to step 2. The new decision vector $^2\mathbf{d}$, along with the associated transition probability matrix ^2P and the reward vector $^2\mathbf{q}$, are shown below:

$$^2\mathbf{d} = \begin{bmatrix} 2 \\ 1 \\ 1 \\ 2 \end{bmatrix}, \quad ^2\mathbf{P} = \begin{matrix} 1 \\ 2 \\ 3 \\ 4 \end{matrix}\begin{bmatrix} 0 & 0 & 0 & 1 \\ 0.45 & 0.55 & 0 & 0 \\ 0 & 0.35 & 0.65 & 0 \\ 0 & 0 & 0.20 & 0.80 \end{bmatrix}, \quad ^2\mathbf{q} = \begin{bmatrix} 62 \\ 423.54 \\ 847.09 \\ 2042.15 \end{bmatrix}.$$

Note that all four states belong to the same recurrent closed set.

Step 2. VD operation
Use p_{ij} and q_i for the second policy, $^2\mathbf{d}=[2\ 1\ 1\ 2]^T$, to solve the VDEs (4.62) for the relative values v_i and the gain g.

$$g + v_1 = 62 + v_4$$
$$g + v_2 = 423.54 + 0.45v_1 + 0.55v_2$$
$$g + v_3 = 847.09 + 0.35v_2 + 0.65v_3$$
$$g + v_4 = 2042.15 + 0.20v_3 + 0.80v_4.$$

Setting $v_4=0$, the solution of the VDEs is

$$g = 1230.5946, \quad v_1 = -1168.5946, \quad v_2 = -2962.0494, \quad v_3 = -4057.7769, \quad v_4 = 0$$

Step 3. IM routine
The second IM routine is shown in Table 5.27.

Step 4. Stopping rule
The new policy is $^3\mathbf{d} = [2\ 2\ 2\ 2]^T$, which is different from the previous policy. Therefore, go to step 2. The new decision vector $^3\mathbf{d}$, along with the associated transition probability matrix ^3P and the reward vector $^3\mathbf{q}$, are shown below:

$$^3\mathbf{d} = \begin{bmatrix} 2 \\ 2 \\ 2 \\ 2 \end{bmatrix}, \quad ^3\mathbf{P} = \begin{matrix} 1 \\ 2 \\ 3 \\ 4 \end{matrix}\begin{bmatrix} 0 & 0 & 0 & 1 \\ 0.55 & 0.45 & 0 & 0 \\ 0 & 0.50 & 0.50 & 0 \\ 0 & 0 & 0.20 & 0.80 \end{bmatrix}, \quad ^3\mathbf{q} = \begin{bmatrix} 62 \\ 468.62 \\ 796.21 \\ 2042.15 \end{bmatrix}. \quad (5.44)$$

Once again all four states belong to the same recurrent closed set.

TABLE 5.27

Second IM for Unichain Model of an Experimental Production Process

State i	Decision Alternative k	Test Quantity $q_i^k + p_{i1}^k v_1 + p_{i2}^k v_2 + p_{i3}^k v_3 + p_{i4}^k v_4$ $= q_i^k + p_{i1}^k(-1168.5946) + p_{i1}^k(-2962.0494)$ $+ p_{i1}^k(-4057.7769)) + p_{i4}^k(0)$	Maximum Value of Test Quantity	Decision $d_i = k^*$
1	1	$56 + 1(-1168.5946) = -1112.5946$		
1	2	$62 + 1(0) = 62 \leftarrow$	62	$d_1 = 2$
2	1	$423.54 + 0.45(-1168.5946)$ $+0.55(-2962.0494) = -1731.4547$		
2	2	$468.62 + 0.55(-1168.5946)$ $+0.45(-2962.0494) = -1507.0293 \leftarrow$	-1507.029	$d_2 = 2$
3	1	$847.09 + 0.35(-2962.0494)$ $+0.65(-4057.7769) = -2827.1823$		
3	2	$796.21 + 0.50(-2962.0494)$ $+0.50(-4057.7769) = -2713.7032 \leftarrow$	-2713.703	$d_3 = 2$
4	1	$1935.36 + 0.25(-4057.7769)$ $+0.75(0) = 920.9158$		
4	2	$2042.15 + 0.20(-4057.7769)$ $+0.80(0) = 1230.5946 \leftarrow$	1230.5946	$d_4 = 2$

Step 2. VD operation

Use p_{ij} and q_i for the third policy, $^3\mathbf{d} = [2\ 2\ 2\ 2]^T$, to solve the VDEs (4.62)

$$g + v_i = q_i + \sum_{j=1}^{4} p_{ij}v_j, \quad i = 1,2,3,4$$

for the relative values v_i and the gain g.

$$g + v_1 = 62 + v_4$$
$$g + v_2 = 468.62 + 0.55v_1 + 0.45v_2$$
$$g + v_3 = 796.21 + 0.50v_2 + 0.50v_3$$
$$g + v_4 = 2042.15 + 0.20v_3 + 0.80v_4.$$

Setting $v_4 = 0$, the solution of the VDEs is

$$g = 1295.2710, \quad v_1 = -1233.2710, \quad v_2 = -2736.2729, \quad v_3 = -3734.3949, \quad v_4 = 0.$$
$$(5.45)$$

Step 3. IM routine
The third IM routine is shown in Table 5.28.

Step 4. Stopping rule
Stop because the new policy, given by the vector $^4d \equiv {}^3d = [2\ 2\ 2\ 2]^T$, is identical to the previous policy. Therefore, this policy is optimal. Under this policy, all states belong to the same recurrent closed class. The relative values and the gain for the optimal policy are shown in Equation (5.45). The recurrent

TABLE 5.28

Third IM for Unichain Model of an Experimental Production Process

State i	Decision Alternative k	Test Quantity $q_i^k + p_{i1}^k v_1 + p_{i2}^k v_2 + p_{i3}^k v_3 + p_{i4}^k v_4$ $= q_i^k + p_{i1}^k(-1233.2710) + p_{i2}^k(-2736.2729)$ $+ p_{i3}^k(-3734.3949)) + p_{i4}^k(0)$	Maximum Value of Test Quantity	Decision $d_i = k^*$
1	2	$56 + 1(-1233.2710) = -1177.271$		
1	2	$62 + 1(0) = 62 \leftarrow$	62	$d_1 = 2$
2	1	$423.54 + 0.45(-1233.2710)$ $+ 0.55(-2736.2729) = -1636.38?$		
2	2	$468.62 + 0.55(-1233.2710)$ $+ 0.45(-2736.2729) = -1441.0019 \leftarrow$	-1441.002	$d_2 = 2$
3	1	$847.09 + 0.35(-2736.2729)$ $+ 0.65(-3734.3949) = -2537.962?$		
3	2	$796.21 + 0.50(-2736.2729)$ $+ 0.50(-3734.3949) = -2439.1239 \leftarrow$	-2439.124	$d_3 = 2$
4	1	$1935.36 + 0.25(-3734.3949)$ $+ 0.75(0) = 1001.7613$		
4	2	$2042.15 + 0.20(-3734.3949)$ $+ 0.80(0) = 1295.271 \leftarrow$	1295.271	$d_4 = 2$

transition probability matrix ^3P and the reward vector ^3q for the optimal policy are shown in Equation (5.44).

Note that the relative values obtained by PI are identical to the corresponding dual variables obtained by LP in Table 5.30 of Section 5.1.3.2.2. The gain obtained by PI is also equal to the gain obtained by LP. By solving Equation (2.12), the recurrent transition probability matrix ^3P in Equation (5.44) has the steady-state probability vector

$$^3\pi = [0.1019 \quad 0.1852 \quad 0.2037 \quad 0.5093].$$ (5.46)

5.1.3.2 Linear Programming

Recall from Section 5.1.3.1 that all states in a unichain MDP have the same gain, irrespective of whether or not the set of transient states is empty. Since the objective of an LP formulation for a unichain MDP is to maximize the gain, the LP formulation is not affected by the presence of transient states. Therefore, the LP formulation in this section for a unichain MDP with transient states is the same as the one for a recurrent MDP. However, there is a distinction [3] in the properties of the feasible solutions to the LP formulations for both models. The distinction is that exactly one stationary policy corresponds to one feasible solution to the LP formulation for a recurrent MDP, but that more than one stationary policy may correspond to a particular feasible solution of the LP formulation for a unichain MDP with transient states. The optimal solution to the LP formulation for a unichain MDP does not specify decisions in transient states. The value of the gain of an optimal policy is not affected by decisions made in transient states. This distinction will be illustrated in Section 5.1.3.2.4 by the solution of a modified unichain MDP model of an experimental production line with transient states.

5.1.3.2.1 *LP Formulation of MDP Model of an Experimental Production Process*

The unichain MDP model of an experimental production process with transient states specified in Table 5.25 will be formulated as an LP. Table 5.29 contains Table 5.25 augmented by a right-hand side column of LP decision variables.

In this example $N = 4$ states, and $K_1 = K_2 = K_3 = K_4 = K = 2$ decisions in every state.

Objective Function

The objective function for the LP is

$$\text{Maximize } g = \sum_{i=1}^{4} \sum_{k=1}^{2} q_i^k \; y_i^k = \sum_{k=1}^{2} q_1^k \; y_1^k + \sum_{k=1}^{2} q_2^k \; y_2^k + \sum_{k=1}^{2} q_3^k \; y_3^k + \sum_{k=1}^{2} q_4^k \; y_4^k$$

$$= (q_1^1 y_1^1 + q_1^2 y_1^2) + (q_2^1 y_2^1 + q_2^2 y_2^2) + (q_3^1 y_3^1 + q_3^2 y_3^2) + (q_4^1 y_4^1 + q_4^2 y_4^2)$$

TABLE 5.29

Data for Unichain Model of an Experimental Production Process

State i	Operation	Decision k	Transition Probability				Reward q_i^k	LP Variable y_i^k
			p_{i1}^k	p_{i2}^k	p_{i3}^k	p_{i4}^k		
1	Training	$1 = I$	1	0	0	0	56	y_1^1
		$2 = C$	0	0	0	1	62	y_1^2
2	Stage 3	$1 = I$	0.45	0.55	0	0	423.54	y_2^1
		$2 = C$	0.55	0.45	0	0	468.62	y_2^2
3	Stage 2	$1 = I$	0	0.35	0.65	0	847.09	y_3^1
		$2 = C$	0	0.50	0.50	0	796.21	y_3^2
4	Stage 1	$1 = I$	0	0	0.25	0.75	1935.36	y_4^1
		$2 = C$	0	0	0.20	0.80	2042.15	y_4^1

$$= (56y_1^1 + 62y_1^2) + (423.54y_2^1 + 468.62y_2^2) + (847.09y_3^1 + 796.21y_3^2)$$
$$+ (1935.36y_4^1 + 2042.15y_4^2).$$

Constraints

State 1 has $K_1 = 2$ possible decisions. The constraint associated with a transition to state $j = 1$ is

$$\sum_{k=1}^{2} y_1^k - \sum_{i=1}^{4} \sum_{k=1}^{2} y_i^k \, p_{i1}^k = 0$$

$$(y_1^1 + y_1^2) - \left(\sum_{k=1}^{2} y_1^k p_{11}^k + \sum_{k=1}^{2} y_2^k p_{21}^k + \sum_{k=1}^{2} y_3^k p_{31}^k + \sum_{k=1}^{2} y_4^k p_{41}^k \right) = 0$$

$$(y_1^1 + y_1^2) - (y_1^1 p_{11}^1 + y_1^2 p_{11}^2) - (y_2^1 p_{21}^1 + y_2^2 p_{21}^2) - (y_3^1 p_{31}^1 + y_3^2 p_{31}^2) - (y_4^1 p_{41}^1 + y_4^2 p_{41}^2) = 0$$

$$(0y_1^1 + y_1^2) - (0.45y_2^1 + 0.55y_2^2) = 0.$$

State 2 has $K_2 = 2$ possible decisions. The constraint associated with a transition to state $j = 2$ is

$$\sum_{k=1}^{2} y_2^k - \sum_{i=1}^{4} \sum_{k=1}^{2} y_i^k p_{i2}^k = 0$$

$$(y_2^1 + y_2^2) - (\sum_{k=1}^{2} y_1^k p_{12}^k + \sum_{k=1}^{2} y_2^k p_{22}^k + \sum_{k=1}^{2} y_3^k p_{32}^k + \sum_{k=1}^{2} y_4^k p_{42}^k) = 0$$

$$(y_2^1 + y_2^2) - (y_1^1 p_{12}^1 + y_1^2 p_{12}^2) - (y_2^1 p_{22}^1 + y_2^2 p_{22}^2) - (y_3^1 p_{32}^1 + y_3^2 p_{32}^2) - (y_4^1 p_{42}^1 + y_4^2 p_{42}^2) = 0$$

$$(0.45 y_2^1 + 0.55 y_2^2) - (0.35 y_3^1 + 0.50 y_3^2) = 0.$$

State 3 has $K_3 = 2$ possible decisions. The constraint associated with a transition to state $j = 3$ is

$$\sum_{k=1}^{2} y_3^k - \sum_{i=1}^{4} \sum_{k=1}^{2} y_i^k p_{i3}^k = 0$$

$$(y_3^1 + y_3^2) - (\sum_{k=1}^{2} y_1^k p_{13}^k + \sum_{k=1}^{2} y_2^k p_{23}^k + \sum_{k=1}^{2} y_3^k p_{33}^k + \sum_{k=1}^{2} y_4^k p_{43}^k) = 0$$

$$(y_3^1 + y_3^2) - (y_1^1 p_{13}^1 + y_1^2 p_{13}^2) - (y_2^1 p_{23}^1 + y_2^2 p_{23}^2) - (y_3^1 p_{33}^1 + y_3^2 p_{33}^2) - (y_4^1 p_{43}^1 + y_4^2 p_{43}^2) = 0$$

$$(0.35 y_3^1 + 0.50 y_3^2) - (0.25 y_4^1 + 0.20 y_4^2) = 0.$$

The redundant constraint associated with the state $j = 4$ will be omitted. The normalizing equation for the four-state MDP is

$$\sum_{i=1}^{4} \sum_{k=1}^{2_i} y_i^k = (y_1^1 + y_1^2) + (y_2^1 + y_2^2) + (y_3^1 + y_3^2) + (y_4^1 + y_4^2) = 1$$

The complete LP formulation for the unichain MDP model of an experimental production process with transient states, with the redundant constraint associated with the state $j = 4$ omitted, is

Maximize $g = (56 y_1^1 + 62 y_1^2) + (423.54 y_2^1 + 468.62 y_2^2)$

$$+ (847.09 y_3^1 + 796.21 y_3^2) + (1935.36 y_4^1 + 2042.15 y_4^2)$$

subject to

(1) $(0y_1^1 + y_1^2) - (0.45y_2^1 + 0.55y_2^2) = 0$
(2) $(0.45y_2^1 + 0.55y_2^2) - (0.35y_3^1 + 0.50y_3^2) = 0$
(3) $(0.35y_3^1 + 0.50y_3^2) - (0.25y_4^1 + 0.20y_4^2) = 0$
(4) $(y_1^1 + y_1^2 + y_1^3) + (y_2^1 + y_2^2) + (y_3^1 + y_3^2) + (y_4^1 + y_4^2) = 1$
$y_1^1 \geq 0, \ y_1^2 \geq 0, \ y_2^2 \geq 0, \ y_3^1 \geq 0, \ y_3^2 \geq 0, \ y_4^1 \geq 0, \ y_4^2 \geq 0.$

5.1.3.2.2 Solution by LP of MDP Model of an Experimental Production Process The LP formulation in Section 5.1.3.2.3 of the unichain MDP model of an experimental production process with transient states is solved to find an optimal policy by using LP software on a personal computer. The output of the LP software is summarized in Table 5.30.

Table 5.30 shows that $y_1^2 = 0.1019$, $y_2^2 = 0.1852$, $y_3^2 = 0.2037$, $y_4^2 = 0.5093$, and the remaining y_i^k equal zero. These nonzero y_i^k values are the steady-state probabilities shown in Equation (5.46) for the recurrent transition probability matrix ^3P associated with the optimal policy. The optimal policy is $^3\mathbf{d} = [2 \ 2 \ 2 \ 2]^T$, which was previously found by PI in Equation (5.45) of Section 5.1.3.1.2. Under this policy, all states are recurrent so that the LP solution prescribes a decision in every state. The associated conditional probabilities,

$$P(\text{decision} = k | \text{state} = i) = y_i^k \Big/ \sum_{k=1}^{K_i} y_i^k,$$

are

$$P(\text{decision} = 2 | \text{state} = 1) = P(\text{decision} = 2 | \text{state} = 2) =$$
$$P(\text{decision} = 2 | \text{state} = 3) = P(\text{decision} = 2 | \text{state} = 4) = 1.$$

TABLE 5.30

LP Solution of MDP Model of an Experimental Production Process

Objective i	Function Value k	$g = 1295.2710$ y_i^k	Row	Dual Variable
1	1	0	Row	Dual Variable
	2	0.1019	Constraint 1	−1233.2710
2	1	0	Constraint 2	−2726.2739
	2	0.1852	Constraint 3	−3734.3949
3	1	0	Constraint 4	1295.2710
	2	0.2037		
4	1	0		
	2	0.5093		

TABLE 5.31

Data for Modified Unichain Model of an Experimental Production Process

State i	Operation	Decision k	Transition Probability				Reward q_i^k	LP Variable y_i^k
			p_{i1}^k	p_{i2}^k	p_{i3}^k	p_{i4}^k		
1	Training	1 = I	1	0	0	0	1300	y_1^1
		2 = C	0	0	0	1	62	y_1^2
2	Stage 3	1 = I	0.45	0.55	0	0	423.54	y_2^1
		2 = C	0.55	0.45	0	0	468.62	y_2^2
3	Stage 2	1 = I	0	0.35	0.65	0	847.09	y_3^1
		2 = C	0	0.50	0.50	0	796.21	y_3^2
4	Stage 1	1 = I	0	0	0.25	0.75	1935.36	y_4^1
		2 = C	0	0	0.20	0.80	2042.15	y_4^2

All the remaining conditional probabilities are zero. The dual variables associated with constraints 1, 2, and 3 are the relative values v_1, v_2, and v_3, respectively, obtained by PI in Equation (5.33) of Section 5.1.3.1.2. The dual variable associated with constraint 4, the normalizing constraint, is the gain, $g = 1295.2710$, which is also the value of the objective function.

5.1.3.2.3 LP Formulation of a Modified MDP Model of an Experimental Production Process

Consider the unichain MDP model of an experimental production process with transient states given in Table 5.29 of Section 5.1.3.2.1. The model is modified so that the reward received by the training center per unit of industrial product is increased from 56 to 1300. Thus, the only difference between the original model and the modified model is that $q_1^1=56$ in the first model is replaced by $q_1^1=1300$ in the modified model. The data for the modified model are given in Table 5.31.

The modified model will be formulated as an LP. The only difference in the LP formulation is that the term $56y_1^1$ in the objective function of the original model is replaced by $1300y_1^1$ in the objective function of the modified model. The LP formulation for the modified unichain MDP model of an experimental production process with transient states, with the redundant constraint associated with the state $j = 4$ omitted, is

Maximize $g = (1300y_1^1 + 62y_1^2) + (423.54y_2^1 + 468.62y_2^2)$
$$+ (847.09y_3^1 + 796.21y_3^2) + (1935.36y_4^1 + 2042.15$$

subject to

(1) $(0y_1^1 + y_1^2) - (0.45y_2^1 + 0.55y_2^2) = 0$
(2) $(0.45y_2^1 + 0.55y_2^2) - (0.35y_3^1 + 0.50y_3^2) = 0$
(3) $(0.35y_3^1 + 0.50y_3^2) - (0.25y_4^1 + 0.20y_4^2) = 0$
(4) $(y_1^1 + y_1^2 + y_1^3) + (y_2^1 + y_2^2) + (y_3^1 + y_3^2) + (y_4^1 + y_4^2) = 1$
   $y_1^1 \geq 0, y_1^2 \geq 0, y_2^1 \geq 0, y_2^2 \geq 0, y_3^1 \geq 0, y_3^2 \geq 0, y_4^1 \geq 0, y_4^2 \geq 0.$

5.1.3.2.4 Solution by LP of a Modified MDP Model of an Experimental Production Process

The LP formulation of the modified unichain MDP model of an experimental production process with transient states is solved by using LP software to find an optimal policy. The output of the LP software is summarized in Table 5.32.

Table 5.32 shows that $y_1^1 = 1$, and all the other y_i^k equal zero. State 1 is absorbing (and also recurrent), and the other three states are transient. Thus, the optimal solution to the LP prescribes a decision in the recurrent state only, in this case state 1, but does not prescribe decisions in the transient states. The gain, which is not affected by decisions in the transient states, equals 1300. In this example, with three transient states and two possible decisions per transient state, $2^3 = 8$ stationary policies correspond to the feasible solution obtained for the LP formulation.

TABLE 5.32

LP Solution of a Modified MDP Model of an
Experimental Production Process

Objective i	Function Value k	$g = 1300$ y_i^t
1	1	1
	2	0
2	1	0
	2	0
3	1	0
	2	0
4	1	0
	2	0

5.1.3.3 Examples of Unichain MDP Models

Examples of unichain MDP models over an infinite planning horizon will be constructed for an inventory system and for component replacement. The secretary problem, which is concerned with optimal stopping, will be modeled over a finite horizon.

5.1.3.3.1 Unichain MDP Model of an Inventory System

Consider the inventory system of a retail operation modeled as a four-state regular Markov chain in Section 1.10.1.1.3. In Section 4.2.3.4.1, this model was enlarged by adding a reward vector to produce a four-state recurrent MCR. The retailer followed a (2, 3) inventory ordering policy for which an expected profit per period, or gain, of 115.6212 was calculated in Equation (4.98). However, she does not know whether a (2, 3) inventory ordering policy will maximize her expected profit per period. The retailer will model her inventory system as a four-state MDP that will allow her to choose among all the ordering decisions which are feasible in every state [1].

Since the retailer can store at most three computers in her shop, she can order from zero computers up to three computers minus the number of computers currently in stock. That is, if her inventory at the beginning of a period is $X_{n-1} = i \leq 3$ computers, she can order from 0 computers up to $(3-i)$ computers. Her feasible ordering decisions corresponding to every possible level of entering inventory are shown in Table 5.33.

TABLE 5.33

Feasible Ordering Decisions Associated with Beginning Inventory Levels

Beginning Inventory $X_{n-1} = i$	Order Quantity $c_{n-1} = k$
If $i = 0$ computers,	$\begin{cases} \text{order } k = 0 \text{ computers, or } k = 1 \text{ computer, or } k = 2 \text{ computers,} \\ \text{or } k = 3 \text{ computers; order up to } k = 3 = (3-0) \text{ computers} \end{cases}$
If $i = 1$ computers,	$\begin{cases} \text{order } k = 0 \text{ computers, or } k = 1 \text{ computer, or } k = 2 \text{ computers;} \\ \text{order up to } k = 2 = (3-1) \text{ computers} \end{cases}$
If $i = 2$ computers,	$\begin{cases} \text{order } k = 0 \text{ computers, or } k = 1 \text{ computer;} \\ \text{order up to } k = 2 = (3-2) \text{ computers} \end{cases}$
If $i = 3$ computers,	order $k = 0$ computers; order up to $k = 0 = (3-3)$ computers

Table 5.34 shows the decisions allowed in every state and the associated transition probabilities expressed as a function of the demand.

Table 5.35 shows the decisions allowed in every state and the associated numerical transition probabilities.

Observe that in state $X_{n-1} = 0$, when decision $c_{n-1} = 0$ is made, $p_{00}^0 = 1$, so that state 0 is absorbing and the other states are transient. In addition, when decision $c_{n-1} = 1$ is made in state $X_{n-1} = 0$, and decision $c_{n-1} = 0$ is made in state $X_{n-1} = 1$, then states 0 and 1 are recurrent and the remaining states are transient. Therefore, the MDP model of the inventory system is unichain with transient states.

The expected reward or profit earned by every decision in every state will be computed next as the expected revenue minus the expected cost. The number of computers sold during a period equals $\min[(X_{n-1} + c_{n-1}), d_n]$. The

TABLE 5.34

Allowable Decisions in Every State and Associated Transition Probabilities Expressed as a Function of the Demand

State $X_{n-1} = 0$	Decision $c_{n-1} = k$	Transition Probability			
		p_{00}^k	p_{01}^k	p_{02}^k	p_{04}^k
	0	$P(d_n \geq 0)$	0	0	0
	1	$P(d_n \geq 1)$	$P(d_n \geq 0)$	0	0
	2	$P(d_n \geq 2)$	$P(d_n \geq 1)$	$P(d_n \geq 0)$	0
	3	$P(d_n \geq 3)$	$P(d_n \geq 2)$	$P(d_n \geq 1)$	$P(d_n \geq 0)$
$X_{n-1} = 1$	$c_{n-1} = k$	p_{10}^k	p_{11}^k	p_{12}^k	p_{13}^k
	0	$P(d_n \geq 1)$	$P(d_n = 0)$	0	0
	1	$P(d_n \geq 2)$	$P(d_n = 1)$	$P(d_n = 0)$	0
	2	$P(d_n = 3)$	$P(d_n = 2)$	$P(d_n = 1)$	$P(d_n = 0)$
$X_{n-1} = 2$	$c_{n-1} = k$	p_{20}^k	p_{21}^k	p_{22}^k	p_{23}^k
	0	$P(d_n \geq 2)$	$P(d_n = 1)$	$P(d_n = 0)$	0
	1	$P(d_n = 3)$	$P(d_n = 2)$	$P(d_n = 1)$	$P(d_n = 0)$
$X_{n-1} = 3$	$c_{n-1} = k$	p_{30}^k	p_{31}^k	p_{32}^k	p_{33}^k
	0	$P(d_n = 3)$	$P(d_n = 2)$	$P(d_n = 1)$	$P(d_n = 0)$

TABLE 5.35

Allowable Decisions in Every State and Associated Numerical Transition Probabilities

			Transition Probability		
State $X_{n-1} = i$	Decision $c_{n-1} = k$	p_{i0}^k	p_{i1}^k	p_{i2}^k	p_{i3}^k
0	0	1	0	0	0
	1	0.7	0.3	0	0
	2	0.3	0.4	0.3	0
	3	0.2	0.1	0.4	0.3
1	0	0.7	0.3	0	0
	1	0.3	0.4	0.3	0
	2	0.2	0.1	0.4	0.3
2	0	0.3	0.4	0.3	0
	1	0.2	0.1	0.1	0.3
3	0	0.2	0.1	0.1	0.3

expected number sold is given by

$$E\{\min[(X_{n-1}+c_{n-1}),d_n]\} = \sum_{d_n=0}^{3} \min[(X_{n-1}+c_{n-1}),d_n]p(d_n)$$

$$= \sum_{k=0}^{3} \min[(X_{n-1}+c_{n-1}),d_n = k]P(d_n = k).$$

The retailer sells computers for \$300 each. The expected revenue equals \$300 per computer sold times the expected number sold. The expected revenue is

$$\$300 \sum_{k=0}^{3} \min[(X_{n-1}+c_{n-1}),d_n = k]P(d_n = k).$$

For example, in state $X_{n-1} = 0$, when decision $c_{n-1} = 2$ is made, the expected revenue is equal to

$$\$300 \sum_{d_n=0}^{3} \min[(X_{n-1}+c_{n-1}),d_n]p(d_n)$$

$$= \$300 \sum_{d_n=0}^{3} \min[(0+2),d_n]p(d_n)$$

$$= \$300[\min(2,0)p(0)+\min(2,1)p(1)+\min(2,2)p(2)+\min(2,3)p(3)]$$

$$= \$300[0p(0)+1p(1)+2p(2)+2p(3)]$$

$$= \$300\{0p(0)+1p(1)+2[p(2)+p(3)]\} = \$300[0(0.3)+1(0.4)+2(0.1+0.2)] = \$300.$$

The expected revenue generated in every state by every decision is calculated in Table 5.36.

The expected cost is equal to the ordering cost plus the expected holding cost plus the expected shortage cost. If $c_{n-1}>0$ computers are ordered at the beginning of a period, the ordering cost for the period is $\$20+\$120c_{n-1}$. If no computers are ordered, the ordering cost is $\$0$. The ordering costs associated with every decision in every state are calculated in Table 5.37.

The expected holding cost equals $\$50$ times the expected number of computers not sold. The number not sold during a period, that is, the surplus of computers at the end of the period, equals $\max[(X_{n-1}+c_{n-1}-d_n), 0]$. The expected number of computers not sold is given by

$$E\{\max[(X_{n-1}+c_{n-1}-d_n),0]\} = \sum_{d_n=0}^{3} \max[(X_{n-1}+c_{n-1}-d_n),0]p(d_n)$$

$$= \sum_{k=0}^{3} \max[(X_{n-1}+c_{n-1}-k),0]P(d_n = k).$$

The expected holding cost is

$$\$50\sum_{k=0}^{3} \max[(X_{n-1}+c_{n-1}-k),0]P(d_n = k).$$

For example, in state $X_{n-1}= 0$, when decision $c_{n-1}= 2$ is made, the expected holding cost is equal to

$$\$50\sum_{d_n=0}^{3} \max[(X_{n-1}+c_{n-1}-d_n),0]p(d_n) = \$50\sum_{d_n=0}^{3} \max[(0+2-d_n),0]p(d_n)$$

$$= \$50[\max(2-0,0)p(0)+\max(2-1,0)p(1)+\max(2-2,0)p(2)+\max(2-3,0)p(3)]$$

$$= \$50\{(2-0)p(0)+(2-1)p(1)+0p(2)+0p(3)\}$$

$$= \$50\{(2-0)p(0)+(2-1)p(1)+0[p(2)+p(3)]\} = \$50\{2p(0)+1p(1)+0[p(2)+p(3)]\}$$

$$= \$50[2(0.3)+1(0.4)+0(0.1+0.2)]= \$50.$$

The expected holding costs generated in every state by every decision are calculated in Table 5.38.

The expected shortage cost equals $\$40$ times the expected number of computers not available to satisfy demand during a period. The number not available to satisfy demand during a period, that is, the shortage of computers at the end of the period, equals $\max[(d_n - X_{n-1} - c_{n-1}),0]$. The expected number of computers not available to satisfy demand during a period is given by

TABLE 5.36

Expected Revenue

State $X_{n-1} = i$	Decision $c_{n-1} = k$	Revenue $\$300\min[(X_{n-1}+c_{n-1}),d_{nl}]$	Expected Revenue $\$300\sum_{d_n=0}^{3}\min[(X_{n-1}+c_{n-1}),d_n]p(d_n)$
0	0	$\$300\min(0, d_n)$	$\$300(0) = \$0$
	1	$\$300\min(1, d_n)$	$\$300\{0 p(0)+1[p(1)+p(2)+p(3)]\} = \$300[0(0.3)+1(0.4+0.1+0.2)]=\$210$
	2	$\$300\min(2, d_n)$	$\$300\{0 p(0)+1 p(1)+2[p(2)+p(3)]\} = \$300[0(0.3)+1(0.4)+2(0.1+0.2)]=\$300$
	3	$\$300\min(3, d_n)$	$\$300[0 p(0)+1 p(1)+2 p(2)+3 p(3)] = \$300[0(0.3)+1(0.4)+2(0.1)+3(0.2)]=\$360$
1	0	$\$300\min(1, d_n)$	$\$300\{0 p(0) + 1[p(1) + p(2) + p(3)]\} = \$210$
	1	$\$300\min(2, d_n)$	$\$300\{0 p(0) + 1 p(1) + 2[p(2) + p(3)]\} = \$300$
	2	$\$300\min(3, d_n)$	$\$300[0 p(0) + 1 p(1) + 2 p(2) + 3 p(3)] = \$360$
2	0	$\$300\min(2, d_n)$	$\$300\{0 p(0) + 1 p(1) + 2[p(2) + p(3)]\} = \$300$
	1	$\$300\min(3, d_n)$	$\$300[0 p(0) + 1 p(1) + 2 p(2) + 3 p(3)] = \$360$
3	0	$\$300\min(3, d_n)$	$\$300[0 p(0) + 1 p(1) + 2 p(2) + 3 p(3)] = \$360$

TABLE 5.37

Ordering Costs

State	Decision	Ordering Cost
X_{n-1}	c_{n-1}	$\$20 + \$120c_{n-1}$ if $c_{n-1} > 0$
0	0	\$0
	1	$\$20 + \$120(1) = \$140$
	2	$\$20 + \$120(2) = \$260$
	3	$\$20 + \$120(3) = \$380$
1	0	\$0
	1	$\$20 + \$120(1) = \$140$
	2	$\$20 + \$120(2) = \$260$
2	0	\$0
	1	$\$20 + \$120(1) = \$140$
3	0	\$0

$$E\{\max[(d_n - X_{n-1} - c_{n-1}), 0]\} = \sum_{d_n=0}^{3} \max[(d_n - X_{n-1} - c_{n-1}), 0]p(d_n)$$

$$= \sum_{k=0}^{3} \max[(k - X_{n-1} - c_{n-1}), 0]P(d_n = k).$$

The expected shortage cost is

$$\$40 \sum_{k=0}^{3} \max[(k - X_{n-1} - c_{n-1}), 0]P(d_n = k).$$

For example, in state $X_{n-1} = 0$, when decision $c_{n-1} = 1$ is made, the expected shortage cost is equal to

$$\$40 \sum_{d_n=0}^{3} \max[(d_n - X_{n-1} - c_{n-1}), 0]p(d_n) = \$40 \sum_{d_n=0}^{3} \max[(d_n - 0 - 1), 0]p(d_n)$$

$$= \$40[\max(0-1,0)p(0) + \max(1-1,0)p(1) + \max(2-1,0)p(2) + \max(3-1,0)p(3)]$$

$$= \$40[0p(0) + (1-1)p(1) + (2-1)p(2) + (3-1)p(3)] = \$40[0p(0) + 0p(1) + 1p(2) + 2p(3)]$$

$$= \$40[0(0.3) + 0(0.4) + 1(0.1) + 2(0.2)] = \$20.$$

The expected shortage costs generated in every state by every decision are calculated in Table 5.39.

TABLE 5.38

Expected Holding Costs

State	Decision	Holding Cost	Expected Holding Cost
X_{n-1}	c_{n-1}	$\$50\max[(X_{n-1}+c_{n-1}-d_n),0]$	$\$50\sum_{d_n=0}^{3}\min[(X_{n-1}+c_{n-1}),d_n]p(d_n)$
0	0	$\$50\max(0-d_n,0)$	$\$50(0)=\$0$
	1	$\$50\max(1-d_n,0)$	$\begin{cases}\$50\{1p(0)+0[p(1)+p(2)+p(3)]\}\\=\$50[1(0.3)+0(0.4+0.1+0.2)]=\$15\end{cases}$
	2	$\$50\max(2-d_n,0)$	$\begin{cases}\$50\{2p(0)+1p(1)+0[p(2)+p(3)]\}\\=\$50[2(0.3)+1(0.4)+0(0.1+0.2)]=\$50\end{cases}$
	3	$\$50\max(3-d_n,0)$	$\begin{cases}\$50[3p(0)+2p(1)+1p(2)+0p(3)]\\=\$50[3(0.3)+2(0.4)+1(0.1)+0(0.2)]=\$90\end{cases}$
1	0	$\$50\max(1-d_n,0)$	$\$50[1p(0)+0[p(1)+p(2)+p(3)]]=\$15$
	1	$\$50\max(2-d_n,0)$	$\$50[2p(0)+1p(1)+0[p(2)+p(3)]]=\$50$
	2	$\$50\max(3-d_n,0)$	$\$50[3p(0)+2p(1)+1p(2)+p(3)]]=\$90$
2	0	$\$50\max(2-d_n,0)$	$\$50[2p(0)+1p(1)+0[p(2)+p(3)]]=\$50$
	1	$\$50\max(3-d_n,0)$	$\$50[3p(0)+2p(1)+1p(2)+0p(3)]]=\$90$
3	0	$\$50\max(3-d_n,0)$	$\$50[3p(0)+2p(1)+1p(2)+0p(3)]]=\$90$

TABLE 5.39

Expected Shortage Costs

State X_{n-1}	Decision c_{n-1}	Shortage Cost $\$40\max[(d_n - X_{n-1} - c_{n-1}), 0]$	Expected Shortage Cost $\$40\sum\limits_{d_n=0}^{3}\min[(X_{n-1}+c_{n-1}), d_n]p(d_n)$
0	0	$\$40\max(d_n - 0, 0)$	$\{\$40[0p(0)+1p(1)+2p(2)+3p(3)]$ $=\$40[0(0.3)+1(0.4)+2(0.1)+3(0.2)]=\$48$
	1	$\$40\max(d_n - 1, 0)$	$\{\$40\{0[p(0)+p(1)]+1p(2)+2p(3)\}$ $=\$40[0(0.3+0.4)+1(0.1)+2(0.2)]=\$20$
	2	$\$40\max(d_n - 2, 0)$	$\{\$40\{0[p(0)+p(1)+p(2)]+1p(3)\}$ $=\$40[0(0.3+0.4+0.1)+1(0.2)]=\$8$
	3	$\$40\max(d_n - 3, 0)$	$\$40(0) = \$0$
1	0	$\$40\max(d_n - 1, 0)$	$\$40\{0[p(0) + p(1)] + 1p(2) + 2p(3)\} = \$20$
	1	$\$40\max(d_n - 2, 0)$	$\$40\{0[p(0) + p(1) + p(2)] + 1p(3)\} = \$8$
	2	$\$40\max(d_n - 3, 0)$	$\$40(0) = \$0$
2	0	$\$40\max(d_n - 2, 0)$	$\$40\{0[p(0) + p(1) + p(2)] + 1p(3)\} = \$8$
	1	$\$40\max(d_n - 3, 0)$	$\$40(0) = \$0$
3	0	$\$40\max(d_n - 3, 0)$	$\$40(0) = \$0$

In Table 5.40 the expected rewards corresponding to every state and decision are calculated.

The complete unichain MDP model of the inventory system is summarized in Table 5.41.

5.1.3.3.1.1 Formulation of a Unichain MDP Model of an Inventory System as an LP   The unichain MDP model of an inventory system will be formulated as an LP. In this example $N = 4$ states. The number of decisions in each state is $K_0 = 4$, $K_1 = 3$, $K_2 = 2$, and $K_3 = 1$. The objective function for the LP is

$$\text{Maximize } g = (-48y_0^0 + 35y_0^1 - 18y_0^2 - 110y_0^3) + (175y_1^0 + 102y_1^1 + 10y_1^2)$$
$$+ (242y_2^0 + 130y_2^1) + 270y_3^0.$$

The constraints are given below:

(1) $(y_0^0 + y_0^1 + y_0^2 + y_0^3) - (y_0^0 + 0.7y_0^1 + 0.3y_0^2 + 0.2y_0^3)$
$- (0.7y_1^0 + 0.3y_1^1 + 0.2y_1^2) - (0.3y_2^0 + 0.2y_2^1) - 0.2y_3^0$

(2) $(y_1^0 + y_1^1 + y_1^2) - (0.3y_0^1 + 0.4y_0^2 + 0.1y_0^3) - (0.3y_1^0 + 0.4y_1^1 + 0.1y_1^2)$
$- (0.4y_2^0 + 0.1y_2^1) - 0.1y_3^0 = 0$

(3) $(y_2^0 + y_2^1) - (0.3y_0^2 + 0.4y_0^3) - (0.3y_1^1 + 0.4y_1^2) - (0.3y_2^0 + 0.4y_2^1)$
$- 0.4y_3^0 = 0$

(4) $y_3^0 - (0.3y_0^3 + 0.3y_1^2 + 0.3y_2^1 + 0.3y_3^0) = 0.$

TABLE 5.40

Expected Rewards

State	Decision	Expected Reward
X_{n-1}	C_{n-1}	E(Revenue) – Ordering Cost – E(Holding Cost) – E (Shortage Cost)
0	0	$0 – $0 – $0 – $48 = –$48
	1	$210 – $140 – $15 – $20 = $35
	2	$300 – $260 – $50 – $8 = $18
	3	$360 – $380 – $90 – $0 = $110
1	0	$210 – $0 – $15 – $20 = $175
	1	$300 – $140 – $50 – $8 = $102
	2	$360 – $260 – $90 – $0 = $10
2	0	$300 – $0 – $50 – $8 = $242
	1	$360 – $140 – $90 – $0 = $130
3	0	$360 – $0 – $90 – $0 = $270

TABLE 5.41

Unichain MDP Model of Inventory System

State $X_{n-1} = i$	Decision $c_{n-1} = k$	\multicolumn{4}{c}{Transition Probability}	Expected Reward q_i^k	LP Variable y_i^k
		$j = 0$	$j = 1$	$j = 2$	$j = 3$		
0	0	1	0	0	0	−$48	y_0^0
	1	0.7	0.3	0	0	$35	y_0^1
	2	0.3	0.4	0.3	0	−$18	y_0^2
	3	0.2	0.1	0.4	0.3	−$110	y_0^3
1	0	0.7	0.3	0	0	$175	y_1^0
	1	0.3	0.4	0.3	0	$102	y_1^1
	2	0.2	0.1	0.4	0.3	$10	y_1^2
2	0	0.3	0.4	0.3	0	$242	y_2^0
	1	0.2	0.1	0.4	0.3	$130	y_2^1
3	0	0.2	0.1	0.4	0.3	$270	y_3^0

(5) $(y_0^0 + y_0^1 + y_0^2 + y_0^3) + (y_1^0 + y_1^1 + y_1^2) + (y_2^0 + y_2^1) + y_3^0 = 1$

$y_0^0 \geq 0,\ y_0^1 \geq 0,\ y_0^2 \geq 0,\ y_0^3 \geq 0,\ y_1^0 \geq 0,\ y_1^1 \geq 0,\ y_1^2 \geq 0,\ y_2^0 \geq 0,\ y_2^1 \geq 0,\ y_3^0 \geq 0.$

The complete LP appears below:

$$\text{Maximize } g = (-48y_0^0 + 35y_0^1 - 18y_0^2 - 110y_0^3) + (175y_1^0 + 102y_1^1 + 10y_1^2)$$
$$+ (242y_2^0 + 130y_2^1) + 270y_3^0$$

subject to

(1) $(0.3y_0^1 + 0.7y_0^2 + 0.8y_0^3) - (0.7y_1^0 + 0.3y_1^1 + 0.2y_1^2) - (0.3y_2^0 + 0.2y_2^1) - 0.2y_3^0 = 0$

(2) $-(0.3y_0^1 + 0.4y_0^2 + 0.1y_0^3) + (0.7y_1^0 + 0.6y_1^1 + 0.9y_1^2) - (0.4y_2^0 + 0.1y_2^1) - 0.1y_3^0 = 0$

(3) $-(0.3y_0^2 + 0.4y_0^3) - (0.3y_1^1 + 0.4y_1^2) + (0.7y_2^0 + 0.6y_2^1) - 0.4y_3^0 = 0$

(4) $-(0.3y_0^3 + 0.3y_1^2 + 0.3y_2^1) + 0.7y_3^0 = 0$

(5) $(y_0^0 + y_0^1 + y_0^2 + y_0^3) + (y_1^0 + y_1^1 + y_1^2) + (y_2^0 + y_2^1) + y_3^0 = 1$

$y_0^0 \geq 0, y_0^1 \geq 0, y_0^2 \geq 0, y_0^3 \geq 0, y_1^0 \geq 0, y_1^1 \geq 0, y_1^2 \geq 0, y_2^0 \geq 0, y_2^1 \geq 0, y_3^0 \geq 0.$

5.1.3.3.1.2 Solution by LP of Unichain MDP Model of an Inventory System After omitting one constraint which is redundant, the LP formulation of the unichain MDP model of an inventory system is solved to find an optimal policy by using LP software on a personal computer. The output of the LP software is summarized in Table 5.42.

Table 5.42 shows that the value of the objective function is 115.6364, which differs slightly from the gain calculated in Equation (4.98) because of roundoff error. The output also shows that all the y_i^k equal zero, except for $y_0^3 = 0.2364$, $y_1^2 = 0.2091$, $y_2^0 = 0.3636$, and $y_3^0 = 0.1909$. These nonzero values are the steady-state probabilities, calculated in Equation (2.27), for the policy given by $\mathbf{d} = [3\ 2\ 0\ 0]^T$, which is therefore an optimal policy. The conditional probabilities,

$$P(\text{order} = k| \text{state} = i) = \frac{y_i^k}{\sum\limits_{k=1}^{K_i} y_i^k}$$

are calculated below:

$P(\text{order} = 3| \text{state} = 0) = P(\text{order} = 2| \text{state} = 1) = P(\text{order} = 0| \text{state} = 2)$
  $= P(\text{order} = 0| \text{state} = 3) = 1.$

TABLE 5.42

LP Solution of Unichain MDP Model of
an Inventory System

		g = 115.6364
Objective i	Function Value k	y_i^k
0	0	0
	1	0
	2	0
	3	0.2364
1	0	0
	1	0
	2	0.2091
2	0	0.3636
	1	0
3	0	0.1909

All the remaining conditional probabilities are zero. Observe that the retailer's expected average profit per period is maximized by the same (2, 3) policy under which she operated the recurrent MCR inventory model in Section 4.2.3.4.1.

5.1.3.3.2 Unichain MDP Model of Component Replacement

The MCR model of component replacement constructed in Equation (4.101) of Section 4.1.3.4.2 is associated with the policy of replacing a surviving component every 4 weeks, its maximum service life. This recurrent MCR model can be transformed into a unichain MDP model by adding two decision alternatives per state in states 0, 1, and 2. The decision alternatives are whether to keep (decision K) or replace (decision R) a surviving component. More precisely, the decision for a component, which has survived i weeks, is whether to keep it for another week (unless it breaks down during the current week), or replace the surviving component with a new component at the end of the current week. If a decision is made to keep a surviving component of age i for another week, then either it will fail during the current week with probability p_{i0} and be replaced, or it will survive to age $i+1$ with probability $p_{i,i+1} = 1 - p_{i0}$. A component of age i, which is replaced, becomes a component of age 0 at the end of the current week. A component of age 3 weeks must be replaced at the end of its fourth week of life [4].

As Section 4.2.3.4.2 indicates, the cost incurred by the decision to keep a component of age i is

$$q_i^K = \$2 + \$10 p_{i0}.$$

A surviving component that is replaced at age 0, 1, or 2 weeks is assumed to have a trade-in or salvage value of \$4. The cost incurred by the decision to replace a surviving component is \$2 for inspecting the component, minus \$4 for trading it in, plus \$10 for replacing it with a new component of age 0. Letting q_i^R denote the cost incurred by the decision to replace a surviving component of age i gives the result that

$$q_i^R = (\text{Cost of inspection}) + (\text{Cost of replacement}) - (\text{Trade-in value})$$

$$q_0^R = q_1^R = q_2^R = \$2 + \$10 - \$4 = \$8.$$

Since a component has no salvage value at the end of its fourth week of life,

$$q_3^R = \$2 + \$10 = \$12.$$

Table 5.43 contains the transition probabilities and cost data for the four-state unichain MDP model of component replacement.

5.1.3.3.2.1 Solution by Enumeration of MDP Model of Component Replacement A component that fails will be replaced at the end of the week in which it has failed. A component that has not failed can be replaced after 1 week of service, or after 2 weeks, 3 weeks, or 4 weeks. Hence there are four replacement policies, which correspond to the four alternative replacement intervals. Enumeration of the MCRs associated with each of these four alternative replacement intervals can be used to determine a least cost replacement interval by finding which MCR has the smallest expected average cost per week, or negative gain. Equation (4.47) is used to calculate the negative gain, which will simply be called the gain, g, associated with each replacement policy. A decision to replace a component of age i is effective 1 year later when the component reaches age $i+1$. For example, a policy of replacing a component every 2 weeks means that you decide at age 0 to keep it until it reaches age 1, and decide at age 1 to replace it when it reaches age 2. The four replacement policies, the corresponding decision vectors, and the corresponding sequence of actions produced by each decision vector are enumerated in Table 5.44.

TABLE 5.43

Data for Unichain MDP Model of Component Replacement

State i	Decision k	Transition Probability				Cost q_i^k	LP Variable y_i^k
		p_{i0}^k	p_{i1}^k	p_{i2}^k	p_{i3}^k		
0	1 = K	0.2	0.8	0	0	$4 = 2 + 0.2(10$	y_0^1
	2 = R	1	0	0	0	$8 = 2 + 10 - 4$	y_0^2
1	1 = K	0.375	0	0.625	0	$5.75 = 2 + 0.375(10)$	y_1^1
	2 = R	1	0	0	0	$8 = 2 + 10 - 4$	y_1^2
2	1 = K	0.8	0	0	0.2	$10 = 2 + 0.8(10)$	y_2^1
	2 = R	1	0	0	0	$8 = 2 + 10 - 4$	y_2^2
3	2 = R	1	0	0	0	$12 = 2 + 10$	y_3^2

TABLE 5.44

Four Alternative Component Replacement Policies

Replacement Policy	Decision Vector at Age $\Rightarrow$ $i = \{0, \quad 1, \quad 2, \quad 3\} \Rightarrow$	Sequence of Actions at Age $i + 1 = \{1, \quad 2, \quad 3, \quad 4\}$
Replace every week	$^1d = [R \quad R \quad R \quad R]^T \Rightarrow$	$[K \quad R \quad R \quad R]^T$
Replace every 2 weeks	$^2d = [K \quad R \quad R \quad R]^T \Rightarrow$	$[K \quad K \quad R \quad R]^T$
Replace every 3 weeks	$^3d = [K \quad K \quad R \quad R]^T \Rightarrow$	$[K \quad K \quad K \quad R]^T$
Replace every 4 weeks	$^4d = [K \quad K \quad K \quad R]^T \Rightarrow$	$[K \quad K \quad K \quad K]^T$

The MCR associated with a replacement interval less than the 4-week service life is a unichain with transient states. If a replacement interval of $t < 4$ weeks is chosen, then states $0, 1, \ldots, t$ form a recurrent closed class, and the remaining states are transient. The MCRs associated with each of the four replacement intervals are shown below, accompanied by their respective gains.

Policy 1, Replace a component every week:

$$^1d = [R \quad R \quad R \quad R]^T \Rightarrow action = [K \quad R \quad R \quad R]^T$$

State	Decision	0	1	2	3		Cost
0	2 = R	1	0	0	0	0	$8
1	2 = R	1	0	0	0	1	8
2	2 = R	1	0	0	0	2	8
3	2 = R	1	0	0	0	3	8

$P =$ ... , $q = $ ... , $\pi = [1 \quad 0 \quad 0 \quad 0]$, $g = \pi q = \$8$.

Under this policy, state 0 is absorbing, and states 1, 2, and 3 are transient.

Policy 2, Replace a component every 2 weeks:

$$^2d = [K \quad R \quad R \quad R]^T \Rightarrow action = [K \quad K \quad R \quad R]^T$$

State	Decision	0	1	2	3		Cost
0	1 = K	0.2	0.8	0	0	0	$4
1	2 = R	1	0	0	0	1	8
2	2 = R	1	0	0	0	2	8
3	2 = R	1	0	0	0	3	8

$P =$ ... , $q = $ ...

$$\pi = [0.5556 \quad 0.4444 \quad 0 \quad 0], \quad g = \pi q = \$5.78.$$

Under this policy, states $\{0, 1\}$ form a recurrent closed class, while states 2 and 3 are transient.

Policy 3, Replace a component every 3 weeks:

$$^3\mathbf{d} = [K \quad K \quad R \quad R]^T \Rightarrow \text{action} = [K \quad K \quad K \quad R]^T$$

State	Decision	0	1	2	3			Cost
0	1 = K	0.2	0.8	0	0		0	$4
$P =$ 1	1 = K	0.375	0	0.625	0	$q = 1$	5.75
2	2 = R	1	0	0	0		2	8
3	2 = R	1	0	0	0		3	8

$$\pi = [0.4348 \quad 0.3478 \quad 0.2174 \quad 0], \quad g = \pi q = \$5.48.$$

Under this policy, states {0, 1, 2} form a recurrent closed class, while state 3 is transient.

Policy 4, Replace a component every 4 weeks:

$$^4\mathbf{d} = [K \quad K \quad K \quad R]^T \Rightarrow \text{action} = [K \quad K \quad K \quad K]^I$$

State	Decision	0	1	2	3			Cost
0	1 = K	0.2	0.8	0	0		0	$4
$P =$ 1	1 = K	0.375	0	0.625	0	$q = 1$	5.75
2	1 = K	0.8	0	0	0.2		2	10
3	2 = R	1	0	0	0		3	12

$$\pi = [0.4167 \quad 0.3333 \quad 0.2083 \quad 0.0417], \quad g = \pi q = \$6.17.$$

Under this policy, all states are recurrent. The minimum cost policy,

$$^3\mathbf{d} = [K \quad K \quad R \quad R]^T \Rightarrow \text{action} = [K \quad K \quad K \quad R]^T, \tag{5.47}$$

is to replace a component every 3 weeks at an expected average cost per week of $5.48.

5.1.3.3.2.2 Solution by LP of MDP Model of Component Replacement The unichain MDP model of component replacement specified in Table 5.43 is formulated as the following LP:

Minimize $g = (4y_0^1 + 8y_0^2) + (5.75y_1^1 + 8y_1^2) + (10y_2^1 + 8y_2^2) + 12y_3^2$

subject to

(1) $(y_0^1 + y_0^2) - (0.2y_0^1 + y_0^2) - (0.375y_1^1 + y_1^2) - (0.8y_2^1 + y_2^2) - y_3^2 = 0$

(2) $(y_1^1 + y_1^2) - 0.8y_0^1 = 0$

(3) $(y_2^1 + y_2^2) - 0.625y_1^1 = 0$

(4) $y_3^2 - 0.2y_2^1 = 0$

(5) $(y_0^1 + y_0^2) + (y_1^1 + y_1^2) + (y_2^1 + y_2^2) + y_3^2 = 1$

$$y_0^1 \ge 0,\ y_0^2 \ge 0,\ y_1^1 \ge 0,\ y_1^2 \ge 0,\ y_2^1 \ge 0,\ y_2^2 \ge 0,\ y_3^2 \ge 0.$$

After omitting one constraint which is redundant, the LP is solved on a personal computer using LP software to find an optimal policy. The output of the LP software is summarized in Table 5.45.

The output of the LP software shows that all the y_i^k equal zero, except for $y_0^1 = 0.4348$, $y_1^1 = 0.3478$, and $y_2^2 = 0.2174$. These values are the steady-state probabilities for the optimal policy $^3\mathbf{d} = [1 \quad 1 \quad 2 \quad 2]^T = [K \quad K \quad R \quad R]^T$, which was previously found by enumeration in Equation (5.47) of Section 5.1.3.3.2.2. The LP solution has confirmed that the minimum cost policy is to replace a component every 3 weeks at an expected average cost per week of \$5.48.

5.1.3.3.3 Optimal Stopping Over a Finite Planning Horizon

An optimal stopping problem is concerned with determining when to stop a sequence of trials with random outcomes. The objective is to maximize an expected reward. This book treats the simplest kind of optimal stopping problem, which is to decide when to stop a sequence of independent trials. The planning horizon can be finite or infinite. In Section 5.1.3.3.3.1, the secretary problem, an example of an optimal stopping problem, is formulated as a unichain MDP over a finite planning horizon, and solved by executing value iteration. In Section 5.2.2.4.2, LP will be used to find an optimal policy for a discounted MDP model of a modified secretary problem over an infinite horizon [1, 3, 4].

In the simplest kind of optimal stopping problem, a system evolves as an independent trials process. As Section 1.10.1.1.6 indicates, a sequence

TABLE 5.45

LP Solution of Unichain MDP Model of Component Replacement

Objective	Function = 5.4783	
i	k	y_i^k
0	1	0.4348
	2	
1	1	0.3478
	2	0
2	1	0
	2	0.2174
3	2	0

of independent trials can be modeled as a Markov chain. At each epoch, there are two decisions in every state: to stop or to continue. If the decision maker stops in state i at epoch n, a reward $q_i(n)$ is received. If the decision maker continues in state i at epoch n, no reward is received. If the decision maker decides to continue in state i at epoch n, then with transition probability p_{ij} the system reaches state j at epoch $n + 1$. If the planning horizon is finite, of length T periods, and the decision maker reaches a state h at epoch T, then a reward $q_h(T)$ is received, and the process stops. The objective is to choose a policy to maximize the expected reward.

5.1.3.3.3.1 Formulation of the Secretary Problem as a Unichain MDP In this section, a colorful example of an optimal stopping problem, called the secretary problem, will be modeled as a unichain MDP. The secretary problem is a continuation of the example of an independent trials process involving the arrival of candidates for a secretarial position, which was described in Section 1.10.1.1.6. To transform that example into the secretary problem, suppose that the executive who is interviewing candidates must hire a secretary within the next 6 days. The executive will interview up to six consecutive candidates at a rate of one applicant per day. After interviewing each candidate or applicant, the executive will assign the current candidate one of the four numerical scores listed in Table 1.8. At the end of an interview, the executive must decide immediately whether to hire or reject the current applicant. A rejected candidate is excluded from further consideration. When a candidate is hired as a secretary, the interviews stop. When an applicant is rejected, the interviews continue. However, if the first five applicants have been rejected, the sixth applicant must be hired. The objective is to maximize the expected score of the secretary.

As Equation (1.51) in Section 1.10.1.1.6 indicates, an independent trials process can be modeled as a Markov chain. The secretary problem will be formulated as a unichain MDP and solved by value iteration over a 6-day planning horizon. To formulate the problem as an MDP, note that there are two decisions: to hire (decision H) or reject (decision R) the nth candidate. There are six alternative policies that correspond to hiring the first, second, third, fourth, fifth, or sixth candidate, and ending the interviews. The state, denoted by X_n, is the numerical score assigned to the nth candidate, for $n = 1$, 2, 3, 4, 5, 6. The sequence $\{X_1, X_2, X_3, X_4, X_5, X_6\}$ is a collection of six independent, identically distributed random variables. The probability distribution of X_n is shown in Table 1.8.

When the nth applicant is rejected, the state X_{n+1} of the next applicant is independent of the state X_n of the current applicant. Hence, when the nth candidate is rejected, the transition probability is $p_{ij} = P(X_{n+1} = j | X_n = i) = P(X_{n+1} = j)$. The state space is augmented by an absorbing state Δ that is reached with probability 1 when an applicant is hired, and with probability 0

TABLE 5.46

Data for Unichain MDP Model of a Secretary Problem Over a Finite Planning Horizon

State i	Decision k	Transition Probability					Reward q_i^k
		$p_{i,15}^k$	$p_{i,20}^k$	$p_{i,25}^k$	$p_{i,30}^k$	$p_{i,\Delta}^k$	
15	1 = H	0	0	0	0	1	15
	2 = R	0.3	0.4	0.2	0.1	0	0
20	1 = H	0	0	0	0	1	20
	2 = R	0.3	0.4	0.2	0.1	0	0
25	1 = H	0	0	0	0	1	25
	2 = R	0.3	0.4	0.2	0.1	0	0
30	1 = H	0	0	0	0	1	30
	2 = R	0.3	0.4	0.2	0.1	0	0
Δ	1 = H	0	0	0	0	1	0
	2 = R	0	0	0	0	1	0

when an applicant is rejected. The augmented state space is $E = \{15, 20, 25, 30, \Delta\}$. If the nth applicant is hired, the daily reward is X_n, equal to the score assigned to the nth applicant. When an applicant is hired, the process goes to the absorbing state Δ, where it remains because no more candidates will be interviewed. If the nth applicant is rejected, the daily reward is zero. Since all the transition probability matrices corresponding to every possible decision contain the single absorbing state Δ, the MDP is unichain. The unichain MDP model of the secretary problem over a finite planning horizon is shown in Table 5.46.

5.1.3.3.3.2 Solution of the Secretary Problem by Value Iteration Value iteration will be executed to determine whether to hire or reject the nth applicant on the basis of the applicant's score, X_n. The length of the planning horizon is 6 days, numbered 1 through 6, corresponding to six candidates who may be interviewed. The calculations will produce a decision rule for each day of the 6-day horizon. To conduct value iteration, let $v_i(n)$ denote the maximum expected score if the nth candidate interviewed is assigned a score of $X_n = i$, and an optimal policy is followed from day n until the end of the planning horizon.

To solve the secretary problem, the backward recursive equations of value iteration (5.6) are modified to have the following form.

$$v_i(n) = \max\{[x_n \text{ if hire}], \ [q_i^2 + \sum_{j=15}^{30} p_{ij}^2 v_j(n+1) \text{ if reject}]\},$$

$$\text{for } n = 1,2,\dots, 5, \text{ and } i = 15, 20, 25, 30$$

Substituting $q_i^2 = 0$,

$$v_i(n) = \max\{[x_n \text{ if hire}], \ [p_{i,15}^2 v_{15}(n+1) + p_{i,20}^2 v_{20}(n+1)$$

$$+ p_{i,25}^2 v_{25}(n+1) + p_{i,30}^2 v_{30}(n+1) \text{ if reject}]\},$$ (5.48)

$$\text{for } n = 1,2,\dots, 5, \text{ and } i = 15, 20, 25, 30$$

$$v_i(n) = \max\{[x_n \text{ if hire}], \ [0.3v_{15}(n+1) + 0.4v_{20}(n+1)$$

$$+ 0.2v_{25}(n+1) + 0.1v_{30}(n+1) \text{ if reject}]\},$$

$$\text{for } n = 1,2,\dots, 5.$$

To begin the backward recursion, the terminal values on day 6 are set equal to the possible scores of the applicant on day 6 because a sixth or final applicant must be hired. Therefore, for $n = 6$,

$$v_i(6) = X_6 = i, \quad \text{for } i = 15, 20, 25, 30$$

$$v_{15}(6) = 15, \quad v_{20}(6) = 20, \quad v_{25}(6) = 25, \quad v_{30}(6) = 30.$$

The calculations for day 5, denoted by $n = 5$, are indicated below:

$$v_i(5) = \max\{[x_5 \text{ if hire}], \ [0.3v_{15}(6) + 0.4v_{20}(6) + 0.2v_{25}(6) + 0.1v_{30}(6) \text{ if reject}]\},$$

for $i = 15, 20, 25, 30$

$$v_i(5) = \max\{[x_5 \text{ if hire}], \ [0.3(15) + 0.4(20) + 0.2(25) + 0.1(30) \text{ if reject}]\},$$

for $i = 15, 20, 25, 30$

$$v_i(5) = \max\{[x_5 \text{ if hire}], \ [20.5 \text{ if reject}]\}, \text{ for } i = 15, 20, 25, 30$$

If $X_5 = i = 15$, then $v_{15}(5) = \max\{[15 \text{ if hire}], \ [20.5 \text{ if reject}]\} = 20.5$, reject

If $X_5 = i = 20$, then $v_{20}(5) = \max\{[20 \text{ if hire}], \ [20.5 \text{ if reject}]\} = 20.5$, reject

If $X_5 = i = 25$, then $v_{25}(5) = \max\{[25 \text{ if hire}], \ [20.5 \text{ if reject}]\} = 25$, hire

If $X_5 = i = 30$, then $v_{30}(5) = \max\{[30 \text{ if hire}], \ [20.5 \text{ if reject}]\} = 30$, hire

The calculations for day 4, denoted by $n = 4$, are indicated below:

$v_i(4) = \max\{[x_4 \text{ if hire}], \ [0.3v_{15}(5) + 0.4v_{20}(5) + 0.2v_{25}(5) + 0.1v_{30}(5) \text{ if reject}]\}$,

for $i = 15, 20, 25, 30$

$v_i(4) = \max\{[x_4 \text{ if hire}], \ [0.3(20.5) + 0.4(20.5) + 0.2(25) + 0.1(30) \text{ if reject}]\}$,

for $i = 15, 20, 25, 30$

$v_i(4) = \max\{[x_4 \text{ if hire}], \ [22.35 \text{ if reject}]\}$, for $i = 15, 20, 25, 30$

If $X_4 = i = 15$, then $v_{15}(4) = \max\{[15 \text{ if hire}], \ [22.35 \text{ if reject}]\} = 22.35$, reject

If $X_4 = i = 20$, then $v_{20}(4) = \max\{[20 \text{ if hire}], \ [22.35 \text{ if reject}]\} = 22.35$, reject

If $X_4 = i = 25$, then $v_{25}(4) = \max\{[25 \text{ if hire}], \ [22.35 \text{ if reject}]\} = 25$, hire

If $X_4 = i = 30$, then $v_{30}(4) = \max\{[30 \text{ if hire}], \ [22.35 \text{ if reject}]\} = 30$, hire

The calculations for day 3, denoted by $n = 3$, are indicated below:

$v_i(3) = \max\{[x_3 \text{ if hire}], \ [0.3v_{15}(4) + 0.4v_{20}(4) + 0.2v_{25}(4) + 0.1v_{30}(4) \text{ if reject}]\}$,

for $i = 15, 20, 25, 30$

$v_i(3) = \max\{[x_3 \text{ if hire}], \ [0.3(22.35) + 0.4(22.35) + 0.2(25) + 0.1(30) \text{ if reject}]\}$,

for $i = 15, 20, 25, 30$

$v_i(3) = \max\{[x_3 \text{ if hire}], \ [23.645 \text{ if reject}]\}$, for $i = 15, 20, 25, 30$

If $X_3 = i = 15$, then $v_{15}(3) = \max\{[15 \text{ if hire}], \ [23.645 \text{ if reject}]\} = 23.645$, reject

If $X_3 = i = 20$, then $v_{20}(3) = \max\{[20 \text{ if hire}], \ [23.645 \text{ if reject}]\} = 23.645$, reject

If $X_3 = i = 25$, then $v_{25}(3) = \max\{[25 \text{ if hire}], \ [23.645 \text{ if reject}]\} = 25$, hire

If $X_3 = i = 30$, then $v_{30}(3) = \max\{[30 \text{ if hire}], \ [23.645 \text{ if reject}]\} = 30$, hire

The calculations for day 2, denoted by $n = 2$, are indicated below.

$v_i(2) = \max\{[x_2 \text{ if hire}], \ [0.3v_{15}(3) + 0.4v_{20}(3) + 0.2v_{25}(3) + 0.1v_{30}(3) \text{ if reject}]\}$,

for $i = 15, 20, 25, 30$

$v_i(2) = \max\{[x_2 \text{ if hire}], \ [0.3(23.645) + 0.4(23.645) + 0.2(25) + 0.1(30) \text{ if reject}]\}$,

for $i = 15, 20, 25, 30$

$v_i(2) = \max\{[x_2 \text{ if hire}], \ [24.5515 \text{ if reject}]\}$, for $i = 15, 20, 25, 30$

If $X_2 = i = 15$, then $v_{15}(2) = \max\{[15 \text{ if hire}], \ [24.5515 \text{ if reject}]\} = 24.5515$, reject

If $X_2 = i = 20$, then $v_{20}(2) = \max\{[20 \text{ if hire}], \ [24.5515 \text{ if reject}]\} = 24.5515$, reject

If $X_2 = i = 25$, then $v_{25}(2) = \max\{[25 \text{ if hire}], \ [24.5515 \text{ if reject}]\} = 25$, hire

If $X_2 = i = 30$, then $v_{30}(2) = \max\{[30 \text{ if hire}], \ [24.5515 \text{ if reject}]\} = 30$, hire.

Finally, the calculations for day 1, denoted by $n = 1$, are indicated below:

$v_i(1) = \max\{[x_1 \text{ if hire}], \ [0.3v_{15}(2)+0.4v_{20}(2)+0.2v_{25}(2)+0.1v_{30}(2) \text{ if reject}]\}$,
for $i = 15, 20, 25, 30$

$v_i(1) = \max\{[x_1 \text{ if hire}], \ [0.3(24.5515)+0.4(24.5515)+0.2(25)+0.1(30) \text{ if reject}]\}$,
for $i = 15, 20, 25, 30$

$v_i(1) = \max\{[x_1 \text{ if hire}], \ [25.18605 \text{ if reject}]\}$, for $i = 15, 20, 25, 30$

If $X_1 = i = 15$, then $v_{15}(1) = \max\{[15 \text{ if hire}], \ [25.18605 \text{ if reject}]\} = 25.18605$, reject

If $X_1 = i = 20$, then $v_{20}(1) = \max\{[20 \text{ if hire}], \ [25.18605 \text{ if reject}]\} = 25.18605$, reject

If $X_1 = i = 25$, then $v_{25}(1) = \max\{[25 \text{ if hire}], \ [25.18605 \text{ if reject}]\} = 25.18605$, reject

If $X_1 = i = 30$, then $v_{30}(1) = \max\{[30 \text{ if hire}], \ [25.18605 \text{ if reject}]\} = 30$, hire.

The results of these calculations are summarized in the form of decision rules given in Table 5.47. By following these decision rules, the executive will maximize the expected score of the secretary.

TABLE 5.47

Decision Rules for Maximizing the Expected Score of the Secretary

Day, n	Candidate Score, X_n	Decision	Minimum Rating of Candidate to Be Hired	Interviews Will
1	If X_1 = 15, 20, or 25, then	Reject		Continue
	If X_1 = 30, then	Hire	$X_1 \geq 25.18605$	Stop
2	If X_2 = 15 or 20, then	Reject		Continue
	If X_2 = 25 or 30, then	Hire	$X_2 \geq 24.5515$	Stop
3	If X_3 = 15 or 20, then	Reject		Continue
	If X_3 = 25 or 30, then	Hire	$X_3 \geq 23.645$	Stop
4	If X_4 = 15 or 20, then	Reject		Continue
	If X_4 = 25 or 30, then	Hire	$X_4 \geq 22.35$	Stop
5	If X_5 = 15 or 20, then	Reject		Continue
	If X_5 = 25 or 30, then	Hire	$X_5 \geq 20.5$	Stop
6	If X_6 = 15, 20, 25, or 30, then	Reject		Continue
		Hire	$X_6 \geq 15$	Stop

Suppose that the optimal policy specified by the decision rules in Table 5.47 is followed. Observe that if the score of the first candidate is greater than or equal to 25.18605, that is, 30, the first candidate should be hired, and the interviews should be stopped. Otherwise, the interviews should be continued. If X_1 equals 30, the maximum expected score is 30 because the first applicant is hired. If X_1 is less than 30, the maximum expected score is 25.18605 because the first applicant is rejected. Prior to knowing X_1, the maximum expected score is $(0.1)(30) + (0.9)(25.18605) = 25.667445$ because there is a probability of 0.1 that the first candidate will be excellent and receive a score of 30, and a probability of 0.9 that the first candidate will not be excellent and the interviews will have to be continued. Note that if the fifth candidate receives a score greater than or equal to 20.5, that is, 25 or 30, the fifth candidate should be hired, and the interviews should be stopped. Finally, if the fifth candidate is rejected, the sixth candidate must be hired.

5.1.4 A Multichain MDP

Recall that an MDP is said to be multichain if the transition matrix associated with at least one stationary policy contains two or more closed communicating sets of recurrent states plus a possibly empty set of transient states. In Section 5.1.4.1, a multichain MDP model of a flexible production process will be constructed. In Section 5.1.4.2, the PI algorithm of Section 5.1.2.3.3.2 will be extended so that it can be used to find an optimal policy for a multichain MDP. An LP formulation for a multichain MDP will be given in Section 5.1.4.3.

5.1.4.1 Multichain Model of a Flexible Production System

To see how PI can be applied to determine an optimal policy for a multichain MDP model, consider the following simplified multichain MDP model of a flexible production system. (A smaller unichain MDP model of an experimental production process was constructed in Section 5.1.3.1.1.) Suppose that a flexible production system consists of three manufacturing stages in series. The three-stage line can manufacture two types of products: industrial (I) or consumer (C). Each stage can be programmed to make either type of product independently of the other stages. The output of each manufacturing stage is inspected. Output with a minor defect is reworked at the current stage. Output from stage 1 or stage 2 that is not defective is passed on to the next stage. Output from any stage with a major defect is discarded as scrap. Nondefective output from stage three is sent to a training center. At the training center, technicians are trained to maintain either the product software (S) or the product hardware (H). The probability of defective output from a stage is different for industrial and commercial output. Output sent from stage 3 to the training center to provide training on software will remain in the training center. Output sent from stage 3 to the training center to provide training on hardware will be disassembled, and returned as input to stage 1 in the following period.

By assigning a state to represent each operation, and adding a reward vector, the flexible production system is modeled as a five-state multichain MDP. The states are indexed in Table 5.48.

Observe that production stage i is represented by transient state $(6-i)$. In states 2, 3, 4, and 5, two decisions are possible. The decision at a manufacturing stage is the type of customer, either industrial (I) or consumer (C), for whom the product is intended. The decision at the training center is whether to focus on hardware (H) or software (S). Since both industrial and consumer output with a major defect at any stage is always discarded as scrap, state 1 is always absorbing. States 3, 4, and 5 are always transient. In contrast, state 2 is absorbing when the training center is dedicated to software, and is transient when the training center is dedicated to hardware. With one decision in state 1 and two decisions in each of the other four states, the number of possible policies is equal to $(1)2^4 = 16$. Table 5.49 contains the data for the five-state multichain MDP model of the flexible production system.

The passage of an item through the flexible production system is shown in Figure 5.4.

TABLE 5.48

States for Flexible Production Process

State	Operation
1	Scrap
2	Training center
3	Stage 3
4	Stage 2
5	Stage 1

TABLE 5.49

Data for Multichain Model of a Flexible Production System

State i	Operation	Decision k	Transition Probability					Reward q_i^k
			p_{i1}^k	p_{i2}^k	p_{i3}^k	p_{i4}^k	p_{i5}^k	
1	Scrap	1	1	0	0	0	0	62
2	Training	1 = S	0	1	0	0	0	100
		2 = H	0	0	0	0	1	22.58
3	Stage 3	1 = I	0.25	0.20	0.55	0	0	423.54
		2 = C	0.20	0.35	0.45	0	0	468.62
4	Stage 2	1 = I	0.15	0	0.20	0.65	0	847.09
		2 = C	0.20	0	0.30	0.50	0	796.21
5	Stage 1	1 = I	0.10	0	0	0.15	0.75	1935.36
		2 = C	0.08	0	0	0.12	0.80	2042.15

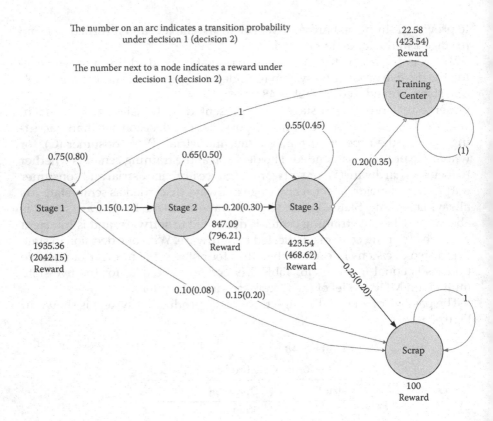

FIGURE 5.4
Passage of an item through a three-stage flexible production system.

5.1.4.2 PI for a Multichain MDP

The PI algorithm [2] for finding an optimal stationary policy for a multi-chain MDP is an extension of the one (described in Section 5.1.2.3.3.2) created for a unichain MDP. The PI algorithm for a multichain MDP has two main steps: the policy evaluation (PE) operation and the IM routine. The algorithm begins by arbitrarily choosing an initial policy. The PE operation for a multichain MDP is analogous to the reward evaluation operation for a multichain MCR described in Section 4.2.5.3. During the PE operation, two sets of N simultaneous linear equations are solved. The first system of equations is called the gain state equations (GSEs). The second system is called the set of VDEs. The two systems of equations, the GSEs plus the VDEs, for a multichain MDP are together called the Policy Evaluation Equations (PEEs). (In Section 4.2.5.3 the GSEs plus the VDEs for a multichain MCR are together called the Reward Evaluation Equations, or REEs.) The PEEs associated with the current policy for a multichain MDP can be solved by following the four-step procedure given in Section 4.2.5.3.3 for solving the REEs for a multichain MCR. Following the completion of the PE operation to obtain the gains and the relative values associated with the current policy, the IM routine attempts to find a better policy. If a better policy is found, the PE operation is repeated using the new policy to identify the appropriate transition probabilities, rewards, and VDEs. The algorithm stops when two successive iterations lead to identical policies.

5.1.4.2.1 Policy Improvement

The IM routine is motivated by the value iteration equation (5.6) of Section 5.1.2.2.1 for a recurrent MDP over a finite planning horizon,

$$v_i(n) = \max_k \left[q_i^k + \sum_{j=1}^{N} p_{ij}^k v_j(n+1) \right], \quad \text{for } n = 0, 1, ..., T-1, \text{ and } i = 1, 2, ..., N, \quad (5.6)$$

where the salvage values $v_i(T)$ are specified for all states $i = 1, 2, ..., N$. The value iteration equation indicates that if an optimal policy is known over a planning horizon starting at epoch $n+1$ and ending at epoch T, then the best decision in state i at epoch n can be found by maximizing a test quantity,

$$q_i^k + \sum_{j=1}^{N} p_{ij}^k v_j(n+1), \qquad (5.49)$$

over all decisions in state i. Therefore, if an optimal policy is known over a planning horizon starting at epoch 1 and ending at epoch T, then at epoch 0, the beginning of the planning horizon, the best decision in state i can be found by maximizing a test quantity,

$$q_i^k + \sum_{j=1}^{N} p_{ij}^k v_j(1), \qquad (5.50)$$

over all decisions in state i. Let g_j denote the gain of state j. Recall from Section 4.2.3.2.2 that when T, the length of the horizon, is very large, $(T-1)$ is also very large, so that

$$v_j(1) \approx (T-1)g_j + v_j. \qquad (5.51)$$

Substituting this expression for $v_j(1)$ in the test quantity produces the result

$$q_i^k + \sum_{j=1}^{N} p_{ij}^k v_j(1) = q_i^k + \sum_{j=1}^{N} p_{ij}^k [(T-1)g_j + v_j] = q_i^k + (T-1)\sum_{j=1}^{N} p_{ij}^k g_j + \sum_{j=1}^{N} p_{ij}^k v_j \quad (5.52)$$

as the test quantity to be maximized with respect to all alternatives in every state. When $(T-1)$ is very large, the term $(T-1)\sum_{j=1}^{N}p_{ij}^{k}g_{j}$ dominates the test quantity, so that the test quantity is maximized by the alternative that maximizes

$$\sum_{j=1}^{N}p_{ij}^{k}g_{j}, \tag{5.53}$$

called the gain test quantity, using the gains of the previous policy. However, when two or more alternatives have the same maximum value of the gain test quantity, there is a tie, and the gain test fails. In that case the decision must be made on the basis of relative values rather than gains. The tie is broken by choosing the alternative that maximizes

$$q_{i}^{k}+\sum_{j=1}^{N}p_{ij}^{k}v_{j}, \tag{5.54}$$

called the value test quantity, by using the relative values of the previous policy.

5.1.4.2.2 PI Algorithm

The detailed steps of the PI algorithm [2] for a multichain MDP are given below:

PI Algorithm

Step 1. Initial policy

Arbitrarily choose an initial policy by selecting for each state i a decision $d_{i} = k$.

Step 2. PE operation

Use p_{ij} and q_{i} for a given policy to solve the set of GSEs

$$g_{i} = \sum_{j=1}^{N}p_{ij}g_{j}, \quad \text{for } i = 1, 2, ..., N. \tag{4.176}$$

and the set of VDEs

$$g_{i}+v_{i} = q_{i}+\sum_{j=1}^{N}p_{ij}v_{j}, \quad i = 1, 2, ..., N, \tag{4.177}$$

for all the relative values v_{i} and gains g_{i} by executing the four-step procedure given in Section 4.2.5.3.3.

Step 3. IM routine

For each state i, find the decision $k*$ that maximizes the gain test quantity

$$\sum_{j=1}^{N} p_{ij}^k g_j \tag{5.53}$$

by using the gains of the previous policy. Then $k*$ becomes the new decision in state i, so that $d_i = k*$, q_i^{k*} becomes q_i, and p_{ij}^{k*} becomes p_{ij}.

If two or more alternatives have the same maximum value of the gain test quantity, the tie is broken by choosing the decision $k*$ that maximizes the value test quantity

$$q_i^k + \sum_{j=1}^{N} p_{ij}^k v_j, \tag{5.54}$$

by using the relative values of the previous policy. Then $k*$ becomes the new decision in state i, so that $d_i = k*$, q_i^{k*} becomes q_i, and p_{ij}^{k*} becomes p_{ij}.

Regardless of whether the IM test is based on gains or values, if the previous decision in state i yields as high a value of the test quantity as any other alternative, leave the previous decision unchanged to assure convergence in the case of equivalent policies.

Step 4. Stopping rule

When the IM test has been completed for all states, a new policy has been determined. A new **P** matrix and **q** vector have been obtained. If the new policy is the same as the previous one, the algorithm has converged, and an optimal policy has been found. If the new policy is different from the previous ous policy in at least one state, go to step 2.

5.1.4.2.3 Solution by PI of Multichain Model of a Flexible Production System

To see how PI can be applied to find an optimal policy for a multichain MDP, PI will be executed to find an optimal policy for the multichain MDP model of a flexible production system specified in Table 5.49 of Section 5.1.4.1.

First iteration

Step 1. Initial policy

Arbitrarily choose an initial policy which consists of making decision 1 in every state. Thus, in state 2 the training center focuses on software. In states 3, 4, and 5, the product is designed for an industrial customer. The initial decision vector $^1\mathbf{d}$, along with the associated transition probability matrix $^1\mathbf{P}$

and the reward vector $^1\mathbf{q}$, are shown below:

$$^1\mathbf{d} = \begin{bmatrix} 1 \\ 1 \\ 1 \\ 1 \\ 1 \end{bmatrix}, \quad ^1\mathbf{P} = \begin{matrix} 1 \\ 2 \\ 3 \\ 4 \\ 5 \end{matrix} \begin{bmatrix} 1 & 0 & 0 & 0 & 0 \\ 0 & 1 & 0 & 0 & 0 \\ 0.25 & 0.20 & 0.55 & 0 & 0 \\ 0.15 & 0 & 0.20 & 0.65 & 0 \\ 0.10 & 0 & 0 & 0.15 & 0.75 \end{bmatrix}, \quad ^1\mathbf{q} = \begin{bmatrix} 62 \\ 100 \\ 423.54 \\ 847.09 \\ 1935.36 \end{bmatrix}.$$

Observe that the initial policy generates an absorbing multichain in which states 1 and 2 are absorbing states and the other three states are transient.

Step 2. PE operation
Use p_{ij} and q_i for the initial policy, $^1\mathbf{d} = [1 \ 1 \ 1 \ 1 \ 1]^T$, to solve the set of GSEs (4.176) and the set of VDEs (4.177) for all the relative values v_i and gains g_i, by setting the value of one v_i in each closed class of recurrent states equal to zero.

Step 1 of PE:
The two recurrent chains are the absorbing states 1 and 2. After setting the relative value $v_1 = 0$ for the highest numbered state in the first recurrent chain, the VDE for the first recurrent chain is

$$g_1 + v_1 = 62 + v_1$$
$$g_1 = 62.$$

After setting the relative value $v_2 = 0$ for the highest numbered state in the second recurrent chain, the VDE for the second recurrent chain is

$$g_2 + v_2 = 100 + v_2$$
$$g_2 = 100.$$

Hence, the independent gains of the two recurrent states are

$$g_1 = 62$$
$$g_2 = 100.$$

Step 2 of PE:
The GSEs for the transient states are

$$g_3 = 0.25g_1 + 0.20g_2 + 0.55g_3$$
$$g_4 = 0.15g_1 + 0.20g_3 + 0.65g_4$$
$$g_5 = 0.10g_1 + 0.15g_4 + 0.75g_5$$

After algebraic simplification, the gains of the three transient states are expressed as weighted averages of the independent gains of the two closed classes of recurrent states

$$g_3 = 0.5556g_1 + 0.4444g_2 = 0.5556(62) + 0.4444(100) = 78.8872$$
$$g_4 = 0.7461g_1 + 0.2539g_2 = 0.7461(62) + 0.2539(100) = 71.6482$$
$$g_5 = 0.8477g_1 + 0.1523g_2 = 0.8477(62) + 0.1523(100) = 67.7874.$$

Step 3 of PE:
The VDEs for the transients states are

$$g_3 + v_3 = 423.54 + 0.25v_1 + 0.20v_2 + 0.55v_3$$
$$g_4 + v_4 = 847.09 + 0.15v_1 + 0.20v_3 + 0.65v_4$$
$$g_5 + v_5 = 1935.36 + 0.10v_1 + 0.15v_4 + 0.75v_5.$$

After substituting the gains of the transient states obtained in step 2 of PE, and the relative values of the recurrent states obtained in step 1 of PE, the VDEs for the transient states are

$$78.8872 + v_3 = 423.54 + 0.25(0) + 0.20(0) + 0.55v_3$$
$$71.6482 + v_4 = 847.09 + 0.15(0) + 0.20(v_3) + 0.65v_4$$
$$67.7874 + v_5 = 1935.36 + 0.10(0) + 0.15v_4 + 0.75v_5.$$

Step 4 of PE:
The VDEs for the transient states are solved to obtain the relative values of the transient states: $v_3 = 765.8951$, $v_4 = 2653.2023$, and $v_5 = 9062.2118$.
   The solutions for the gain vector and the vector of relative values for the initial policy are summarized below:

$$^1g = \begin{matrix} 1 \\ 2 \\ 3 \\ 4 \\ 5 \end{matrix} \begin{bmatrix} 62 \\ 100 \\ 78.8872 \\ 71.6482 \\ 67.7874 \end{bmatrix}, \quad ^1v = \begin{matrix} 1 \\ 2 \\ 3 \\ 4 \\ 5 \end{matrix} \begin{bmatrix} 0 \\ 0 \\ 765.8951 \\ 2653.2023 \\ 9062.2118 \end{bmatrix}.$$

Step 3. IM routine
For each state i, find the decision $k*$ that maximizes the gain test quantity

$$\sum_{j=1}^{5} p_{ij}^k g_j = p_{i1}^k g_1 + p_{i2}^k g_2 + (p_{i3}^k g_3 + p_{i4}^k g_4 + p_{i5}^k g_5).$$

TABLE 5.50

First IM for Model of a Flexible Production System

State i	Decision k	Gain Test Quantity $\sum_{j=1}^{5} p_{ij}^{k} g_j$	Value Test Quantity $q_i^k + \sum_{j=1}^{5} p_{ij}^{k} v_j$
1	1	$1(62) = 62 \leftarrow$	
2	1	$1(100) = 100 \leftarrow$	
	2	$1(67.7874) = 67.7874$	
3	1	$0.25(62) + 0.20(100) + 0.55(78.8872)$ $= 78.8880$	
	2	$0.20(62) + 0.35(100) + 0.45(78.8872)$ $= 82.8992 \leftarrow$	
4	1	$0.15(62) + 0.20(78.8872) + 0.65(71.6482) = 71.6488$	
	2	$0.20(62) + 0.30(78.8872) + 0.50(71.6482) = 71.8903 \leftarrow$	
5	1	67.7878	$1935.36 + (7194.6392) = 9129.9992$
	2	67.7878	$2042.15 + (7568.1537) = 9610.3037 \leftarrow$

by using the gains of the previous policy. Then $k*$ becomes the new decision in state i, so that $d_i = k*$, q_i^{k*} becomes q_i, and p_{ij}^{k*} becomes p_{ij}. The first IM routine is executed in Table 5.50.

Step 4. Stopping rule
The new policy is given by the vector $^2\mathbf{d} = [1 \ \ 1 \ \ 2 \ \ 2 \ \ 2]^T$, which is different from the initial policy. Therefore, go to step 2. The new decision vector $^2\mathbf{d}$, along with the associated transition probability matrix $^2\mathbf{P}$ and the reward vector $^2\mathbf{q}$, are shown below.

Second iteration

$$
^2\mathbf{d} = \begin{bmatrix} 1 \\ 1 \\ 2 \\ 2 \\ 2 \end{bmatrix}, \quad
^2\mathbf{P} = \begin{matrix} 1 \\ 2 \\ 3 \\ 4 \\ 5 \end{matrix}
\begin{bmatrix}
1 & 0 & 0 & 0 & 0 \\
0 & 1 & 0 & 0 & 0 \\
0.20 & 0.35 & 0.45 & 0 & 0 \\
0.20 & 0 & 0.30 & 0.50 & 0 \\
0.08 & 0 & 0 & 0.12 & 0.80
\end{bmatrix}, \quad
^2\mathbf{q} = \begin{bmatrix} 62 \\ 100 \\ 468.62 \\ 796.21 \\ 2042.15 \end{bmatrix}.
\tag{5.55}
$$

Step 2. PE operation
Use p_{ij} and q_i for the second policy, $^2\mathbf{d} = [1 \ \ 1 \ \ 2 \ \ 2 \ \ 2]^T$, to solve the set of GSEs (4.176) and the set of VDEs (4.177) for all the relative values v_i and gains g_i, by setting the value of one v_i in each closed class of recurrent states equal to zero.

Step 1 of PE:
After setting the relative values $v_1 = v_2 = 0$ in the VDEs for the highest numbered states in the two recurrent closed classes, the independent gains of the two recurrent states are once again

$$g_1 = 62$$
$$g_2 = 100.$$

Step 2 of PE:
The GSEs for the transient states are

$$g_3 = 0.20g_1 + 0.35g_2 + 0.45g_3$$
$$g_4 = 0.20g_1 + 0.30g_3 + 0.50g_4$$
$$g_5 = 0.08g_1 + 0.12g_4 + 0.80g_5.$$

After algebraic simplification, the gains of the three transient states expressed as weighted averages of the independent gains of the two closed classes of recurrent states are

$$g_3 = 0.3636g_1 + 0.6364g_2 = 0.3636(62) + 0.6364(100) = 86.1818$$
$$g_4 = 0.6182g_1 + 0.3818g_2 = 0.6182(62) + 0.3818(100) = 76.5091$$
$$g_5 = 0.7709g_1 + 0.2291g_2 = 0.7709(62) + 0.2291(100) = 70.7055.$$

Step 3 of PE:
The VDEs for the transients states are

$$g_3 + v_3 = 468.62 + 0.20v_1 + 0.35v_2 + 0.45v_3$$
$$g_4 + v_4 = 796.21 + 0.20v_1 + 0.30v_3 + 0.50v_4$$
$$g_5 + v_5 = 2042.15 + 0.08v_1 + 0.12v_4 + 0.80v_5.$$

After substituting the gains of the transient states obtained in step 2 of PE, and the relative values of the recurrent states obtained in step 1 of PE, the VDEs for the transient states are

$$86.1818 + v_3 = 468.62 + 0.20(0) + 0.35(0) + 0.45v_3$$
$$76.5091 + v_4 = 796.21 + 0.20(0) + 0.30(v_3) + 0.50v_4$$
$$70.7055 + v_5 = 2042.15 + 0.08(0) + 0.12v_4 + 0.80v_5.$$

Step 4 of PE:
The VDEs for the transient states are solved to obtain the relative values of the transient states: $v_3 = 695.3422$, $v_4 = 1856.6071$, and $v_5 = 10971.187$.

The solutions for the gain vector and the vector of relative values for the second policy are summarized below:

$$
{}^2g = \begin{matrix} 1 \\ 2 \\ 3 \\ 4 \\ 5 \end{matrix}\begin{bmatrix} 62 \\ 100 \\ 86.1818 \\ 76.5091 \\ 70.7055 \end{bmatrix}, \quad {}^2v = \begin{matrix} 1 \\ 2 \\ 3 \\ 4 \\ 5 \end{matrix}\begin{bmatrix} 0 \\ 0 \\ 695.3422 \\ 1856.6071 \\ 10971.187 \end{bmatrix}. \tag{5.56}
$$

Step 3. IM routine
For each state i, find the decision $k*$ that maximizes the gain test quantity

$$
\sum_{j=1}^{5} p_{ij}^k g_j = p_{i1}^k g_1 + p_{i2}^k g_2 + (p_{i3}^k g_3 + p_{i4}^k g_4 + p_{i5}^k g_5)
$$

by using the gains of the previous policy. Then $k*$ becomes the new decision in state i, so that $d_i = k*$, q_i^{k*} becomes q_i, and p_{ij}^{k*} becomes p_{ij}. The second IM routine is executed in Table 5.51.

Step 4. Stopping rule
Stop because the new policy, given by the vector ${}^3d \equiv {}^2d = [1 \quad 1 \quad 2 \quad 2 \quad 2]^T$, is identical to the previous policy. Therefore, this policy is optimal. The firm

TABLE 5.51

Second IM for Model of a Flexible Production System

State	Decision	Gain Test Quantity $\sum_{j=1}^{5} p_{ij}^k g_j$	Value Test Quantity $q_i^k + \sum_{j=1}^{5} p_{ij}^k v_j$
i	k		
1	1	$1(62) = 62 \leftarrow$	
2	1	$1(100) = 100 \leftarrow$	
	2	$1(70.7055) = 70.7055$	
3	1	$0.25(62) + 0.20(100) + 0.55(86.1818)$	
		$= 82.9000$	
	2	$0.20(62) + 0.35(100) + 0.45(86.1818)$	
		$= 86.1818 \leftarrow$	
4	1	$0.15(62) + 0.20(86.1818) + 0.65(76.5091) = 76.2673$	
	2	$0.20(62) + 0.30(86.1818) + 0.50(76.5091) = 76.5091 \leftarrow$	
5	1	70.7056	$1935.36 + (8506.8813) = 10442.241$
	2	70.7056	$2042.15 + (8999.7425) = 11041.892 \leftarrow$

will manufacture a consumer product at stages 1, 2, and 3, and train employees on software in state 2. The optimal policy generates an absorbing multichain MCR with two recurrent closed classes, absorbing states 1 and 2,
respectively, and a set of transient states {3, 4, 5}. The gain vector ^2g, and the
vector of relative values ^2v, for the optimal policy are shown in Equation
(5.56). The optimal decision vector $^2\mathbf{d}$, the associated transition probability
matri, $^2\mathbf{P}$, and the reward vector $^2\mathbf{q}$, are shown in Equation (5.55).

5.1.4.3 Linear Programming

Puterman [3] constructs the LP formulation for a multichain MDP from the
LP formulation in Section 5.1.2.3.4.1 for a unichain MDP by adding an additional set of variables, x_i^k, and an additional set of constraints which general-
izes the single normalizing constraint for a unichain MDP. Since a multichain
MDP may not have a unique gain g, the value of the objective function for
the LP is denoted by z instead of g. The LP formulation for a multichain MDP
with N states and K_i decisions in each state i is

$$\text{Maximize } z = \sum_{i=1}^{N} \sum_{k=1}^{K_i} q_i^k \, y_i^k$$

subject to

$$\sum_{k=1}^{K_i} y_j^k - \sum_{i=1}^{N} \sum_{k=1}^{K_i} y_i^k \, p_{ij}^k = 0, \quad \text{for } j = 1, 2, ..., N.$$

$$\sum_{k=1}^{K_i} y_j^k + \sum_{k=1}^{K_i} x_j^k - \sum_{i=1}^{N} \sum_{k=1}^{K_i} x_i^k \, p_{ij}^k = b_j, \quad \text{for } j = 1, 2, ..., N,$$

where each constant $b_j > 0$ and $\sum_{j=1}^{N} b_j = 1$ and $y_i^k \geq 0$, $x_i^k \geq 0$, for $i = 1, 2, ..., N$,

and $k = 1, 2, 3, ..., K_i$. At least one of the equations in the first set of constraints
is redundant.

The optimal stationary policy generated by the solution of the LP for a unichain MDP is deterministic. A deterministic policy is one for which

$$P(\text{decision} = k | \text{state} = i) = \begin{cases} 1 & \text{for a single decision } k \text{ in state } i \\ 0 & \text{for all other decisions in state } i. \end{cases}$$

The optimal LP solution for a multichain MDP may generate a deterministic
stationary policy or a randomized stationary policy. In a randomized policy,

$P(\text{decision} = k|\text{state} = i)$ is a conditional probability distribution for the decision to be made in state i, so that $0 \le P(\text{decision} = k|\text{state} = i) \le 1$. A randomized stationary policy generated by a feasible solution to the LP formulation for a multichain MDP is characterized by

$$P(\text{decision} = k| \text{state} = i) = \begin{cases} y_i^k / \sum_{k=1}^{K_i} y_i^k & \text{for states } i \text{ in which } \sum_{k=1}^{K_i} y_i^k > 0 \\ x_i^k / \sum_{k=1}^{K_i} x_i^k & \text{for states } i \text{ in which } \sum_{k=1}^{K_i} x_i^k > 0. \end{cases}$$

In the final simplex tableau, under a deterministic policy, the dual variable associated with each equation in the second set of constraints is equal to the gain in the corresponding state. That is, for an N-state MDP, under a deterministic policy, the dual variable associated with constraint equation i, for $i > N$, is equal to the gain in state $(i - N)$.

5.1.4.3.1 LP Formulation of Multichain Model of a Flexible Production System

The multichain MDP model of a flexible production system constructed in Section 5.1.4.1 will be formulated as an LP. Table 5.49 from Section 5.1.4.1 is repeated in Table 5.52, augmented by two right-hand side columns of LP decision variables.

In this example $N = 5$ states, $K_1 = 1$ decision in state 1, and $K_2 = K_3 = K_4 = K_5 = 2$ decisions in every other state.

TABLE 5.52

Data for Multichain Model of a Flexible Production System

State i	Operation	Decision k	p_{i1}^k	p_{i2}^k	p_{i3}^k	p_{i4}^k	p_{i5}^k	Reward q_i^k	LP y_i^k	Variable x_i^k
1	Scrap	1	1	0	0	0	0	62	y_1^1	x_1^1
2	Training	1 = S	0	1	0	0	0	100	y_2^1	x_2^1
		2 = H	0	0	0	0	1	22.58	y_2^2	x_2^2
3	Stage 3	1 = I	0.25	0.20	0.55	0	0	423.54	y_3^1	x_3^1
		2 = C	0.20	0.35	0.45	0	0	468.62	y_3^2	x_3^2
4	Stage 2	1 = I	0.15	0	0.20	0.65	0	847.09	y_4^1	x_4^1
		2 = C	0.2	0	0.30	0.50	0	796.21	y_4^2	x_4^2
5	Stage 1	1 = I	0.10	0	0	0.15	0.75	1935.36	y_5^1	x_5^1
		2 = C	0.08	0	0	0.12	0.80	2042.15	y_5^2	x_5^2

Objective Function

The objective function for the LP is

$$
\text{Maximize } z = \sum_{k=1}^{1} q_1^k\, y_1^k + \sum_{i=2}^{5} \sum_{k=1}^{2} q_i^k\, y_i^k
$$

$$
= \sum_{k=1}^{1} q_1^k\, y_1^k + \sum_{k=1}^{2} q_2^k\, y_2^k + \sum_{k=1}^{2} q_3^k\, y_3^k + \sum_{k=1}^{2} q_4^k\, y_4^k + \sum_{k=1}^{2} q_5^k\, y_5^k
$$

$$
= q_1^1 y_1^1 + (q_2^1 y_2^1 + q_2^2 y_2^2) + (q_3^1 y_3^1 + q_3^2 y_3^2) + (q_4^1 y_4^1 + q_4^2 y_4^2) + (q_5^1 y_5^1 + q_5^2 y_5^2)
$$

$$
= 62y_1^1 + (100y_2^1 + 22.58y_2^2) + (423.54y_3^1 + 468.62y_3^2)
$$
$$
+ (847.09y_4^1 + 796.21y_4^2) + (1935.36y_5^1 + 2042.15y_5^2).
$$

Form the first set of constraints in terms of the variable y_i^k.
State 1 has $K_1 = 1$ decision. In the first set of constraints, the equation associated with a transition to state $j = 1$ is

$$
\sum_{k=1}^{1} y_1^k - \left(\sum_{k=1}^{1} y_1^k p_{11}^k + \sum_{i=2}^{5} \sum_{k=1}^{2} y_i^k\, p_{i1}^k \right) = 0
$$

$$
\sum_{k=1}^{1} y_1^k - \left(\sum_{k=1}^{1} y_1^k p_{11}^k + \sum_{k=1}^{2} y_2^k p_{21}^k + \sum_{k=1}^{2} y_3^k p_{31}^k + \sum_{k=1}^{2} y_4^k p_{41}^k + \sum_{k=1}^{2} y_5^k p_{51}^k \right) = 0
$$

$$
y_1^1 - (y_1^1 p_{11}^1) - (y_2^1 p_{21}^1 + y_2^2 p_{21}^2) - (y_3^1 p_{31}^1 + y_3^2 p_{31}^2)
$$
$$
- (y_4^1 p_{41}^1 + y_4^2 p_{41}^2) - (y_5^1 p_{51}^1 + y_5^2 p_{51}^2) = 0
$$

$$
y_1^1 - (1 y_1^1) - (0 y_2^1 + 0 y_2^2) - (0.25 y_3^1 + 0.20 y_3^2)
$$
$$
- (0.15 y_4^1 + 0.20 y_4^2) - (0.10 y_5^1 + 0.08 y_5^2) = 0.
$$

State 2 has $K_2 = 2$ possible decisions. In the first set of constraints, the equation associated with a transition to state $j = 2$ is

$$
\sum_{k=1}^{2} y_2^k - \left(\sum_{k=1}^{1} y_1^k p_{12}^k + \sum_{i=2}^{5} \sum_{k=1}^{2} y_i^k\, p_{i2}^k \right) = 0
$$

$$
\sum_{k=1}^{2} y_2^k - \left(\sum_{k=1}^{1} y_1^k p_{12}^k + \sum_{k=1}^{2} y_2^k p_{22}^k + \sum_{k=1}^{2} y_3^k p_{32}^k + \sum_{k=1}^{2} y_4^k p_{42}^k + \sum_{k=1}^{2} y_5^k p_{52}^k \right) = 0
$$

$$
(y_2^1 + y_2^2) - (y_1^1 p_{12}^1) - (y_2^1 p_{22}^1 + y_2^2 p_{22}^2) - (y_3^1 p_{32}^1 + y_3^2 p_{32}^2)
$$
$$
- (y_4^1 p_{42}^1 + y_4^2 p_{42}^2) - (y_5^1 p_{52}^1 + y_5^2 p_{52}^2) = 0
$$

$$(y_2^1 + y_2^2) - (0y_1^1) - (1y_2^1 + 0y_2^2) - (0.20y_3^1 + 0.35y_3^2)$$
$$- (0y_4^1 + 0y_4^2) - (0y_5^1 + 0y_5^2) = 0.$$

State 3 has $K_3 = 2$ possible decisions. In the first set of constraints, the equation associated with a transition to state $j = 3$ is

$$\sum_{k=1}^{2} y_3^k - \left(\sum_{k=1}^{1} y_1^k p_{13}^k + \sum_{i=2}^{5} \sum_{k=1}^{2} y_i^k \ p_{i3}^k \right) = 0$$

$$\sum_{k=1}^{2} y_3^k - \left(\sum_{k=1}^{1} y_1^k p_{13}^k + \sum_{k=1}^{2} y_2^k p_{23}^k + \sum_{k=1}^{2} y_3^k p_{33}^k + \sum_{k=1}^{2} y_4^k p_{43}^k + \sum_{k=1}^{2} y_5^k p_{53}^k \right) = 0$$

$$(y_3^1 + y_3^2) - (y_1^1 p_{13}^1) - (y_2^1 p_{23}^1 + y_2^2 p_{23}^2) - (y_3^1 p_{33}^1 + y_3^2 p_{33}^2)$$
$$- (y_4^1 p_{43}^1 + y_4^2 p_{43}^2) - (y_5^1 p_{53}^1 + y_5^2 p_{53}^2) = 0$$

$$(y_3^1 + y_3^2) - (0y_1^1) - (0y_2^1 + 0y_2^2) - (0.55y_3^1 + 0.45y_3^2)$$
$$- (0.20y_4^1 + 0.30y_4^2) - (0y_5^1 + 0y_5^2) = 0.$$

State 4 has $K_4 = 2$ possible decisions. In the first set of constraints, the equation associated with a transition to state $j = 4$ is

$$\sum_{k=1}^{2} y_4^k - \left(\sum_{k=1}^{1} y_1^k p_{14}^k + \sum_{i=2}^{5} \sum_{k=1}^{2} y_i^k \ p_{i4}^k \right) = 0$$

$$\sum_{k=1}^{2} y_4^k - \left(\sum_{k=1}^{1} y_1^k p_{14}^k + \sum_{k=1}^{2} y_2^k p_{24}^k + \sum_{k=1}^{2} y_3^k p_{34}^k + \sum_{k=1}^{2} y_4^k p_{44}^k + \sum_{k=1}^{2} y_5^k p_{54}^k \right) = 0$$

$$(y_4^1 + y_4^2) - (y_1^1 p_{14}^1) - (y_2^1 p_{24}^1 + y_2^2 p_{24}^2) - (y_3^1 p_{34}^1 + y_3^2 p_{34}^2)$$
$$- (y_4^1 p_{44}^1 + y_4^2 p_{44}^2) - (y_5^1 p_{54}^1 + y_5^2 p_{54}^2) = 0$$

$$(y_4^1 + y_4^2) - (0y_1^1) - (0y_2^1 + 0y_2^2) - (0y_3^1 + 0y_3^2)$$
$$- (0.65y_4^1 + 0.50y_4^2) - (0.15y_5^1 + 0.12y_5^2) = 0.$$

State 5 has $K_5 = 2$ possible decisions. In the first set of constraints, the equation associated with a transition to state $j = 5$ is

$$\sum_{k=1}^{2} y_5^k - \left(\sum_{k=1}^{1} y_1^k p_{15}^k + \sum_{i=2}^{5} \sum_{k=1}^{2} y_i^k \ p_{i5}^k \right) = 0$$

$$\sum_{k=1}^{2} y_5^k - \left(\sum_{k=1}^{1} y_1^k p_{15}^k + \sum_{k=1}^{2} y_2^k p_{25}^k + \sum_{k=1}^{2} y_3^k p_{35}^k + \sum_{k=1}^{2} y_4^k p_{45}^k + \sum_{k=1}^{2} y_5^k p_{55}^k \right) = 0$$

$$(y_5^1 + y_5^2) - (y_1^1 p_{15}^1) - (y_2^1 p_{25}^1 + y_2^2 p_{25}^2) - (y_3^1 p_{35}^1 + y_3^2 p_{35}^2)$$
$$- (y_4^1 p_{45}^1 + y_4^2 p_{45}^2) - (y_5^1 p_{55}^1 + y_5^2 p_{55}^2) = 0$$

$$(y_5^1 + y_5^2) - (0y_1^1) - (0y_2^1 + 1y_2^2) - (0y_3^1 + 0y_3^2)$$
$$- (0y_4^1 + 0y_4^2) - (0.75y_5^1 + 0.80y_5^2) = 0.$$

Form the second set of constraints in terms of the variable x_i^k by setting the constants $b_1 = b_2 = b_3 = b_4 = b_5 = 0.2$.

State 1 has $K_1 = 1$ decision. In the second set of constraints, the equation associated with a transition to state $j = 1$ is

$$\sum_{k=1}^{1} y_1^k + \sum_{k=1}^{1} x_1^k - \left(\sum_{k=1}^{1} x_1^k p_{11}^k + \sum_{i=2}^{5} \sum_{k=1}^{2} x_i^k \, p_{i1}^k \right) = b_1 = 0.2$$

$$y_1^1 + x_1^1 - \left(\sum_{k=1}^{1} x_1^k p_{11}^k + \sum_{k=1}^{2} x_2^k p_{21}^k + \sum_{k=1}^{2} x_3^k p_{31}^k + \sum_{k=1}^{2} x_4^k p_{41}^k + \sum_{k=1}^{2} x_5^k p_{51}^k \right) = 0.2$$

$$y_1^1 + x_1^1 - (x_1^1 p_{11}^1) - (x_2^1 p_{21}^1 + x_2^2 p_{21}^2) - (x_3^1 p_{31}^1 + x_3^2 p_{31}^2)$$
$$- (x_4^1 p_{41}^1 + x_4^2 p_{41}^2) - (x_5^1 p_{51}^1 + x_5^2 p_{51}^2) = 0.2$$

$$y_1^1 + x_1^1 - (1x_1^1) - (0x_2^1 + 0x_2^2) - (0.25x_3^1 + 0.20x_3^2)$$
$$- (0.15x_4^1 + 0.20x_4^2) - (0.10x_5^1 + 0.08x_5^2) = 0.2.$$

State 2 has $K_2 = 2$ possible decisions. In the second set of constraints, the equation associated with a transition to state $j = 2$ is

$$\sum_{k=1}^{2} y_2^k + \sum_{k=1}^{2} x_2^k - \left(\sum_{k=1}^{1} x_1^k p_{11}^k + \sum_{i=2}^{5} \sum_{k=1}^{2} x_i^k \, p_{i2}^k \right) = b_2 = 0.2$$

$$(y_2^1 + y_2^2) + (x_2^1 + x_2^2) - \left(\sum_{k=1}^{1} x_1^k p_{12}^k + \sum_{k=1}^{2} x_2^k p_{22}^k + \sum_{k=1}^{2} x_3^k p_{32}^k + \sum_{k=1}^{2} x_4^k p_{42}^k + \sum_{k=1}^{2} x_5^k p_{52}^k \right) = 0.2$$

$$(y_2^1 + y_2^2) + (x_2^1 + x_2^2) - (x_1^1 p_{12}^1) - (x_2^1 p_{22}^1 + x_2^2 p_{22}^2) - (x_3^1 p_{32}^1 + x_3^2 p_{32}^2)$$
$$- (x_4^1 p_{42}^1 + x_4^2 p_{42}^2) - (x_5^1 p_{52}^1 + x_5^2 p_{52}^2) = 0.2$$

$$(y_2^1 + y_2^2) + (x_2^1 + x_2^2) - (0x_1^1) - (1x_2^1 + 0x_2^2) - (0.20x_3^1 + 0.35x_3^2)$$
$$- (0x_4^1 + 0x_4^2) - (0x_5^1 + 0x_5^2) = 0.2.$$

State 3 has $K_3 = 2$ possible decisions. In the second set of constraints, the equation associated with a transition to state $j = 3$ is

$$\sum_{k=1}^{2} y_3^k + \sum_{k=1}^{2} x_3^k - \left(\sum_{k=1}^{1} x_1^k p_{11}^k + \sum_{i=2}^{5} \sum_{k=1}^{2} x_i^k \ p_{i3}^k \right) = b_3 = 0.2$$

$$(y_3^1 + y_3^2) + (x_3^1 + x_3^2) - \left(\sum_{k=1}^{1} x_1^k p_{13}^k + \sum_{k=1}^{2} x_2^k p_{23}^k + \sum_{k=1}^{2} x_3^k p_{33}^k + \sum_{k=1}^{2} x_4^k p_{43}^k + \sum_{k=1}^{2} x_5^k p_{53}^k \right) = 0.2$$

$$(y_3^1 + y_3^2) + (x_3^1 + x_3^2) - (x_1^1 p_{13}^1) - (x_3^1 p_{23}^1 + x_2^2 p_{23}^2) - (x_3^1 p_{33}^1 + x_3^2 p_{33}^2)$$
$$- (x_4^1 p_{43}^1 + x_4^2 p_{43}^2) - (x_5^1 p_{53}^1 + x_5^2 p_{53}^2) = 0.2$$

$$(y_3^1 + y_3^2) + (x_3^1 + x_3^2) - (0x_1^1) - (0x_2^1 + 0x_2^2) - (0.55x_3^1 + 0.45x_3^2)$$
$$- (0.20x_4^1 + 0.30x_4^2) - (0x_5^1 + 0x_5^2) = 0.2.$$

State 4 has $K_4 = 2$ possible decisions. In the second set of constraints, the equation associated with a transition to state $j = 4$ is

$$\sum_{k=1}^{2} y_4^k + \sum_{k=1}^{2} x_4^k - \left(\sum_{k=1}^{1} x_1^k p_{11}^k + \sum_{i=2}^{5} \sum_{k=1}^{2} x_i^k \ p_{i4}^k \right) = b_4 = 0.2$$

$$(y_4^1 + y_4^2) + (x_4^1 + x_4^2) - \left(\sum_{k=1}^{1} x_1^k p_{14}^k + \sum_{k=1}^{2} x_2^k p_{24}^k + \sum_{k=1}^{2} x_3^k p_{34}^k + \sum_{k=1}^{2} x_4^k p_{44}^k + \sum_{k=1}^{2} x_5^k p_{54}^k \right) = 0.2$$

$$(y_4^1 + y_4^2) + (x_4^1 + x_4^2) - (x_1^1 p_{14}^1) - (x_3^1 p_{24}^1 + x_2^2 p_{24}^2) - (x_3^1 p_{34}^1 + x_3^2 p_{34}^2)$$
$$- (x_4^1 p_{44}^1 + x_4^2 p_{44}^2) - (x_5^1 p_{54}^1 + x_5^2 p_{54}^2) = 0.2$$

$$(y_4^1 + y_4^2) + (x_4^1 + x_4^2) - (0x_1^1) - (0x_2^1 + 0x_2^2) - (0x_3^1 + 0x_3^2)$$
$$- (0.65x_4^1 + 0.50x_4^2) - (0.15x_5^1 + 0.12x_5^2) = 0.2.$$

State 5 has $K_5 = 2$ possible decisions. In the second set of constraints, the equation associated with a transition to state $j = 5$ is

$$\sum_{k=1}^{2} y_5^k + \sum_{k=1}^{2} x_5^k - \left(\sum_{k=1}^{1} x_1^k p_{11}^k + \sum_{i=2}^{5} \sum_{k=1}^{2} x_i^k \ p_{i5}^k \right) = b_5 = 0.2$$

$$(y_5^1 + y_5^2) + (x_5^1 + x_5^2) - \left(\sum_{k=1}^{1} x_1^k p_{15}^k + \sum_{k=1}^{2} x_2^k p_{25}^k + \sum_{k=1}^{2} x_3^k p_{35}^k + \sum_{k=1}^{2} x_4^k p_{45}^k + \sum_{k=1}^{2} x_5^k p_{55}^k \right) = 0.2$$

$$(y_5^1 + y_5^2) + (x_5^1 + x_5^2) - (x_1^1 p_{15}^1) - (x_3^1 p_{25}^1 + x_2^2 p_{25}^2) - (x_3^1 p_{35}^1 + x_3^2 p_{35}^2)$$
$$- (x_4^1 p_{45}^1 + x_4^2 p_{45}^2) - (x_5^1 p_{55}^1 + x_5^2 p_{55}^2) = 0.2$$

$$(y_5^1 + y_5^2) + (x_5^1 + x_5^2) - (0x_1^1) - (0x_2^1 + 1x_2^2) - (0x_3^1 + 0x_3^2)$$
$$- (0x_4^1 + 0x_4^2) - (0.75x_5^1 + 0.80x_5^2) = 0.2.$$

The complete LP formulation for the multichain MDP model of a flexible production system is

$$\text{Maximize } z = 62y_1^1 + (100y_2^1 + 22.58y_2^1) + (423.54y_3^1 + 468.62y_3^2)$$
$$+ (847.09y_4^1 + 796.21y_4^2) + (1935.36y_5^1 + 2042.15y_5^2)$$

subject to

(1) $-(0.25y_3^1 + 0.20y_3^2) - (0.15y_4^1 + 0.20y_4^2) - (0.10y_5^1 + 0.08y_5^2) = 0$

(2) $(y_2^2) - (0.20y_3^1 + 0.35y_3^2) = 0$

(3) $(0.45y_3^1 + 0.55y_3^2) - (0.20y_4^1 + 0.30y_4^2) = 0$

(4) $(0.35y_4^1 + 0.50y_4^2) - (0.15y_5^1 + 0.12y_5^2) = 0$

(5) $-(y_2^2) + (0.25y_5^1 + 0.20y_5^2) = 0$

(6) $(y_1^1) - (0.25x_3^1 + 0.20x_3^2) - (0.15x_4^1 + 0.20x_4^2) - (0.10x_5^1 + 0.08x_5^2) = 0.2$

(7) $(y_2^1 + y_2^2) + (x_2^2) - (0.20x_3^1 + 0.35x_3^2) = 0.2$

(8) $(y_3^1 + y_3^2) + (0.45x_3^1 + 0.55x_3^2) - (0.20x_4^1 + 0.30x_4^2) = 0.2$

(9) $(y_4^1 + y_4^2) + (0.35x_4^1 + 0.50x_4^2) - (0.15x_5^1 + 0.12x_5^2) = 0.2$

(10) $(y_5^1 + y_5^2) - (x_2^2) + (0.25x_5^1 + 0.20x_5^2) = 0.2$

$y_1^1 \geq 0, \ y_1^2 \geq 0, \ y_2^1 \geq 0, \ y_2^2 \geq 0, \ y_3^1 \geq 0, \ y_3^2 \geq 0, \ y_4^1 \geq 0, \ y_4^2 \geq 0, \ y_5^1 \geq 0, \ y_5^2 \geq 0.$

$x_1^1 \geq 0, \ x_1^2 \geq 0, \ x_2^1 \geq 0, \ x_2^2 \geq 0, \ x_3^1 \geq 0, \ x_3^2 \geq 0, \ x_4^1 \geq 0, \ x_4^2 \geq 0, \ x_5^1 \geq 0, \ x_5^2 \geq 0.$

5.1.4.3.2 Solution by LP of the Multichain MDP Model of a Flexible Production System

The LP formulation of the multichain MDP model of a flexible production system is solved on a personal computer using LP software to find an optimal policy. The output of the LP software is summarized in Table 5.53.

The objective function value is 79.0793. The optimal policy, given by the vector $\mathbf{d} = \begin{bmatrix} 1 & 1 & 2 & 2 & 2 \end{bmatrix}^T$, is deterministic, and was obtained by PI in Equation (5.55) of Section 5.1.4.2.3. Observe that the gains in states 1, 2, 3, 4, and 5, respectively, obtained by PI and shown in Equation (5.56), are equal to the dual variables associated with constraints 6, 7, 8, 9, and 10, respectively, shown in Table 5.53 of the LP solution. For the five-state MDP, the dual variable associated with constraint equation i, for $i > 5$, is equal to the gain in state $(i-5)$. For example, the dual variable associated with constraint equation 9 is equal to $g_4 = 76.5091$, the gain in state $(9-5) = 4$.

5.1.4.4 A Multichain MDP Model of Machine Maintenance

Consider the unichain MCR model of machine maintenance introduced as a Markov chain in Section 1.10.1.2.2.1, and enlarged to produce an MCR in Section 4.2.4.1.3. The four states of the machine are listed in Table 1.9 of Section 1.10.1.2.1.2. The unichain MCR model will be expanded to create a multichain MDP model by allowing choices among the four alternative maintenance actions described in Table 1.10 of Section 1.10.1.2.2.1. The maintenance policy followed for the unichain MCR model in Section 4.2.4.2.3 was $d = [2 \quad 1 \quad 1 \quad 1]^T$, which calls for overhauling the machine in state 1, and doing nothing in states 2, 3, and 4. (This is called the modified maintenance policy shown in Equation (4.117) of Sections 1.10.1.2.2.1 and 4.2.4.1.3.)

Suppose that the engineer responsible for maintaining the machine has determined that the decisions listed in Table 5.54 are feasible in the different states.

The daily revenue earned from production in every state is given in Table 4.8 of Section 4.2.4.1.3. The daily maintenance costs in every state are summarized in Table 4.9 of Section 4.2.4.1.3. The daily rewards for the decisions that are feasible in the different states are indicated in Table 5.55.

TABLE 5.53

LP Solution of Multichain MDP Model of a Flexible Production System

i	k	y_i^k	x_i^k			
1	1	0.5505	0	Objective	Function	Value = 79.0793
2	1	0.4495	0	Row		Dual Variable
	2	0	0	Constraint 6		$62 = g_1$
3	1	0	0	Constraint 7		$100 = g_2$
	2	0	0.7127	Constraint 8		$86.1818 = g_3$
4	1	0	0	Constraint 9		$76.5091 = g_4$
	2	0	0.6400	Constraint 10		$70.7055 = g_5$
5	1	0	0			
	2	0	1			

TABLE 5.54

Decisions Feasible in the Different States

State	Description	Feasible Decision
1	Not Working (NW)	Overhaul (OV) Repair (RP) Replace (RL)
2	Working, with a Major Defect (WM)	Do Nothing (DN) Repair (RP)
3	Working, with a Minor Defect (Wm)	Do Nothing (DN) Overhaul (OV)
4	Working Properly (WP)	Do Nothing (DN) Overhaul (OV)

TABLE 5.55

Daily Rewards for the Decisions That Are Feasible in the Different States

State\Decision	Do Nothing	Overhaul	Repair	Replace
Not Working	Not Feasible	−$300	−$700	−$1,200
Working, with Major Defect	$200	Not Feasible	−$700	Not Feasible
Working, with Minor Defect	$500	−$300	Not Feasible	Not Feasible
Working Properly	$1,000	−$300	Not Feasible	Not Feasible

TABLE 5.56

The 24 Stationary Policies for Machine Maintenance

Policy	State 1	State 2	State 3	State 4	Structure
^1d	OV	DN	OV	DN	U
^2d	OV	DN	OV	OV	M
^3d	OV	RP	OV	DN	I
^4d	OV	RP	OV	OV	U
^5d	RP	DN	OV	DN	I
^6d	RP	DN	OV	OV	U
^7d	RP	RP	OV	DN	I
^8d	RP	RP	OV	OV	U
^9d	RL	DN	OV	DN	I
^{10}d	RL	DN	OV	OV	U
^{11}d	RL	RP	OV	DN	I
^{12}d	RL	RP	OV	OV	U
^{13}d	OV	DN	DN	DN	U
^{14}d	OV	DN	DN	OV	M
^{15}d	OV	RP	DN	DN	I
^{16}d	OV	RP	DN	OV	U
^{17}d	RP	DN	DN	DN	U
^{18}d	RP	DN	DN	OV	M
^{19}d	RP	RP	DN	DN	I
^{20}d	RP	RP	DN	OV	U
^{21}d	RL	DN	DN	DN	I
^{22}d	RL	DN	DN	OV	U
^{23}d	RL	RP	DN	DN	I
^{24}d	RL	RP	DN	OV	U

Recall that under a stationary policy the same decision is made each time that the process returns to a particular state. With three feasible decisions in state 1, and two decisions in states 2, 3, and 4, the number of stationary policies is equal to (3)(2)(2)(2) = 24. The 24 stationary policies, labeled $^1\mathbf{d}$ through $^{24}\mathbf{d}$, are enumerated in Table 5.56. The right most column indicates whether the associated Markov chain is irreducible (I), unichain (U), or multichain (M).

TABLE 5.57

Data for Multichain Model of Machine Maintenance

State i	Decision k	Transition Probability				Reward q_i^k	LP y_i^k	Variable x_i^k
		p_{i1}^k	p_{i2}^k	p_{i3}^k	p_{i4}^k			
1 = NW	2 = OV	0.2	0.8	0	0	−$300	y_1^2	x_1^2
	3 = RP	0.3	0	0.7	0	−$700	y_1^3	x_1^3
	4 = RL	0	0	0	1	−$1,200	y_1^4	x_1^4
2 = WM	1 = DN	0.6	0.4	0	0	$200	y_2^1	x_2^1
	3 = RP	0	0.3	0	0.7	−$700	y_2^3	x_2^3
3 = Wm	1 = DN	0.2	0.3	0.5	0	$500	y_3^1	x_3^1
	2 = OV	0	0	0.2	0.8	−$300	y_3^2	x_3^2
4 = WP	1 = DN	0.3	0.2	0.1	0.4	$1,000	y_4^1	x_4^1
	2 = OV	0	0	0	1	−$300	y_4^2	x_4^2

Policies ^{2}**d**, ^{14}**d**, and ^{18}**d** produce recurrent multichains, each with two recurrent chains and no transient states. Table 5.57 contains the data for the four-state multichain MDP model of machine maintenance. The variables for an LP formulation appear in the two right-most columns.

5.1.4.4.1 *LP Formulation of Multichain Model of Machine Maintenance*

The multichain MDP model of machine maintenance will be formulated as an LP. In this example $N = 4$ states, $K_1 = 3$ decisions in state 1, and $K_2 = K_3 = K_4 = 2$ decisions in states 2, 3, and 4.

Objective Function

The objective function for the LP is

$$\text{Maximize } z = (q_1^2 y_1^2 + q_1^3 y_1^3 + q_1^4 y_1^4) + (q_2^1 y_2^1 + q_2^3 y_2^3) + (q_3^1 y_3^1 + q_3^2 y_3^2) + (q_4^1 y_4^1 + q_4^2 y_4^2)$$

$$= (-300y_1^2 - 700y_1^3 - 1200y_1^4) + (200y_2^1 - 700y_2^3)$$
$$+ (500y_3^1 - 300y_3^2) + (1000y_4^1 - 300y_4^2).$$

Form the first set of constraints.

State 1 has $K_1 = 3$ decisions. In the first set of constraints, the equation associated with a transition to state $j = 1$ is

$$(y_1^2 + y_1^3 + y_1^4) - (y_1^2 p_{11}^2 + y_1^3 p_{11}^3 + y_1^4 p_{11}^4) - (y_2^1 p_{21}^1 + y_2^3 p_{21}^3)$$
$$- (y_3^1 p_{31}^1 + y_3^2 p_{31}^2) - (y_4^1 p_{41}^1 + y_4^2 p_{41}^2) = 0$$

$$(y_1^2 + y_1^3 + y_1^4) - (0.2y_1^2 + 0.3y_1^3 + 0y_1^4) - (0.6y_2^1 + 0y_2^3) - (0.2y_3^1 + 0y_3^2) - (0.3y_4^1 + 0y_4^2) = 0.$$

State 2 has $K_2 = 2$ decisions. In the first set of constraints, the equation associated with a transition to state $j = 2$ is

$$(y_2^1 + y_2^3) - (y_1^2 p_{12}^2 + y_1^3 p_{12}^3 + y_1^4 p_{12}^4) - (y_2^1 p_{22}^1 + y_2^3 p_{22}^3)$$
$$- (y_3^1 p_{32}^1 + y_3^2 p_{32}^2) - (y_4^1 p_{42}^1 + y_4^2 p_{42}^2) = 0$$

$$(y_2^1 + y_2^3) - (0.8y_1^2 + 0y_1^3 + 0y_1^4) - (0.4y_2^1 + 0.3y_2^3) - (0.3y_3^1 + 0y_3^2) - (0.2y_4^1 + 0y_4^2) = 0.$$

State 3 has $K_3 = 2$ decisions. In the first set of constraints, the equation associated with a transition to state $j = 3$ is

$$(y_3^1 + y_3^2) - (y_1^2 p_{13}^2 + y_1^3 p_{13}^3 + y_1^4 p_{13}^4) - (y_2^1 p_{23}^1 + y_2^3 p_{23}^3)$$
$$- (y_3^1 p_{33}^1 + y_3^2 p_{33}^2) - (y_4^1 p_{43}^1 + y_4^2 p_{43}^2) = 0$$

$$(y_3^1 + y_3^2) - (0y_1^2 + 0.7y_1^3 + 0y_1^4) - (0y_2^1 + 0y_2^3) - (0.5y_3^1 + 0.2y_3^2) - (0.1y_4^1 + 0y_4^2) = 0.$$

State 4 has $K_4 = 2$ decisions. In the first set of constraints, the equation associated with a transition to state $j = 4$ is

$$(y_4^1 + y_4^2) - (y_1^2 p_{14}^2 + y_1^3 p_{14}^3 + y_1^4 p_{14}^4) - (y_2^1 p_{24}^1 + y_2^3 p_{24}^3)$$
$$- (y_3^1 p_{34}^1 + y_3^2 p_{34}^2) - (y_4^1 p_{44}^1 + y_4^2 p_{44}^2) = 0$$

$$(y_4^1 + y_4^2) - (0y_1^2 + 0y_1^3 + y_1^4) - (0y_2^1 + 0.7y_2^3) - (0y_3^1 + 0.8y_3^2) - (0.4y_4^1 + y_4^2) = 0.$$

Form the second set of constraints by setting $b_1 = b_2 = b_3 = b_4 = 0.25$.
State 1 has $K_1 = 3$ decisions. In the second set of constraints, the equation associated with a transition to state $j = 1$ is

$$(y_1^2 + y_1^3 + y_1^4) + (x_1^2 + x_1^3 + x_1^4) - (x_1^2 p_{11}^2 + x_1^3 p_{11}^3 + x_1^4 p_{11}^4) - (x_2^1 p_{21}^1 + x_2^3 p_{21}^3)$$
$$- (x_3^1 p_{31}^1 + x_3^2 p_{31}^2) - (x_4^1 p_{41}^1 + x_4^2 p_{41}^2) = 0.25$$

$$(y_1^2 + y_1^3 + y_1^4) + (x_1^2 + x_1^3 + x_1^4) - (0.2x_1^2 + 0.3x_1^3 + 0x_1^4) - (0.6x_2^1 + 0x_2^3)$$
$$- (0.2x_3^1 + 0x_3^2) - (0.3x_4^1 + 0x_4^2) = 0.25.$$

State 2 has $K_2 = 2$ decisions. In the second set of constraints, the equation associated with a transition to state $j = 2$ is

$$(y_2^1 + y_2^3) + (x_2^1 + x_2^3) - (x_1^2 p_{12}^2 + x_1^3 p_{12}^3 + x_1^4 p_{12}^4) - (x_2^1 p_{22}^1 + x_2^3 p_{22}^3)$$
$$- (x_3^1 p_{32}^1 + x_3^2 p_{32}^2) - (x_4^1 p_{42}^1 + x_4^2 p_{42}^2) = 0.25$$

$$(y_2^1 + y_2^3) + (x_2^1 + x_2^3) - (0.8x_1^2 + 0x_1^3 + 0x_1^4) - (0.4x_2^1 + 0.3x_2^3)$$
$$- (0.3x_3^1 + 0x_3^2) - (0x_4^1 + 0.2x_4^2) = 0.25.$$

State 3 has $K_3 = 2$ decisions. In the second set of constraints, the equation associated with a transition to state $j = 3$ is

$$(y_3^1 + y_3^2) + (x_3^1 + x_3^2) - (x_1^2 p_{13}^2 + x_1^3 p_{13}^3 + x_1^4 p_{13}^4) - (x_2^1 p_{23}^1 + x_2^3 p_{23}^3)$$
$$- (x_3^1 p_{33}^1 + x_3^2 p_{33}^2) - (x_4^1 p_{43}^1 + x_4^2 p_{43}^2) = 0.25$$

$$(y_3^1 + y_3^2) + (x_3^1 + x_3^2) - (0x_1^2 + 0.7x_1^3 + 0x_1^4) - (0x_2^1 + 0x_2^3)$$
$$- (0.5x_3^1 + 0.2x_3^2) - (0.1x_4^1 + 0x_4^2) = 0.25.$$

State 4 has $K_4 = 2$ decisions. In the second set of constraints, the equation associated with a transition to state $j = 4$ is

$$(y_4^1 + y_4^2) + (x_4^1 + x_4^2) - (x_1^2 p_{14}^2 + x_1^3 p_{14}^3 + x_1^4 p_{14}^4) - (x_2^1 p_{24}^1 + x_2^3 p_{24}^3)$$
$$- (x_3^1 p_{34}^1 + x_3^2 p_{34}^2) - (x_4^1 p_{44}^1 + x_4^2 p_{44}^2) = 0.25$$

$$(y_4^1 + y_4^2) + (x_4^1 + x_4^2) - (0x_1^2 + 0x_1^3 + x_1^4) - (0x_2^1 + 0.7x_2^3)$$
$$- (0x_3^1 + 0.8x_3^2) - (0.4x_4^1 + x_4^2) = 0.25.$$

The complete LP formulation for the multichain MDP model of machine maintenance is

$$\text{Maximize } z = (-300y_1^2 - 700y_1^3 - 1200y_1^4) + (200y_2^1 - 700y_2^3)$$
$$+ (500y_3^1 - 300y_3^2) + (1000y_4^1 - 300y_4^4)$$

subject to

(1) $(0.8y_1^2 + 0.7y_1^3 + y_1^4) - 0.6y_2^1 - 0.2y_3^1 - 0.3y_4^1 = 0$

(2) $-(0.8y_1^2) + (0.6y_2^1 + 0.7y_2^3) - 0.3y_3^1 - (0.2y_4^1) = 0$

(3) $-(0.7y_1^3) + (0.5y_3^1) + (0.8y_3^2) - (0.1y_4^1) = 0$

(4) $-(y_1^4) - (0.7y_2^3) - (0.8y_3^2) + (0.6y_4^1) = 0$

(5) $(y_1^2 + y_1^3 + y_1^4) + (0.8x_1^2 + 0.7x_1^3 + x_1^4) - 0.6x_2^1 - 0.2x_3^1 - 0.3x_4^1 = 0.25$

(6) $(y_2^1 + y_2^3) - (0.8x_1^2) + (0.6x_2^1 + 0.7x_2^3) - (0.3x_3^1) - (0.2x_4^1) = 0.25$

(7) $(y_3^1 + y_3^2) - (0.7x_1^3) + (0.5x_3^1) + (0.8x_3^2) - (0.1x_4^1) = 0.25$

(8) $(y_4^1 + y_4^2) - (x_1^4) - (0.7x_2^3) - (0.8x_3^2) + (0.6x_4^1) = 0.25$

$$y_1^2 \geq 0,\ y_1^3 \geq 0,\ y_1^4 \geq 0,\ y_2^1 \geq 0,\ y_2^3 \geq 0,\ y_3^1 \geq 0,\ y_3^2 \geq 0,\ y_4^1 \geq 0,\ y_4^2 \geq 0.$$

$$x_1^2 \geq 0,\ x_1^3 \geq 0,\ x_1^4 \geq 0,\ x_2^1 \geq 0,\ x_2^3 \geq 0,\ x_3^1 \geq 0,\ x_3^2 \geq 0,\ x_4^1 \geq 0,\ x_4^2 \geq 0.$$

5.1.4.4.2 Solution by LP of the Multichain Model of Machine Maintenance

The LP formulation of the multichain MDP model of machine maintenance is solved on a personal computer using LP software to find an optimal policy. The output of the LP software is summarized in Table 5.58.

The optimal policy, given by the decision vector $^{17}\mathbf{d} = [3 \quad 1 \quad 1 \quad 1]^T$ in Table 5.56, is deterministic. The engineer will repair the machine when it is in state 1, not working, and will do nothing when the machine is in states 2, 3, and 4. The optimal policy generates a unichain MCR with one recurrent closed class consisting of states 1, 2, and 3. State 4 is transient. The unichain transition probability matrix $^{17}\mathbf{P}$, and the reward vector $^{17}\mathbf{q}$, generated by the optimal decision vector $^{17}\mathbf{d}$ are shown in Table 5.59.

Since the optimal policy has produced a unichain MCR, all states have the same gain. The gain in every state is $g_1 = g_2 = g_3 = g_4 = g = 45.1613$, equal to the optimal value of the LP objective function. The gain is also equal to the optimal value of the dual variables associated with constraints 5, 6, 7, and 8.

TABLE 5.58

LP Solution of Multichain Model of Machine Maintenance

i	k	y_i^k	x_i^k		
1	2	0	0		
	3	0.3226	0.2285	Objective	Function Value = 45.1613
	4	0	0	Constraint	Dual Variable
2	1	0.2258	0.1792	5	$45.1613 = g_1 = g$
	3	0	0	6	$45.1613 = g_2 = g$
3	1	0.4516	0	7	$45.1613 = g_3 = g$
	2	0	0	8	$45.1613 = g_4 = g$
4	1	0	0.4167		
	2	0	0		

TABLE 5.59

Unichain MCR Corresponding to Optimal Policy for Machine Maintenance

State, $X_{n-1} = i$	Decision, k	State	1	2	3	4		Reward
1, Not Working	3, Repair	1	0.3	0	0.7	0		$-\$700 = q_1$
2, Working, Major Defect	1, Do Nothing $^{17}P =$	2	0.6	0.4	0	0	$,\ ^{17}q =$	$\$200 = q_2$
3, Working, Minor Defect	1, Do Nothing	3	0.2	0.3	0.5	0		$\$500 = q_3$
4, Working Properly	1, Do Nothing	4	0.3	0.2	0.1	0.4		$\$1,000 = q_4$

5.2 A Discounted MDP

A discounted MDP is one for which a dollar earned one year from now has a present value of α dollars now, where $0 < \alpha < 1$. As Section 4.3 indicates, α is called a discount factor. As is true for a discounted MCR treated in Section 4.3, chain structure is not relevant when analyzing an MDP with discounting.

5.2.1 Value Iteration over a Finite Planning Horizon

An optimal policy for a discounted MDP is defined as one which maximizes the expected total discounted reward earned in every state. When the planning horizon is finite, an optimal policy can be found by using value iteration, which was applied to a discounted MCR in Section 4.3.2 [2, 3].

5.2.1.1 Value Iteration Equation

When value iteration is applied to a discounted MDP, the definition of $v_i(n)$ given in Section 4.3.2.1 for a discounted MCR is changed by adding the requirement that an optimal policy is followed from epoch n to the end of the planning horizon. Hence, $v_i(n)$ now denotes the value, at epoch n, of the expected total discounted reward earned during the $T-n$ periods from epoch n to epoch T, the end of the planning horizon, if the system is in state i at epoch n, and an optimal policy is followed. The value iteration equation (4.243) for a discounted MCR is modified to obtain the following value iteration equation in algebraic form for a discounted MDP:

$$\left. \begin{array}{l} v_i(n) = \max_k \left[q_i^k + \alpha \sum_{j=1}^{N} p_{ij}^k v_j(n+1) \right], \\[2ex] \text{for } n = 0, 1,..., T-1, \text{ and } i = 1, 2, ..., N, \text{ where } v_i(T) \text{ is specified for all states } i. \end{array} \right\} \quad (5.57)$$

5.2.1.2 Value Iteration for Discounted MDP Model of Monthly Sales

Consider the MDP model of monthly sales specified in Table 5.3 of Section 5.1.2.1, and solved by value iteration without discounting over a seven month horizon in Section 5.1.2.2.2. In this section, value iteration will be applied to a discounted MDP model of monthly sales over a seven month horizon. A discount factor of $\alpha = 0.9$ is chosen. To begin, the following salvage values are specified at the end of month 7.

$$v_1(7) = v_2(7) = v_3(7) = v_4(7) = 0. \tag{5.58}$$

For the four-state discounted MDP, the value iteration equations are

$$v_i(7) = 0, \quad \text{for } i = 1, 2, 3, 4 \tag{5.59}$$

$$\left. \begin{aligned} v_i(n) &= \max_k \left[q_i^k + \alpha \sum_{j=1}^{4} p_{ij}^k v_j(n+1) \right], \quad \text{for } n = 0, 1, ..., 6, \quad \text{and} \quad i = 1, 2, 3, 4 \\ &= \max_k \{ q_i^k + 0.9[p_{i1}^k v_1(n+1) + p_{i2}^k v_2(n+1) + p_{i3}^k v_3(n+1) + p_{i4}^k v_4(n+1)] \}, \\ &\quad \text{for } n = 0, 1, ..., 6. \end{aligned} \right\} \tag{5.60}$$

The calculations for month 6, denoted by $n = 6$, are indicated in Table 5.60.

$$\left. \begin{aligned} v_i(6) &= \max_k \{ q_i^k + \alpha[p_{i1}^k v_1(7) + p_{i2}^k v_2(7) + p_{i3}^k v_3(7) + p_{i4}^k v_4(7)] \}, \quad \text{for } i = 1, 2, 3, 4 \\ &= \max_k \{ q_i^k + 0.9[p_{i1}^k(0) + p_{i2}^k(0) + p_{i3}^k(0) + p_{i4}^k(0)] \}, \quad \text{for } i = 1, 2, 3, 4 \\ &= \max_k [q_i^k], \quad \text{for } i = 1, 2, 3, 4. \end{aligned} \right\}$$

$$\tag{5.61}$$

At the end of month 6, the optimal decision is to select the alternative which will maximize the expected total discounted reward in every state. Thus, the decision vector at the end of month 6 is $\mathbf{d(6)} = [3\ 2\ 2\ 1]^T$.

The calculations for month 5, denoted by $n = 5$, are indicated in Table 5.61.

$$\left. \begin{aligned} v_i(5) &= \max_k \{ q_i^k + \alpha[p_{i1}^k v_1(6) + p_{i2}^k v_2(6) + p_{i3}^k v_3(6) + p_{i4}^k v_4(6)] \}, \quad \text{for } i = 1, 2, 3, 4 \\ &= \max_k \{ q_i^k + 0.9[p_{i1}^k(-20) + p_{i2}^k(10) + p_{i3}^k(-5) + p_{i4}^k(35)] \}, \quad \text{for } i = 1, 2, 3, 4. \end{aligned} \right\}$$

$$\tag{5.62}$$

TABLE 5.60
Value Iteration for $n = 6$

$n=6$	q_i^1	q_i^2	q_i^3	Expected Total Discounted Reward	Decision k
i	$k=1$	$k=2$	$k=3$	$v_i(6) = \max_k[q_i^1, q_i^2, q_i^3]$, for $i = 1, 2, 3, 4$	
1	−30	−25	−20	$v_1(6) = \max[-30, -25, -20]$	3
2	5	10	—	$v_2(6) = \max[5, 10] = 10$	2
3	−10	−5	—	$v_3(6) = \max[-10, -5] = -5$	2
4	35	25	—	$v_4(6) = \max[35, 25] = 35$	1

TABLE 5.61

Value Iteration for $n = 5$

i k	$\begin{aligned} &q_i^k + a[p_{i1}^k v_1(6) + p_{i2}^k v_2(6) + p_{i3}^k v_3(6) \\ &+ p_{i4}^k v_4(6)] = q_i^k + 0.9[p_{i1}^k(-20) + p_{i2}^k(10) \\ &+ p_{i3}^k(-5) + p_{i4}^k(35)] \end{aligned}$	$v_i(5) =$ Expected Total Discounted Reward	Decision $d_i(5) = k$
1 1	$-30 + 0.9[0.15(-20) + 0.40(10) + 0.35(-5)$ $+ 0.10(35)] = -27.525$		
1 2	$-25 + 0.9[0.45(-20) + 0.05(10) + 0.20(-5)$ $+ 0.30(35)] = -24.1 \leftarrow$	$\max[-27.525,$ $-24.1, -28.55\,]$ $= -24.1 = v_1(5)$	$d_1(5) = 2$
1 3	$-20 + 0.9[0.60(-20) + 0.30(10) + 0.10(-5)$ $+ 0(35)] = -28.55$		
2 1	$5 + 0.9[0.25(-20) + 0.30(10) + 0.35(-5)$ $+ 0.10(35)] = 4.775$		
2 2	$10 + 0.9[0.30(-20) + 0.40(10) + 0.25(-5)$ $+ 0.05(35)] = 8.65 \leftarrow$	$\max[4.775, 8.65]$ $= 8.65 = v_2(5)$	$d_2(5) = 2$
3 1	$-10 + 0.9[0.05(-20) + 0.65(10) + 0.25(-5)$ $+ 0.05(35)] = -4.6$		
3 2	$-5 + 0.9[0.05(-20) + 0.25(10) + 0.50(-5)$ $+ 0.20(35)] = 0.4 \leftarrow$	$\max[-4.6, 0.4]$ $= 0.4 = v_3(5)$	$d_3(5) = 2$
4 1	$35 + 0.9[0.05(-20) + 0.20(10) + 0.40(-5)$ $+ 0.35(35)] = 45.125 \leftarrow$	$\max[45.125, 43.45]$ $= 45.125 = v_4(5)$	$d_4(5) = 1$
4 2	$25 + 0.9[0(-20) + 0.10(10) + 0.30(-5)$ $+ 0.60(35)] = 43.45$		

At the end of month 5, the optimal decision is to select the second alternative in states 1, 2, and 3, and the first alternative in state 4. Thus, the decision vector at the end of month 5 is $\mathbf{d}(5) = [2\ 2\ 2\ 1]^T$. The calculations for month 4, denoted by $n = 4$, are indicated in Table 5.62.

TABLE 5.62

Value Iteration for $n = 4$

i k	$q_i^k + \alpha[p_{i1}^k v_1(5) + p_{i2}^k v_2(5) + p_{i3}^k v_3(5)$ $+ p_{i4}^k v_4(5)] = q_i^k + 0.9[p_{i1}^k(-24.1)$ $+ p_{i2}^k(8.65) + p_{i3}^k(0.4) + p_{i4}^k(45.125)]$	$v_i(4)$ = Expected Total Discounted Reward	Decision $d_i(4) = k$
1 1	$-30 + 0.9[0.15(-24.1) + 0.40(8.65)$ $+ 0.35(0.4) + 0.10(45.125)] = -25.95225$		
1 2	$-25 + 0.9[0.45(-24.1) + 0.05(8.65)$ $+ 0.20(0.4) + 0.30(45.125)] = -22.1155 \leftarrow$	max $[-25.9523,$ $-22.1155, -30.6435]$ $= -22.1155 = v_1(4)$	$d_1(4) = 2$
1 3	$-20 + 0.9[0.60(-24.1) + 0.30(8.65)$ $+ 0.10(0.4) + 0(45.125)] = -30.6425$		
2 1	$5 + 0.9[0.25(-24.1) + 0.30(8.65)$ $+ 0.35(0.4) + 0.10(45.125)] = 6.1003$		
2 2	$10 + 0.9[0.30(-24.1) + 0.40(8.65)$ $+ 0.25(0.4) + 0.05(45.125)] = 8.7276 \leftarrow$	max $[6.1003, 8.7276]$ $= 8.7276 = v_2(4)$	$d_2(4) = 2$
3 1	$-10 + 0.9[0.05(-24.1) + 0.65(8.65)$ $+ 0.25(0.4) + 0.05(45.125)] = -3.9063$		
3 2	$-5 + 0.9[0.05(-24.1) + 0.25(8.65)$ $+ 0.50(0.4) + 0.20(45.125)] = 4.1643 \leftarrow$	max $[-3.9063,$ $4.1643] = 4.1643$ $= v_3(4)$	$d_3(4) = 2$
4 1	$35 + 0.9[0.05(-24.1) + 0.20(8.65)$ $+ 0.40(0.4) + 0.35(45.125)] = 49.8309$		
4 2	$25 + 0.9[0(-24.1) + 0.10(8.65) + 0.30(0.4)$ $+ 0.60(45.125)] = 50.2540 \leftarrow$	max $[49.8309, 50.2540]$ $= 50.2540 = v_4(4)$	$d_4(4) = 2$

$$v_i(4) = \max_k \{q_i^k + \alpha[p_{i1}^k v_1(5) + p_{i2}^k v_2(5) + p_{i3}^k v_3(5) + p_{i4}^k v_4(5)]\} \text{ for } i = 1, 2, 3, 4$$
$$= \max_k \{q_i^k + 0.9[p_{i1}^k(-24.1) + p_{i2}^k(8.65) + p_{i3}^k(0.4) + p_{i4}^k(45.125)]\}$$
$$\text{for } i = 1, 2, 3, 4$$

$$(5.63)$$

At the end of month 4, the optimal decision is to select the second alternative in every state. Thus, the decision vector at the end of month 4 is $\mathbf{d}(4) = [2\ 2\ 2\ 2]^{\mathrm{T}}$.

The calculations for month 3, denoted by $n = 3$, are indicated in Table 5.63.

$$
\begin{aligned}
v_i(3) &= \max_k \{q_i^k + \alpha[p_{i1}^k v_1(4) + p_{i2}^k v_2(4) + p_{i3}^k v_3(4) + p_{i4}^k v_4(4)]\} \text{ for } i = 1,2,3,4 \\
&= \max_k \{q_i^k + 0.9[p_{i1}^k(-22.1155) + p_{i2}^k(8.7276) + p_{i3}^k(4.1643) + p_{i4}^k(50.2540)]\} \\
&\quad \text{for } i = 1,2,3,4.
\end{aligned}
$$

$$(5.64)$$

TABLE 5.63

Value Iteration for $n = 3$

i k	$\begin{aligned} q_i^k + \alpha[p_{i1}^k v_1(4) + p_{i2}^k v_2(4) + p_{i3}^k v_3(4) \\ + p_{i4}^k v_4(4)] = q_i^k + 0.9[p_{i1}^k(-22.1155) \\ + p_{i2}^k(8.7276) + p_{i3}^k(4.1643) + p_{i4}^k(50.2540)] \end{aligned}$	$v_i(3) =$ Expected Total Discounted Reward	Decision $d_i(3) = k$
1 1	$-30 + 0.9[0.15(-22.1155) + 0.40(8.7276)$ $+ 0.35(4.1643) + 0.10(50.2540)] = -24.0090$		
1 2	$-25 + 0.9[0.45(-22.1155) + 0.05(8.7276)$ $+ 0.20(4.1643) + 0.30(50.2540)]$ $= -19.2459 \leftarrow$	max $[-24.0090,$ $-19.2459, -29.2111]$ $= -19.2459 = v_1(3)$	$d_1(3) = 2$
1 3	$-20 + 0.9[0.60(-22.1155) + 0.30(8.7276)$ $+ 0.10(4.1643) + 0(50.2540)] = -29.2111$		
2 1	$5 + 0.9[0.25(-22.1155) + 0.30(8.7276)$ $+ 0.35(4.1643) + 0.10(50.2540)] = 8.2151$		
2 2	$10 + 0.9[0.30(-22.1155) + 0.40(8.7276)$ $+ 0.25(4.1643) + 0.05(50.2540)] = 10.3691 \leftarrow$	max $[8.2151,$ $10.3691] = 10.3691 = v_2(3)$	$d_2(3) = 2$
3 1	$-10 + 0.9[0.05(-22.1155) + 0.65(8.7276)$ $+ 0.25(4.1643) + 0.05(50.2540)] = -2.6912$		
3 2	$-5 + 0.9[0.05(-22.1155) + 0.25(8.7276)$ $+ 0.50(4.1643) + 0.20(50.2540)] = 6.8882 \leftarrow$	max $[-2.6912,$ $6.8882] = 6.8882 = v_3(3)$	$d_3(3) = 2$
4 1	$35 + 0.9[0.05(-22.1155) + 0.20(8.7276)$ $+ 0.40(4.1643) + 0.35(50.2540)] = 52.9049$		
4 2	$25 + 0.9[0(-22.1155) + 0.10(8.7276)$ $+ 0.30(4.1643) + 0.60(50.2540)] = 54.0470 \leftarrow$	max $[52.9049,$ $54.0470] = 54.0470$ $= v_4(3)$	$d_4(3) = 2$

At the end of month 3, the optimal decision is to select the second alternative in every state. Thus, the decision vector at the end of month 3 is $\mathbf{d}(3) = [2\ 2\ 2\ 2]^T$.

The calculations for month 2, denoted by $n = 2$, are indicated in Table 5.64.

$$
\begin{aligned}
v_i(2) &= \max_k \{q_i^k + \alpha[p_{i1}^k v_1(3) + p_{i2}^k v_2(3) + p_{i3}^k v_3(3) + p_{i4}^k v_4(3)]\} \text{ for } i = 1, 2, 3, 4 \\
&= \max_k \{q_i^k + 0.9[p_{i1}^k(-19.2459) + p_{i2}^k(10.3691) + p_{i3}^k(6.8882) + p_{i4}^k(54.0470)]\} \\
&\quad \text{for } i = 1, 2, 3, 4.
\end{aligned}
$$

(5.65)

At the end of month 2, the optimal decision is to select the second alternative in every state. Thus, the policy at the end of month 2 is $\mathbf{d}(2) = [2\ 2\ 2\ 2]^T$.

The calculations for month 1, denoted by $n = 1$, are indicated in Table 5.65.

TABLE 5.64

Value Iteration for $n = 2$

i k	$q_i^k + \alpha[p_{i1}^k v_1(3) + p_{i2}^k v_2(3) + p_{i3}^k v_3(3)$ $+ p_{i4}^k v_4(3)] = q_i^k + 0.9[p_{i1}^k(-19.2459)$ $+ p_{i2}^k(10.3691) + p_{i3}^k(6.8882) + p_{i4}^k(54.0470)]$	$v_i(2)$ = Expected Total Discounted Reward	Decision $d_i(2) = k$
1 1	$-30 + 0.9[0.15(-19.2459) + 0.40(10.3691)$ $+ 0.35(6.8882) + 0.10(54.0470)] = -21.8313$		
1 2	$-25 + 0.9[0.45(-19.2459) + 0.05(10.3691)$ $+ 0.20(6.8882) + 0.30(54.0470)]$ $= -16.4954 \leftarrow$	$\max[-21.8313,$ $-16.4954, -26.9732]$ $= -16.4954 = v_1(2)$	$d_1(2) = 2$
1 3	$-20 + 0.9[0.60(-19.2459) + 0.30(10.3691)$ $+ 0.10(6.8882) + 0(54.0470)] = -26.9732$		
2 1	$5 + 0.9[0.25(-19.2459) + 0.30(10.3691)$ $+ 0.35(6.8882) + 0.10(54.0470)] = 10.5033$		
2 2	$10 + 0.9[0.30(-19.2459) + 0.40(10.3691)$ $+ 0.25(6.8882) + 0.05(54.0470)]$ $= 12.5184 \leftarrow$	$\max[10.5033,$ $12.5184] = 12.5184$ $= v_2(2)$	$d_2(2) = 2$
3 1	$-10 + 0.9[0.05(-19.2459) + 0.65(10.3691)$ $+ 0.25(6.8882) + 0.05(54.0470)] = -0.8182$		
3 2	$-5 + 0.9[0.05(-19.2459) + 0.25(10.3691)$ $+ 0.50(6.8882) + 0.20(54.0470)]$ $= 9.2951 \leftarrow$	$\max[-0.8182,$ $9.2951] = 9.2951$ $= v_3(2)$	$d_3(2) = 2$

continued

TABLE 5.64 (continued)

Value Iteration for $n = 2$

4 1	$35 + 0.9[0.05(-19.2459) + 0.20(10.3691)$	
	$+ 0.40(6.8882) + 0.35(54.0470)] = 55.5049$	
4 2	$25 + 0.9[0(-19.2459) + 0.10(10.3691)$	max $[55.5049,$ $d_4(2) = 2$
	$+ 0.30(6.8882) + 0.60(54.0470)]$	$56.9784] = 56.9784$
	$= 56.9784 \leftarrow$	$= v_4(2)$

TABLE 5.65

Value Iteration for $n = 1$

$$q_i^k + \alpha[p_{i1}^k v_1(2) + p_{i2}^k v_2(2) + p_{i3}^k v_3(2)$$
$$+ p_{i4}^k v_4(2)] = q_i^k + 0.9[p_{i1}^k(-16.4954)$$
$$+ p_{i2}^k(12.5184) + p_{i3}^k(9.2951) + p_{i4}^k(56.9784)]$$

i k		$v_i(1) =$ Expected Total Discounted Reward	Decision $d_i(1) = k$
1 1	$-30 + 0.9[0.15(-16.4954) + 0.40(12.5184)$ $+ 0.35(9.2951) + 0.10(56.9784)] = -19.6642$		
1 2	$-25 + 0.9[0.45(-16.4954) + 0.05(12.5184)$ $+ 0.20(9.2951) + 0.30(56.9784)]$ $= -14.0600 \leftarrow$	max$[-19.6642,$ $-14.0060, -24.6910]$ $= -14.0060 = v_1(1)$	$d_1(1) = 2$
1 3	$-20 + 0.9[0.60(-16.4954) + 0.30(12.5184)$ $+ 0.10(9.2951) + 0(56.9784)] = -24.6910$		
2 1	$5 + 0.9[0.25(-16.4954) + 0.30(12.5184)$ $+ 0.35(9.2951) + 0.10(56.9784)] = 12.7245$		
2 2	$10 + 0.9[0.30(-16.4954) + 0.40(12.5184)$ $+ 0.25(9.2951) + 0.05(56.9784)]$ $= 14.7083 \leftarrow$	max$[12.7245,$ $14.7083] = 14.7083$ $= v_2(1)$	$d_2(1) = 2$
3 1	$-10 + 0.9[0.05(-16.4954) + 0.65(12.5184)$ $+ 0.25(9.2951) + 0.05(56.9784)] = 1.2364$		
3 2	$-5 + 0.9[0.05(-16.4954)$ $+ 0.25(12.5184) + 0.50(9.2951)$ $+ 0.20(56.9784)] = 11.5133 \leftarrow$	max$[1.2364,$ $11.5133] = 11.5133$ $= v_3(1)$	$d_3(1) = 2$

TABLE 5.65 (continued)

Value Iteration for $n = 1$

4	1	$35 + 0.9[0.05(-16.4954) + 0.20(12.5184)$
		$+ 0.40(9.2951) + 0.35(56.9784)] = 55.5049$

4 2 $25 + 0.9[0(-16.4954) + 0.10(12.5184)$ 　　　　$\max[57.8055, \quad d_4(1)$

　　　　$+ 0.30(9.2951) + 0.60(56.9784)]$　　　　$59.4047] = 59.4047 \quad = 2$

　　　　$= 59.4047 \leftarrow$　　　　　　　$= v_4(1)$

$$v_i(1) = \max_k \{q_i^k + \alpha[p_{i1}^k v_1(2) + p_{i2}^k v_2(2) + p_{i3}^k v_3(2) + p_{i4}^k v_4(2)]\} \text{ for } i = 1,2,3,4$$
$$= \max_k \{q_i^k + 0.9[p_{i1}^k(-16.4954) + p_{i2}^k(12.5184) + p_{i3}^k(9.2951) + p_{i4}^k(56.9784)]\}$$
$$\text{for } i = 1,2,3,4.$$

$$(5.66)$$

At the end of month 1, the optimal decision is to select the second alternative in every state. Thus, the decision vector at the end of month 1 is $d(1) = [2\,2\,2\,2]^T$. The calculations for month 0, denoted by $n = 0$, are indicated in Table 5.66.

$$v_i(0) = \max_k \{q_i^k + \alpha[p_{i1}^k v_1(1) + p_{i2}^k v_2(1) + p_{i3}^k v_3(1) + p_{i4}^k v_4(1)]\} \text{ for } i = 1,2,3,4$$
$$\max_k \{q_i^k + 0.9[p_{i1}^k(-14.0600) + p_{i2}^k(14.7083) + p_{i3}^k(11.5133) + p_{i4}^k(59.4047)]\}$$
$$\text{for } i = 1,2,3,4.$$

$$(5.67)$$

At the end of month 0, which is the beginning of month 1, the optimal decision is to select the second alternative in every state. Thus, the decision vector at the beginning of month 1 is $d(0) = [2\,2\,2\,2]^T$.

The results of these calculations for the expected total discounted rewards and the optimal decisions at the end of each month of the seven month planning horizon are summarized in Table 5.67.

If the process starts in state 4, the expected total discounted reward is 61.5109, the highest for any state. On the other hand, the lowest expected total discounted reward is –11.9208, which is the negative of an expected total discounted cost of 11.9208, if the process starts in state 1.

TABLE 5.66

Value Iteration for $n = 0$

	$\begin{aligned} q_i^k + \alpha\,[&p_{i1}^k v_1(1) + p_{i2}^k v_2(1) + p_{i3}^k v_3(1) \\ &+ p_{i4}^k v_4(1)\,] = q_i^k + 0.9[p_{i1}^k(-14.0600) \\ &+ p_{i2}^k(14.7083) + p_{i3}^k(11.5133) + p_{i4}^k(59.4047)] \end{aligned}$	$v_i(0)$ = Expected Total Discounted Reward	Decision $d_i(0) = k$
i k			
1 1	$-30 + 0.9[0.15(-14.0600) + 0.40(14.7083)$ $+\,0.35(11.5133) + 0.10(59.4047)] = -17.6300$		
1 2	$-25 + 0.9[0.45(-14.0600) + 0.05(14.7083)$ $+\,0.20(11.5133) + 0.30(59.4047)]$ $= -11.9208 \leftarrow$	$\max[\,-17.6300,$ $-11.9208,$ $-22.5850\,]$ $= -11.9208 = v_1(0)$	$d_1(0) = 2$
1 3	$-20 + 0.9[0.60(-14.0600) + 0.30(14.7083)$ $+\,0.10(11.5133) + 0(59.4047)] = -22.5850$		
2 1	$5 + 0.9[0.25(-14.0600) + 0.30(14.7083)$ $+\,0.35(11.5133) + 0.10(59.4047)] = 14.7809$		
2 2	$10 + 0.9[0.30(-14.0600)$ $+\,0.40(14.7083) + 0.25(11.5133)$ $+\,0.05(59.4047)] = 16.7625 \leftarrow$	$\max[14.7809,$ $16.7625]$ $= 16.7625 = v_2(0)$	$d_2(0) = 2$
3 1	$-10 + 0.9[0.05(-14.0600) + 0.65(14.7083)$ $+\,0.25(11.5133) + 0.05(59.4047)] = 3.2354$		
3 2	$-5 + 0.9[0.05(-14.0600) + 0.25(14.7083)$ $+\,0.50(11.5133) + 0.20(59.4047)]$ $= 13.5505 \leftarrow$	$\max[3.2354,$ $13.5505] = 13.5505$ $= v_3(0)$	$d_3(0) = 2$
4 1	$35 + 0.9[0.05(-14.0600) + 0.20(14.7083)$ $+\,0.40(11.5133) + 0.35(59.4047)] = 59.8721$		
4 2	$25 + 0.9[0(-14.0600) + 0.10(14.7083)$ $+\,0.30(11.5133) + 0.60(59.4047)]$ $= 61.5109 \leftarrow$	$\max[59.8721,$ $61.5109] = 61.5109$ $= v_4(0)$	$d_4(0) = 2$

5.2.2 An Infinite Planning Horizon

When an optimal stationary policy is to be determined for a discounted MDP over an infinite planning horizon, either value iteration, PI, or LP can

TABLE 5.67

Expected Total Discounted Rewards and Optimal Decisions for a Planning Horizon of 7 Months

Epoch	0	1	2	3	4	5	6	7
$v_1(n)$	−11.9208	−14.0600	−16.4954	−19.2459	−22.1155	−24.10	−20	0
$v_2(n)$	16.7625	14.7083	12.5184	10.3691	8.7276	8.65	10	0
$v_3(n)$	13.5505	11.5133	9.2951	6.8882	4.1643	0.40	−5	0
$v_4(n)$	61.5109	59.4047	56.9784	54.0470	50.254	45.125	35	0
$d_1(n)$	2	2	2	2	2	2	3	—
$d_2(n)$	2	2	2	2	2	2	2	—
$d_3(n)$	2	2	2	2	2	2	2	—
$d_4(n)$	2	2	2	2	2	1	1	—

Table spans n = 0 to 7.

be applied. Value iteration imposes the smallest computational burden, but yields only an approximate solution for the expected total discounted rewards. Both PI and LP produce exact solutions.

5.2.2.1 Value Iteration

In Section 4.3.3.2, value iteration was used to find the expected total discounted rewards received by an MCR over an infinite planning horizon. In Section 5.1.2.3.2, value iteration was used to find an optimal stationary policy plus the expected relative rewards received by an undiscounted MDP over an infinite horizon. In this section, value iteration will be applied to find an optimal stationary policy plus the expected total discounted rewards received by an MDP over an infinite horizon.

5.2.2.1.1 Value Iteration Algorithm

By extending the value iteration algorithm for an MCR in Section 4.3.3.2.2, the following value iteration algorithm in its most basic form (adapted from Puterman [3]) can be applied to a discounted MDP over an infinite horizon. This value iteration algorithm assumes that epochs are numbered as consecutive negative integers from −n at the beginning of the horizon to 0 at the end.

Step 1. Select arbitrary salvage values for

$v_i(0)$, for $i = 1, 2, ..., N$. For simplicity, set $v_i(0) = 0$. Specify $\varepsilon > 0$. Set $n = -1$.

Step 2. For each state i, use the value iteration equation to compute

$$v_i(n) = \max_k \left[q_i^k + \alpha \sum_{j=1}^N p_{ij}^k v_j(n+1) \right], \quad \text{for } i = 1, 2, ..., N.$$

Step 3. If $\max_{i=1,2,\ldots,N} |v_i(n) - v_i(n+1)| < \varepsilon$, go to step 4. Otherwise, decrement n by 1 and return to step 2.

Step 4. For each state i, choose the decision $d_i = k$ that yields the maximum value of $v_i(n)$, and stop.

5.2.2.1.2 Solution by Value Iteration of Discounted MDP Model of Monthly Sales

In Section 5.2.1.2, value iteration was executed to find an optimal policy for the four-state discounted MDP model of monthly sales over a 7 month planning horizon. The solution is summarized in Table 5.67, and is repeated in Table 5.68. In accordance with the treatment of a discounted MCR model in Section 4.3.3.2.3, the last eight epochs of an infinite horizon are numbered sequentially in Table 5.68 as −7, −6, −5, −4, −3, −2, −1, and 0. The absolute value of a negative epoch represents the number of months remaining in the horizon.

Table 5.68 shows that when 1 month remains in an infinite horizon, the optimal policy is to maximize the reward received in every state. When 2 months remain, the expected total discounted rewards are maximized by making decision 2 in states 1, 2, and 3, and decision 1 in state 4. When 3 or more months remain, the optimal policy is to make decision 2 in every state. Thus, as value iteration proceeds backward from the end of the horizon, convergence to an optimal policy given by the decision vector $\mathbf{d}(-3) = [2\ 2\ 2\ 2]^T$ appears to have occurred at $n = -3$.

Table 5.69 gives the absolute values of the differences between the expected total discounted rewards earned over planning horizons, which differ in length by one period. Note that the maximum absolute differences become progressively smaller for each epoch added to the planning horizon.

TABLE 5.68

Expected Total Discounted Rewards and Optimal Decisions during the Last 7 Months of an Infinite Planning Horizon

				n				
Epoch	−7	−6	−5	−4	−3	−2	−1	−0
$v_1(n)$	−11.9208	−14.0600	−16.4954	−19.2459	−22.1155	−24.10	−20	0
$v_2(n)$	16.7625	14.7083	12.5184	10.3691	8.7276	8.65	10	0
$v_3(n)$	13.5505	11.5133	9.2951	6.8882	4.1643	0.40	−5	0
$v_4(n)$	61.5109	59.4047	56.9784	54.0470	50.254	45.125	35	0
$d_1(n)$	2	2	2	2	2	2	3	–
$d_2(n)$	2	2	2	2	2	2	2	–
$d_3(n)$	2	2	2	2	2	2	2	–
$d_4(n)$	2	2	2	2	2	1	1	–

TABLE 5.69

Absolute Values of the Differences between the Expected Total Discounted Rewards Earned Over Planning Horizons, Which Differ in Length by One Period

	n						
	-7	-6	-5	-4	-3	-2	-1
Epoch	$\begin{matrix}\lvert v_i(-7)\rvert \\ \lvert -v_i(-6)\rvert\end{matrix}$	$\begin{matrix}\lvert v_i(-6)\rvert \\ \lvert -v_i(-5)\rvert\end{matrix}$	$\begin{matrix}\lvert v_i(-5)\rvert \\ \lvert -v_i(-4)\rvert\end{matrix}$	$\begin{matrix}\lvert v_i(-4)\rvert \\ \lvert -v_i(-3)\rvert\end{matrix}$	$\begin{matrix}\lvert v_i(-3)\rvert \\ \lvert -v_i(-2)\rvert\end{matrix}$	$\begin{matrix}\lvert v_i(-2)\rvert \\ \lvert -v_i(-1)\rvert\end{matrix}$	$\begin{matrix}\lvert v_i(-1)\rvert \\ \lvert -v_i(0)\rvert\end{matrix}$
i							
1	2.1392U	2.4354U	2.7505	2.8696	1.9845	4.1	20
2	2.0542	2.1899	2.1493	1.6415	0.0776	1.35	5
3	2.0372	2.2181	2.4069	2.7239	3.7643	5.4	5
4	2.1062	2.4263	2.9314U	3.7930U	5.129U	10.125U	35U
$\mathrm{Max}\left(\begin{matrix}\lvert v_i(n)\rvert \\ \lvert -v_i(n+1)\rvert\end{matrix}\right)$	2.1392	2.4354	2.9314	3.7930	5.129	10.125	35

In Table 5.69, a suffix U identifies the maximum absolute difference for each epoch. The maximum absolute differences obtained for all seven epochs are listed in the bottom row of Table 5.69.

Table 5.68 shows that under the optimal policy $d(-7) = [2\,2\,2\,2]^T$, the approximate expected total discounted rewards,

$$v_1 = -11.9208, \quad v_2 = 16.7625, \quad v_3 = 13.5505, \quad v_4 = 61.5109,$$

obtained after seven repetitions of value iteration are significantly different from the actual expected total discounted rewards,

$$v_1 = 6.8040, \quad v_2 = 35.4613, \quad v_3 = 32.2190, \quad v_4 = 80.1970,$$

obtained by PI in Equation (5.71) of Section 5.2.2.2.3, and by LP as dual variables in Table 5.73 of Section 5.2.2.3.4. Thus, while value iteration appears to have converged quickly to an optimal policy, it has not converged after seven repetitions to the actual expected total discounted rewards.

5.2.2.2 Policy Iteration

The PI algorithm created by Ronald A. Howard [2] for finding an optimal stationary policy for a discounted MDP over an infinite planning horizon is similar to the one (described in Section 5.1.2.3.3.2) that he created for an undiscounted, recurrent MDP. Since the gain is not meaningful for a discounted process, an optimal policy is one which maximizes the expected total discounted rewards received in all states. Howard proved that after a finite number of iterations, an optimal policy will be determined. Once again, the PI algorithm has two main steps: the VD operation and the IM routine. The algorithm begins by arbitrarily choosing an initial policy. During the VD operation, the VDEs corresponding to the current policy are solved for the expected total discounted rewards received in all states. The IM routine attempts to find a better policy. If a better policy is found, the VD operation is repeated using the new policy to identify a new system of associated VDEs. The algorithm stops when two successive iterations lead to identical policies.

5.2.2.2.1 Policy Improvement

The IM routine was motivated by the value iteration equation (5.57) for a discounted MDP in Section 5.2.1.1. The value iteration equation indicates that if an optimal policy is known over a planning horizon starting at epoch $n + 1$ and ending at epoch T, then the best decision in state i at epoch n can be found by maximizing a test quantity

$$q_i^k + \alpha \sum_{j=1}^{N} p_{ij}^k v_j(n+1) \tag{5.68}$$

over all decisions in state i. Therefore, if an optimal policy is known over a planning horizon starting at epoch 1 and ending at epoch T, then at epoch 0, which marks the beginning of period 1, the best decision in state i can be found by maximizing a test quantity

$$q_i^k + \alpha \sum_{j=1}^{N} p_{ij}^k v_j(1) \tag{5.69}$$

over all decisions in state i. Recall from Section 4.3.3.1 that when T, the length of the planning horizon, is very large, $(T-1)$ is also very large.

Substituting v_j in Equation (4.62) for $v_j(1)$ in the test quantity produces the result

$$q_i^k + \alpha \sum_{j=1}^{N} p_{ij}^k v_j(1) = q_i^k + \alpha \sum_{j=1}^{N} p_{ij}^k v_j \tag{5.70}$$

as the test quantity to be maximized with respect to all alternatives when making decisions in state i, for $i = 1, 2, \ldots, N$. The expected total discounted rewards produced by solving the VDEs associated with the most recent policy can be used.

5.2.2.2.2 PI Algorithm

The detailed steps of the PI algorithm are given below.

Step 1. Initial policy
Arbitrarily choose an initial policy by selecting for each state i a decision $d_i = k$.

Step 2. VD operation
Use p_{ij} and q_i for a given policy to solve the VDEs (4.260)

$$v_i = q_i + \alpha \sum_{j=1}^{N} p_{ij}v_j, \quad i = 1, 2, ..., N$$

for all expected total discounted rewards v_i.

Step 3. IM routine
For each state i, find the decision $k*$ that maximizes the test quantity

$$q_i^k + \alpha \sum_{j=1}^{N} p_{ij}^k v_j$$

using the expected total discounted rewards v_i of the previous policy. Then $k*$ becomes the new decision in state i, so that $d_i = k*$, q_i^{k*} becomes q_i, and p_{ij}^{k*} becomes p_{ij}.

Step 4. Stopping rule
When the policies on two successive iterations are identical, the algorithm stops because an optimal policy has been found. Leave the old d_i unchanged if the test quantity for that d_i is equal to that of any other alternative in the new policy determination. If the new policy is different from the previous policy in at least one state, go to step 2.

Howard proved that, for each policy, the expected total discounted rewards received in every state will be greater than or equal to their respective values for the previous policy. The algorithm will terminate after a finite number of iterations.

5.2.2.2.3 Solution by PI of MDP Model of Monthly Sales

Consider a discounted MDP model of monthly sales. This model was specified in Table 5.3 of Section 5.1.2.1. In Table 5.68 of Section 5.2.2.1.2, using a discount factor of $\alpha = 0.9$, value iteration was executed to find an optimal policy over an infinite horizon. In this section, PI will be executed to find an optimal policy using the same discount factor.

First iteration

Step 1. Initial policy
Arbitrarily choose the initial policy $^{23}\mathbf{d} = [3\ 2\ 2\ 1]^T$ by making decision 3 in state 1, decision 2 in states 2 and 3, and decision 1 in state 4. Thus, $d_1 = 3$, $d_2 = d_3 = 2$, and $d_4 = 1$. The initial decision vector $^{23}\mathbf{d}$, along with the associated transition probability matrix $^{23}\mathbf{P}$, and the reward vector $^{23}\mathbf{q}$, are shown below.

$$
{}^{23}\mathbf{d} = \begin{bmatrix} 3 \\ 2 \\ 2 \\ 1 \end{bmatrix}, \quad
{}^{23}\mathbf{P} = \begin{matrix} 1 \\ 2 \\ 3 \\ 4 \end{matrix}\begin{bmatrix} 0.60 & 0.30 & 0.10 & 0 \\ 0.30 & 0.40 & 0.25 & 0.05 \\ 0.05 & 0.25 & 0.50 & 0.20 \\ 0.05 & 0.20 & 0.40 & 0.35 \end{bmatrix}, \quad
{}^{23}\mathbf{q} = \begin{matrix} 1 \\ 2 \\ 3 \\ 4 \end{matrix}\begin{bmatrix} -20 \\ 10 \\ -5 \\ 35 \end{bmatrix}.
$$

Step 2. VD operation

Use p_{ij} and q_i for the initial policy, ${}^{23}\mathbf{d} = [3\ 2\ 2\ 1]^T$, to solve the VDEs (4.257)

$$
\begin{aligned}
v_1 &= q_1 + \alpha(p_{11}v_1 + p_{12}v_2 + p_{13}v_3 + p_{14}v_4) \\
v_2 &= q_2 + \alpha(p_{21}v_1 + p_{22}v_2 + p_{23}v_3 + p_{24}v_4) \\
v_3 &= q_3 + \alpha(p_{31}v_1 + p_{32}v_2 + p_{33}v_3 + p_{34}v_4) \\
v_4 &= q_4 + \alpha(p_{41}v_1 + p_{42}v_2 + p_{43}v_3 + p_{44}v_4)
\end{aligned}
$$

for all the expected total discounted rewards v_i

$$
\begin{aligned}
v_1 &= -20 + 0.9(0.60v_1 + 0.30v_2 + 0.10v_3 + 0v_4) \\
v_2 &= 10 + 0.9(0.30v_1 + 0.40v_2 + 0.25v_3 + 0.05v_4) \\
v_3 &= -5 + 0.9(0.05v_1 + 0.25v_2 + 0.50v_3 + 0.20v_4) \\
v_4 &= 35 + 0.9(0.05v_1 + 0.20v_2 + 0.40v_3 + 0.35v_4).
\end{aligned}
$$

The solution of the VDEs is

$$
v_1 = -38.2655, \quad v_2 = 6.1707, \quad v_3 = 8.1311, \quad v_4 = 54.4759.
$$

Step 3. IM routine

For each state i, find the decision $k*$ that maximizes the test quantity

$$
q_i^k + \alpha \sum_{j=1}^{4} p_{ij}^k v_j
$$

using the expected total discounted rewards v_i of the initial policy. Then $k*$ becomes the new decision in state i, so that $d_i = k*$, q_i^{k*} becomes q_i, and p_{ij}^{k*} becomes p_{ij}. The first IM routine is executed in Table 5.70.

Step 4. Stopping rule

The new policy is ${}^{16}\mathbf{d} = [2\ 2\ 2\ 2]^T$, which is different from the initial policy. Therefore, go to step 2. The new decision vector, ${}^{16}\mathbf{d}$, along with the associated transition probability matrix, ${}^{16}\mathbf{P}$, and the reward vector, ${}^{16}\mathbf{q}$, are shown below.

TABLE 5.70

First IM for Monthly Sales Example

State i	Decision Alternative k	Test Quantity $q_i^k + \alpha(p_{i1}^k v_1 + p_{i2}^k v_2 + p_{i3}^k v_3 + p_{i4}^k v_4)$ $= q_i^k + 0.9[p_{i1}^k(-38.2655) + p_{i2}^k(6.1707)$ $+ p_{i3}^k(8.1311) + p_{i4}^k(54.4759)]$	Maximum Value of Test Quantity	Decision $d_i = k*$
1	1	$-30 + 0.9[0.15(-38.2655)$ $+ 0.40(6.1707) + 0.35(8.1311)$ $+ 0.10(54.4759)] = -25.4803$		
1	2	$-25 + 0.9[0.45(-38.2655)$ $+ 0.05(6.1707) + 0.20(8.1311)$ $+ 0.30(54.4759)] = -24.0478 \leftarrow$	max[-25.4803, -24.0478, -38.2655] $= -24.0478$	$d_1 = 2$
1	3	$-20 + 0.9[0.60(-38.2655)$ $+ 0.30(6.1707) + 0.10(8.1311)$ $+ 0(54.4759)] = -38.2655$		
2	1	$5 + 0.9[0.25(-38.2655)$ $+ 0.30(6.1707) + 0.35(8.1311)$ $+ 0.10(54.4759)] = 5.5205$		
2	2	$10 + 0.9[0.30(-38.2655)$ $+ 0.40(6.1707) + 0.25(8.1311)$ $+ 0.05(54.4759)] = 6.1707 \leftarrow$	max[5.5205, 6.1707] $= 6.1707$	$d_2 = 2$
3	1	$-10 + 0.9[0.05(-38.2655)$ $+ 0.65(6.1707) + 0.25(8.1311)$ $+ 0.05(54.4759)] = -3.8312$		
3	2	$-5 + 0.9[0.05(-38.2655)$ $+ 0.25(6.1707) + 0.50(8.1311)$ $+ 0.20(54.4759)] = 8.1311 \leftarrow$	max[-3.8312, 8.1311] $= 8.1311$	$d_3 = 2$

continued

TABLE 5.70 (continued)

First IM for Monthly Sales Example

4	1	$35 + 0.9[0.05(-38.2655)$
		$+ 0.20(6.1707) + 0.40(8.1311)$
		$+ 0.35(54.4759)] = 54.4759$

4	2	$25 + 0.9[0(-38.2655)$	$\max[54.4759,\ d_4 = 2$
		$+ 0.10(6.1707) + 0.30(8.1311)$	$57.1677]$
		$+ 0.60(54.4759)] = 57.1677 \leftarrow$	$= 57.1677$

$$^{16}\mathbf{d} = \begin{bmatrix} 2 \\ 2 \\ 2 \\ 2 \end{bmatrix}, \quad ^{16}\mathbf{P} = \begin{matrix} 1 \\ 2 \\ 3 \\ 4 \end{matrix}\begin{bmatrix} 0.45 & 0.05 & 0.20 & 0.30 \\ 0.30 & 0.40 & 0.25 & 0.05 \\ 0.05 & 0.25 & 0.50 & 0.20 \\ 0 & 0.10 & 0.30 & 0.60 \end{bmatrix}, \quad ^{16}\mathbf{q} = \begin{matrix} 1 \\ 2 \\ 3 \\ 4 \end{matrix}\begin{bmatrix} -25 \\ 10 \\ -5 \\ 25 \end{bmatrix}. \quad (5.71)$$

Step 2. VD operation

Use p_{ij} and q_i for the new policy, $^{16}\mathbf{d} = [2\ 2\ 2\ 2]^T$, to solve the VDEs (4.257) for all the expected total discounted rewards v_i.

$$v_1 = -25 + 0.9[0.45v_1 + 0.05v_2 + 0.20v_3 + 0.30v_4]$$
$$v_2 = 10 + 0.9[0.30v_1 + 0.40v_2 + 0.25v_3 + 0.05v_4]$$
$$v_3 = -5 + 0.9[0.05v_1 + 0.25v_2 + 0.50v_3 + 0.20v_4]$$
$$v_4 = 25 + 0.9[0v_1 + 0.10v_2 + 0.30v_3 + 0.60v_4].$$

The solution of the VDEs is

$$v_1 = 6.8040, \quad v_2 = 35.4613, \quad v_3 = 32.2190, \quad v_4 = 80.1970. \quad (5.72)$$

Step 3. IM routine

For each state i, find the decision $k*$ that maximizes the test quantity

$$q_i^k + \alpha \sum_{j=1}^{4} p_{ij}^k v_j$$

using the expected total discounted rewards v_i of the previous policy. Then $k*$ becomes the new decision in state i, so that $d_i = k*$, q_i^{k*} becomes q_i, and p_{ij}^{k*} becomes p_{ij}. The second IM routine is s executed in Table 5.71.

Step 4. Stopping rule

Stop because the new policy, given by the vector $^{16}\mathbf{d} = [2\ 2\ 2\ 2]^T$, is identical to the previous policy. Therefore, this policy is optimal. The expected

TABLE 5.71

Second IM for Monthly Sales Example

State i	Decision Alternative k	Test Quantity $q_i^k + \alpha(p_{i1}^k v_1 + p_{i2}^k v_2 + p_{i3}^k v_3 + p_{i4}^k v_4)$ $= q_i^k + 0.9[p_{i1}^k(6.8040) + p_{i2}^k(35.4613)$ $+ p_{i3}^k(32.2190) + p_{i4}^k(80.1970)]$	Maximum Value of Test Quantity	Decision $d_i = k*$
1	1	$-30 + 0.9[0.15(6.8040)$ $+0.40(35.4613) + 0.35(32.2190)$ $+0.10(80.1970)] = 1.0513$		
1	2	$-25 + 0.9[0.45(6.8040)$ $+0.05(35.4613) + 0.20(32.2190)$ $+0.30(80.1970)] = 6.8040 \leftarrow$	max[1.0513, 6.8040, -38.5158] $= 6.8040$	$d_1 = 2$
1	3	$-20 + 0.9[0.60(6.8040)$ $+0.30(35.4613) + 0.10(32.2190)$ $+0(80.1970)] = -38.5158$		
2	1	$5 + 0.9[0.25(6.8040)$ $+0.30(35.4613) + 0.35(32.2190)$ $+0.10(80.1970)] = 33.4722$		
2	2	$10 + 0.9[0.30(6.8040)$ $+0.40(35.4613) + 0.25(32.2190)$ $+0.05(80.1970)] = 35.4613 \leftarrow$	max[33.4722, 35.4613] $= 35.4613$	$d_2 = 2$
3	1	$-10 + 0.9[0.05(6.8040)$ $+0.65(35.4613) + 0.25(32.2190)$ $+0.05(80.1970)] = 21.9092$		
3	2	$-5 + 0.9[0.05(6.8040)$ $+0.25(35.4613) + 0.50(32.2190)$ $+0.20(80.1970)] = 32.2190 \leftarrow$	max[21.9092, 32.2190] $= 32.21901$	$d_3 = 2$

continued

TABLE 5.71 (continued)

Second IM for Monthly Sales Example

4	1	$35 + 0.9[0.05(6.8040)$	
		$+0.20(35.4613) + 0.40(32.2190)$	
		$+0.35(80.1970)] = 78.5501$	
4	2	$25 + 0.9[0(6.8040)$	$\max[78.5501, \ d_4 = 2$
		$+0.10(35.4613) + 0.30(32.2190)$	$80.1970]$
		$+0.60(80.1970)] = 80.1970 \leftarrow$	$= 80.1970$

total discounted rewards for the optimal policy are calculated in Equation (5.77). The transition probability matrix ^{16}P and the reward vector ^{16}q for the optimal policy are given in Equation (5.71). Note that the optimal policy is the same as the one obtained without discounting in Equation (5.24) of Section 5.1.2.3.3.3. Note also that the expected total discounted rewards obtained by PI are identical to the corresponding dual variables obtained by LP in Table 5.73 of Section 5.2.2.3.4.

5.2.2.2.4 *Optional Insight: IM for a Two-State Discounted MDP*

To gain insight (adapted from Howard [2]) into why the IM routine works for a discounted MDP, it is instructive to examine the special case of a generic two-state process with two decisions per state. (The special case of a two-state recurrent MDP without discounting was addressed in Section 5.1.2.3.3.4.) The two-state discounted MDP is shown in Table 5.20 in which μ replaces a. The discount factor is denoted by α, where $0 < \alpha < 1$.

Since the process has two states with two decisions per state, there are $2^2 = 4$ possible policies, identified by the letters A, B, C, and D. The decision vectors for the four policies are

$$d^A = \frac{1}{2}\begin{bmatrix} 1 \\ 1 \end{bmatrix}, \quad d^B = \frac{1}{2}\begin{bmatrix} 1 \\ 2 \end{bmatrix}, \quad d^C = \frac{1}{2}\begin{bmatrix} 2 \\ 1 \end{bmatrix}, \quad d^D = \frac{1}{2}\begin{bmatrix} 2 \\ 2 \end{bmatrix}.$$

Suppose that policy A is the current policy. Policy A is characterized by the following decision vector, transition probability matrix, reward vector, and expected total discounted reward vector:

$$\mathbf{d}^A = \begin{bmatrix} 1 \\ 1 \end{bmatrix}, \quad \mathbf{P}^A = \frac{1}{2}\begin{bmatrix} 1-u & u \\ c & 1-c \end{bmatrix}, \quad \mathbf{q}^A = \frac{1}{2}\begin{bmatrix} e \\ s \end{bmatrix}, \quad \mathbf{v}^A = \frac{1}{2}\begin{bmatrix} v_1^A \\ v_2^A \end{bmatrix}.$$

The VDEs for states 1 and 2 under policy A are given below:

$$v_i^A = q_i^A + \alpha \sum_{j=1}^{2} p_{ij}^A v_j^A, \quad i = 1, 2$$

For $i = 1$,

$$v_1^A = q_1^A + \alpha p_{11}^A v_1^A + \alpha p_{12}^A v_2^A$$
$$v_1^A = e + \alpha(1-u)v_1^A + \alpha u v_2^A$$

For $i = 2$,

$$v_2^A = q_2^A + \alpha p_{21}^A v_1^A + \alpha p_{22}^A v_2^A$$
$$v_2^A = s + \alpha c v_1^A + \alpha(1-c)v_2^A$$

Suppose that the evaluation of policy A by the IM routine has produced a new policy, policy D. Policy D is characterized by the following decision vector, transition probability matrix, reward vector, and expected total discounted reward vector:

$$\mathbf{d}^D = \begin{bmatrix} 2 \\ 2 \end{bmatrix}, \quad \mathbf{P}^D = \frac{1}{2}\begin{bmatrix} 1-b & b \\ d & 1-d \end{bmatrix}, \quad \mathbf{q}^D = \frac{1}{2}\begin{bmatrix} f \\ h \end{bmatrix}, \quad \mathbf{v}^D = \frac{1}{2}\begin{bmatrix} v_1^D \\ v_2^D \end{bmatrix}.$$

The VDEs for states 1 and 2 under policy D are given below.

$$v_i^D = q_i^D + \alpha \sum_{j=1}^{2} p_{ij}^D v_j^D, \quad i = 1, 2.$$

For $i = 1$,

$$v_1^D = q_1^D + \alpha p_{11}^D v_1^D + \alpha p_{12}^D v_2^D$$
$$v_1^D = f + \alpha(1-b)v_1^D + \alpha b v_2^D.$$

For $i = 2$,

$$v_2^D = q_2^D + \alpha p_{21}^D v_1^D + \alpha p_{22}^D v_2^D$$
$$v_2^D = h + \alpha d v_1^D + \alpha(1-d)v_2^D.$$

Since the IM routine has chosen policy D over A, the test quantity for policy D must be greater than or equal to the test quantity for A in both states.

Therefore,

For $i = 1$,

$$q_1^D + \alpha(p_{11}^D v_1^A + p_{12}^D v_2^A) \geq q_1^A + \alpha(p_{11}^A v_1^A + p_{12}^A v_2^A)$$

$$f + \alpha(1-b)v_1^A + \alpha b v_2^A \geq e + \alpha(1-u)v_1^A + \alpha u v_2^A.$$

Let γ_1 denote the improvement in the test quantity achieved in state 1.

$$\gamma_1 = f + \alpha(1-b)v_1^A + \alpha b v_2^A - e - \alpha(1-u)v_1^A - \alpha u v_2^A \geq 0.$$

For $i = 2$,

$$q_2^D + \alpha(p_{21}^D v_1^A + p_{22}^D v_2^A) \geq q_2^A + \alpha(p_{21}^A v_1^A + p_{22}^A v_2^A)$$

$$h + \alpha d v_1^A + \alpha(1-d)v_2^A \geq s + \alpha c v_1^A + \alpha(1-c)v_2^A.$$

Let γ_2 denote the improvement in the test quantity achieved in state 2.

$$\gamma_2 = h + \alpha d v_1^A + \alpha(1-d)v_2^A - s - \alpha c v_1^A - \alpha(1-c)v_2^A \geq 0.$$

The objective of this insight is to show that the IM routine must increase the expected total discounted rewards of one or both states.

For $i = 1$,

$$\begin{aligned}
v_1^D - v_1^A &= [f + \alpha(1-b)v_1^D + \alpha b v_2^D] - [e + \alpha(1-u)v_1^A + \alpha u v_2^A] \\
&= [(f-e) + \alpha(1-b)v_1^A + \alpha b v_2^A - \alpha(1-u)v_1^A - \alpha u v_2^A] \\
&\quad - \alpha(1-b)v_1^A - \alpha b v_2^A + \alpha(1-b)v_1^D + \alpha b v_2^D \\
&= \gamma_1 - \alpha(1-b)v_1^A - \alpha b v_2^A + \alpha(1-b)v_1^D + \alpha b v_2^D \\
&= \gamma_1 + \alpha p_{11}^D(v_1^D - v_1^A) + \alpha p_{12}^D(v_2^D - v_2^A).
\end{aligned}$$

For $i = 2$,

$$\begin{aligned}
v_2^D - v_2^A &= [h + \alpha d v_1^D + \alpha(1-d)v_2^D] - [s + \alpha c v_1^A + \alpha(1-c)v_2^A] \\
&= [(h-s) + \alpha d v_1^A + \alpha(1-d)v_2^A - \alpha c v_1^A - \alpha(1-c)v_2^A] \\
&\quad - \alpha d v_1^A - \alpha(1-d)v_2^A + \alpha d v_1^D + \alpha(1-d)v_2^D \\
&= \gamma_2 - \alpha d v_1^A - \alpha(1-d)v_2^A + \alpha d v_1^D + \alpha(1-d)v_2^D \\
&= \gamma_2 + \alpha p_{21}^D(v_1^D - v_1^A) + \alpha p_{22}^D(v_2^D - v_2^A)
\end{aligned}$$

Observe that the pair of equations for the increase in the expected total discounted rewards has the same form as the pair of equations for the expected total discounted rewards,

$$v_1^D = q_1^D + \alpha p_{11}^D v_1^D + \alpha p_{12}^D v_2^D$$
$$v_2^D = q_2^D + \alpha p_{21}^D v_1^D + \alpha p_{22}^D v_2^D.$$

In matrix form, the pair of equations for the expected total discounted rewards is

$$\mathbf{v}^D = \mathbf{q}^D + \alpha \mathbf{P}^D \mathbf{v}^D.$$

The solution for the vector of the expected total discounted rewards is

$$\mathbf{v}^D = (\mathbf{I} - \alpha \mathbf{P}^D)^{-1} \mathbf{q}^D.$$

Similarly, the matrix form of the pair of equations for the increase in the expected total discounted rewards is

$$\mathbf{v}^D - \mathbf{v}^A = \gamma + \alpha \mathbf{P}^D (\mathbf{v}^D - \mathbf{v}^A),$$

where $\mathbf{v}^D - \mathbf{v}^A$ is the vector of the increase in the expected total discounted rewards in both states, and $\gamma = [\gamma_1 \quad \gamma_2]^T$), is the vector of improvement of the test quantity in both states. Therefore, the solution for the vector of the increase in the expected total discounted rewards is

$$\mathbf{v}^D - \mathbf{v}^A = (\mathbf{I} - \alpha \mathbf{P}^D)^{-1} \gamma.$$

For the two-state process under policy D,

$$\alpha \mathbf{P}^D = \begin{bmatrix} \alpha(1-b) & \alpha b \\ \alpha d & \alpha(1-d) \end{bmatrix}$$

$$\mathbf{I} - \alpha \mathbf{P}^D = \begin{bmatrix} 1 - \alpha + \alpha b & -\alpha b \\ -\alpha d & 1 - \alpha + \alpha d \end{bmatrix}$$

$$(\mathbf{I} - \alpha \mathbf{P}^D)^{-1} = \frac{1}{(1-\alpha+\alpha b)(1-\alpha+\alpha d) - \alpha^2 bd} \begin{bmatrix} 1 - \alpha + \alpha d & \alpha b \\ \alpha d & 1 - \alpha + \alpha b \end{bmatrix}.$$

Note that $\alpha \geq 0$, $b \geq 0$, and $d \geq 0$. Hence, $\alpha b \geq 0$ and $\alpha d \geq 0$. Also, $1 - \alpha \geq 0$. Hence, $1 - \alpha + \alpha b \geq 0$ and $1 - \alpha + \alpha d \geq 0$. Thus, all entries in the numerator of the matrix $(\mathbf{I} - \alpha \mathbf{P}^D)^{-1}$ are nonnegative. The denominator of all entries in $(\mathbf{I}-\alpha\mathbf{P}^D)^{-1}$ is

$$(1-\alpha+\alpha b)(1-\alpha+\alpha d)-\alpha^2 bd$$
$$= (1-\alpha+\alpha d-\alpha+\alpha^2-\alpha^2 d+\alpha b-\alpha^2 b+\alpha^2 bd)-\alpha^2 bd$$
$$= (1-\alpha-\alpha+\alpha^2)+(\alpha d-\alpha^2 d)+(\alpha b-\alpha^2 b)$$
$$= [(1-\alpha)-\alpha(1-\alpha)]+\alpha d(1-\alpha)+\alpha b(1-\alpha)$$
$$= (1-\alpha)(1-\alpha+\alpha d+\alpha b)\geq 0.$$

The denominator of the elements of $(\mathbf{I}-\alpha\mathbf{P}^D)^{-1}$ is also nonnegative because it is the product of two factors which are both greater than or equal to zero. Therefore, all entries in the matrix $(\mathbf{I}-\alpha\mathbf{P}^D)^{-1}$ are greater than or equal to zero. Since both elements of the vector γ are also greater than or equal to zero, the elements of the vector

$$\mathbf{v}^D - \mathbf{v}^A = (\mathbf{I}-\alpha\mathbf{P}^D)^{-1}\gamma$$

must be nonnegative. Therefore, the IM routine cannot decrease the expected total discounted rewards of either state. Howard proves that for a discounted MDP with N states, the IM routine must increase the expected total discounted rewards of at least one state.

5.2.2.3 LP for a Discounted MDP

When a linear program is formulated for a discounted MDP, Markov chain structure is not relevant.

5.2.2.3.1 Formulation of a LP for a Discounted MDP

Informally, an LP formulation for a discounted MDP can be obtained by modifying the LP formulation given in Equations (5.43) of Section 5.1.2.3.4.1 for an undiscounted, recurrent MDP in the following manner. The transition probability p_{ij} in the undiscounted MDP is replaced with αp_{ij} in the discounted process. Note that αp_{ij} is not a transition probability because the row sums of the discounted transition matrix, $\alpha\mathbf{P}$, equal α, which is less than one. With discounting, no constraint is redundant, so that no constraint (5.42) is discarded. In the absence of transition probabilities, steady-state probabilities are not meaningful, so that there is no normalizing constraint equation (5.40) for steady-state probabilities. Thus, the LP formulation for an N-state discounted process will have N independent constraints. The LP objective function is unchanged from the one for an undiscounted MDP because its coefficients do not contain p_{ij}.

When p_{ij} is replaced by αp_{ij}, the constraint equations (5.39),

$$\sum_{k=1}^{K_j} y_j^k - \sum_{i=1}^{N}\sum_{k=1}^{K_i} y_i^k\, p_{ij}^k = 0, \quad \text{for } j = 1, 2, \ldots, N, \tag{5.39}$$

which do contain p_{ij}, are transformed into the inequality constraints

$$\sum_{k=1}^{K_i} y_j^k - \sum_{i=1}^{N} \sum_{k=1}^{K_i} y_i^k \, (\alpha p_{ij}^k) = \sum_{k=1}^{K_i} y_j^k - \alpha \sum_{i=1}^{N} \sum_{k=1}^{K_i} y_i^k \, p_{ij}^k > 0, \quad \text{for } j = 1, 2, ..., N, \quad (5.73)$$

because, for $0 < \alpha < 1$,

$$\sum_{k=1}^{K_i} y_j^k > \alpha \sum_{i=1}^{N} \sum_{k=1}^{K_i} y_i^k \, p_{ij}^k. \quad (5.74)$$

The inequality constraints are expressed as equations by transposing all terms to the left hand sides, and making the right-hand sides arbitrary positive constants, denoted by b_j. The equality constraints are

$$\sum_{k=1}^{K_i} y_j^k - \alpha \sum_{i=1}^{N} \sum_{k=1}^{K_i} y_i^k \, p_{ij}^k = b_j, \quad (5.75)$$

for $j = 1, 2, ..., N$, where $b_j > 0$.

The complete LP formulation for a discounted MDP is [1, 3]

$$\text{Maximize } \sum_{i=1}^{N} \sum_{k=1}^{K_i} q_i^k \, y_i^k$$

subject to

$$\sum_{k=1}^{K_i} y_j^k - \alpha \sum_{i=1}^{N} \sum_{k=1}^{K_i} y_i^k \, p_{ij}^k = b_j, \quad (5.76)$$

for $j = 1, 2, ..., N$, where $b_j > 0$.

$$y_i^k \geq 0, \quad \text{for } i = 1, 2, ..., N, \quad \text{and} \quad k = 1, 2, ..., K_i$$

When LP software has found an optimal solution for the decision variables, the conditional probabilities,

$$P(\text{decision} = k | \text{state} = i) = \frac{y_i^k}{\sum_{k=1}^{K_i} y_i^k}, \quad (5.30)$$

can be calculated to determine the optimal decision in every state, thereby specifying an optimal policy. The expected total discounted reward, v_j, earned by starting in state j, is equal to the dual variable associated with row j in the final LP tableau. Since the LP constraints are equations, the dual variables, and hence the expected total discounted rewards, v_j, are unrestricted in sign. Both an optimal policy and the expected total discounted rewards are independent of the values assigned to the arbitrary positive constants, b_j.

Suppose that

$$\sum_{j=1}^{N} b_j = 1. \tag{5.77}$$

Since b_j is the right-hand side constant of the jth constraint in the LP, b_j is also the objective function coefficient of the jth dual variable, v_j. The optimal dual objective function is equal to $\sum_{j=1}^{N} b_j v_j$, which is also the optimal value of the primal LP objective function.

5.2.2.3.2 *Optional Insight: Informal Derivation of the LP Formulation for a Discounted MDP*

The following informal derivation of the LP formulation for a discounted MDP is adapted from Puterman [3]. This informal derivation begins by defining the nonnegative decision variable y_i^k as the following infinite series:

$$
\begin{aligned}
y_i^k &= \alpha^0 P(\text{at epoch 0, state} = i \text{ and decision} = k) + \alpha^1 P(\text{at epoch 1,} \\
&\quad \text{state} = i \text{ and decision} = k) + \alpha^2 P(\text{at epoch 2, state} = i \text{ and} \\
&\quad \text{decision} = k) + \alpha^3 P(\text{at epoch 3, state} = i \text{ and decision} = k) + \cdots \\
&= \alpha^0 P(X_0 = i, d_i = k) + \alpha^1 P(X_1 = i, d_i = k) + \alpha^2 P(X_2 = i, d_i = k) \\
&\quad + \alpha^3 P(X_3 = i, d_i = k) + \cdots \\
&= \sum_{n=0}^{\infty} \alpha^n P(X_n = i, d_i = k).
\end{aligned}
$$

If the positive constants denoted by b_j are chosen such that $\sum_{j=1}^{N} b_j = 1$, then b_j can be interpreted as the probability that the system starts in state j. That is,

$$b_j = P(X_0 = j). \tag{5.78}$$

If $\sum_{j=1}^{N} b_j = 1$, then the variable y_i^k can be interpreted as the joint discounted probability of being in state i and making decision k.

Objective Function

The objective function for the LP formulation for a discounted MDP is

$$\text{Maximize } \sum_{n=0}^{\infty} \alpha^n \sum_{i=1}^{N} \sum_{k=1}^{K_i} q_i^k P(X_n = i, d_i = k)$$

$$= \sum_{i=1}^{N} \sum_{k=1}^{K_i} q_i^k \sum_{n=0}^{\infty} \alpha^n P(X_n = i, d_i = k)$$

$$= \sum_{i=1}^{N} \sum_{k=1}^{K_i} q_i^k y_i^k$$

which is the same objective function as the one for the LP formulation for an undiscounted, recurrent MDP.

Constraints

The constraints can be obtained by starting with the expression $\sum_{i=1}^{N} \sum_{k=1}^{K_i} \alpha p_{ij}^k y_i^k$.

$$\sum_{i=1}^{N} \sum_{k=1}^{K_i} \alpha p_{ij}^k y_i^k = \sum_{i=1}^{N} \sum_{k=1}^{K_i} \alpha p_{ij}^k \sum_{n=0}^{\infty} \alpha^n P(X_n = i, d_i = k)$$

$$= \sum_{k=1}^{K_i} \sum_{n=0}^{\infty} \alpha^{n+1} \sum_{i=1}^{N} p_{ij}^k P(X_n = i, d_i = k)$$

$$= \sum_{k=1}^{K_j} \sum_{n=0}^{\infty} \alpha^{n+1} P(X_{n+1} = j, d_j = k)$$

$$= \sum_{k=1}^{K_j} [\alpha^1 P(X_1 = j, d_j = k) + \alpha^2 P(X_2 = j, d_j = k)$$

$$+ \alpha^3 P(X_3 = j, d_j = k) + \cdots]$$

$$= \sum_{k=1}^{K_j} \{\alpha^0 P(X_0 = j, d_j = k) + [\alpha^1 P(X_1 = j, d_j = k) + \alpha^2 P(X_2 = j, d_j = k)$$

$$+ \alpha^3 P(X_3 = j, d_j = k) + \cdots] - \alpha^0 P(X_0 = j, d_j = k)\}$$

$$= \sum_{k=1}^{K_j} [\alpha^0 P(X_0 = j, d_j = k) + \alpha^1 P(X_1 = j, d_j = k) + \alpha^2 P(X_2 = j, d_j = k)$$

$$+ \alpha^3 P(X_3 = j, d_j = k) + \cdots] - \sum_{k=1}^{K_j} \alpha^0 P(X_0 = j, d_j = k)$$

$$= \sum_{k=1}^{K_j} \sum_{n=0}^{\infty} \alpha^n P(X_n = j, d_j = k) - \sum_{k=1}^{K_j} \alpha^0 P(X_0 = j, d_j = k)$$

$$= \sum_{k=1}^{K_j} \sum_{n=0}^{\infty} \alpha^n P(X_n = j, d_j = k) - P(X_0 = j)$$

$$= \sum_{k=1}^{K_j} y_j^k - b_j.$$

Hence, the constraints are

$$\sum_{k=1}^{K_j} y_j^k - \alpha \sum_{i=1}^{N} \sum_{k=1}^{K_i} y_i^k \, p_{ij}^k = b_j.$$

for $j = 1, 2, \dots, N$.

5.2.2.3.3 Formulation of a Discounted MDP Model of Monthly Sales as an LP

The discounted MDP model of monthly sales, specified in Table 5.3 of Section 5.1.2.1, was solved over an infinite horizon by value iteration in Table 5.68 of Section 5.2.2.1.2, and by PI in Section 5.2.2.3. Using a discount factor of $\alpha = 0.9$, the discounted MDP model will be formulated in this section as an LP, and solved in Section 5.2.2.3.4 to find an optimal policy. Table 5.3, augmented by a right hand column of LP variables, is repeated as Table 5.72.

In this example $N = 4$ states, $K_1 = 3$ decisions in state 1, and $K_2 = K_3 = K_4 = 2$ decisions in states 2, 3, and 4, respectively. The positive constants are arbitrarily chosen to be $b_1 = 0.1$, $b_2 = 0.2$, $b_3 = 0.3$, and $b_4 = 0.4$. Note that

$$\sum_{j=1}^{N} b_j = 1.$$

TABLE 5.72

Data for Discounted MDP Model of Monthly Sales, $\alpha = 0.9$

State i	Decision k		Transition Probability				Reward q_i^k	LP Variable
			p_{i1}^k	p_{i2}^k	p_{i3}^k	p_{i4}^k		
1	1	Sell noncore assets	0.15	0.40	0.35	0.10	−30	y_1^1
	2	Take firm private	0.45	0.05	0.20	0.30	−25	y_1^2
	3	Offer employee buyouts	0.60	0.30	0.10	0	−20	y_1^3
2	1	Reduce management salaries	0.25	0.30	0.35	0.10	5	y_2^1
	2	Reduce employee benefits	0.30	0.40	025	0.05	10	y_2^2
3	1	Design more appealing products	0.05	0.65	0.25	0.05	−10	y_3^1
	2	Invest in new technology	0.05	0.25	0.50	0.20	−5	y_3^2
4	1	Invest in new projects	0.05	0.20	0.40	0.35	35	y_4^1
	2	Make strategic acquisitions	0	0.10	0.30	0.60	25	y_4^2

Objective Function
The objective function for the LP is

$$\text{Maximize} \sum_{i=1}^{4} \sum_{k=1}^{K_i} q_i^k \, y_i^k$$

$$\text{Maximize} \sum_{k=1}^{3} q_1^k \, y_1^k + \sum_{k=1}^{2} q_2^k \, y_2^k + \sum_{k=1}^{3} q_3^k \, y_3^k + \sum_{k=1}^{4} q_4^k \, y_4^k$$

$$= (q_1^1 y_1^1 + q_1^2 y_1^2 + q_1^3 y_1^3) + (q_2^1 y_2^1 + q_2^2 y_2^2) + (q_3^1 y_3^1 + q_3^2 y_3^2) + (q_4^1 y_4^1 + q_4^2 y_4^2)$$

$$= (-30y_1^1 - 25y_1^2 - 20y_1^3) + (5y_2^1 + 10y_2^2) + (-10y_3^1 - 5y_3^2) + (35y_4^1 + 25y_4^2)$$

Constraints
State 1 has $K_1 = 3$ possible decisions. The constraint associated with a transition to state $j = 1$ is

$$\sum_{k=1}^{3} y_1^k - \alpha \sum_{i=1}^{4} \sum_{k=1}^{K_{ii}} y_i^k \, p_{i1}^k = b_1$$

$$(y_1^1 + y_1^2 + y_1^3) - \alpha \left(\sum_{k=1}^{3} y_1^k p_{11}^k + \sum_{k=1}^{2} y_2^k p_{21}^k + \sum_{k=1}^{2} y_3^k p_{31}^k + \sum_{k=1}^{2} y_4^k p_{41}^k \right) = b_1$$

$$(y_1^1 + y_1^2 + y_1^3) - \alpha(y_1^1 p_{11}^1 + y_1^2 p_{11}^2 + y_1^3 p_{11}^3) - \alpha(y_2^1 p_{21}^1 + y_2^2 p_{21}^2) - \alpha(y_3^1 p_{31}^1 + y_3^2 p_{31}^2)$$
$$- \alpha(y_4^1 p_{41}^1 + y_4^2 p_{41}^2) = b_1$$

$$(y_1^1 + y_1^2 + y_1^3) - 0.9(0.15y_1^1 + 0.45y_1^2 + 0.60y_1^3) - 0.9(0.25y_2^1 + 0.30y_2^2)$$
$$- 0.9(0.05y_3^1 + 0.05y_3^2) - 0.9(0.05y_4^1 + 0y_4^2) = 0.1.$$

State 2 has $K_2 = 2$ possible decisions. The constraint associated with a transition to state $j = 2$ is

$$\sum_{k=1}^{2} y_2^k - \alpha \sum_{i=1}^{4} \sum_{k=1}^{K_i} y_i^k \, p_{i2}^k = b_2$$

$$(y_2^1 + y_2^2) - \alpha \left(\sum_{k=1}^{3} y_1^k p_{12}^k + \sum_{k=1}^{2} y_2^k p_{22}^k + \sum_{k=1}^{2} y_3^k p_{32}^k + \sum_{k=1}^{2} y_4^k p_{42}^k \right) = b_2$$

$$(y_2^1 + y_2^2) - \alpha(y_1^1 p_{12}^1 + y_1^2 p_{12}^2 + y_1^3 p_{12}^3) - \alpha(y_2^1 p_{22}^1 + y_2^2 p_{22}^2) - \alpha(y_3^1 p_{32}^1 + y_3^2 p_{32}^2)$$
$$- \alpha(y_4^1 p_{42}^1 + y_4^2 p_{42}^2) = b_2$$

$$(y_2^1 + y_2^2) - 0.9(0.40y_1^1 + 0.05y_1^2 + 0.30y_1^3) - 0.9(0.30y_2^1 + 0.40y_2^2) - 0.9(0.65y_3^1 + 0.25y_3^2)$$
$$- 0.9(0.20y_4^1 + 0.10y_4^2) = 0.2$$

State 3 has $K_3 = 2$ possible decisions. The constraint associated with a transition to state $j = 3$ is

$$\sum_{k=1}^{2} y_3^k - \alpha \sum_{i=1}^{4} \sum_{k=1}^{K_i} y_i^k \, p_{i3}^k = b_3$$

$$(y_3^1 + y_3^2) - \alpha \left(\sum_{k=1}^{3} y_1^k p_{13}^k + \sum_{k=1}^{2} y_2^k p_{23}^k + \sum_{k=1}^{2} y_3^k p_{33}^k + \sum_{k=1}^{2} y_4^k p_{43}^k \right) = b_3$$

$$(y_3^1 + y_3^2) - \alpha(y_1^1 p_{13}^1 + y_1^2 p_{13}^2 + y_1^3 p_{13}^3) - \alpha(y_2^1 p_{23}^1 + y_2^2 p_{23}^2) - \alpha(y_3^1 p_{33}^1 + y_3^2 p_{33}^2)$$
$$- \alpha(y_4^1 p_{43}^1 + y_4^2 p_{43}^2) = b_3$$

$$(y_3^1 + y_3^2) - 0.9(0.35 y_1^1 + 0.20 y_1^2 + 0.10 y_1^3) - 0.9(0.35 y_2^1 + 0.25 y_2^2)$$
$$- 0.9(0.25 y_3^1 + 0.50 y_3^2) - 0.9(0.40 y_4^1 + 0.30 y_4^2) = 0.3.$$

State 4 has $K_4 = 2$ possible decisions. The constraint associated with a transition to state $j = 4$ is

$$\sum_{k=1}^{2} y_4^k - \alpha \sum_{i=1}^{4} \sum_{k=1}^{K_i} y_i^k \, p_{i4}^k = b_4$$

$$(y_4^1 + y_4^2) - \alpha \left(\sum_{k=1}^{3} y_1^k p_{14}^k + \sum_{k=1}^{2} y_2^k p_{24}^k + \sum_{k=1}^{2} y_3^k p_{34}^k + \sum_{k=1}^{2} y_4^k p_{44}^k \right) = b_4$$

$$(y_4^1 + y_4^2) - \alpha(y_1^1 p_{14}^1 + y_1^2 p_{14}^2 + y_1^3 p_{14}^3) - \alpha(y_2^1 p_{24}^1 + y_2^2 p_{24}^2) - \alpha(y_3^1 p_{34}^1 + y_3^2 p_{34}^2)$$
$$- \alpha(y_4^1 p_{44}^1 + y_4^2 p_{44}^2) = b_4$$

$$(y_4^1 + y_4^2) - 0.9(0.10 y_1^1 + 0.30 y_1^2 + 0 y_1^3) - 0.9(0.10 y_2^1 + 0.05 y_2^2)$$
$$- 0.9(0.05 y_3^1 + 0.20 y_3^2) - 0.9(0.35 y_4^1 + 0.60 y_4^2) = 0.4.$$

The complete LP formulation for the discounted MDP model of monthly sales is

Maximize $(-30 y_1^1 - 25 y_1^2 - 20 y_1^3) + (5 y_2^1 + 10 y_2^2) + (-10 y_3^1 - 5 y_3^2) + (35 y_4^1 + 25 y_4^2)$

subject to

(1) $(0.865y_1^1 + 0.595y_1^2 + 0.46y_1^3) - (0.225y_2^1 + 0.27y_2^2) - (0.045y_3^1 + 0.045y_3^2)$
$\quad - (0.045y_4^1 + 0y_4^2) = 0.1$

2) $-(0.36y_1^1 + 0.045y_1^2 + 0.27y_1^3) + (0.73y_2^1 + 0.64y_2^2) - (0.585y_3^1 + 0.225y_3^2)$
$\quad - (0.18y_4^1 + 0.09y_4^2) = 0.2$

3) $-(0.315y_1^1 + 0.18y_1^2 + 0.09y_1^3) - (0.315y_2^1 + 0.225y_2^2) + (0.775y_3^1 + 0.55y_3^2)$
$\quad - (0.36y_4^1 + 0.27y_4^2) = 0.3$

4) $-(0.09y_1^1 + 0.27y_1^2 + 0y_1^3) - (0.09y_2^1 + 0.045y_2^2) - (0.045y_3^1 + 0.18y_3^2)$
$\quad + (0.685y_4^1 + 0.46y_4^2) = 0.4$

$y_1^1 \geq 0,\ y_1^2 \geq 0,\ y_1^3 \geq 0,\ y_2^1 \geq 0,\ y_2^2 \geq 0,\ y_3^1 \geq 0,\ y_3^2 \geq 0,\ y_4^1 \geq 0,\ y_4^2 \geq 0.$

5.2.2.3.4 *Solution by LP of a Discounted MDP Model of Monthly Sales*

The LP formulation of the discounted MDP model of monthly sales is solved on a personal computer using LP software to find an optimal policy. The output of the LP software is summarized in Table 5.73.

The output of the LP software shows that all the y_i^k equal zero, except for $y_1^2 = 1.3559$, $y_2^2 = 2.0515$, $y_3^2 = 3.3971$, and $y_4^2 = 3.1954$. The optimal policy is given by $^{16}\mathbf{d} = [2\ 2\ 2\ 2]^T$. This optimal policy was previously found by value iteration in Table 5.68 of Section 5.2.2.1.2, and by PI in Equation (5.71) of Section 5.2.2.2.3. This is also the same optimal policy obtained for the

TABLE 5.73

LP Solution of Discounted MDP Model of Monthly Sales

		Function value = 49.5172		
Objective i	k	y_i^k		
1	1	0		
	2	1.3559	Row	Dual Variable
	3	0	Constraint 1	6.8040
2	1	0	Constraint 2	35.4613
	2	2.0515	Constraint 3	32.2190
3	1	0	Constraint 4	80.1970
	2	3.3971		
4	1	0		
	2	3.1954		

undiscounted, recurrent MDP model of monthly sales in Sections 5.1.2.3.2.2, 5.1.2.3.3.3, and 5.1.2.3.4.3. The conditional probabilities,

$$P(\text{decision} = k | \text{state} = i) = y_i^k / \sum_{k=1}^{K_i} y_i^k,$$

are calculated below:

$$P(\text{decision} = 2 | \text{state} = 1) = P(\text{decision} = 2 | \text{state} = 2) = P(\text{decision}$$
$$= 2 | \text{state} = 3) = P(\text{decision} = 2 | \text{state} = 4) = 1.$$

All the remaining conditional probabilities are zero.

The dual variables associated with constraints 1, 2, 3, and 4, respectively, are the expected total discounted rewards, $v_1 = 6.8040$, $v_2 = 35.4613$, $v_3 = 32.2190$, and $v_4 = 80.1970$, respectively. The expected total discounted rewards obtained as dual variables by LP are identical to those obtained by PI in Equation (5.72) of Section 5.2.2.2.3. Since $\sum_{j=1}^{N} b_j = 1$, the

optimal value of the LP objective function, 49.5172, equals $\sum_{j=1}^{4} b_j v_j$, as is stated at the end of Section 5.2.2.3.1.

That is,

$$\sum_{j=1}^{4} b_j v_j = 49.5172 = 0.1(6.8040) + (0.2)(35.4613) + (0.3)(32.2190) + (0.4)(80.1970)$$

5.2.2.4 Examples of Discounted MDP Models

Discounted MDP models over an infinite horizon will be constructed for an inventory system and for a modified form of the secretary problem.

5.2.2.4.1 Inventory System

The inventory system modeled as a unichain MDP in Section 5.1.3.3.1 will be treated in this section as a discounted MDP [1]. Using a discount factor of $\alpha = 0.9$, the discounted MDP model will be formulated as an LP in Section 5.2.2.4.1.1, and solved in Section 5.2.2.4.1.2 to find an optimal policy. Data for the discounted MDP model of an inventory system are given in Table 5.41 of Section 5.1.3.3.1.

5.2.2.4.1.1 Formulation of a Discounted MDP Model of an Inventory System as an LP As Section 5.1.3.3.1 indicates, the LP formulations for both the unichain and discounted models have $N = 4$ states, $K_0 = 4$ decisions in state 0, $K_1 = 3$ decisions in state 1, $K_2 = 2$ decisions in state 2, and $K_3 = 1$ decision in state 3. The positive constants for the right hand sides of the constraints in the discounted model are arbitrarily chosen to be $b_1 = 0.1$, $b_2 = 0.2$, $b_3 = 0.3$, and $b_4 = 0.4$. Note that $\sum_{j=1}^{N} b_j = 1$.

Objective Function

The objective function for the LP for the discounted model, unchanged from the objective function for the unichain model, is

$$\text{Maximize} \quad (-48y_0^0 + 35y_0^1 - 18y_0^2 - 110y_0^3) + (175y_1^0 + 102y_1^1 + 10y_1^2)t$$
$$+ (242y_2^0 + 130y_2^1) + 270y_3^0.$$

The constraints are given below:

(1) $(y_0^0 + y_0^1 + y_0^2 + y_0^3) - 0.9(y_0^0 + 0.7y_0^1 + 0.3y_0^2 + 0.2y_0^3) - 0.9(0.7y_1^0 + 0.3y_1^1$
$+ 0.2y_1^2) - 0.9(0.3y_2^0 + 0.2y_2^1) - 0.9(0.2y_3^0) = 0.1$

(2) $(y_1^0 + y_1^1 + y_1^2) - 0.9(0.3y_0^0 + 0.4y_0^2 + 0.1y_0^3) - 0.9(0.3y_1^0 + 0.4y_1^1 + 0.1y_1^2)$
$- 0.9(0.4y_2^0 + 0.1y_2^1) - 0.9(0.1)y_3^0 = 0.2$

(3) $(y_2^0 + y_2^1) - 0.9(0.3y_0^0 + 0.4y_0^2) - 0.9(0.3y_1^1 + 0.4y_1^1) - 0.9(0.3y_2^0 + 0.4y_2^1)$
$- 0.9(0.4)y_3^0 = 0.3$

(4) $y_3^0 - 0.9(0.3y_0^3 + 0.3y_1^2 + 0.3y_2^1 + 0.3y_3^0) = 0.4$

$y_0^0 \geq 0$, $y_0^1 \geq 0$, $y_0^2 \geq 0$, $y_0^3 \geq 0$, $y_1^0 \geq 0$, $y_1^1 \geq 0$, $y_1^2 \geq 0$, $y_2^0 \geq 0$, $y_2^1 \geq 0$, $y_3^0 \geq 0$.

The complete LP appears below:

$$\text{Maximize} \quad (-48y_0^0 + 35y_0^1 - 18y_0^2 - 110y_0^3) + (175y_1^0 + 102y_1^1 + 10y_1^2)$$
$$+ (242y_2^0 + 130y_2^1) + 270y_3^0$$

subject to

(1) $(0.1y_0^0 + 0.37y_0^1 + 0.73y_0^2 + 0.82y_0^3) - (0.63y_1^0 + 0.27y_1^1 + 0.18y_1^2)$
$-(0.27y_2^0 + 0.18y_2^1) - 0.18y_3^0 = 0.1$

(2) $-(0.27y_0^1 + 0.36y_0^2 + 0.09y_0^3) + (0.73y_1^0 + 0.64y_1^1 + 0.91y_1^2)$
$-(0.36y_2^0 + 0.09y_2^1) - 0.09y_3^0 = 0.2$

(3) $-(0.27y_0^2 + 0.36y_0^3) - (0.27y_1^1 + 0.36y_1^2) + (0.73y_2^0 + 0.64y_2^1)$
$-0.36y_3^0 = 0.3$

(4) $-(0.27y_0^3 + 0.27y_1^2 + 0.27y_2^1) + 0.73y_3^0 = 0.4$

$y_0^0 \geq 0,\ y_0^1 \geq 0,\ y_0^2 \geq 0,\ y_0^3 \geq 0,\ y_1^0 \geq 0,\ y_1^1 \geq 0,\ y_1^2 \geq 0,\ y_2^0 \geq 0,\ y_2^1 \geq 0, y_3^0 \geq 0.$

5.2.2.4.1.2 Solution by LP of Discounted MDP Model of an Inventory System   The
LP formulation of the discounted MDP model of an inventory system is
solved on a personal computer using LP software to find an optimal policy.
The output of the LP software is summarized in Table 5.74.

Table 5.74 shows that the value of the objective function is 1218.2572. The
output also shows that all the y_i^k equal zero, except for $y_0^3 = 2.2220$, $y_1^2 = 2.0661$,
$y_2^0 = 3.5780$, and $y_3^0 = 2.1339$. The optimal policy is given by $\mathbf{d} = [3\ 2\ 0\ 0]^T$,
which is the same optimal policy obtained for the undiscounted, unichain

TABLE 5.74

LP Solution of Discounted MDP Model of an Inventory System

Objective i	k	y_i^k	Row	Dual Variable
		Function Value = 1218.2572		
0	0	0		
	1	0	Row	Dual Variable
	2	0	Constraint 1	964.7156
	3	2.2220	Constraint 2	1084.7156
1	0	0	Constraint 3	1223.2477
	1	0	Constraint 4	1344.7156
	2	2.0661		
2	0	3.5780		
	1	0		
3	0	2.1339		

MDP model of an inventory system in Table 5.42 of Section 5.1.3.3.1.2. The conditional probabilities,

$$P(\text{order} = k| \text{state} = i) = \frac{y_i^k}{\sum\limits_{k=1}^{K_i} y_i^k},$$

are calculated below:

$$P(\text{order} = 3| \text{state} = 0) = P(\text{order} = 2| \text{state} = 1) = P(\text{order} = 0| \text{state} = 2)$$
$$= P(\text{order} = 0| \text{state} = 3) = 1.$$

All the remaining conditional probabilities are zero. The retailer's expected total discounted profit is maximized by the same (2, 3) inventory policy that maximized her expected average profit per period without discounting in Section 5.1.3.3.1.2.

The dual variables associated with constraints 1, 2, 3, and 4, respectively, are the expected total discounted rewards, $v_0 = 964.7156$, $v_1 = 1084.7156$, $v_2 = 1223.2477$, and $v_3 = 1344.7156$, respectively. Since $\sum_{j=1}^{N} b_j = 1$, the optimal value of the LP objective function, 1218.2752, equals $\sum_{j=0}^{3} b_j v_j$, as is stated at the end of Section 5.2.2.3.1. That is,

$$\sum_{j=0}^{3} b_j v_j = 1218.2752 = 0.1(964.7156) + (0.2)(1084.7156) + (0.3)(1223.2477)$$

$$+ (0.4)(1344.7156).$$

5.2.2.4.2 Optimal Stopping Over an Infinite Planning Horizon: The Secretary Problem

In Section 5.1.3.3.3.1, the secretary problem was treated as an optimal stopping problem over a finite planning horizon by formulating it as an undiscounted, unichain MDP. Recall that the state space is augmented with an absorbing state Δ that is reached with probability 1 when a candidate is hired and with probability 0 when an applicant is rejected. Value iteration was executed to find an optimal policy. A modified form of an optimal stopping problem can be analyzed over an infinite planning horizon by formulating it as a discounted MDP. When rewards are discounted, the secretary problem can be analyzed over an infinite horizon by making the following two modifications in the undiscounted model analyzed over a finite horizon. (1) There is no limit to the number of candidates who may be interviewed. (2) If the nth applicant is rejected, the executive must pay a continuation cost to interview the next candidate. Recall that the state, denoted by X_n, is the numerical score assigned to the nth candidate. The continuation cost is equivalent to a numerical score of -2. Hence, if the nth applicant is hired, the daily reward is X_n, the applicant's numerical score. If the nth applicant is rejected, the daily reward is $X_n = -2$, the continuation cost. A daily reward of zero is associated with the absorbing state Δ because when state Δ is reached, no more interviews will be conducted [1, 3, 4].

5.2.2.4.2.1 Formulation of the Secretary Problem as a Discounted MDP   The discounted MCR model of the secretary problem analyzed over an infinite planning horizon is shown in Table 5.75. LP decision variables y_i^k are shown in the right-hand-column.

There are four alternative policies, which correspond to hiring a candidate rated poor (15), fair (20), good (25), or excellent (30).

5.2.2.4.2.2 Solution of the Secretary Problem by LP   The discounted MDP model of the secretary problem over an infinite planning horizon is formulated as the following LP. In this example $N = 5$ states, and $K_{15} = K_{20} = K_{25} = K_{30} = K_\Delta = K = 2$ decisions in every state. The LP decision variables associated with the absorbing state Δ, and the constraint associated with a transition to the absorbing state Δ can be omitted from the LP formulation. A daily discount factor of $\alpha = 0.9$ is specified. The positive constants are arbitrarily chosen to be $b_1 = b_2 = b_3 = b_4 = 0.25$. Note that they sum to one.

TABLE 5.75

Data for Discounted MDP Model of a Secretary Problem Over an Infinite Planning Horizon

State i	Decision k	Transition Probility					Reward q_i^k	LP Variable y_i^k
		$p_{i,15}^k$	$p_{i,20}^k$	$p_{i,25}^k$	$p_{i,30}^k$	$p_{i,\Delta}^k$		
15	1 = H	0	0	0	0	1	15	y_{15}^1
	2 = R	0.3	0.4	0.2	0.1	0	-2	y_{15}^2
20	1 = H	0	0	0	0	1	20	y_{20}^1
	2 = R	0.3	0.4	0.2	0.1	0	-2	y_{20}^2
25	1 = H	0	0	0	0	1	25	y_{25}^1
	2 = R	0.3	0.4	0.2	0.1	0	-2	y_{25}^2
30	1 = H	0	0	0	0	1	30	y_{30}^1
	2 = R	0.3	0.4	0.2	0.1	0	-2	y_{30}^2
Δ	1 = H	0	0	0	0	1	0	y_Δ^1
	2 = R	0	0	0	0	1	0	y_Δ^2

Objective Function
The objective function for the LP is

$$\text{Maximize } \sum_{i=15}^{30} \sum_{k=1}^{2} q_i^k \, y_i^k$$

$$= \sum_{k=1}^{2} q_{15}^k \, y_{15}^k + \sum_{k=1}^{2} q_{20}^k \, y_{20}^k + \sum_{k=1}^{2} q_{25}^k \, y_{25}^k + \sum_{k=1}^{2} q_{30}^k \, y_{30}^k$$

$$= (q_{15}^1 y_{15}^1 + q_{15}^2 y_{15}^2) + (q_{20}^1 y_{20}^1 + q_{20}^2 y_{20}^2) + (q_{25}^1 y_{25}^1 + q_{25}^2 y_{25}^2) + (q_{30}^1 y_{30}^1 + q_{30}^2 y_{30}^2)$$

$$= (15 y_{15}^1 - 2 y_{15}^2) + (20 y_{20}^1 - 2 y_{20}^2) + (25 y_{25}^1 - 2 y_{25}^2) + (30 y_{30}^1 - 2 y_{30}^2).$$

Constraints
State 15 has $K_{15} = 2$ possible decisions. The constraint associated with a transition to state $j = 15$ is

$$\sum_{k=1}^{2} y_{15}^k - \alpha \sum_{i=15}^{30} \sum_{k-1}^{2} y_i^k \, p_{i,15}^k = b_1$$

$$(y_{15}^1 + y_{15}^2) - \alpha \left(\sum_{k=1}^{2} y_{15}^k p_{15,15}^k + \sum_{k=1}^{2} y_{20}^k p_{20,15}^k + \sum_{k=1}^{2} y_{25}^k p_{25,15}^k + \sum_{k=1}^{2} y_{30}^k p_{30,15}^k \right) = b_1$$

$$(y_{15}^1 + y_{15}^2) - \alpha(y_{15}^1 p_{15,15}^1 + y_{15}^2 p_{15,15}^2) - \alpha(y_{20}^1 p_{20,15}^1 + y_{20}^2 p_{20,15}^2) - \alpha(y_{25}^1 p_{25,15}^1 + y_{25}^2 p_{25,15}^2)$$
$$- \alpha(y_{30}^1 p_{30,15}^1 + y_{30}^2 p_{30,15}^2) = b_1$$

$$(y_{15}^1 + y_{15}^2) - 0.9(0 y_{15}^1 + 0.3 y_{15}^2) - 0.9(0 y_{20}^1 + 0.3 y_{20}^2) - 0.9(0 y_{25}^1 + 0.3 y_{25}^2)$$
$$- 0.9(0 y_{30}^1 + 0.3 y_{30}^2) = 0.25.$$

State 20 has $K_{20} = 2$ possible decisions. The constraint associated with a transition to state $j = 20$ is

$$\sum_{k=1}^{2} y_{20}^k - \alpha \sum_{i=15}^{30} \sum_{k=1}^{2} y_i^k \, p_{i,20}^k = b_2$$

$$(y_{20}^1 + y_{20}^2) - \alpha(\sum_{k=1}^{2} y_{15}^k p_{15,20}^k + \sum_{k=1}^{2} y_{20}^k p_{20,20}^k + \sum_{k=1}^{2} y_{25}^k p_{25,20}^k + \sum_{k=1}^{2} y_{30}^k p_{30,20}^k) = b_2$$

$$(y_{20}^1 + y_{20}^2) - \alpha(y_{15}^1 p_{15,20}^1 + y_{15}^2 p_{15,20}^2) - \alpha(y_{20}^1 p_{20,20}^1 + y_{20}^2 p_{20,20}^2) - \alpha(y_{25}^1 p_{25,20}^1 + y_{25}^2 p_{25,20}^2)$$
$$- \alpha(y_{30}^1 p_{30,20}^1 + y_{30}^2 p_{30,20}^2) = b_2$$

$$(y_{20}^1 + y_{20}^2) - 0.9(0 y_{15}^1 + 0.4 y_{15}^2) - 0.9(0 y_{20}^1 + 0.4 y_{20}^2) - 0.9(0 y_{25}^1 + 0.4 y_{25}^2)$$
$$- 0.9(0 y_{30}^1 + 0.4 y_{30}^2) = 0.25.$$

State 25 has $K_{25} = 2$ possible decisions. The constraint associated with a transition to state $j = 25$ is

$$\sum_{k=1}^{2} y_{25}^{k} - \alpha \sum_{i=15}^{30} \sum_{k=1}^{2} y_{i}^{k} \, p_{i,25}^{k} = b_3$$

$$(y_{25}^{1} + y_{25}^{2}) - \alpha \left(\sum_{k=1}^{2} y_{15}^{k} p_{15,25}^{k} + \sum_{k=1}^{2} y_{20}^{k} p_{20,25}^{k} + \sum_{k=1}^{2} y_{25}^{k} p_{25,25}^{k} + \sum_{k=1}^{2} y_{30}^{k} p_{30,25}^{k} \right) = b_3$$

$$(y_{25}^{1} + y_{25}^{2}) - \alpha(y_{15}^{1} p_{15,25}^{1} + y_{15}^{2} p_{15,25}^{2}) - \alpha(y_{20}^{1} p_{20,25}^{1} + y_{20}^{2} p_{20,25}^{2}) - \alpha(y_{25}^{1} p_{25,25}^{1} + y_{25}^{2} p_{25,25}^{2})$$
$$- \alpha(y_{30}^{1} p_{30,25}^{1} + y_{30}^{2} p_{30,25}^{2}) = b_3$$

$$(y_{25}^{1} + y_{25}^{2}) - 0.9(0y_{15}^{1} + 0.2y_{15}^{2}) - 0.9(0y_{20}^{1} + 0.2y_{20}^{2}) - 0.9(0y_{25}^{1} + 0.2y_{25}^{2})$$
$$- 0.9(0y_{30}^{1} + 0.2y_{30}^{2}) = 0.25.$$

State 30 has $K_{30} = 2$ possible decisions. The constraint associated with a transition to state $j = 30$ is

$$\sum_{k=1}^{2} y_{30}^{k} - \alpha \sum_{i=15}^{30} \sum_{k=1}^{2} y_{i}^{k} \, p_{i,30}^{k} = b_4$$

$$(y_{30}^{1} + y_{30}^{2}) - \alpha \left(\sum_{k=1}^{2} y_{15}^{k} p_{15,30}^{k} + \sum_{k=1}^{2} y_{20}^{k} p_{20,30}^{k} + \sum_{k=1}^{2} y_{25}^{k} p_{25,30}^{k} + \sum_{k=1}^{2} y_{30}^{k} p_{30,30}^{k} \right) = b_4$$

$$(y_{30}^{1} + y_{30}^{2}) - \alpha(y_{15}^{1} p_{15,30}^{1} + y_{15}^{2} p_{15,30}^{2}) - \alpha(y_{20}^{1} p_{20,30}^{1} + y_{20}^{2} p_{20,30}^{2}) - \alpha(y_{25}^{1} p_{25,30}^{1} + y_{25}^{2} p_{25,30}^{2})$$
$$- \alpha(y_{30}^{1} p_{30,30}^{1} + y_{30}^{2} p_{30,30}^{2}) = b_4$$

$$(y_{30}^{1} + y_{30}^{2}) - 0.9(0y_{15}^{1} + 0.1y_{15}^{2}) - 0.9(0y_{20}^{1} + 0.1y_{20}^{2}) - 0.9(0y_{25}^{1} + 0.1y_{25}^{2})$$
$$- 0.9(0y_{30}^{1} + 0.1y_{30}^{2}) = 0.25.$$

The complete LP formulation for the discounted MDP model of the secretary problem analyzed over an infinite planning horizon is

Maximize $(15y_{15}^{1} - 2y_{15}^{2}) + (20y_{20}^{1} - 2y_{20}^{2}) + (25y_{25}^{1} - 2y_{25}^{2}) + (30y_{30}^{1} - 2y_{30}^{2})$

subject to

(1) $(y_{15}^{1} + 0.73y_{15}^{2}) - (0.27y_{20}^{2} + 0.27y_{25}^{2} + 0.27y_{30}^{2}) = 0.25$

(2) $(y_{20}^{1} + 0.64y_{20}^{2}) - (0.36y_{15}^{2} + 0.36y_{25}^{2} + 0.36y_{30}^{2}) = 0.25$

(3) $(y_{25}^{1} + 0.82y_{25}^{2}) - (0.18y_{15}^{2} + 0.18y_{20}^{2} + 0.18y_{30}^{2}) = 0.25$

(4) $(y_{30}^{1} + 0.91y_{30}^{2}) - (0.09y_{15}^{2} + 0.09y_{20}^{2} + 0.09y_{25}^{2}) = 0.25$

$y_1^2 \geq 0, \ y_1^2 \geq 0, \ y_1^3 \geq 0, \ y_2^1 \geq 0, \ y_2^2 \geq 0, \ y_3^1 \geq 0, \ y_3^2 \geq 0, \ y_4^1 \geq 0, \ y_4^2 \geq 0.$

The LP formulation of the discounted MDP model of the secretary problem is solved on a personal computer using LP software to find an optimal policy. The output of the LP software is summarized in Table 5.76.

The output of the LP software shows that $y_{15}^2 = 0.3425$, $y_{20}^1 = 0.3733$, $y_{25}^1 = 0.3116$, $y_{30}^1 = 0.2808$, and the remaining y_i^k equal zero. The optimal policy, given by $d = [2\ 1\ 1\ 1]^T$, is to reject a candidate rated 15, and hire a candidate rated 20 or 25 or 30. The associated conditional probabilities,

$$P(\text{decision} = k|\text{state} = i)\ y_i^k / \sum_{k=1}^{K_i} y_i^k,$$

are calculated below:

$$P(\text{decision} = 2|\text{ state} = 15) = P(\text{decision} = 1|\text{ state} = 20) = P(\text{decision}$$
$$= 1|\text{ state} = 25) = P(\text{decision} = 1|\text{ state} = 30) = 1.$$

All the remaining conditional probabilities are zero.

The dual variables associated with constraints 1, 2, 3, and 4, respectively, are the expected total discounted rewards, $v_{15} = 16.9863$, $v_{20} = 20$, $v_{25} = 25$, and $v_{30} = 30$, respectively. Since $\sum_{j=1}^{N} b_j = 1$, the optimal value of the LP objective function, 22.9966, equals $\sum_{j=1}^{4} b_j v_j$, as is stated at the end of Section 5.2.2.3.1. That is,

$$\sum_{j=1}^{4} b_j v_j = 22.9966 = 0.25(16.9863) + (0.25)(20) + (0.25)(25) + (0.25)(30).$$

TABLE 5.76

LP Solution of Discounted MDP Model of the Secretary Problem

Function = 22.9966				
Objective i	k	y_i^k		
15	1	0	Row	Dual Variable
	2	0.3425	Constraint 1	16.9863
20	1	0.3733	Constraint 2	20
	2	0	Constraint 3	25
25	1	0.3116	Constraint 4	30
	2	0		
30	1	0.2808		
	2	0		

It is instructive to note that the MCR associated with the optimal policy given by $\mathbf{d} = [2\ 1\ 1\ 1]^T$, with the absorbing state Δ restored, is

$$
P = \begin{array}{c} 15 \\ 20 \\ 25 \\ 30 \\ \Delta \end{array}
\begin{bmatrix}
0.3 & 0.4 & 0.2 & 0.1 & 0 \\
0 & 0 & 0 & 0 & 1 \\
0 & 0 & 0 & 0 & 1 \\
0 & 0 & 0 & 0 & 1 \\
0 & 0 & 0 & 0 & 1
\end{bmatrix}, \quad
q = \begin{array}{c} 15 \\ 20 \\ 25 \\ 30 \\ \Delta \end{array}
\begin{bmatrix}
-2 \\
20 \\
25 \\
30 \\
0
\end{bmatrix}.
$$

Also,

$$
\alpha P = \begin{array}{c} 15 \\ 20 \\ 25 \\ 30 \\ \Delta \end{array}
\begin{bmatrix}
0.27 & 0.36 & 0.18 & 0.09 & 0 \\
0 & 0 & 0 & 0 & 0.9 \\
0 & 0 & 0 & 0 & 0.9 \\
0 & 0 & 0 & 0 & 0.9 \\
0 & 0 & 0 & 0 & 0.9
\end{bmatrix}
$$

$$
(I - \alpha P) = \begin{array}{c} 15 \\ 20 \\ 25 \\ 30 \\ \Delta \end{array}
\begin{bmatrix}
0.73 & -0.36 & -0.18 & -0.09 & 0 \\
0 & 1 & 0 & 0 & -0.9 \\
0 & 0 & 1 & 0 & -0.9 \\
0 & 0 & 0 & 1 & -0.9 \\
0 & 0 & 0 & 0 & 0.10
\end{bmatrix}.
$$

Recall that the matrix equation (4.253) in Section 4.3.3.1 is an alternate form of the VDEs. Solving the matrix equation (4.253) for the vector of expected total discounted rewards,

$$
v = (I - \alpha P)^{-1} q = \begin{array}{c} 15 \\ 20 \\ 25 \\ 30 \\ \Delta \end{array}
\begin{bmatrix}
1.369863 & 0.493151 & 0.246575 & 0.123288 & 7.767123 \\
0 & 1 & 0 & 0 & 9 \\
0 & 0 & 1 & 0 & 9 \\
0 & 0 & 0 & 1 & 9 \\
0 & 0 & 0 & 0 & 10
\end{bmatrix}
\begin{bmatrix}
-2 \\
20 \\
25 \\
30 \\
0
\end{bmatrix}
$$

$$
= \begin{array}{c} 15 \\ 20 \\ 25 \\ 30 \\ \Delta \end{array}
\begin{bmatrix}
16.9863 \\
20 \\
25 \\
30 \\
0
\end{bmatrix}.
$$

(5.79)

Substituting the components of vector v into the equations of the IM routine will confirm that $\mathbf{d} = [2\ 1\ 1\ 1]^T$ is an optimal policy. Table 5.76 shows, once again, that under an optimal policy, the expected total discounted reward in every state is equal to the corresponding dual variable in the optimal tableau of the LP formulation.

PROBLEMS

5.1 A woman wishes to sell a car within the next 4 weeks. She expects to receive one bid or offer each week from a prospective buyer. The weekly offer is a random variable, which has the following stationary probability distribution:

Offer per week	$16,000	$18,000	$22,000	$24,000
Probability	0.40	0.10	0.30	0.20

Once the woman accepts an offer, the bidding stops. Her objective is to maximize her expected total income over a 4-week planning horizon.
(a) Formulate this optimal stopping problem as a unichain MDP.
(b) Use value iteration to find an optimal policy which will specify when to accept or reject an offer during each week of the planning horizon.

5.2 In Problem 3.10, suppose that the investor receives a monthly dividend of $1 per share. No dividend is received when the stock is sold. The investor uses a monthly discount factor of $\alpha = 0.9$. She wants to determine when to sell and when to hold the stock. Her objective is to maximize her expected total discounted reward.
(a) Treat this problem as an optimal stopping problem over an infinite planning horizon. Formulate this optimal stopping problem as a discounted, unichain MDP.
(b) Formulate the discounted, unichain MDP as a LP, and solve it to find an optimal policy.
(c) Use PI to find an optimal policy.

5.3 A consumer electronics retailer can place orders for flat panel TVs at the beginning of each day. All orders are delivered immediately. Every time an order is placed for one or more TVs, the retailer pays a fixed cost of $60. Every TV ordered costs the retailer $100. The daily holding cost per unsold TV is $10. A daily shortage cost of $180 is incurred for each TV that is not available to satisfy demand. The retailer can accommodate a maximum inventory of two TVs. The daily demand for TVs is an independent, identically distributed random variable which has the following stationary probability distribution:

Daily demand d	0	1	2
Probability, $p(d)$	0.5	0.4	0.1

(a) Formulate this model as an MDP.
(b) Determine an optimal ordering policy that will minimize the expected total cost over the next 3 days.
(c) Use LP to determine an optimal ordering policy that will minimize the expected average cost per day over an infinite planning horizon.
(d) Use PI to determine an optimal ordering policy over an infinite planning horizon.
(e) Use LP with a discount factor of $\alpha = 0.9$ to find an optimal ordering policy over an infinite planning horizon.
(f) Use PI with a discount factor of $\alpha = 0.9$ to find an optimal ordering policy and the associated expected total discounted cost vector over an infinite planning horizon.

5.4   At the start of each day, the condition of a machine is classified as either excellent, acceptable, or poor. The daily behavior of the machine is modeled as a three-state absorbing unichain with the following transition probability matrix:

$P = [p_{ij}] =$ State	E	A	P
Excellent (E)	0.1	0.7	0.2
Acceptable (A)	0	0.4	0.6
Poor (P)	0	0	1

A machine in excellent condition earns revenue of $600 per day. A machine in acceptable condition earns daily revenue of $300, and a machine in poor condition earns daily revenue of $100. At the start of each day, the engineer responsible for the machine can make one of the following three maintenance decisions: do nothing, repair the machine, or replace it with a new machine. One day is needed to repair the machine at a cost of $500, or to replace it at a cost of $1,000. A machine which is repaired starts the following day in excellent condition with probability 0.6, in acceptable condition with probability 0.3, or in poor condition with probability 0.1. A machine that is replaced always starts the following day in excellent condition.
(a) Formulate this model as an MDP.
(b) Determine an optimal maintenance policy that will maximize the expected total profit over the next 3 days.
(c) Use LP to determine an optimal maintenance policy that will maximize the expected average profit per day over an infinite planning horizon.
(d) Use PI to determine an optimal maintenance policy over an infinite planning horizon.
(e) Use LP with a discount factor of $\alpha = 0.9$ to find an optimal maintenance policy over an infinite planning horizon.
(f) Use PI with a discount factor of $\alpha = 0.9$ to find an optimal maintenance policy and the associated expected total discounted reward vector over an infinite planning horizon.

5.5 Suppose that the condition of a machine can be described by one of the following three states:

State	Condition
NW	Not Working
WI	Working Intermittently
WP	Working Properly

The daily behavior of the machine when it is left alone for 1 day is modeled as an absorbing unichain with the following transition probability matrix:

$$P = [p_{ij}] = \begin{array}{l|ccc} \text{State} & \text{NW} & \text{WI} & \text{WP} \\ \hline \text{Not Working (NW)} & 1 & 0 & 0 \\ \text{Working Intermittently (WI)} & 0.8 & 0.2 & 0 \\ \text{Working Properly (WP)} & 0.2 & 0.5 & 0.3 \end{array}$$

The machine is observed at the start of each day. Suppose initially that the machine is in state WP at the beginning of the day. If, with probability 0.3, it works properly throughout the day, it earns $1,000 in revenue. If, with probability 0.5, the machine deteriorates during the day, it earns $500 in revenue and enters state WI, working intermittently. If, with probability 0.2, the machine fails during the day, it earns zero revenue and enters state NW, not working. Suppose next that the machine starts the day in state WI. If, with probability 0.2, it works intermittently throughout the day, it earns $500 in revenue. If, with probability 0.8, the machine fails during the day, it earns zero revenue and enters state NW. Daily revenue is summarized in the table below:

State	Condition	Daily Revenue
NW	Not Working	$0
WI	Working Intermittently	$500
WP	Working Properly	$1000

At the beginning of the day, the engineer responsible for the machine can make one of the following four maintenance decisions: do nothing (DN), perform preventive maintenance (PM), repair the machine (RP), or replace the machine (RL). The costs of the maintenance actions are summarized in the table below:

Maintenance Action	Cost
Do Nothing (DN)	$0
Preventive Maintenance (PM)	$100
Repair (RP)	$400
Replace (RL)	$1600

All maintenance actions are completed instantly. However, neither preventive maintenance nor a repair are always successful. The following maintenance actions are feasible in each state:

State	Maintenance Action
Not Working (NW)	Repair (RP), Replace (RL)
Working Intermittently (WI)	Repair (RP), Preventive Maintenance (PM)
Working Properly (WP)	Preventive Maintenance (PM), Do Nothing (DN)

Replacement, and a repair which is completely successful, always produces a machine that works properly throughout the day. A partly successful repair when the machine is in state NW enables a machine to work intermittently throughout the day. An unsuccessful repair when the machine is in state NW leaves the machine not working throughout the day. An unsuccessful repair when the machine is in state WI leaves the machine working intermittently throughout the day. Preventive maintenance which is completely successful enables a machine to remain in its current state throughout the day.

When the machine begins the day in state NW, a repair is completely successful with probability 0.5, partly successful with probability 0.4, and unsuccessful with probability 0.1. When the machine starts the day in state WI, a repair is successful with probability 0.6, and unsuccessful with probability 0.4. When the machine begins the day in state WP, preventive maintenance is completely successful with probability 0.7, partly successful with probability 0.2, and unsuccessful with probability 0.1. When the machine starts the day in state WI, preventive maintenance is successful with probability 0.8, and unsuccessful with probability 0.2.

(a) Construct the vector of expected immediate rewards associated with each state and decision.
(b) Formulate this model as an MDP.
(c) Determine an optimal maintenance policy that will maximize the expected total profit over the next 3 days.
(d) Use LP to determine an optimal maintenance policy that will maximize the expected average profit per day over an infinite planning horizon.
(e) Use PI to determine an optimal maintenance policy over an infinite planning horizon.
(f) Use LP with a discount factor of $\alpha = 0.9$ to find an optimal maintenance policy over an infinite planning horizon.
(g) Use PI with a discount factor of $\alpha = 0.9$ to find an optimal maintenance policy and the associated expected total discounted reward vector over an infinite planning horizon.

5.6 In Problem 4.1, suppose that the engineer responsible for the machine has the options of bringing a machine with a major defect or a minor defect to the repair process. When the engineer elects to bring a machine with a major defect or a minor defect to the repair process, the machine makes a transition with probability 1 to state 1 (NW). The daily costs to bring a machine in state 2 (MD) and state 3 (mD) to the repair process are $80 and $40, respectively. A repair takes one day to complete. No revenue is earned on days during which a machine is repaired. When the machine is in state 1 (NW), it is always under repair. Hence, the only feasible action in state 1 is to repair (RP) the machine. Since a machine in state 4 (WP) is never brought to the repair process, the only feasible action in state 4 (WP) is to do nothing (DN). The following four policies are feasible:

State	Policy 1	Policy 2	Policy 3	Policy 4
1 (NW)	RP	RP	RP	RP
2 (MD)	DN	DN	RP	RP
3 (mD)	DN	RP	DN	RP
4 (WP)	DN	DN	DN	DN

The daily revenue earned in every state under policies 1 and 4 is calculated in the table below:

State	Daily revenue	Repair Cost	Policy 1	Daily Reward of Policy 1	Policy 4	Daily Reward of Policy
1 (NW)	$0	$0	Repair	$0	Repair	$0
2 (MD)	$200	$80	DN	$200	Repair	−$80
3 (mD)	$400	$40	DN	$400	Repair	−$40
4 (WP)	$600		DN	$600	DN	$600

(a) Model this machine repair problem as a recurrent MDP.
(b) Find an optimal repair policy and its expected total reward over a three day planning horizon.
(c) Use LP to find an optimal repair policy and its expected average reward, or gain, over an infinite planning horizon.
(d) Use PI to find an optimal repair policy and its expected average reward, or gain, over an infinite planning horizon.
(e) Use LP with a discount factor of $\alpha = 0.9$ to find an optimal repair policy over an infinite planning horizon.
(f) Use PI with a discount factor of $\alpha = 0.9$ to find an optimal repair policy and the associated expected total discounted reward vector over an infinite planning horizon.

5.7 Consider the following stochastic shortest route problem over a three-period planning horizon for the network shown below:

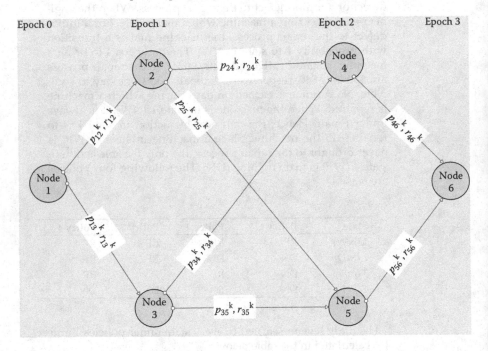

Network through which a stochastic shortest route is to be found

The network has six nodes numbered 1 through 6. One time period is needed to traverse an arc between two adjacent nodes. The origin, node 1, is entered at epoch 0. Nodes 2 and 3 can be reached at epoch 1 in one step from node 1 by traversing arcs (1,2) and (1, 3), respectively. Nodes 4 and 5 can be reached at epoch 2 in one step from node 2 by traversing arcs (2, 4) and (2, 5), respectively. Nodes 4 and 5 can also be reached at epoch 2 in one step from node 3 by traversing arcs (3, 4) and (3, 5), respectively. The final destination, node 6, can be reached at epoch 3 in one step from nodes 4 and 5 by traversing arcs (4, 6) and (5, 6), respectively. The network has no cycles. That is, once a node has been has been reached, it will not be visited again. A decision k at a node i identifies the node k chosen to be reached at the next epoch. The network is stochastic because a decision k made at a node i can result in the process moving to another node instead of the one chosen. Also, the distance, or arc length, between adjacent nodes is a function of the decision made at the current node. The objective is to find a path of minimum expected length from node 1, the origin, to node 6, the final destination.

To model this network problem as an MDP, the following variables are defined.

$d_i(n) = k$ is the decision k made at node i at epoch n.

$p^k_{ij} = P(X_{n+1} = j \mid X_n = i \cap d_i(n) = k) =$ the probability of moving from node i to node j when decision k is made at node i at epoch n.

$r_{ij}^k =$ the actual distance from node i to node j when decision k is made at node i at epoch n and node j is reached at epoch $n + 1$.

$$q^k_i = \sum_{j=i+1}^{6} p^k_{ij} r^k_{ij}, \quad i = 1, 2, ..., 5,$$

is the expected distance from node i to node j when decision k is made at node i at epoch n.

The network is Markovian because the node j reached at epoch $n+1$ depends only on the present node i and the decision k made at epoch n. The transition probabilities and the actual distances corresponding to the decisions made at every node are shown in the table below.

Transition probabilities and actual distances for stochastic shortest route problem

State	Decision	Transition				Probability		Actual				Distance		
i	k	p^k_{i1}	p^k_{i2}	p^k_{i3}	p^k_{i4}	p^k_{i5}	p^k_{i6}	r^k_{i1}	r^k_{i2}	r^k_{i3}	r^k_{i4}	r^k_{i5}	r^k_{i6}	q^k_i
1	2	0	0.7	0.3	0	0	0	0	23	25	0	0	0	
	3	0	0.2	0.8	0	0	0	0	26	22	0	0	0	
2	4	0	0	0	0.4	0.6	0	0	0	0	37	33	0	
	5	0	0	0	0.65	0.35	0	0	0	0	35	36	0	
3	4	0	0	0	0.75	0.25	0	0	0	0	31	34	0	
	5	0	0	0	0.15	0.85	0	0	0	0	28	32	0	
4	6	0	0	0	0	0	1	0	0	0	0	0	46	
5	6	0	0	0	0	0	1	0	0	0	0	0	42	

(a) Calculate the expected distances q_i^k from node i to node j when decision k is made at every node i at epoch n.

(b) Model this stochastic shortest route problem as a unichain MDP.

(c) Use value iteration to find a path of minimum expected length from node 1, the origin, to node 6, the final destination, over a 3 day planning horizon.

5.8   At the beginning of every month, a woman executes diagnostic software on her personal computer to determine whether it is virus-free or is infected with a malignant or benign virus. She models the condition of her computer as a Markov chain with the following three states:

If her computer is virus-free at the start of the current month, the probability that it will be virus-free at the start of the next month is 0.45, the probability that it will be infected with a benign virus is 0.40, and the probability that it will be infected

State	Condition
F	Virus-free
B	Benign virus
M	Malignant virus

with a malignant virus is 0.15. If her computer is infected with a virus, she can hire either virus-removal consultant A or virus-removal consultant B. Both consultants require a full month in which to attempt to remove a virus, but are not always successful. Consultant A charges higher monthly fees than consultant B because she has higher probabilities of success. To remove a benign virus, consultant A charges $160 and consultant B charges $110. To remove a malignant virus, consultant A charges $280 and consultant B charges $200. Consultant A has a 0.64 probability of removing a malignant virus and a 0.28 probability of reducing it to a benign virus. She has a 0.92 probability of removing a benign virus and a 0.08 probability of leaving it unchanged. Consultant B has a 0.45 probability of removing a malignant virus and a 0.38 probability of reducing it to a benign virus. She has a 0.62 probability of removing a benign virus, and a 0.04 probability of transforming it into a malignant virus.

(a) Model this virus-removal problem as a recurrent MDP.
(b) Find an optimal virus-removal policy and its expected total cost over a 3 month planning horizon.
(c) Use LP to find an optimal virus-removal policy and its expected average cost over an infinite planning horizon.
(d) Use PI to find an optimal virus-removal policy and its expected average cost over an infinite planning horizon.
(e) Use LP with a discount factor of $\alpha = 0.9$ to find an optimal virus-removal policy over an infinite planning horizon.
(f) Use PI with a discount factor of $\alpha = 0.9$ to find an optimal virus-removal policy and the associated expected total discounted cost vector over an infinite planning horizon.

5.9   In Problem 4.8, suppose that the manager of the dam is considering the following two policies for releasing water at the beginning of every week: she can release at most 2 units of water, or she can release at most 3 units. The implementation of the decision to release at most 2 units of water was described in Problem 4.8.

Suppose that the decision is made to release at most 3 units of water at the beginning of every week. If the volume of water stored in the dam plus the volume flowing into the dam at the beginning of the week exceeds 3 units, then 3 units of water are released. The first unit of water released is used to generate electricity, which is sold for $5. The second and third units released are used for irrigation which earns $4 per unit. If the volume of water stored in the dam plus the volume flowing into the dam at the beginning of the week equals 3 units, then

2 units of water are released. The first unit of water released is used to generate electricity, which is sold for $5. The second unit released is used for irrigation which earns $4. If the volume of water stored in the dam plus the volume flowing into the dam at the beginning of the week equals 2 units, then only 1 unit of water is released which is used to generate electricity that is sold for $5. If the volume of water stored in the dam plus the volume flowing into the dam at the beginning of the week is less than 2 units, no water is released.

(a) Model the dam problem as a four-state unichain MDP. For each state and decision, calculate the associated transition probabilities and the reward.

(b) Use LP to find a release policy which maximizes the expected average reward, or gain, over an infinite planning horizon.

(c) Use PI to find a release policy which maximizes the expected average reward, or gain, over an infinite planning horizon.

(d) Use LP with a discount factor of $\alpha = 0.9$ to find a release policy which maximizes the vector of expected total discounted rewards.

(e) Use PI with a discount factor of $\alpha = 0.9$ to find an optimal release policy and the associated expected total discounted reward vector over an infinite planning horizon.

5.10 Suppose that in a certain state, every registered professional engineer (P.E.) is required to complete 15 h of continuing professional development (CPD) every year. CPD hours may be earned by successfully completing continuing education courses offered by a professional or trade organization, or offered in-house by a corporation. All registrants need to keep a yearly log showing the type of courses attended. Each year the State Board of Registration for Professional Engineers will audit randomly selected registrants. If selected for an audit, a registrant must submit a CPD activity log and supporting documentation.

A consulting engineer plans to construct a two-state Markov chain model of the possibility that the Board will audit a registrant's record of continuing engineering education. The state variable is

$X_n = 1$ if a registrant was audited at the end of year n,

$X_n = 0$ if a registrant was not audited at the end of year n.

Historical data indicates that an engineer who was not audited last year has a 0.3 probability of being audited this year. An engineer who was audited last year has a 0.2 probability of being audited this year.

The consulting engineer is concerned that a portion of her CPD education consists of in-house courses offered by a corporation which provides engineering products and services that she uses and recommends to her clients. Critics have said that such courses are often little more than company marketing in the guise of education. If she is audited, she must demonstrate that the course provider did not bias the course material in favor of its own products and services. To prepare her CPD log and assemble supporting documentation for a possible audit at the end of the current year, she is considering hiring a prominent engineering educator as an audit consultant. The audit consultant will charge her a fee of $2,000. Experience indicates that the Board often imposes a fine for deficiencies in CPD courses taken in prior years when a registrant is audited. Past records show that when a CPD log is prepared by an audit consultant, the average fine is $600 after a log is audited. However, when a CPD log is prepared by the registrant herself, the average fine is $8,000. The consulting engineer must decide whether or not to hire an audit consultant to prepare her CPD log for a possible audit at the end of the current year. Hiring an audit consultant appears to reduce by 0.05 the probability that a registrant will be audited in the following year:

(a) Formulate the consulting engineer's decision alternatives as a two-state recurrent MDP. For each state and decision, specify the associated transition probabilities and calculate the expected immediate cost.

(b) Execute value iteration to find a policy which minimizes the vector of expected total costs that will be incurred in both states after 3 years. Assume zero terminal costs at the end of year 3.

(c) Use exhaustive enumeration to find a policy that minimizes the expected average cost, or negative gain, over an infinite planning horizon.

(d) Use PI to find an optimal policy over an infinite horizon.

References

1. Hillier, F. S. and Lieberman G. J., *Introduction to Operations Research*, 8th ed., McGraw-Hill, New York, 2005.
2. Howard, R. A., *Dynamic Programming and Markov Processes*, M.I.T. Press, Cambridge, MA, 1960.
3. Puterman, M. L., *Markov Decision Processes: Discrete Stochastic Dynamic Programming*, Wiley, New York, 1994.
4. Wagner, H. M., *Principles of Operations Research*, 2nd ed., Prentice-Hall, Englewood Cliffs, NJ, 1975.

6

Special Topics: State Reduction and Hidden Markov Chains

This chapter introduces two interesting special topics: state reduction in Section 6.1, and hidden Markov chains in Section 6.2. State reduction is an iterative procedure which reduces a Markov chain to a smaller chain from which a solution to the original chain can be found. Roundoff error can be reduced when subtractions are avoided. A hidden Markov chain is one for which the states cannot be observed. When a hidden Markov chain enters a state, only an observation symbol, not the state itself, can be detected. Hidden Markov models have been constructed for various applications such as speech recognition and bioinformatics.

6.1 State Reduction

State reduction is an alterative procedure for finding various quantities for a Markov chain, such as steady-state probabilities, mean first passage times (MFPTs), and absorption probabilities. State reduction has two steps: matrix reduction and back substitution. Each iteration of matrix reduction produces a reduced matrix one state smaller than its predecessor, resulting in a final reduced matrix from which the solution to the original problem can be obtained by back substitution.

State reduction is based on a theorem in Kemeny and Snell [2], and is equivalent to Gaussian elimination. Kemeny and Snell showed that a partitioned Markov chain can be reduced in size by performing matrix operations to produce a smaller Markov chain. They showed that if the steady-state probabilities for the original Markov chain are known, then the steady-state probabilities for the smaller chain are proportional to those for the corresponding states of the original chain. Three state reduction algorithms will be described. The first two compute steady-state probabilities and MFPT, respectively, for a regular Markov chain. The third computes absorption probabilities for a reducible chain.

6.1.1 Markov Chain Partitioning Algorithm for Computing Steady-State Probabilities

The first state reduction algorithm, called the Markov chain partitioning algorithm [4, 1], or MCPA, computes the vector of steady-state probabilities for a regular Markov chain. The MCPA contain two steps: matrix reduction and back substitution. The matrix reduction step successively partitions the transition probability matrix. Each partition of a transition matrix produces four submatrices: a square matrix, a column vector, a single element in the upper left hand corner, and a row vector. When combined in accordance with the theorem in Kemeny and Snell, the four submatrices create a reduced transition matrix that also represents a regular Markov chain. The number of states in the reduced matrix is one less than the number in the original matrix. The reduced matrix equals the sum of the square matrix and the product of the column vector, the reciprocal of one minus the single element, and the row vector. To improve numerical accuracy, subtractions may be eliminated by replacing one minus the single element by the sum of the remaining elements in the top row. Each reduced matrix is partitioned in the same manner as the original matrix, creating a sequence of successively smaller reduced matrices. Each reduced matrix overwrites its predecessor, beginning with the diagonal element in the upper left hand corner. Matrix reduction ends when the last reduced matrix has two states. Recall that the steady-state probability vector for a generic two-state regular Markov chain is known, as it was computed in Equation (2.16).

Matrix reduction is followed by back substitution, which begins by using the ratio of the steady-state probabilities for the final two-state reduced matrix to express the steady-state probability for the next to last state as a constant times the steady-state probability for the last state. Then the first equation for steady-state probabilities from the next larger reduced matrix is used to express the steady-state probability for the third from last state as a function of the steady-state probabilities for the last two states. The steady-state probability for the next to last state is substituted in this equation, expressed in terms of the steady-state probability for the last state. This recursive procedure is repeated, using the first equation for steady-state probabilities from each reduced matrix next larger in size, until the steady-state probabilities for all states have been expressed as constants times the steady-state probability for the last state. Then the normalizing equation (2.6) for the original Markov chain is solved to find the steady-state probability for the last state. All other steady-state probabilities are obtained by multiplying the constants previously found by the steady-state probability for the last state.

6.1.1.1 Matrix Reduction of a Partitioned Markov Chain

The partitioning algorithm for recursively computing steady-state probabilities is motivated by the following theorem, in modified form, of Kemeny and

Snell (pp. 114–116). Consider a regular Markov chain with N states, indexed $1, 2, \ldots, N$. For simplicity, assume $N = 4$. The transition probability matrix is denoted by P. The states are divided into two subsets. The first subset contains state 1. The second subset contains the remaining states, and is denoted by $S = \{2, 3, 4\}$. The transition matrix P is partitioned into four submatrices called p_{11}, u, v, and T, as shown below:

$$P = [p_{ij}] = \begin{array}{c} 1 \\ 2 \\ 3 \\ 4 \end{array} \left[\begin{array}{c|ccc} p_{11} & p_{12} & p_{13} & p_{14} \\ p_{21} & p_{22} & p_{23} & p_{24} \\ p_{31} & p_{32} & p_{33} & p_{34} \\ p_{41} & p_{42} & p_{43} & p_{44} \end{array} \right]$$

$$= \begin{array}{c} 1 \\ S \end{array} \left[\begin{array}{cc} p_{11} & u \\ v & T \end{array} \right], \tag{6.1}$$

where p_{11} is the single element in the upper left-hand corner of P,

$$u = [p_{12} \quad p_{13} \quad p_{14}] \text{ is a } 1\times3 \text{ row vector,} \tag{6.1}$$

$$v = [p_{21} \quad p_{31} \quad p_{41}]^T \text{ is a } 3\times1 \text{ column vector, and} \tag{6.1}$$

$$T = [p_{ij}] = \begin{array}{c} 2 \\ 3 \\ 4 \end{array} \left[\begin{array}{ccc} p_{22} & p_{23} & p_{24} \\ p_{32} & p_{33} & p_{34} \\ p_{42} & p_{43} & p_{44} \end{array} \right] \text{ is a } 3\times3 \text{ square matrix.} \tag{6.1}$$

A reduced matrix, $\overline{P}$, which represents a regular three-state Markov chain, is computed according to the formula

$$\overline{P} = T + v(1 - p_{11})^{-1}u. \tag{6.2}$$

Since roundoff error can be reduced by eliminating subtractions [1], the factor $(1 - p_{11})^{-1}$ is replaced by $(p_{12} + p_{13} + p_{14})^{-1}$. The modified formula, called the matrix reduction formula, is

$$\overline{P} = T + v(p_{12} + p_{13} + p_{14})^{-1}u. \tag{6.3}$$

The reduced transition matrix, denoted by $\overline{P}$, is calculated below using the matrix reduction formula (6.3):

$$\overline{P} = \begin{matrix} 2 \\ 3 \\ 4 \end{matrix} \begin{bmatrix} p_{22} & p_{23} & p_{24} \\ p_{32} & p_{33} & p_{34} \\ p_{42} & p_{43} & p_{44} \end{bmatrix} + \begin{bmatrix} p_{21} \\ p_{31} \\ p_{41} \end{bmatrix} [p_{12} + p_{13} + p_{14}]^{-1} [p_{12} \quad p_{13} \quad p_{14}]$$

$$= \begin{matrix} 2 \\ 3 \\ 4 \end{matrix} \begin{bmatrix} p_{22} & p_{23} & p_{24} \\ p_{32} & p_{33} & p_{34} \\ p_{42} & p_{43} & p_{44} \end{bmatrix} + [p_{12} + p_{13} + p_{14}]^{-1} \begin{bmatrix} p_{21}p_{12} & p_{21}p_{13} & p_{21}p_{14} \\ p_{31}p_{12} & p_{31}p_{13} & p_{31}p_{14} \\ p_{41}p_{12} & p_{41}p_{13} & p_{41}p_{14} \end{bmatrix}$$

$$= \begin{matrix} 2 \\ 3 \\ 4 \end{matrix} \left[\begin{matrix} p_{22} + \dfrac{p_{21}p_{12}}{p_{12}+p_{13}+p_{14}} & p_{23} + \dfrac{p_{21}p_{13}}{p_{12}+p_{13}+p_{14}} & p_{24} + \dfrac{p_{21}p_{14}}{p_{12}+p_{13}+p_{14}} \\[2ex] p_{32} + \dfrac{p_{31}p_{12}}{p_{12}+p_{13}+p_{14}} & p_{33} + \dfrac{p_{31}p_{13}}{p_{12}+p_{13}+p_{14}} & p_{34} + \dfrac{p_{31}p_{14}}{p_{12}+p_{13}+p_{14}} \\[2ex] p_{42} + \dfrac{p_{41}p_{12}}{p_{12}+p_{13}+p_{14}} & p_{43} + \dfrac{p_{41}p_{13}}{p_{12}+p_{13}+p_{14}} & p_{44} + \dfrac{p_{41}p_{14}}{p_{12}+p_{13}+p_{14}} \end{matrix} \right\} \quad (6.4)$$

$$= \begin{matrix} 2 \\ 3 \\ 4 \end{matrix} \begin{bmatrix} \overline{p_{22}} & \overline{p_{23}} & \overline{p_{24}} \\ \overline{p_{32}} & \overline{p_{33}} & \overline{p_{34}} \\ \overline{p_{42}} & \overline{p_{43}} & \overline{p_{44}} \end{bmatrix}.$$

The foregoing process of applying the matrix reduction formula (6.3) to the original matrix, P, to form the reduced matrix, $\overline{P}$, is called matrix reduction.

6.1.1.2 Optional Insight: Informal Justification of the Formula for Matrix Reduction

An informal justification of the formula for matrix reduction is instructive. Assume that the original four-state Markov chain is observed only when it is in the subset $S = \{2, 3, 4\}$ of the last three states. A new three-state Markov chain, called the reduced chain, is produced, with a transition probability matrix denoted by $\overline{P}$. A single step in the reduced chain corresponds, in the original chain, to the transition, not necessarily in one step, from a state in S to another state in S. Consider two states of S, say states 2 and 4. The transition probability, $\overline{p_{24}}$, for the reduced matrix is equal to the probability that the original chain, starting in state 2, enters S for the first time at state 4. This is the probability that the original chain moves from state 2 to state 4 in one step, plus the probability that it moves from state 2 to state 1 in one step, and enters S from state 1 for the first time at state 4. Therefore,

$$
\left.\begin{aligned}
\overline{P_{24}} &= P_{24} + P_{21}P_{14} + P_{21}P_{11}P_{14} + P_{21}P_{11}^2P_{14} + P_{21}P_{11}^3P_{14} + \cdots \\
&= P_{24} + P_{21}(1 + P_{11} + P_{11}^2 + P_{11}^3 + \cdots)P_{14} \\
&= P_{24} + P_{21}(1 - P_{11})^{-1}P_{14} \\
&= P_{24} + P_{21}(P_{12} + P_{13} + P_{14})^{-1}P_{14}.
\end{aligned}\right\}
\tag{6.5}
$$

By repeating this argument for the remaining pairs of states in S, the following eight additional transition probabilities for the reduced matrix, $\overline{P}$, shown in equation (6.4), can be obtained.

$$
\overline{P_{22}} = P_{22} + P_{21}(P_{12} + P_{13} + P_{14})^{-1}P_{12}
\tag{6.6}
$$

$$
\overline{P_{23}} = P_{23} + P_{21}(P_{12} + P_{13} + P_{14})^{-1}P_{13}
\tag{6.7}
$$

$$
\overline{P_{32}} = P_{32} + P_{31}(P_{12} + P_{13} + P_{14})^{-1}P_{12}
\tag{6.8}
$$

$$
\overline{P_{33}} = P_{33} + P_{31}(P_{12} + P_{13} + P_{14})^{-1}P_{13}
\tag{6.9}
$$

$$
\overline{P_{34}} = P_{34} + P_{31}(P_{12} + P_{13} + P_{14})^{-1}P_{14}
\tag{6.10}
$$

$$
\overline{P_{42}} = P_{42} + P_{41}(P_{12} + P_{13} + P_{14})^{-1}P_{12}
\tag{6.11}
$$

$$
\overline{P_{43}} = P_{43} + P_{41}(P_{12} + P_{13} + P_{14})^{-1}P_{13}
\tag{6.12}
$$

$$
\overline{P_{44}} = P_{44} + P_{41}(P_{12} + P_{13} + P_{14})^{-1}P_{14}.
\tag{6.13}
$$

The matrix form of the nine formulas (6.5) through (6.13) is the matrix reduction formulas (6.3).

6.1.1.3 Optional Insight: Informal Derivation of the MCPA

In this optional section, an informal derivation of the MCPA will be given. For simplicity, consider the transition probability matrix for a regular four-state Markov chain partitioned as shown in Equation (6.1). The first step of the MCPA is matrix reduction using Equation (6.3). The original transition probability matrix is identified by superscript one. The superscript is incremented by one for each reduced matrix.

$$
P^{(1)} = \begin{array}{c} 1 \\ 2 \\ 3 \\ 4 \end{array}
\left[\begin{array}{cccc}
p_{11}^{(1)} & p_{12}^{(1)} & p_{13}^{(1)} & p_{14}^{(1)} \\
p_{21}^{(1)} & p_{22}^{(1)} & p_{23}^{(1)} & p_{24}^{(1)} \\
p_{31}^{(1)} & p_{32}^{(1)} & p_{33}^{(1)} & p_{34}^{(1)} \\
p_{41}^{(1)} & p_{42}^{(1)} & p_{43}^{(1)} & p_{44}^{(1)}
\end{array}\right]
= \begin{array}{c} 1 \\ S \end{array}
\left[\begin{array}{cc}
p_{11}^{(1)} & u^{(1)} \\
v^{(1)} & T^{(1)}
\end{array}\right]
\tag{6.14}
$$

$$P^{(2)} = T^{(1)} + v^{(1)} \left[p_{12}^{(1)} + p_{13}^{(1)} + p_{14}^{(1)} \right]^{-1} u^{(1)} \tag{6.15}$$

$$
\begin{aligned}
&=
\begin{array}{c} 2 \\ 3 \\ 4 \end{array}
\begin{bmatrix} p_{22}^{(1)} & p_{23}^{(1)} & p_{24}^{(1)} \\ p_{32}^{(1)} & p_{33}^{(1)} & p_{34}^{(1)} \\ p_{42}^{(1)} & p_{43}^{(1)} & p_{44}^{(1)} \end{bmatrix}
+
\begin{bmatrix} p_{21}^{(1)} \\ p_{31}^{(1)} \\ p_{41}^{(1)} \end{bmatrix}
\left[p_{12}^{(1)} + p_{13}^{(1)} + p_{14}^{(1)} \right]^{-1}
\begin{bmatrix} p_{12}^{(1)} & p_{13}^{(1)} & p_{14}^{(1)} \end{bmatrix}
\\[4em]
&=
\begin{array}{c} 2 \\ 3 \\ 4 \end{array}
\left[\begin{array}{c|cc} p_{22}^{(2)} & p_{23}^{(2)} & p_{24}^{(2)} \\ \hline p_{32}^{(2)} & p_{33}^{(2)} & p_{34}^{(2)} \\ p_{42}^{(2)} & p_{43}^{(2)} & p_{44}^{(2)} \end{array} \right]
=
\begin{bmatrix} p_{22}^{(2)} & u^{(2)} \\ v^{(2)} & T^{(2)} \end{bmatrix}
\end{aligned}
\tag{6.15}
$$

$$
P^{(3)} = T^{(2)} + v^{(2)} \left[p_{23}^{(2)} + p_{24}^{(2)} \right]^{-1} u^{(2)} =
\begin{array}{c} 3 \\ 4 \end{array}
\begin{bmatrix} p_{33}^{(2)} & p_{34}^{(2)} \\ p_{43}^{(2)} & p_{44}^{(2)} \end{bmatrix}
+
\begin{bmatrix} p_{32}^{(2)} \\ p_{42}^{(2)} \end{bmatrix}
\left(p_{23}^{(2)} + p_{24}^{(2)} \right)^{-1}
\begin{bmatrix} p_{23}^{(2)} & p_{24}^{(2)} \end{bmatrix}
$$

$$
=
\begin{array}{c} 3 \\ 4 \end{array}
\left[\begin{array}{c|c} p_{33}^{(3)} & p_{34}^{(3)} \\ \hline p_{43}^{(3)} & p_{44}^{(3)} \end{array} \right]
=
\begin{array}{c} 3 \\ 4 \end{array}
\begin{bmatrix} p_{33}^{(3)} & u^{(3)} \\ v^{(3)} & T^{(3)} \end{bmatrix} .
\tag{6.16}
$$

The matrix reduction step has ended with a two-state Markov chain, which has a transition probability matrix denoted by $P^{(3)}$. Using Equation (2.16), the steady-state probability vector for $P^{(3)}$, denoted by $\pi^{(3)}$, is known, and is shown below with its two components.

$$
\pi^{(3)} = \begin{bmatrix} \pi_3^{(3)} & \pi_4^{(3)} \end{bmatrix} = \begin{bmatrix} \dfrac{p_{43}^{(3)}}{p_{34}^{(3)} + p_{43}^{(3)}} & \dfrac{p_{34}^{(3)}}{p_{34}^{(3)} + p_{43}^{(3)}} \end{bmatrix} .
\tag{6.17}
$$

The second step of the MCPA is back substitution, which begins by solving for $\pi_3^{(3)}$ as a constant k_3 times $\pi_4^{(3)}$.

$$
\pi_3^{(3)} = \frac{p_{43}^{(3)}}{p_{34}^{(3)}} \pi_4^{(3)} = (p_{34}^{(3)})^{-1} p_{43}^{(3)} \pi_4^{(3)} = k_3 \pi_4^{(3)} ,
\tag{6.18}
$$

where the constant

$$
k_3 = (p_{34}^{(3)})^{-1} p_{43}^{(3)} .
\tag{6.19}
$$

The steady-state probability vector for $P^{(2)}$ is denoted by

$$
\pi^{(2)} = \begin{bmatrix} \pi_2^{(2)} & \pi_3^{(2)} & \pi_4^{(2)} \end{bmatrix} .
\tag{6.20}
$$

The first steady-state equation of the system

$$\pi^{(2)} = \pi^{(2)} P^{(2)} \tag{6.21}$$

is

$$\pi_2^{(2)} = \pi_2^{(2)} p_{22}^{(2)} + \pi_3^{(2)} p_{32}^{(2)} + \pi_4^{(2)} p_{42}^{(2)}$$
$$(1 - p_{22}^{(2)})\pi_2^{(2)} = \pi_3^{(2)} p_{32}^{(2)} + \pi_4^{(2)} p_{42}^{(2)}. \tag{6.22}$$

By the theorem of Kemeny and Snell,

$$\pi_3^{(3)} = \frac{\pi_3^{(2)}}{\pi_3^{(2)} + \pi_4^{(2)}}, \quad \text{and} \quad \pi_4^{(3)} = \frac{\pi_4^{(2)}}{\pi_3^{(2)} + \pi_4^{(2)}} \tag{6.23}$$

Observe that

$$\frac{\pi_3^{(2)}}{\pi_4^{(2)}} = \frac{\pi_3^{(3)}}{\pi_4^{(3)}} = \frac{k_3 \pi_4^{(3)}}{\pi_4^{(3)}} = k_3. \tag{6.24}$$

Hence, $\pi_3^{(3)} = k_3 \pi_4^{(3)}$ implies that $\pi_3^{(2)} = k_3 \pi_4^{(2)}$. Substituting $\pi_3^{(2)} = k_3 \pi_4^{(2)}$ in the first steady-state equation of the system (6.21),

$$(1 - p_{22}^{(2)})\pi_2^{(2)} = p_{32}^{(2)} k_3 \pi_4^{(2)} + p_{42}^{(2)} \pi_4^{(2)} = (p_{32}^{(2)} k_3 + p_{42}^{(2)})\pi_4^{(2)}$$
$$\pi_2^{(2)} = (1 - p_{22}^{(2)})^{-1}(p_{32}^{(2)} k_3 + p_{42}^{(2)})\pi_4^{(2)} = (p_{23}^{(2)} + p_{24}^{(2)})^{-1}(p_{32}^{(2)} k_3 + p_{42}^{(2)})\pi_4^{(2)}, \tag{6.25}$$
$$= k_2 \pi_4^{(2)},$$

where the constant

$$k_2 = (p_{23}^{(2)} + p_{24}^{(2)})^{-1}(p_{32}^{(2)} k_3 + p_{42}^{(2)}), \tag{6.26}$$

and where $(p_{23}^{(2)} + p_{24}^{(2)})^{-1}$ is substituted for $(1 - p_{22}^{(2)})^{-1}$ to avoid subtractions. The steady-state probability vector for $P^{(1)}$ is denoted by

$$\pi^{(1)} = \begin{bmatrix} \pi_1^{(1)} & \pi_2^{(1)} & \pi_3^{(1)} & \pi_4^{(1)} \end{bmatrix}. \tag{6.27}$$

The first steady-state equation of the system

$$\pi^{(1)} = \pi^{(1)} P^{(1)} \tag{6.28}$$

is

$$\left. \begin{array}{l} \pi_1^{(1)} = \pi_1^{(1)} p_{11}^{(1)} + \pi_2^{(1)} p_{21}^{(1)} + \pi_3^{(1)} p_{31}^{(1)} + \pi_4^{(1)} p_{41}^{(1)} \\ (1 - p_{11}^{(1)})\pi_1^{(1)} = \pi_2^{(1)} p_{21}^{(1)} + \pi_3^{(1)} p_{31}^{(1)} + \pi_4^{(1)} p_{41}^{(1)} \end{array} \right\}. \tag{6.29}$$

By the theorem of Kemeny and Snell,

$$\pi_2^{(2)} = \frac{\pi_2^{(1)}}{\pi_2^{(1)} + \pi_3^{(1)} + \pi_4^{(1)}}, \quad \pi_3^{(2)} = \frac{\pi_3^{(1)}}{\pi_2^{(1)} + \pi_3^{(1)} + \pi_4^{(1)}}, \quad \text{and} \quad \pi_4^{(2)} = \frac{\pi_4^{(1)}}{\pi_2^{(1)} + \pi_3^{(1)} + \pi_4^{(1)}} \quad (6.30)$$

Observe that

$$\frac{\pi_2^{(1)}}{\pi_4^{(1)}} = \frac{\pi_2^{(2)}}{\pi_4^{(2)}} = \frac{k_2 \pi_4^{(2)}}{\pi_4^{(2)}} = k_2. \quad (6.31)$$

Hence, $\pi_2^{(2)} = k_2 \pi_4^{(2)}$ implies that $\pi_2^{(1)} = k_2 \pi_4^{(1)}$.
Similarly, $\pi_3^{(2)} = k_3 \pi_4^{(2)}$ implies that $\pi_3^{(1)} = k_3 \pi_4^{(1)}$.
Substituting $\pi_2^{(1)} = k_2 \pi_4^{(1)}$ and $\pi_3^{(1)} = k_3 \pi_4^{(1)}$ in the first steady-state equation of the system (6.28),

$$\left.\begin{aligned}
(1 - p_{11}^{(1)})\pi_1^{(1)} &= p_{21}^{(1)} k_2 \pi_4^{(1)} + p_{31}^{(1)} k_3 \pi_4^{(1)} + p_{41}^{(1)} \pi_4^{(1)} = (p_{21}^{(1)} k_2 + p_{31}^{(1)} k_3 + p_{41}^{(1)})\pi_4^{(1)} \\
\pi_1^{(1)} &= (1 - p_{11}^{(1)})^{-1}(p_{21}^{(1)} k_2 + p_{31}^{(1)} k_3 + p_{41}^{(1)})\pi_4^{(1)} \\
&= (p_{12}^{(1)} + p_{13}^{(1)} + p_{14}^{(1)})^{-1}(p_{21}^{(1)} k_2 + p_{31}^{(1)} k_3 + p_{41}^{(1)})\pi_4^{(1)} = k_1 \pi_4^{(1)},
\end{aligned}\right\} \quad (6.32)$$

where the constant

$$k_1 = (p_{12}^{(1)} + p_{13}^{(1)} + p_{14}^{(1)})^{-1}(p_{21}^{(1)} k_2 + p_{31}^{(1)} k_3 + p_{41}^{(1)}).$$

Solving the normalizing equation (2.6) for the system (6.28) for $\pi_4^{(1)}$ as a function of the constants,

$$\left.\begin{aligned}
\pi_1^{(1)} + \pi_2^{(1)} + \pi_3^{(1)} + \pi_4^{(1)} &= 1 \\
k_1 \pi_4^{(1)} + k_2 \pi_4^{(1)} + k_3 \pi_4^{(1)} + \pi_4^{(1)} &= 1 \\
(k_1 + k_2 + k_3 + 1)\pi_4^{(1)} &= 1
\end{aligned}\right\} \quad (6.33)$$

yields

$$\pi_4^{(1)} = (1 + k_1 + k_2 + k_3)^{-1}. \quad (6.34)$$

Solving for the steady-state probabilities for the original Markov chain,

$$\left.\begin{aligned}
\pi_4 &= \pi_4^{(1)} \\
\pi_3 &= \pi_3^{(1)} = k_3 \pi_4^{(1)} \\
\pi_2 &= \pi_2^{(1)} = k_2 \pi_4^{(1)} \\
\pi_1 &= \pi_1^{(1)} = k_1 \pi_4^{(1)}
\end{aligned}\right\}. \quad (6.35)$$

6.1.1.4 Markov Chain Partitioning Algorithm

The MCPA will calculate the steady-state probability vector for a regular Markov chain with N states indexed $1, 2, \ldots, N$, and a transition probability matrix, P. The detailed steps of the MCPA are given below:

Matrix Reduction

1. Initialize $n = 1$.
2. Let $P^{(n)} = P = [p_{ij}]$ for $n \le i \le N$ and $n \le j \le N$.
3. Partition $P^{(n)}$ as

$$
P^{(n)} = \begin{array}{c} n \\ n+1 \\ \vdots \\ N \end{array} \left[\begin{array}{c|ccc} p_{nn}^{(n)} & p_{n,n+1}^{(n)} & \cdots & p_{n,N}^{(n)} \\ \hline p_{n+1,n}^{(n)} & p_{n+1,n+1}^{(n)} & \cdots & p_{n+1,N}^{(n)} \\ \vdots & \vdots & \cdots & \vdots \\ p_{N,n}^{(n)} & p_{N,n+1}^{(n)} & \cdots & p_{N,N}^{(n)} \end{array} \right] = \left[\begin{array}{cc} p_{nn}^{(n)} & u^{(n)} \\ v^{(n)} & T^{(n)} \end{array} \right].
$$

4. Store the first row and first column, respectively, of $P^{(n)}$ by overwriting the first row and the first column, respectively, of P.
5. Compute $P^{(n+1)} = T^{(n)} + v^{(n)} \left[p_{n,n+1}^{(n)} + \cdots + p_{n,N}^{(n)} \right]^{-1} u^{(n)}$
6. Increment n by 1. If $n < N - 1$, go to step 3. Otherwise, go back substitution.

Back Substitution

1. Initialize $i = N - 1$.
2. $k_{N-1} = \left(p_{N-1,N}^{(N-1)} \right)^{-1} p_{N,N-1}^{(N-1)}$
3. Decrement i by 1
4. $k_i = \left(\sum\limits_{j=i+1}^{N} p_{ij}^{(i)} \right)^{-1} \left(p_{Ni}^{(i)} + \sum\limits_{h=i+1}^{N-1} p_{hi}^{(i)} k_h \right)$, $i = N-2, N-3, \ldots, 1$.
5. If $i > 1$, go to step 3.
6. $i = 1$
7. $\pi_N = \left(1 + \sum\limits_{h=1}^{N-1} k_h \right)^{-1}$.
8. $\pi_h = k_h \pi_N$, $h = 1, 2, \ldots, N-1$.
9. $\pi = [\pi_1 \quad \pi_2 \quad \cdots \quad \pi_N]$

6.1.1.5 Using the MCPA to Compute the Steady-State Probabilities for a Four-State Markov Chain

In this section, the MCPA will be executed to calculate the steady-state probability vector for the regular four-state Markov chain model of the weather for which the transition matrix is shown in Equation (1.9). The steady-state probability vector was previously computed in Equation (2.22). The transition probability matrix is partitioned as shown in Equation (6.36).

Matrix Reduction

$$P = P^{(1)} = \begin{matrix} 1 \\ 2 \\ 3 \\ 4 \end{matrix} \begin{bmatrix} 0.3 & 0.1 & 0.4 & 0.2 \\ \hline 0.2 & 0.5 & 0.2 & 0.1 \\ 0.3 & 0.2 & 0.1 & 0.4 \\ 0 & 0.6 & 0.3 & 0.1 \end{bmatrix} = \begin{matrix} 1 \\ S \end{matrix} \begin{bmatrix} p_{11}^{(1)} & u^{(1)} \\ v^{(1)} & T^{(1)} \end{bmatrix} \tag{6.36}$$

$$\left. \begin{aligned} P^{(2)} &= T^{(1)} + v^{(1)} \left[p_{12}^{(1)} + p_{13}^{(1)} + p_{14}^{(1)} \right]^{-1} u^{(1)} \\[2mm] &= \begin{matrix} 2 \\ 3 \\ 4 \end{matrix} \begin{bmatrix} 0.5 & 0.2 & 0.1 \\ 0.2 & 0.1 & 0.4 \\ 0.6 & 0.3 & 0.1 \end{bmatrix} + \begin{bmatrix} 0.2 \\ 0.3 \\ 0 \end{bmatrix} (0.1+0.4+0.2)^{-1} \begin{bmatrix} 0.1 & 0.4 & 0.2 \end{bmatrix} \\[2mm] &= \begin{matrix} 2 \\ 3 \\ 4 \end{matrix} \begin{bmatrix} 37/70 & 22/70 & 11/70 \\ \hline 17/70 & 19/70 & 34/70 \\ 42/70 & 21/70 & 7/70 \end{bmatrix} = \begin{bmatrix} p_{22}^{(2)} & u^{(2)} \\ v^{(2)} & T^{(2)} \end{bmatrix} \end{aligned} \right\} \tag{6.37}$$

$$\left. \begin{aligned} P^{(3)} &= T^{(2)} + v^{(2)} \left[p_{23}^{(2)} + p_{24}^{(2)} \right]^{-1} u^{(2)} \\[2mm] &= \begin{matrix} 3 \\ 4 \end{matrix} \begin{bmatrix} 19/70 & 34/70 \\ \hline 21/70 & 7/70 \end{bmatrix} + \begin{bmatrix} 17/70 \\ 42/70 \end{bmatrix} (22/70 + 11/70)^{-1} \begin{bmatrix} 22/70 & 11/70 \end{bmatrix} . \\[2mm] &= \begin{matrix} 3 \\ 4 \end{matrix} \begin{bmatrix} 13/30 & 17/30 \\ \hline 21/30 & 9/30 \end{bmatrix} = \begin{bmatrix} p_{33}^{(3)} & u^{(3)} \\ v^{(3)} & T^{(3)} \end{bmatrix} \end{aligned} \right\} \tag{6.38}$$

Back Substitution

$$k_3 = (p_{34}^{(3)})^{-1} p_{43}^{(3)} = (17/30)^{-1}(21/30) = 1617/1309. \tag{6.39}$$

$$k_2 = \left[p_{23}^{(2)} + p_{24}^{(2)} \right]^{-1} \left[p_{32}^{(2)} k_3 + p_{42}^{(2)} \right] \left. \\ = (22/70 + 11/70)^{-1} \left[(17/70)(21/17) + 42/70 \right] = 2499/1309 \right\} \tag{6.40}$$

$$k_1 = \left[p_{12}^{(1)} + p_{13}^{(1)} + p_{14}^{(1)} \right]^{-1} \left[p_{21}^{(1)} k_2 + p_{31}^{(1)} k_3 + p_{41}^{(1)} \right] \left. \\ = (0.1 + 0.4 + 0.2)^{-1} \left[(0.2)(2499/1309) + (0.3)(1617/1309) \right] = 1407/1309 \right\} \tag{6.41}$$

$$\pi_4 = (1 + k_1 + k_2 + k_3)^{-1} = (1 + 1407/1309 + 2499/1309 + 1617/1309)^{-1} = 1309/6832 \tag{6.42}$$

$$\pi_3 = k_3 \pi_4 = (1617/1309)\,(1309/6832) = 1617/6832 \tag{6.43}$$

$$\pi_2 = k_2 \pi_4 = (2499/1309)\,(1309/6832) = 2499/6832 \tag{6.44}$$

$$\pi_1 = k_1 \pi_4 = (1407/1309)\,(1309/6832) = 1407/6832. \tag{6.45}$$

These are the same steady-state probabilities that were obtained in Equation (2.22).

6.1.1.6 Optional Insight: Matrix Reduction and Gaussian Elimination

To see that matrix reduction is equivalent to Gaussian elimination, consider again the transition probability matrix shown in Equation (6.36) for the four-state Markov chain model of the weather. Equations (2.13), with the normalizing equation omitted, can be expressed as

$$\pi = \pi P \Rightarrow \pi^{(1)} = \pi^{(1)} P^{(1)}. \tag{6.46}$$

Equations (2.12), with the normalizing equation omitted, can be expressed as

$$\left. \begin{aligned} \pi_1 &= (0.3)\pi_1 + (0.2)\pi_2 + (0.3)\pi_3 + (0)\pi_4 \\ \pi_2 &= (0.1)\pi_1 + (0.5)\pi_2 + (0.2)\pi_3 + (0.6)\pi_4 \\ \pi_3 &= (0.4)\pi_1 + (0.2)\pi_2 + (0.1)\pi_3 + (0.3)\pi_4 \\ \pi_4 &= (0.2)\pi_1 + (0.1)\pi_2 + (0.4)\pi_3 + (0.1)\pi_4 \end{aligned} \right\}. \tag{6.47}$$

When Gaussian elimination is applied to the linear system (6.47), the first step is to solve the first equation for π_1 as a function of π_2, π_3, and π_4 in the following manner.

$$\left.\begin{array}{l} \pi_1 = (0.3)\,\pi_1 + (0.2)\,\pi_2 + (0.3)\,\pi_3 + (0)\,\pi_4 \\ (1-0.3)\,\pi_1 = (0.2)\,\pi_2 + (0.3)\,\pi_3 + (0)\,\pi_4, \end{array}\right\} \tag{6.48}$$

to obtain

$$\pi_1 = (20/70)\,\pi_2 + (30/70)\,\pi_3 + (0)\,\pi_4. \tag{6.49}$$

To reduce roundoff error by avoiding subtractions, both matrix reduction and Gaussian elimination can be modified by replacing the coefficient $(1-0.3)$ of π_1 in Equation (6.48) with the coefficient $(0.1 + 0.4 + 0.2)$, equal to the sum of the remaining coefficients in the first row of the transition matrix $P^{(1)}$. The expression for π_1 is substituted into the remaining three equations of (6.47) to produce the following reduced system of three equations in three unknowns.

$$\left.\begin{array}{l} \pi_2 = (37/70)\,\pi_2 + (17/70)\,\pi_3 + (42/70)\,\pi_4 \\ \pi_3 = (22/70)\,\pi_2 + (19/70)\,\pi_3 + (21/70)\,\pi_4 \\ \pi_4 = (11/70)\,\pi_2 + (34/70)\,\pi_3 + (7/70)\,\pi_4 \end{array}\right\}. \tag{6.50}$$

Expressed in matrix form equations (6.50) are

$$\bar{\pi} = \overline{\pi P} \Rightarrow \pi^{(2)} = \pi^{(2)} P^{(2)}, \tag{6.51}$$

or

$$\begin{bmatrix} \pi_2^{(2)} & \pi_3^{(2)} & \pi_4^{(2)} \end{bmatrix} = \begin{bmatrix} \pi_2^{(2)} & \pi_3^{(2)} & \pi_4^{(2)} \end{bmatrix} \begin{array}{c} 2 \\ 3 \\ 4 \end{array} \begin{bmatrix} 37/70 & 22/70 & 11/70 \\ 17/70 & 19/70 & 34/70 \\ 42/70 & 21/70 & 7/70 \end{bmatrix}. \tag{6.52}$$

Thus, the first step of Gaussian elimination has produced the first reduced matrix, $P^{(2)}$, which is calculated in Equation (6.37). This procedure can be repeated until only two equations remain, producing the final reduced matrix, $P^{(3)}$. The results of this example can be generalized to conclude that each step of Gaussian elimination produces a reduced coefficient matrix equivalent to the reduced matrix produced by the corresponding step of matrix reduction. To conform to the order in which the equations of a linear system are removed during Gaussian elimination, which begins with the first equation, matrix reduction starts in the upper left-hand corner of a transition probability matrix. Back substitution is the same for both Gaussian elimination and state reduction.

6.1.2 Mean First Passage Times

Recall from Sections 2.2.2.1 and 3.3.1 that when a vector of MFPTs to a target state 0 (or *j*) in a regular Markov chain is to be calculated, the process begins by making the target state 0 an absorbing state. A state reduction algorithm for computing MFPTs also makes the target state 0 an absorbing state [3]. A state reduction algorithm for computing MFPTs has three steps: augmentation, matrix reduction, and back substitution. This algorithm differs from the MCPA of Section 6.1.1 in two major respects. The first difference is that an augmentation step has been added to avoid subtractions. The second difference is that matrix reduction stops when the final reduced matrix has only one row.

6.1.2.1 Forming the Augmented Matrix

To understand the augmentation step, consider the following transition matrix for a generic three-state regular Markov chain:

$$P = \begin{matrix} 0 \\ 1 \\ 2 \end{matrix} \begin{bmatrix} p_{00} & p_{01} & p_{02} \\ p_{10} & p_{11} & p_{12} \\ p_{20} & p_{21} & p_{22} \end{bmatrix}. \tag{6.53}$$

Suppose that MFPTs to target state 0 are desired. When target state 0 is made an absorbing state, the modified transition probability matrix, $\mathbf{P_M}$, is partitioned in the following manner:

$$\mathbf{P_M} = \begin{matrix} 0 \\ 1 \\ 2 \end{matrix} \begin{bmatrix} 1 & 0 & 0 \\ \hline p_{10} & p_{11} & p_{12} \\ p_{20} & p_{21} & p_{22} \end{bmatrix} = \begin{bmatrix} 1 & 0 \\ D & Q \end{bmatrix}, \tag{6.54a}$$

where

$$D = \begin{matrix} 1 \\ 2 \end{matrix} \begin{bmatrix} p_{10} \\ p_{20} \end{bmatrix}, \quad Q = \begin{matrix} 1 \\ 2 \end{matrix} \begin{bmatrix} p_{11} & p_{12} \\ p_{21} & p_{22} \end{bmatrix}. \tag{6.54b}$$

In the augmentation step, a rectangular augmented matrix, denoted by G, is constructed, and is shown below in partitioned form.

$$G = [Q \quad D \quad e], \tag{6.55}$$

where e is a two-component column vector with all entries one.

The augmented matrix, G, is shown below. Note that the columns in the augmented matrix are numbered consecutively from 1 to 4, so that state 0 is placed in column 3, and the vector e is placed in column 4.

$$
\begin{aligned}
G &= \begin{array}{c} 1 \\ 2 \end{array}\!\!\left[\begin{array}{ccc|c} p_{11} & p_{12} & p_{10} & e_1 \\ p_{21} & p_{22} & p_{20} & e_2 \end{array}\right] = [Q \quad D \quad e] = \begin{array}{c} 1 \\ 2 \end{array}\!\!\left[\begin{array}{cc|c|c} g_{11} & g_{12} & g_{13} & 1 \\ g_{21} & g_{22} & g_{23} & 1 \end{array}\right] \\
&= \begin{array}{c} 1 \\ 2 \end{array}\!\!\left[\begin{array}{cc|c|c} g_{11} & g_{12} & g_{13} & g_{14} \\ g_{21} & g_{22} & g_{23} & g_{24} \end{array}\right].
\end{aligned} \tag{6.56}
$$

Matrix Q is augmented with vector D to avoid subtractions. Observe that

$$D = \begin{array}{c} 1 \\ 2 \end{array}\!\!\left[\begin{array}{c} p_{10} \\ p_{20} \end{array}\right] = \begin{array}{c} 1 \\ 2 \end{array}\!\!\left[\begin{array}{c} g_{13} \\ g_{23} \end{array}\right]. \tag{6.57}$$

Since the sum of the entries in the first three columns in each row of the augmented matrix equals one, subtractions can be eliminated by making the substitution

$$(1 - p_{11}) = (p_{12} + p_{10}) = (1 - g_{11}) = (g_{12} + g_{13}). \tag{6.58}$$

6.1.2.2 State Reduction Algorithm for Computing MFPTs

The detailed steps of a state reduction algorithm for computing MFPTs for a regular Markov chain with N states indexed $1, 2, \ldots, N$, and a transition probability matrix, P, are given below. As Section 6.1.2 indicates, the modified transition probability matrix, $\mathbf{P_M}$, is assumed to be partitioned such that target state 0 is made an absorbing state.

A. Augmentation

An $N \times (N + 2)$ rectangular augmented matrix, $\mathbf{G}$, is formed such that $G = [Q, D, e] = [g_{ij}]$, where matrices Q and D are defined as in Equation (2.50) and vector e is an N-component column vector with all entries one.

B. Matrix Reduction

Matrix reduction is applied to the augmented matrix, $\mathbf{G}$.

    1. Initialize $n = 1$

    2. Let $\mathbf{G}^{(n)} = G = [g_{ij}]$ for $n \le i \le N$ and $n \le j \le N + 2$.

3. Partition $G^{(n)}$ as

$$
G^{(n)} = \begin{array}{c} n \\ n+1 \\ \vdots \\ N \end{array} \left[\begin{array}{c|cccc} g_{nn}^{(n)} & g_{n,n+1}^{(n)} & \cdots & g_{n,N+1}^{(n)} & g_{n,N+2}^{(n)} \\ \hline g_{n+1,n}^{(n)} & g_{n+1,n+1}^{(n)} & \cdots & g_{n+1,N+1}^{(n)} & g_{n+1,N+2}^{(n)} \\ \vdots & \vdots & \cdots & \vdots & \vdots \\ g_{N,n}^{(n)} & g_{N,n+1}^{(n)} & \cdots & g_{N,N+1}^{(n)} & g_{N,N+2}^{(n)} \end{array} \right] = \left[\begin{array}{cc} g_{nn}^{(n)} & \mathbf{u}^{(n)} \\ \mathbf{v}^{(n)} & \mathbf{T}^{(n)} \end{array} \right]. \qquad (6.59)
$$

4. Store the first row of $G^{(n)}$.
5. Compute $G^{(n+1)} = \mathbf{T}^{(n)} + \mathbf{v}^{(n)} \left[g_{n,n+1}^{(n)} + \cdots + g_{n,N+1}^{(n)} \right]^{-1} \mathbf{u}^{(n)}$
6. Increment n by 1. If $n < N$, go to step 3. Otherwise, go back substitution.

C. Back Substitution

1. The entry in row N of the vector MFPTs to state 0 is computed.
$m_{N0} = g_{N,N+2}^{(N)} / g_{N,N+1}^{(N)}$.

2. Let $i = N - 1$.

3. For $i \leq 1 < N$, compute the entry in row i of the vector of mean first passage times to state 0.

$$
m_{i0} = \left(g_{i,N+2}^{(i)} + \sum_{h=i+1}^{N} g_{ih}^{(i)} m_{h0} \right) \bigg/ \sum_{h=i+1}^{N+1} g_{ih}^{(i)}.
$$

4. Decrement i by 1.

5. If $i > 0$, go to step 3. Otherwise, stop.

6.1.2.3 Using State Reduction to Compute MFPTs for a Five-State Markov Chain

To demonstrate how the state reduction algorithm is used to compute the MFPTs for a regular Markov chain, consider the following numerical example, treated in Section 2.2.2.1, of a Markov chain with five states, indexed 0, 1, 2, 3, and 4. The vector of MFPTs was calculated in Equations (2.62) and (3.22). The transition probability matrix is shown in Equation (2.57). The vector of MFPTs to target state 0 can be found by making state 0 an absorbing state. The modified transition probability matrix, $\mathbf{P}_M$, is partitioned as shown in Equation (2.58).

A. Augmentation

$$
G = \begin{bmatrix} \mathbf{Q} & \mathbf{D} & \mathbf{e} \end{bmatrix} = \begin{array}{c} 1 \\ 2 \\ 3 \\ 4 \end{array} \left[\begin{array}{cccc|c|c} 0.4 & 0.1 & 0.2 & 0.3 & 0 & 1 \\ 0.1 & 0.2 & 0.3 & 0.2 & 0.2 & 1 \\ 0.2 & 0.1 & 0.4 & 0.2 & 0.1 & 1 \\ 0.3 & 0.4 & 0.2 & 0.1 & 0 & 1 \end{array} \right]. \qquad (6.60)
$$

B. Matrix Reduction

$$\mathbf{G}^{(1)} = \begin{array}{c} 1 \\ 2 \\ 3 \\ 4 \end{array}\left[\begin{array}{c|ccccc} 0.4 & 0.1 & 0.2 & 0.3 & 0 & 1 \\ \hline 0.1 & 0.2 & 0.3 & 0.2 & 0.2 & 1 \\ 0.2 & 0.1 & 0.4 & 0.2 & 0.1 & 1 \\ 0.3 & 0.4 & 0.2 & 0.1 & 0 & 1 \end{array}\right] = \begin{bmatrix} g_{11}^{(1)} & \mathbf{u}^{(1)} \\ \mathbf{v}^{(1)} & \mathbf{T}^{(1)} \end{bmatrix}$$

(6.61)

$$\mathbf{G}^{(2)} = \mathbf{T}^{(1)} + \mathbf{v}^{(1)}\left[g_{12}^{(1)} + g_{13}^{(1)} + g_{14}^{(1)} + g_{15}^{(1)} \right]^{-1} \mathbf{u}^{(1)}$$

$$\mathbf{G}^{(2)} = \begin{array}{c} 2 \\ 3 \\ 4 \end{array}\left[\begin{array}{c|cccc} 0.2167 & 0.3333 & 0.25 & 0.2 & 1.1667 \\ \hline 0.1333 & 0.4667 & 0.30 & 0.1 & 1.3333 \\ 0.4500 & 0.3000 & 0.25 & 0 & 1.5000 \end{array}\right] = \begin{bmatrix} g_{22}^{(2)} & \mathbf{u}^{(2)} \\ \mathbf{v}^{(2)} & \mathbf{T}^{(2)} \end{bmatrix}$$

(6.62)

$$\mathbf{G}^{(3)} = \mathbf{T}^{(2)} + \mathbf{v}^{(2)}\left[g_{23}^{(2)} + g_{24}^{(2)} + g_{25}^{(2)} \right]^{-1} \mathbf{u}^{(2)}$$

$$\mathbf{G}^{(3)} = \begin{array}{c} 3 \\ 4 \end{array}\left[\begin{array}{c|ccc} 0.5234 & 0.3425 & 0.1341 & 1.5138 \\ \hline 0.4915 & 0.3936 & 0.1149 & 2.1703 \end{array}\right] = \begin{bmatrix} g_{33}^{(3)} & \mathbf{u}^{(3)} \\ \mathbf{v}^{(3)} & \mathbf{T}^{(3)} \end{bmatrix}$$

(6.63)

$$\mathbf{G}^{(4)} = \mathbf{T}^{(3)} + \mathbf{v}^{(3)}\left[g_{34}^{(3)} + g_{35}^{(3)} \right]^{-1} \mathbf{u}^{(3)}$$

$$\mathbf{G}^{(4)} = \begin{bmatrix} 0.7468 & 0.2532 & 3.7500 \end{bmatrix} = \begin{bmatrix} g_{44}^{(4)} & g_{45}^{(4)} & g_{46}^{(4)} \end{bmatrix}$$

(6.64)

C. Back Substitution

Back substitution begins by computing the entry in row four of the vector of MFPTs to state *0*.

$$m_{40} = g_{46}^{(4)} / g_{45}^{(4)} = 3.7500/0.2532 = 14.8104.$$

(6.65)

Next, the entry in row three is computed.

$$\left.\begin{aligned} m_{30} &= (g_{36}^{(3)} + g_{34}^{(3)} m_{40}) / (g_{34}^{(3)} + g_{35}^{(3)}) \\ &= (1.5318 + 0.3425(14.8104)) / (0.3425 + 0.1341) = 13.8573. \end{aligned}\right\}$$

(6.66)

Then the entry in row two is computed.

$$\left.\begin{aligned} m_{20} &= (g_{26}^{(2)} + g_{23}^{(2)} m_{30} + g_{24}^{(2)} m_{40}) / (g_{23}^{(2)} + g_{24}^{(2)} + g_{25}^{(2)}) \\ &= (1.1667 + 0.3333(13.8573) + 0.25(14.8104)) / (0.3333 + 0.25 + 0.2) = 12.1128. \end{aligned}\right\}$$

(6.67)

Finally, the entry in row one is computed.

$$\left.\begin{aligned} m_{10} &= (g_{16}^{(1)} + g_{12}^{(1)} m_{20} + g_{13}^{(1)} m_{30} + g_{14}^{(1)} m_{40}) / (g_{12}^{(1)} + g_{13}^{(1)} + g_{14}^{(1)} + g_{15}^{(1)}) \\ &= (1 + 0.1(12.1128) + 0.2(13.8573) + 0.3(14.8104)) / (0.1 + 0.2 + 0.3 + 0) = 15.7098. \end{aligned}\right\}$$

(6.68)

This vector of MFPTs differs slightly from those computed in Equations (2.62) and (3.22). Discrepancies are due to roundoff error because only the first four significant decimal digits were stored.

6.1.3 Absorption Probabilities

In Sections 6.1.1 and 6.1.2, state reduction was applied to calculate quantities for regular Markov chains. In this section, a state reduction algorithm will be developed to compute absorption probabilities for a reducible chain [3]. To demonstrate how state reduction can be used to compute a vector of absorption probabilities, consider the six-state absorbing multichain model of patient flow in a hospital introduced in Section 1.10.2.1.2. States 0 and 5 are absorbing, and the other four states are transient. The transition probability matrix is given in canonical form in Equation (1.59).

$$
P = \begin{array}{c|cc|cccc}
\text{State} & 0 & 5 & 1 & 2 & 3 & 4 \\
\hline
0 & 1 & 0 & 0 & 0 & 0 & 0 \\
5 & 0 & 1 & 0 & 0 & 0 & 0 \\
\hline
1 & 0 & 0 & 0.4 & 0.1 & 0.2 & 0.3 \\
2 & 0.15 & 0.05 & 0.1 & 0.2 & 0.3 & 0.2 \\
3 & 0.07 & 0.03 & 0.2 & 0.1 & 0.4 & 0.2 \\
4 & 0 & 0 & 0.3 & 0.4 & 0.2 & 0.1
\end{array} = \begin{bmatrix} I & 0 \\ D & Q \end{bmatrix}, \qquad (1.59)
$$

where

$$
D = \begin{array}{c}1\\2\\3\\4\end{array}\begin{bmatrix} 0 & 0 \\ 0.15 & 0.05 \\ 0.07 & 0.03 \\ 0 & 0 \end{bmatrix} = \begin{array}{c}1\\2\\3\\4\end{array}\begin{bmatrix} 0 & 0 \\ 0.15 & 0.05 \\ 0.07 & 0.03 \\ 0 & 0 \end{bmatrix} = [D_1 \;\; D_2], \quad Q = \begin{array}{c}1\\2\\3\\4\end{array}\begin{bmatrix} 0.4 & 0.1 & 0.2 & 0.3 \\ 0.1 & 0.2 & 0.3 & 0.2 \\ 0.2 & 0.1 & 0.4 & 0.2 \\ 0.3 & 0.4 & 0.2 & 0.1 \end{bmatrix}. \quad (6.69)
$$

A state reduction algorithm will be constructed to compute the vector of the probabilities of absorption in state 0. This vector of absorption probabilities was previously computed in Equations (3.93) and (3.95). The state reduction algorithm is similar in structure to the one constructed for computing MFPTs in Section 6.1.2. The major difference is in the augmentation step.

6.1.3.1 *Forming the Augmented Matrix*

To compute a vector of absorption probabilities, an $N \times (N + 2)$ rectangular augmented matrix, $\mathbf{G}$, is formed such that

$$
\mathbf{G} = [\mathbf{Q} \quad D_2 \quad D_1] = [g_{ij}], \qquad (6.70)
$$

where column vector D_1 of matrix D governs one-step transitions from transient states to the target absorbing state, state 0, and the entry in each row of column vector D_2 is the sum of the entries in the corresponding row of matrix D, excluding the entry in column D_1. For the absorbing multichain model of patient flow in a hospital, $N = 4$. The augmented matrix is

$$G = [Q \quad D_2 \quad D_1] = \begin{array}{c} 1 \\ 2 \\ 3 \\ 4 \end{array}\left[\begin{array}{cccc|c|c} g_{11} & g_{12} & g_{13} & g_{14} & g_{15} & g_{16} \\ g_{21} & g_{22} & g_{23} & g_{24} & g_{25} & g_{26} \\ g_{31} & g_{32} & g_{33} & g_{34} & g_{35} & g_{36} \\ g_{41} & g_{42} & g_{43} & g_{44} & g_{45} & g_{46} \end{array}\right]$$

$$= \begin{array}{c} 1 \\ 2 \\ 3 \\ 4 \end{array}\left[\begin{array}{cccc|c|c} p_{11} & p_{12} & p_{13} & p_{14} & p_{15} & p_{10} \\ p_{21} & p_{22} & p_{23} & p_{24} & p_{25} & p_{20} \\ p_{31} & p_{32} & p_{33} & p_{34} & p_{35} & p_{30} \\ p_{41} & p_{42} & p_{43} & p_{44} & p_{45} & p_{40} \end{array}\right]. \quad (6.71)$$

To avoid subtractions, matrix Q has been augmented with vectors D_1 and D_2, where

$$D_1 = \begin{bmatrix} p_{10} \\ p_{20} \\ p_{30} \\ p_{40} \end{bmatrix} = \begin{array}{c} 1 \\ 2 \\ 3 \\ 4 \end{array}\begin{bmatrix} 0 \\ 0.15 \\ 0.07 \\ 0 \end{bmatrix}, \quad D_2 = \begin{bmatrix} p_{15} \\ p_{25} \\ p_{35} \\ p_{45} \end{bmatrix} = \begin{array}{c} 1 \\ 2 \\ 3 \\ 4 \end{array}\begin{bmatrix} 0 \\ 0.05 \\ 0.03 \\ 0 \end{bmatrix} \quad (6.72)$$

Since the row sums of the augmented matrix equal one, subtractions are avoided by making the substitution

$$(1 - p_{11}) = (p_{12} + p_{13} + p_{14} + p_{15} + p_{10}) = (1 - g_{11}) = (g_{12} + g_{13} + g_{14} + g_{15} + g_{16}). \quad (6.73)$$

6.1.3.2 State Reduction Algorithm for Computing Absorption Probabilities

The detailed steps of a state reduction algorithm for computing absorption probabilities for a regular Markov chain with N states indexed $1, 2, \ldots, N$, and a transition probability matrix, P, are given below. The transition probability matrix, P, is assumed to be partitioned as in Equations (1.59) and (6.69).

A. Augmentation
An $N \times (N + 2)$ augmented matrix, **G**, is formed such that

$$G = [Q \quad D_2 \quad D_1] = [g_{ij}]. \quad (6.70)$$

B. Matrix Reduction
Matrix reduction is applied to the augmented matrix, **G**.

1. Initialize $n = 1$.
2. Let $\mathbf{G}^{(n)} = \mathbf{G} = [g_{ij}]$ for $n \le i \le N$ and $n \le j \le N + 2$.
3. Partition $\mathbf{G}^{(n)}$ as

$$
\mathbf{G}^{(n)} = \begin{matrix} n \\ n+1 \\ \vdots \\ N \end{matrix}
\left[
\begin{array}{c|cccc}
g_{nn}^{(n)} & g_{n,n+1}^{(n)} & \cdots & g_{n,N+1}^{(n)} & g_{n,N+2}^{(n)} \\
\hline
g_{n+1,n}^{(n)} & g_{n+1,n+1}^{(n)} & \cdots & g_{n+1,N+1}^{(n)} & g_{n+1,N+2}^{(n)} \\
\vdots & \vdots & \cdots & \vdots & \vdots \\
g_{N,n}^{(n)} & g_{N,n+1}^{(n)} & \cdots & g_{N,N+1}^{(n)} & g_{N,N+2}^{(n)}
\end{array}
\right]
$$

$$
= \begin{bmatrix} g_{nn}^{(n)} & \mathbf{u}^{(n)} \\ \mathbf{v}^{(n)} & \mathbf{T}^{(n)} \end{bmatrix}.
$$

4. Store the first row of $\mathbf{G}^{(n)}$.
5. Compute $\mathbf{G}^{(n+1)} = \mathbf{T}^{(n)} + \mathbf{v}^{(n)} \left[g_{n,n+1}^{(n)} + \cdots + g_{n,N+2}^{(n)} \right]^{1} \mathbf{u}^{(n)}$.
6. Increment n by 1. If $n < N$, go to step 3. Otherwise, go back to substitution.

C. Back Substitution

1. Compute the entry in row N of the vector of the probabilities of absorption in state 0.

$$
f_{N0} = g_{N,N+2}^{(N)} / \left(g_{N,N+1}^{(N)} + g_{N,N+2}^{(N)} \right).
$$

2. Let $i = N - 1$.
3. For $1 \le i < N$, Compute the entry in row i of the vector of the probabilities of absorption in state 0.

$$
f_{i0} = \left(g_{i,N+2}^{(i)} + \sum_{h=i+1}^{N} g_{ih}^{(i)} f_{h0} \right) \Big/ \sum_{h=i+1}^{N+2} g_{ih}^{(i)}.
$$

4. Decrement i by 1.
5. If $i > 0$, go to step 3. Otherwise, stop.

6.1.3.3 Using State Reduction to Compute Absorption Probabilities for an Absorbing Multichain Model of Patient Flow in a Hospital

The state reduction algorithm will be executed to compute the vector of the probabilities of absorption in state 0, which indicates that a hospital patient has been discharged.

A. Augmentation

$$
\mathbf{G}=\begin{bmatrix}\mathbf{Q} & \mathbf{D}_2 & \mathbf{D}_1\end{bmatrix}=\begin{matrix}1\\2\\3\\4\end{matrix}\begin{bmatrix}0.4 & 0.1 & 0.2 & 0.3 & 0 & 0\\0.1 & 0.2 & 0.3 & 0.2 & 0.05 & 0.15\\0.2 & 0.1 & 0.4 & 0.2 & 0.03 & 0.07\\0.3 & 0.4 & 0.2 & 0.1 & 0 & 0\end{bmatrix}. \tag{6.74}
$$

B. Matrix Reduction

$$
\mathbf{G}^{(1)}=\begin{matrix}1\\2\\3\\4\end{matrix}\begin{bmatrix}0.4 & 0.1 & 0.2 & 0.3 & 0 & 0\\0.1 & 0.2 & 0.3 & 0.2 & 0.05 & 0.15\\0.2 & 0.1 & 0.4 & 0.2 & 0.03 & 0.07\\0.3 & 0.4 & 0.2 & 0.1 & 0 & 0\end{bmatrix}=\begin{bmatrix}g_{11}^{(1)} & \mathbf{u}^{(1)}\\\mathbf{v}^{(1)} & \mathbf{T}^{(1)}\end{bmatrix} \tag{6.75}
$$

$$
\mathbf{G}^{(2)}=\mathbf{T}^{(1)}+\mathbf{v}^{(1)}\left[g_{12}^{(1)}+g_{13}^{(1)}+g_{14}^{(1)}+g_{15}^{(1)}+g_{16}^{(1)}\right]^{-1}\mathbf{u}^{(1)}
$$

$$
\mathbf{G}^{(2)}=\begin{matrix}2\\3\\4\end{matrix}\begin{bmatrix}0.2167 & 0.3333 & 0.25 & 0.05 & 0.15\\0.1333 & 0.4667 & 0.30 & 0.03 & 0.07\\0.4500 & 0.3000 & 0.25 & 0 & 0\end{bmatrix}=\begin{bmatrix}g_{22}^{(2)} & \mathbf{u}^{(2)}\\\mathbf{v}^{(2)} & \mathbf{T}^{(2)}\end{bmatrix} \tag{6.76}
$$

$$
\mathbf{G}^{(3)}=\mathbf{T}^{(2)}+\mathbf{v}^{(2)}\left[g_{23}^{(2)}+g_{24}^{(2)}+g_{25}^{(2)}+g_{26}^{(2)}\right]^{-1}\mathbf{u}^{(2)}
$$

$$
\mathbf{G}^{(3)}=\begin{matrix}3\\4\end{matrix}\begin{bmatrix}0.5234 & 0.3426 & 0.0385 & 0.0955\\0.4915 & 0.3936 & 0.0287 & 0.0862\end{bmatrix}=\begin{bmatrix}g_{33}^{(3)} & \mathbf{u}^{(3)}\\\mathbf{v}^{(3)} & \mathbf{T}^{(3)}\end{bmatrix} \tag{6.77}
$$

$$
\mathbf{G}^{(4)}=\mathbf{T}^{(3)}+\mathbf{v}^{(3)}\left[g_{34}^{(3)}+g_{35}^{(3)}+g_{36}^{(3)}\right]^{-1}\mathbf{u}^{(3)}
$$

$$
\mathbf{G}^{(4)}=\begin{bmatrix}0.7469 & 0.0684 & 0.1847\end{bmatrix}=\begin{bmatrix}g_{44}^{(4)} & g_{45}^{(4)} & g_{46}^{(4)}\end{bmatrix}. \tag{6.78}
$$

C. Back Substitution
Back substitution begins by computing the entry in row four of the vector of probabilities of absorption in state 0.

$$
f_{40}=(g_{46}^{(4)})/(g_{45}^{(4)}+g_{46}^{(4)})=(0.1847)/(0.0684+0.1847)=0.7298. \tag{6.79}
$$

Next the entry in row three is computed.

$$f_{30} = (g_{36}^{(3)} + g_{34}^{(3)} f_{40})/(g_{34}^{(3)} + g_{35}^{(3)} + g_{36}^{(3)})$$
$$= (0.0955 + 0.3426(0.7298))/(0.3426 + 0.0385 + 0.0955) = 0.7250. \tag{6.80}$$

Then the entry in row two is computed.

$$f_{20} = (g_{26}^{(2)} + g_{23}^{(2)} f_{30} + g_{24}^{(2)} f_{40})/(g_{23}^{(2)} + g_{24}^{(2)} + g_{25}^{(2)} + g_{26}^{(2)})$$
$$= (0.15 + 0.3333(0.7250) + 0.25(0.7298))/(0.3333 + 0.25 + 0.05 + 0.15)$$
$$= 0.7329. \tag{6.81}$$

Finally, the entry in row one is computed.

$$f_{10} = (g_{16}^{(1)} + g_{12}^{(1)} f_{20} + g_{13}^{(1)} f_{30} + g_{14}^{(1)} f_{40})/(g_{12}^{(1)} + g_{13}^{(1)} + g_{14}^{(1)} + g_{15}^{(1)} + g_{16}^{(1)})$$
$$= (0 + 0.1(0.7329) + 0.2(0.7250) + 0.3(0.7298))/(0.1 + 0.2 + 0.3 + 0 + 0) = 0.7287. \tag{6.82}$$

The probability f_{i0} that a patient in a transient state i will eventually be discharged can be expressed as an entry in the vector f_0 of the probabilities of absorption in absorbing state 0.

$$f_0 = [f_{10} \quad f_{20} \quad f_{30} \quad f_{40}]^T = [0.7287 \quad 0.7329 \quad 0.7250 \quad 0.7298]^T. \tag{6.83}$$

These probabilities of absorption in state 0 differ only slightly from those computed in Equation (3.95). Discrepancies are due to roundoff error because only the first four significant digits after the decimal point were stored.

6.2 An Introduction to Hidden Markov Chains

The following introduction to hidden Markov chains is based primarily on the first three sections of an excellent tutorial by Rabiner [4]. In certain applications such as speech recognition, bioinformatics, and musicology, the states of a Markov chain model may be hidden from an observer. To treat such cases, an extension of a Markov chain model, called a hidden Markov model, or HMM, has been developed. When a hidden or underlying Markov chain in an HMM enters a state, only an observation symbol, not the state itself, can be detected. In this section, the underlying Markov chain is assumed to be a regular chain.

6.2.1 HMM of the Weather

Consider the following small example of an HMM for the hourly weather on two remote hidden islands. The weather on the hidden islands cannot be observed by scientists at a distant meteorological station. Suppose that every hour a weather satellite randomly selects one of the two hidden islands and scans the weather on that island. Every hour, a signal indicating dry weather or wet weather on the randomly selected hidden island is relayed by the weather satellite to the distant meteorological station. The two islands are distinguished by the integers 1 and 2.

The hourly weather is treated as an independent trials process by assuming that the weather in one hour does not affect the weather in any other hour. Suppose that dry weather is denoted by D and wet weather by W. The probability that island i will have dry weather is denoted by d_i. Hence, for the ith island, $P(D) = d_i$ and $P(W) = 1 - d_i$. The probabilities of wet and dry weather on each hidden island are indicated symbolically in Table 6.1.

Letting $d_1 = 0.55$, and $d_2 = 0.25$, the probabilities of wet and dry weather on each hidden island are indicated numerically in Table 6.2.

Scientists at the meteorological station can see only the hourly report of the weather on a hidden island, either D or W, but cannot identify the island for which the weather is reported. The random choice of which island is to be scanned each hour by the satellite is governed by a two-state, regular, hidden Markov chain. The state, denoted by X_n, represents the identity of the hidden island scanned in hour n. The state space is $E = \{1, 2\}$. The number of states is denoted by N. In this example $N = 2$. Suppose that the hidden two-state Markov chain has the following transition probability matrix:

$$P = [p_{ij}] = \begin{array}{c|cc} \text{Island = State} & 1 & 2 \\ \hline 1 & p_{11} & p_{12} \\ 2 & p_{21} & p_{22} \end{array} = \begin{array}{c|cc} X_n\backslash X_{n+1} & 1 & 2 \\ \hline 1 & 0.6 & 0.4 \\ 2 & 0.7 & 0.3 \end{array} \qquad (6.84)$$

TABLE 6.1

Symbolic Probabilities of Weather on Islands

	Dry Weather, D	Wet Weather, W
Island 1	$P(D) = d_1$	$P(W) = 1 - d_1$
Island 2	$P(D) = d_2$	$P(W) = 1 - d_2$

TABLE 6.2

Numerical Probabilities of Weather on Islands

	Dry Weather, D	Wet Weather, W
Island 1	$P(D) = 0.55$	$P(W) = 0.45$
Island 2	$P(D) = 0.25$	$P(W) = 0.75$

Suppose that the initial state probability vector for the two-state hidden Markov chain is

$$p^{(0)} = \left[p_1^{(0)} \quad p_2^{(0)} \right] = [P(X_0 = 1) \quad P(X_0 = 2)] = [0.35 \quad 0.65]. \tag{6.85}$$

Each island or state has two mutually exclusive observation symbols, namely, D for dry weather and W for wet weather. The two distinct symbols are members of an alphabet denoted by

$$V = \{v_1, v_2\} = \{D, W\}. \tag{6.86}$$

The number of distinct observation symbols per state, or alphabet size, is denoted by K. In this example $K = 2$. The HMM can generate a sequence of observations denoted by

$$O = \{O_0, O_1, \ldots, O_M\}, \tag{6.87}$$

where each observation, O_n, is one of the symbols from the alphabet V, and $M + 1$ is the number of observations in the sequence. The observation symbols are generated in accordance with an observation symbol probability distribution. The observation symbol probability distribution function in state i is

$$b_i(k) = P(O_n = v_k | X_n = i), \quad \text{for } i = 1, \ldots, N \quad \text{and} \quad k = 1, \ldots, K. \tag{6.88}$$

In the hidden island weather example, $N = 2$ states and $K = 2$ observation symbols per state. In matrix form the observation symbol probability distribution is shown in Table 6.3.

Substituting $\{v_1, v_2\} = \{D, W\}$, the observation symbol probability distribution matrix is shown in Table 6.4.

Letting $d_i = P(O_n = D | X_n = i)$ and $1 - d_i = P(O_n = W | X_n = i)$ for island i, the observation symbol probability distribution matrix is shown in Table 6.5.

Substituting $d_1 = 0.55$, and $d_2 = 0.25$, the observation symbol probability distribution matrix is shown in Equation (6.89).

$$B = [b_i(k)] = \begin{array}{c|cc} & \text{Weather} = D & \text{Weather} = W \\ \hline \text{Island 1} & 0.55 & 0.45 \\ \text{Island 2} & 0.25 & 0.75 \end{array}. \tag{6.89}$$

TABLE 6.3

Observation Symbol Probability Distribution when $V = \{v_1, v_2\}$

	Observation Symbol $v_k = v_1$	Observation Symbol $v_k = v_2$
$B = [b_i(k)] = $ State $i = 1$	$P(O_n = v_1 | X_n = 1)$	$P(O_n = v_2 | X_n = 1)$
State $i = 2$	$P(O_n = v_1 | X_n = 2)$	$P(O_n = v_1 | X_n = 2)$

TABLE 6.4

Observation Symbol Probability Distribution when $V=\{D,W\}$

	Observation Symbol $v_1 = D$	Observation Symbol $v_2 = W$
$B = [b_i(k)] =$ State $i = 1$	$P(O_n = D \mid X_n = 1)$	$P(O_n = W \mid X_n = 1)$
State $i = 2$	$P(O_n = D \mid X_n = 2)$	$P(O_n = W \mid X_n = 2)$

TABLE 6.5

Observation Symbol Probability Distribution as a Function of d_i

	Weather Symbol $v_1 = D$	Weather Symbol $v_2 = W$
$B = [b_i(k)] =$ State 1 = Island 1	$b_1(D) = d_1$	$b_1(W) = 1 - d_1$
State 2 = Island 2	$b_2(D) = d_2$	$b_2(W) = 1 - d_2$

$X_0 = i$	$\rightarrow$	$X_1 = j$	$\rightarrow$	$X_2 = k$	$\cdots$	$X_n = g$	$\rightarrow$	$X_{n+1} = h$	State
O_0		O_1		O_2	$\cdots$	O_n		O_{n+1}	Observation Symbol
$b_i(O_0)$		$b_j(O_1)$		$b_k(O_2)$	$\cdots$	$b_g(O_n)$		$b_h(O_{n+1})$	Symbol Probability
$p_i^{(0)}$	p_{ij}		p_{jk}		$\cdots$		p_{gh}		Transition Probability
0		1		2	$\cdots$	n		$n+1$	Epoch

FIGURE 6.1
State transition and observation symbol generation for a sample path.

In summary, a hidden Markov chain starts at epoch 0 in state $X_0 = i$ with an initial state probability $p_i^{(0)} = P(X_0 = i)$, and generates an observation symbol O_0 with a symbol probability $b_i(O_0) = P(O_0 \mid X_0 = i)$. At epoch 1, with transition probability $p_{ij} = P(X_1 = j \mid X_0 = i)$, the hidden chain moves to state $X_1 = j$, and generates an observation symbol O_1 with symbol probability $b_j(O_1) = P(O_1 \mid X_1 = j)$. Continuing in this manner, a sequence $X = \{X_0, X_1, ...,X_M\}$ of hidden states (a sample path) generates a corresponding sequence $O = \{O_0, O_1, ... ,O_M\}$ of observation symbols. The process of state transition and observation symbol generation for a sample path is illustrated in Figure 6.1.

6.2.2 Generating an Observation Sequence

The following procedure will generate a sequence of $M + 1$ observations, $O = \{O_0, O_1, ... , O_M\}$.

1. Chose a starting state, $X_n = i$, according to the initial state probability vector,

$$p^{(0)} = \begin{bmatrix} p_1^{(0)} & p_2^{(0)} & \cdots & p_N^{(0)} \end{bmatrix} = [P(X_0 = 1) \quad P(X_0 = 2) \quad \cdots \quad P(X_0 = N)].$$

2. Set $n = 0$.

3. Choose $O_n = v_k$ according to the observation symbol probability distribution in state i,

$$b_i(k) = P(O_n = v_k | X_n = i), \quad \text{for } i = 1, \dots, N \quad \text{and} \quad k = 1, \dots, K.$$

4. At epoch $n + 1$ the hidden Markov chain moves to a state $X_{n+1} = j$ according to the state transition probability $p_{ij} = P(X_{n+1} = j | X_n = i)$.

5. Set $n = n + 1$, and return to step 3 if $n < M$. Otherwise, stop.

6.2.3 Parameters of an HMM

An HMM can be completely specified in terms of five parameters: N, K, $p^{(0)}$, P, and B. The first two parameters, are numbers, the third is a vector, and the last two are matrices. Parameter N is the number of states in the hidden Markov chain, while K is the number of distinct observation symbols per state, or alphabet size. For the HMM example of the weather on two hidden islands, $N = 2$ and $K = 2$. Parameter $p^{(0)}$ is the initial state probability vector for the hidden Markov chain. Parameter $P = [p_{ij}]$ is the transition probability matrix for the hidden Markov chain. Finally, parameter $B = [b_i(k)]$ is the observation symbol probability distribution matrix. A compact notation for the set of the three probability parameters used to specify an HMM is $\lambda = \{p^{(0)}, P, B\}$. As Equations (6.85), (6.84), and (6.89) indicate, the set of three probability parameters for the HMM example of the weather on two hidden islands is given by

$$\lambda = \left\{ [0.35 \quad 0.65], \begin{bmatrix} 0.6 & 0.4 \\ 0.7 & 0.3 \end{bmatrix}, \begin{bmatrix} 0.55 & 0.45 \\ 0.25 & 0.75 \end{bmatrix} \right\}. \tag{6.90}$$

6.2.4 Three Basic Problems for HMMs

In order for an HMM to be useful in applications, three basic problems must be addressed. The three problems can be linked together under a probabilistic framework. While an exact solution can be obtained for problem 1, problem 2 can be solved in several ways. No analytical solution can be found for problem 3, which is not treated in this book.

Problem 1 is how to calculate the probability that a particular sequence of observations, represented by $O = \{O_0, O_1, \dots, O_M\}$, is generated by a given model specified by the set of parameters $\lambda = \{p^{(0)}, P, B\}$. This probability is denoted by $P(O|\lambda)$. Problem 1 is useful in evaluating how well a given model matches a particular observation sequence.

In problem 2 a sequence of observations, represented by $O = \{O_0, O_1, \ldots, O_M\}$, is given, and a model is specified by the set of parameters $\lambda = \{p^{(0)}, P, B\}$. The objective is to choose a corresponding hidden sequence of states, $X = \{X_0, X_1, \ldots, X_M\}$, which best fits or explains the given observation sequence. Among several possible optimality criteria for determining a best fit, the one selected is to choose the state sequence that is most likely to have generated the given observation sequence.

Problem 3 is concerned with how to adjust the model parameters, $\lambda = \{p^{(0)}, P, B\}$, so as to maximize the probability of a particular observation sequence, $P(O|\lambda)$. Rabiner [4] and references [5, 6] describe iterative procedures for choosing model parameters such that $P(O|\lambda)$ is locally maximized.

6.2.4.1 Solution to Problem 1

If an HMM is extremely small, problem 1 can be solved by exhaustive enumeration. Otherwise, problem 1 is solved more efficiently by a forward procedure described in Section 6.2.4.1.2, or by a backward procedure described in Section 6.2.4.1.3.

6.2.4.1.1 Exhaustive Enumeration

Given a very small model and a particular observation sequence, the simplest way to solve problem 1 is by exhaustive enumeration. This would involve enumerating every possible state sequence capable of producing the particular observation sequence, calculating the probability of each of these enumerated state sequences, and adding these probabilities together. Recall that the set of parameters $\lambda = \{p^{(0)}, P, B\}$ for the HMM of the weather on two hidden islands is given in Equation (6.90).

Consider the particular sequence, $O = \{O_0, O_1, O_2\} = \{W, D, W\}$, of $M + 1 = 3$ (three) observations. This particular observation sequence can be generated by a corresponding sequence of hidden states denoted by $X = \{X_0, X_1, X_2\}$. The state space for distinguishing the two islands is $E = \{1, 2\}$. Since the observation sequence consists of three observation symbols, and each observation symbol can be generated by one of two states, the number of possible three-state sequences that must be enumerated is $(2)(2)(2) = 2^3 = 8$. The observations are assumed to be independent random variables. Then the probability that the observation sequence $O = \{O_0, O_1, O_2\} = \{W, D, W\}$ is generated by the state sequence $X = \{X_0, X_1, X_2\} = \{i, j, k\}$ is equal to

$$\left. \begin{aligned} &P(X_0 = i)P(O_0 = W| X_0 = i)P(X_1 = j| X_0 = i)P(O_1 = D| X_1 = j) \\ &P(X_2 = k| X_1 = j)P(O_2 = W| X_2 = k) \\ &= p_i^{(0)} b_i(W) p_{ij} b_j(D) p_{jk} b_k(W). \end{aligned} \right\} \tag{6.91}$$

In Table 6.6 below, all eight possible three-state sequences, $X = \{X_0, X_1, X_2\}$, are enumerated. The joint probabilities,

TABLE 6.6

Enumeration of Joint Probabilities for all Eight Possible Three-State Sequences

X_0	X_1	X_2	Joint Probability $P(X_0, X_1, X_2; W, D, W \mid \lambda)$
1	1	1	$p_1^{(0)} b_1(W) p_{11} b_1(D) p_{11} b_1(WR)$
			$= 0.35(0.45)0.6(0.55)0.6(0.45) = 0.0140332$
1	1	2	$p_1^{(0)} b_1(W) p_{11} b_1(D) p_{12} b_2(W)$
			$= 0.35(0.45)0.6(0.55)0.4(0.75) = 0.0155925$
1	2	1	$p_1^{(0)} b_1(W) p_{12} b_2(D) p_{21} b_1(W)$
			$= 0.35(0.45)0.4(0.25)0.7(0.45) = 0.00496125$
1	2	2	$p_1^{(0)} b_1(W) p_{12} b_2(D) p_{22} b_2(W)$
			$= 0.35(0.45)0.4(0.25)0.3(0.75) = 0.00354375$
2	1	1	$p_2^{(0)} b_2(W) p_{21} b_1(D) p_{11} b_1(W)$
			$= 0.65(0.75)0.7(0.55)0.6(0.45) = 0.0506756$
2	1	2	$p_2^{(0)} b_2(W) p_{21} b_1(D) p_{12} b_2(W)$
			$= 0.65(0.75)0.7(0.55)0.4(0.75) = 0.0563062 \leftarrow$ highest
2	2	1	$p_2^{(0)} b_2(W) p_{22} b_2(D) p_{21} b_1(W)$
			$= 0.65(0.75)0.3(0.25)0.7(0.45) = 0.0115171$
2	2	2	$p_2^{(0)} b_2(W) p_{22} b_2(D) p_{22} b_2(W)$
			$= 0.65(0.75)0.3(0.25)0.3(0.75) = 0.0082265625$

$$P(W, D, W \mid \lambda) = \sum_{X_0, X_1, X_2} P(X_0, X_1, X_2; W, D, W \mid \lambda) = 0.164856$$

$$P(X, O \mid \lambda) = P(X_0, X_1, X_2; O_0, O_1, O_2 \mid \lambda) = P(X_0, X_1, X_2; W, D, W \mid \lambda) \quad (6.92)$$

are calculated for each three-state sequence. The sum, taken over all values of X, of these eight joint probabilities is equal to the marginal probability, $P(O \mid \lambda)$. That is,

$$P(O \mid \lambda) = \sum_{\text{all } X} P(X, O \mid \lambda) = \sum_{\text{all } X} P(O \mid X, \lambda) P(X \mid \lambda)$$

$$\left. \begin{aligned} P(O \mid \lambda) &= \sum_{\text{all } X} P(X, O \mid \lambda) = \sum_{\text{all } X} P(O \mid X, \lambda) P(X \mid \lambda) \\ &= P(O_0, O_1, O_2 \mid \lambda) = P(W, D, W \mid \lambda) = \sum_{X_0, X_1, X_2} P(X_0, X_1, X_2; W, D, W \mid \lambda). \end{aligned} \right\} \quad (6.93)$$

In this small example involving a sequence of $M + 1 = 3$ observation symbols and $N = 2$ states, approximately $[2(M + 1) - 1]N^{M+1} = [2(3) - 1]2^3 = 40$ multiplications were needed to calculate

$$P(O|\lambda) = P(O_0, O_1, O_2|\lambda) = P(W, D, W|\lambda) = 0.164856 \qquad (6.94)$$

by using exhaustive enumeration in Table 6.6. However, in larger problems, exhaustive enumeration is not a practical procedure for calculating $P(O|\lambda)$. to solve problem 1. Either one of two alternative procedures, called the forward procedure and the backward procedure, can be used to solve problem 1 far more efficiently than exhaustive enumeration. Calculations in the forward procedure move forward in time, while those in the backward procedure move backward.

6.2.4.1.2 Forward Procedure

The forward procedure defines a forward variable,

$$\alpha_n(i) = P(O_0, O_1, O_2, \ldots, O_n, X_n = i|\lambda) \qquad (6.95)$$

which is the joint probability that the partial sequence $\{O_0, O_1, O_2, \ldots, O_n\}$ of $n + 1$ observations is generated from epoch 0 until epoch n, and the HMM is in state i at epoch n. The forward procedure has three steps, which are labeled initialization, induction, and termination. The three steps will be described in reference to the small HMM of the weather on two hidden islands.

Step 1. Initialization
At epoch 0, the forward variable is

$$\alpha_0(i) = P(O_0, X_0 = i|\lambda) = P(X_0 = i)P(O_0|X_0 = i) = p_i^{(0)}b_i(O_0) \quad \text{for } 1 \leq i \leq N = 2$$

$$\alpha_0(1) = p_1^{(0)}b_1(O_0)$$
$$\alpha_0(2) = p_2^{(0)}b_2(O_0).$$

Step 2. Induction
At epoch 1,

$$\alpha_1(j) = P(O_0, O_1, X_1 = j|\lambda) = P(X_0 = 1)P(O_0|X_0 = 1)P(X_1 = j|X_0 = 1)P(O_1|X_1 = j)$$
$$+ P(X_0 = 2)P(O_0|X_0 = 2)P(X_1 = j|X_0 = 2)P(O_1|X_1 = j) \quad \text{for } 1 \leq j \leq N = 2.$$

$$\alpha_1(j) = p_1^{(0)} b_1(O_0) p_{1j} b_j(O_1) + p_2^{(0)} b_2(O_0) p_{2j} b_j(O_1)$$
$$= [p_1^{(0)} b_1(O_0) p_{1j} + p_2^{(0)} b_2(O_0) p_{2j}] b_j(O_1)$$
$$= [\alpha_0(1) p_{1j} + \alpha_0(2) p_{2j}] b_j(O_1)$$
$$= \left[\sum_{i=1}^{2} \alpha_0(i) p_{ij} \right] b_j(O_1).$$

Hence, at epoch 1, the forward variable is

$$\alpha_1(j) = \left[\sum_{i=1}^{2} \alpha_0(i) p_{ij} \right] b_j(O_1) \quad \text{for} \quad 1 \le j \le N = 2.$$

Similarly, at epoch 2,

$$\alpha_2(j) = P(O_0, O_1, O_2, X_2 = j | \lambda)$$
$$\alpha_2(j) = p_1^{(0)} b_1(O_0) p_{11} b_1(O_1) p_{1j} b_j(O_2) + p_1^{(0)} b_1(O_0) p_{12} b_2(O_1) p_{2j} b_j(O_2)$$
$$+ p_2^{(0)} b_2(O_0) p_{21} b_1(O_1) p_{1j} b_j(O_2) + p_2^{(0)} b_2(O_0) p_{22} b_2(O_1) p_{2j} b_j(O_2)$$
$$= p_1^{(0)} b_1(O_0) p_{11} b_1(O_1) p_{1j} b_j(O_2) + p_2^{(0)} b_2(O_0) p_{21} b_1(O_1) p_{1j} b_j(O_2)$$
$$+ p_1^{(0)} b_1(O_0) p_{12} b_2(O_1) p_{2j} b_j(O_2) + p_2^{(0)} b_2(O_0) p_{22} b_2(O_1) p_{2j} b_j(O_2).$$

Substituting $\alpha_0(1) = p_1^{(0)} b_1(O_0)$ and $\alpha_0(2) = p_2^{(0)} b_2(O_0)$,

$$\alpha_2(j) = \alpha_0(1) p_{11} b_1(O_1) p_{1j} b_j(O_2) + \alpha_0(2) p_{21} b_1(O_1) p_{1j} b_j(O_2)$$
$$+ \alpha_0(1) p_{12} b_2(O_1) p_{2j} b_j(O_2) + \alpha_0(2) p_{22} b_2(O_1) p_{2j} b_j(O_2)$$
$$= [\alpha_0(1) p_{11} + \alpha_0(2) p_{21}] b_1(O_1) p_{1j} b_j(O_2)$$
$$+ [\alpha_0(1) p_{12} + \alpha_0(2) p_{22}] b_2(O_1) p_{2j} b_j(O_2)$$

Substituting $\alpha_1(1) = [\alpha_0(1) p_{11} + \alpha_0(2) p_{21}] b_1(O_1)$ and $\alpha_1(2)$

$$= [\alpha_0(1) p_{12} + \alpha_0(2) p_{22}] b_2(O_1),$$
$$\alpha_2(j) = \alpha_1(1) p_{1j} b_j(O_2) + \alpha_1(2) p_{2j} b_j(O_2)$$
$$= [\alpha_1(1) p_{1j} + \alpha_1(2) p_{2j}] b_j(O_2)$$
$$= \left[\sum_{i=1}^{2} \alpha_1(i) p_{ij} \right] b_j(O_2)$$

Thus, at epoch 2 the forward variable is

$$\alpha_2(j) = [\sum_{i=1}^{2} \alpha_1(i)p_{ij}]b_j(O_2) \quad \text{for } 1 \le j \le N = 2.$$

By induction, one may conclude that in the general case the forward variable is

$$\alpha_{n+1}(j) = [\sum_{i=1}^{N} \alpha_n(i)p_{ij}]b_j(O_{n+1}) \text{ for epochs } 0 \le n \le M-1 \text{ and states } 1 \le j \le N.$$

Step 3. Termination
At epoch $M=2$, the forward procedure ends with the desired probability expressed as the sum of the terminal forward variables.

$$P(O|\lambda) = \sum_{\text{all values of } X_M} P(O, X_M|\lambda)$$

$$= P(O_0, O_1, O_2|\lambda) = P(O_0, O_1, O_2, X_2 = 1|\lambda) + P(O_0, O_1, O_2, X_2 = 2|\lambda)$$

$$= \alpha_2(1) + \alpha_2(2) = \sum_{i=1}^{2} \alpha_2(i).$$

The complete forward procedure is given below:

Step 1. Initialization
At epoch 0, the forward variable is

$$\alpha_0(i) = p_i^{(0)}b_i(O_0) \quad \text{for } 1 \le i \le N.$$

Step 2. Induction

$$\alpha_{n+1}(j) = [\sum_{i=1}^{N} \alpha_n(i)p_{ij}]b_j(O_{n+1}) \text{ for epochs } 0 \le n \le M-1 \text{ and states } 1 \le j \le N.$$

Step 3. Termination

$$P(O|\lambda) = P(O_0, O_1, O_2, \dots, O_M|\lambda) = \sum_{i=1}^{N} \alpha_M(i).$$

The forward procedure will be executed to calculate $P(O|\lambda) = P(O_0, O_1, O_2|\lambda) = P(W,D,W|\lambda)$ for the small example concerning the weather on two hidden islands. Recall that the set of parameters $\lambda = \{p^{(0)}, P, B\}$ for this example is given in Equation (6.90).

Step 1. Initialization
At epoch 0,

$$\alpha_0(i) = p_i^{(0)}b_i(O_0) = p_i^{(0)}b_i(W)$$
$$\alpha_0(1) = p_1^{(0)}b_1(W) = (0.35)(0.45) = 0.1575$$
$$\alpha_0(2) = p_2^{(0)}b_2(W) = (0.65)(0.75) = 0.4875$$

Step 2. Induction
At epoch 1,

$$\alpha_1(j) = \left[\sum_{i=1}^{2}\alpha_0(i)p_{ij}\right]b_j(O_1) = [\alpha_0(1)p_{1j} + \alpha_0(2)p_{2j}]b_j(D)$$
$$\alpha_1(1) = [\alpha_0(1)p_{11} + \alpha_0(2)p_{21}]b_1(D) = [(0.1575)(0.6) + (0.4875)(0.7)](0.55) = 0.2396625$$
$$\alpha_1(2) = [\alpha_0(1)p_{12} + \alpha_0(2)p_{22}]b_2(D) = [(0.1575)(0.4) + (0.4875)(0.3)](0.25) = 0.0523125$$

At epoch 2,

$$\alpha_2(j) = \left[\sum_{i=1}^{2}\alpha_1(i)p_{ij}\right]b_j(O_2) = [\alpha_1(1)p_{1j} + \alpha_1(2)p_{2j}]b_j(W)$$
$$\alpha_2(1) = [\alpha_1(1)p_{11} + \alpha_1(2)p_{21}]b_1(W)$$
$$= [(0.2396625)(0.6) + (0.0523125)(0.7)](0.45) = 0.0811873$$
$$\alpha_2(2) = [\alpha_1(1)p_{12} + \alpha_1(2)p_{22}]b_2(W)$$
$$= [(0.2396625)(0.4) + (0.0523125)(0.3)](0.75) = 0.083669.$$

Step 3. Termination
At epoch $M = 2$,

$$P(O|\lambda) = P(O_0, O_1, O_2|\lambda) = P(W, D, W|\lambda) = \sum_{i=1}^{2}\alpha_2(i) = \alpha_2(1) + \alpha_2(2)$$
$$= 0.0811873 + 0.083669 = 0.1648563 \tag{6.96}$$

in agreement with the result obtained by exhaustive enumeration in Table 6.6.

In contrast to exhaustive enumeration, which required about 40 multiplications to calculate $P(O|\lambda) = P(O_0, O_1, O_2|\lambda)$ for this small HMM involving a sequence of $M + 1 = 3$ observation symbols and $N = 2$ states, the forward procedure required only $N(N + 1)M + N = 2(3)2 + 2 = 14$ multiplications.

6.2.4.1.3 Backward Procedure

Although the forward procedure alone is sufficient for solving problem 1, the backward procedure is introduced as an alternative procedure for solving problem 1 because both forward and backward calculations are needed for solving problem 3. Calculations in the backward procedure move backward in time. The backward procedure defines a backward variable,

$$\beta_n(i) = P(O_{n+1}, O_{n+2}, O_{n+3}, \ldots, O_M | X_n = i, \lambda) \tag{6.97}$$

which is the conditional probability that the partial sequence $\{O_{n+1}, O_{n+2}, O_{n+3}, \ldots, O_M\}$ of $M - n$ observations is generated from epoch $n + 1$ until epoch M, given that the HMM is in state i at epoch n. The backward procedure, when it is applied to solve problem 1, also has three steps called initialization, induction, and termination. The three-step backward procedure used to solve problem 1 is given below.

Step 1. Initialization
At epoch M, the backward variable is

$$\beta_M(i) = 1 \quad \text{for } 1 \le i \le N.$$

Step 2. Induction

$$\beta_n(i) = \sum_{j=1}^{N} p_{ij} b_j(O_{n+1}) \beta_{n+1}(j) \quad \text{for epochs } n = M-1, M-2, \ldots, 0 \quad \text{and} \quad \text{states } 1 \le i \le N.$$

Step 3. Termination
At epoch 0,

$$P(O|\lambda) = P(O_0, O_1, O_2, \ldots O_M | \lambda) = \sum_{i=1}^{N} P(O_0, X_0 = i | \lambda) P(O_1, O_2, O_3, \ldots, O_M | X_0 = i, \lambda)$$

$$= \sum_{i=1}^{N} P(X_0 = i) P(O_0 | X_0 = i) P(O_1, O_2, O_3, \ldots, O_M | X_0 = i, \lambda)$$

$$= \sum_{i=1}^{N} p_i^{(0)} b_i(O_0) \beta_0(i) = \sum_{i=1}^{N} \alpha_0(i) \beta_0(i),$$

after substituting $\alpha_0(i) = p_i^{(0)} b_i(O_0)$ (the initialization step of the forward procedure).

The backward procedure will be executed to calculate $P(O|\lambda) = P(O_0, O_1, O_2 | \lambda) = P(W, D, W | \lambda)$ for the small example involving the weather on two hidden islands.

Step 1. Initialization.

At epoch $M = 2$,

$$\beta_2(1) = 1 \quad \text{and} \quad \beta_2(2) = 1.$$

Step 2. Induction

At epoch 1,

$$\beta_1(i) = \sum_{j=1}^{2} p_{ij} b_j(O_2)\beta_2(j) = p_{i1} b_1(W)\beta_2(1) + p_{i2} b_2(W)\beta_2(2)$$

$$\beta_1(1) = p_{11} b_1(W)\beta_2(1) + p_{12} b_2(W)\beta_2(2) = p_{11} b_1(W)(1) + p_{12} b_2(W)(1)$$
$$= (0.6)(0.45)(1) + (0.4)(0.75)(1) = 0.57$$

$$\beta_1(2) = p_{21} b_1(W)\beta_2(1) + p_{22} b_2(W)\beta_2(2) = p_{21} b_1(W)(1) + p_{22} b_2(W)(1)$$
$$= (0.7)(0.45)(1) + (0.3)(0.75)(1) = 0.54.$$

At epoch 0,

$$\beta_0(i) = \sum_{j=1}^{2} p_{ij} b_j(O_1)\beta_1(j) = p_{i1} b_1(D)\beta_1(1) + p_{i2} b_2(D)\beta_1(2)$$

$$\beta_0(1) = p_{11} b_1(D)\beta_1(1) + p_{12} b_2(D)\beta_1(2) = (0.6)(0.55)(0.57) + (0.4)(0.25)(0.54) = 0.2421$$

$$\beta_0(2) = p_{21} b_1(D)\beta_1(1) + p_{22} b_2(D)\beta_1(2) = (0.7)(0.55)(0.57) + (0.3)(0.25)(0.54) = 0.25995.$$

Step 3. Termination

At epoch 0, after substituting $\alpha_0(i) = p_i^{(0)} b_i(O_0)$,

$$\alpha_0(1) = p_1^{(0)} b_1(W) = (0.35)(0.45) = 0.1575$$
$$\alpha_0(2) = p_2^{(0)} b_2(W) = (0.65)(0.75) = 0.4875,$$

$$P(O|\lambda) = P(O_0, O_1, O_2|\lambda) = P(W, D, W|\lambda) = \sum_{i=1}^{2} \alpha_0(i)\beta_0(i) = \alpha_0(1)\beta_0(1) + \alpha_0(2)\beta_0(2)$$

$$= (0.1575)(0.2421) + (0.4875)(0.25995) = 0.1648563,$$

which is the same result obtained by exhaustive enumeration in Table 6.6 and Equation (6.94), and by the forward procedure in Equation (6.96).

6.2.4.2 Solution to Problem 2

Problem 2 assumes that a sequence of symbols has been observed, and the parameter set λ of the model has been specified. Given the sequence of observation symbols, $O = \{O_0, O_1, \ldots, O_M\}$, the objective of problem 2 is to identify a corresponding sequence of states, $X = \{X_0, X_1, \ldots, X_M\}$, which is most likely to have generated the given observation sequence. In other words, given the observation sequence, $O = \{O_0, O_1, \ldots, O_M\}$, the objective of problem 2 is to find a corresponding state sequence, $X = \{X_0, X_1, \ldots, X_M\}$, which maximizes the probability of having generated the given observation sequence. Mathematically, this objective is to find the state sequence X, which maximizes the conditional probability, $P(X|O, \lambda)$. This objective is expressed by the notation $\text{argmax}_X P(X|O, \lambda)$. The term argmax refers to the argument X which yields the maximum value of $P(X|O, \lambda)$. By the definition of conditional probability,

$$\max_X P(X|O, \lambda) = \max_X P(X, O|\lambda)/P(O|\lambda). \tag{6.98}$$

Since the denominator on the right hand side is not a function of X, maximizing the conditional probability $P(X|O, \lambda)$ with respect to X is equivalent to maximizing the numerator $P(X, O|\lambda)$ with respect to X. Hence an equivalent objective of problem 2 is to find the state sequence X, which maximizes $P(X, O|\lambda)$, the joint probability of state sequence X and observation sequence O.

A formal procedure called the Viterbi algorithm, which is based on dynamic programming, is used to calculate $\max_X P(X, O|\lambda.)$ and find $\text{argmax}_X P(X, O|\lambda)$. The algorithm first calculates $\max_X P(X, O|\lambda.)$. Then the algorithm works backward or backtracks to recover the state sequence X which maximizes $P(X, O|\lambda)$. If this state sequence is not unique, the Viterbi algorithm will find one state sequence, which maximizes $P(X, O|\lambda)$. To find the state sequence X that is most likely to correspond to the given observation sequence O, the algorithm defines, at epoch n, the quantity

$$\delta_n(j) = \max_{X_0, X_1, \ldots, X_{n-1}} P(X_0, X_1, X_2, \ldots, X_n = j, \ O_0, O_1, O_2, \ldots, O_n|\lambda) \tag{6.99a}$$

The quantity $\delta_n(i)$ is the highest probability of the state sequence ending in state X_n at epoch n, which accounts for the first $n + 1$ observations.

At epoch 0, $\delta_n(i)$ is initialized as

$$\delta_0(i) = p_i^{(0)} b_i(O_0). \tag{6.99b}$$

As is true for the forward variable, $\alpha_n(i)$, and the backward variable, $\beta_n(i)$, the probability $\delta_n(i)$ can be calculated by induction. The induction equation for calculating $\delta_n(i)$ is developed informally below.

At epoch 0,

$$\delta_0(i) = p_i^{(0)} b_i(O_0) \quad \text{for } i = 1, 2, \ldots, N.$$

At epoch 1,

$$\delta_1(X_1 = j) = \max_{X_0} p_{X_0}^{(0)} b_{X_0}(O_0) p_{X_0,X_1} b_{X_1}(O_1)$$

$$\delta_1(j) = \max_{i=1,\ldots,N}[p_i^{(0)} b_i(O_0)]p_{ij} b_j(O_1) = \max_{i=1,\ldots,N}[\delta_0(i)p_{ij}]b_j(O_1) \quad \text{for } j = 1,\ldots,N.$$

At epoch 2,

$$\delta_2(X_2 = k) = \max_{X_0,X_1} [\max_{X_0} p_{X_0}^{(0)} b_{X_0}(O_0) p_{X_0,X_1} b_{X_1}(O_1)]p_{X_1,X_2} b_{X_2}(O_2)$$

$$= \max_{i=1,\ldots,N}[p_i^{(0)} b_i(O_0) p_{ij} b_j(O_1)]p_{jk} b_k(O_2)$$

$$\delta_2(k) = \max_{i=1,\ldots,N}[\delta_1(i)p_{ik}]b_k(O_2) \quad \text{for } k = 1,\ldots,N.$$

At epoch $n + 1$, for $n = 0, \ldots, M - 1$, by induction,

$$\delta_{n+1}(j) = \max_{i}[\delta_n(i)p_{ij}]b_j(O_{n+1}) \quad \text{for } j = 1,\ldots,N.$$

Observe that at epoch M,

$$\delta_M(j) = \max_{i}[\delta_{M-1}(i)p_{ij}]b_j(O_M) \quad \text{for } j = 1,\ldots,N.$$

Recall that $\delta_M(i)$ is the highest probability of the state sequence ending in state X_M at epoch M, which accounts for the observation sequence ending in symbol O_M at epoch M. Therefore, the joint probability of this state sequence and the associated observation sequence is

$$\delta_M(j) = \max_{X_0,X_1,\ldots,X_{M-1}} P(X_0,X_1,X_2,\ldots,X_M = j, \; O_0,O_1,O_2,\ldots,O_M|\lambda). \tag{6.100}$$

To recover the state sequence $\{X_0, X_1, X_2,\ldots,X_M\}$, which has the highest probability of generating the observation sequence $\{O_0, O_1, O_2,\ldots,O_M\}$, it is necessary
to keep a record of the argument, which maximizes the induction equation for
$\delta_{n+1}(j)$ for each n and j. To construct this record, an array $\psi_n(j)$ is defined as

$$\psi_n(j) = \operatorname*{argmax}_{i=1,\ldots,N}[\delta_{n-1}(i)p_{ij}] \quad \text{for } n = 1,\ldots,M \quad \text{and} \quad j = 1,\ldots,N. \tag{6.101}$$

The complete Viterbi algorithm is given below:

Step 1. Initialization

$$\delta_0(i) = p_i^{(0)} b_i(O_0) \quad \text{for } 1 \le i \le N$$

$$\psi_0(i) = 0$$

Step 2. Recursion

$$\delta_n(j) = \max_{1 \le i \le N}[\delta_{n-1}(i)p_{ij}]b_j(O_n), \quad \text{for } 1 \le n \le M \text{ and } 1 \le j \le N$$

$$\psi_n(j) = \operatorname*{argmax}_{1 \le i \le N}[\delta_{n-1}(i)p_{ij}], \quad \text{for } 1 \le n \le M \text{ and } 1 \le j \le N.$$

Step 3. Termination

$$P^* = \max_{1 \le i \le N}[\delta_M(i)]$$

$$X_M^* = \operatorname*{argmax}_{1 \le i \le N}[\delta_M(i)].$$

Step 4. Backtracking to recover the state sequence

$$X_n^* = \psi_{n+1}(X_{n+1}^*), \quad \text{for } n = M-1, M-2, \dots, 0.$$

The Viterbi algorithm will be applied to the example HMM of the weather on two hidden islands to recover the state sequence, $X = \{X_0, X_1, X_2\}$, which has the highest probability of generating the weather observation sequence, $O = \{O_0, O_1, O_2\} = \{W, D, W\}$, representing the weather at three consecutive epochs. Recall that the hidden state indicates which island is the source of the observed weather symbol. The algorithm calculates $\max_X P(X, O)$ and retrieves $\operatorname{argmax}_X P(X, O)$. The set of model parameters $f\lambda = \{p^{(0)}, P, B\}$ for this example is given in equation (6.90).

Step 1. Initialization
At epoch 0,

$$\delta_0(i) = p_i^{(0)}b_i(O_0) = p_i^{(0)}b_i(W), \quad \text{for } 1 \le i \le 2$$

$$\delta_0(1) = p_1^{(0)}b_1(W) = (0.35)(0.45) = 0.1575$$

$$\delta_0(2) = p_2^{(0)}b_2(W) = (0.65)(0.75) = 0.4875$$

$$\psi_0(i) = 0 = \psi_0(1) = \psi_0(2).$$

Step 2. Recursion
At epoch 1,

$$\delta_1(j) = \max_{1 \le i \le 2}[\delta_0(i)p_{ij}]b_j(O_1) = \max_{1 \le i \le 2}[\delta_0(i)p_{ij}]b_j(D), \quad \text{for } 1 \le j \le 2$$

$$\delta_1(j) = \max[\delta_0(1)p_{1j}, \ \delta_0(2)p_{2j}]b_j(D), \quad \text{for } 1 \le j \le 2$$

$$\delta_1(1) = \max[\delta_0(1)p_{11}, \ \delta_0(2)p_{21}]b_1(D), \quad \text{for } j = 1$$

$$= \max[(0.1575)(0.6), \ (0.4875)(0.7)](0.55)$$

$$= [(0.4875)(0.7)](0.55) = 0.1876875$$

$$\delta_1(2) = \max[\delta_0(1)p_{12}, \quad \delta_0(2)p_{22}]b_2(D), \quad \text{for } j = 2$$
$$= \max[(0.1575)(0.4), \quad (0.4875)(0.3)](0.25)$$
$$= [(0.4875)(0.3)](0.25) = 0.0365625$$

$$\psi_1(j) = \operatorname*{argmax}_{1 \le i \le 2}[\delta_0(i)p_{ij}], \quad \text{for } 1 \le j \le 2$$
$$= \arg\max[\delta_0(1)p_{1j} \text{ for } i = 1, \, \delta_0(2)p_{2j} \text{ for } i = 2], \text{ for } 1 \le j \le 2$$

$$\psi_1(1) = \arg\max[\delta_0(1)p_{11}, \quad \delta_0(2)p_{21}] \text{ for } j = 1$$
$$= \arg\max[(0.1575)(0.6), \quad (0.4875)(0.7)]$$
$$= \arg\max[0.0945, \quad 0.34125] = 2$$

$$\psi_1(2) = \arg\max[\delta_0(1)p_{12}, \quad \delta_0(2)p_{22}] \text{ for } j = 2,$$
$$= \arg\max[(0.1575)(0.4), \quad (0.4875)(0.3)]$$
$$= \arg\max[0.063, \quad 0.14625] = 2.$$

At epoch 2,

$$\delta_2(j) = \max_{1 \le i \le 2}[\delta_1(i)p_{ij}]b_j(O_2) = \max_{1 \le i \le 2}[\delta_1(i)p_{ij}]b_j(W), \quad \text{for } 1 \le j \le 2$$
$$= \max[\delta_1(1)p_{1j}, \quad \delta_1(2)p_{2j}]b_j(W), \quad \text{for } 1 \le j \le 2$$

$$\delta_2(1) = \max[\delta_1(1)p_{11}, \quad \delta_1(2)p_{21}]b_1(W) \text{ for } j = 1$$
$$= \max[(0.1876875)(0.6), \quad (0.0365625)(0.7)](0.45)$$
$$= [(0.1876875)(0.6)](0.45) = 0.0506756$$

$$\delta_2(2) = \max[\delta_1(1)p_{12}, \quad \delta_1(2)p_{22}]b_2(W) \text{ for } j = 2$$
$$= \max[(0.1876875)(0.4)], \quad (0.0365625)(0.3)](0.75)$$
$$= [(0.1876875)(0.4)](0.75) = 0.0563062$$

$$\psi_2(j) = \operatorname*{argmax}_{1 \le i \le 2}[\delta_1(i)p_{ij}], \quad \text{for } 1 \le j \le 2$$
$$= \arg\max[\delta_1(1)p_{1j} \text{ for } i = 1, \delta_1(2)p_{2j} \text{ for } i = 2], \text{ for } 1 \le j \le 2$$

$$\psi_2(1) = \arg\max[\delta_1(1)p_{11}, \quad \delta_1(2)p_{21}], \text{ for } j = 1.$$
$$= \arg\max[(0.1876875)(0.6), \quad (0.0365625)(0.7)]$$
$$= \arg\max[0.1126125, \quad 0.0255937]$$
$$= 1$$

$$\psi_2(2) = \arg\max[\delta_1(1)p_{12}, \quad \delta_1(2)p_{22}], \text{ for } j = 2$$
$$= \arg\max[(0.1876875)(0.4), \quad (0.0365625)(0.3)]$$
$$= \arg\max[(0.075075, \quad 0.0109687]$$
$$= 1.$$

Step 3. Termination

$$P^* = \max_{1 \le i \le 2}[\delta_2(i)] = \max[\delta_2(1), \ \delta_2(2)] = \max[0.0506756, \ 0.0563062] = 0.0563062 = \delta_2(2)$$

$$X_2^* = \operatorname{argmax}_{1 \le i \le 2}[\delta_2(i)] = \arg\max[\delta_2(1), \ \delta_2(2)] = \arg\max[0.0506756, \ 0.0563062] = 2$$

Step 4. Backtracking to recover the state sequence

$$X_n^* = \psi_{n+1}(X_{n+1}^*), \text{ for } n = (2-1), 0 = 1, 0$$

$$X_1^* = \psi_{1+1}(X_{1+1}^*) = \psi_2(X_2^*) = \psi_2(2) = 1, \text{ for } n = 1$$

$$X_0^* = \psi_{0+1}(X_{0+1}^*) = \psi_1(X_1^*) = \psi_1(1) = 2, \text{ for } n = 0.$$

Note that $P^* = 0.0563062$ is the highest joint probability that the state sequence $\{X_0, X_1, X_2\}$ accounts for the given observation sequence $\{O_0, O_1, O_2\} = \{W, D, W\}$. Backtracking has retrieved the accountable state sequence $X = \{X_0, X_1, X_2\} = \{2,1,2\}$. Hence,

$$P^* = P(X_0, X_1, X_2; \ O_0, O_1, O_2 | \lambda) = P(2,1,2; \ W, D, W | \lambda) = 0.0563062. \quad (6.102)$$

This result was obtained previously by exhaustive enumeration in Table 6.6, where it is flagged by an arrow. The solution to problem 2 has identified the state sequence that is most likely to have occurred, given a particular observation sequence. Other criteria for identifying a state sequence can be imposed.

PROBLEMS

6.1   Consider a regular Markov chain that has the following transition probability matrix:

$$P = \quad
\begin{array}{c|cccc}
\text{State} & 0 & 1 & 2 & 3 \\
\hline
0 & 0.23 & 0.34 & 0.26 & 0.17 \\
1 & 0.08 & 0.42 & 0.18 & 0.32 \\
2 & 0.31 & 0.17 & 0.23 & 0.29 \\
3 & 0.24 & 0.36 & 0.34 & 0.06 \\
\end{array}$$

Use the MCPA to find the vector of steady-state probabilities.

6.2   Consider a regular Markov chain which has the following transition probability matrix:

$$P = \quad
\begin{array}{c|cccc}
\text{State} & 0 & 1 & 2 & 3 \\
\hline
0 & 0.23 & 0.34 & 0.26 & 0.17 \\
1 & 0.08 & 0.42 & 0.18 & 0.32 \\
2 & 0.31 & 0.17 & 0.23 & 0.29 \\
3 & 0.24 & 0.36 & 0.34 & 0.06 \\
\end{array}$$

Use state reduction to find the vector of MFPTs to state 0.

6.3 Consider an absorbing multichain which has the following transition probability matrix expressed in canonical form:

$$P = \begin{array}{c|ccc|ccc}
\text{State} & 0 & 4 & 5 & 1 & 2 & 3 \\
\hline
0 & 1 & 0 & 0 & 0 & 0 & 0 \\
4 & 0 & 1 & 0 & 0 & 0 & 0 \\
5 & 0 & 0 & 1 & 0 & 0 & 0 \\
\hline
1 & 0.15 & 0.05 & 0.20 & 0.2 & 0.3 & 0.1 \\
2 & 0.09 & 0.03 & 0.18 & 0.1 & 0.4 & 0.2 \\
3 & 0.12 & 0.02 & 0.06 & 0.5 & 0.2 & 0.1
\end{array} = \begin{bmatrix} I & 0 \\ D & Q \end{bmatrix}, \text{ where}$$

$$I = \begin{bmatrix} 1 & 0 & 0 \\ 0 & 1 & 0 \\ 0 & 0 & 1 \end{bmatrix}, D = \begin{array}{c}1\\2\\3\end{array}\begin{bmatrix} 0.15 & 0.05 & 0.20 \\ 0.09 & 0.03 & 0.18 \\ 0.12 & 0.02 & 0.06 \end{bmatrix}, Q = \begin{array}{c}1\\2\\3\end{array}\begin{bmatrix} 0.2 & 0.3 & 0.1 \\ 0.1 & 0.4 & 0.2 \\ 0.5 & 0.2 & 0.1 \end{bmatrix}$$

Use state reduction to find the vector of probabilities of absorption in state 0.

6.4 Consider a HMM specified by the set of parameters $\lambda = \{p^{(0)}, P, B\}$. The following parameters are given.

$N = 3$ states in a hidden Markov chain, $E = \{1,2,3\}$ is the state space.

$K = 4$ distinct observation symbols, $V = \{v_1,v_2,v_3,v_4\} = \{c,d,e,f\}$ is the alphabet.

The initial state probability vector for the three-state hidden Markov chain is

$$p^{(0)} = \begin{bmatrix} p_1^{(0)} & p_2^{(0)} & p_3^{(0)} \end{bmatrix} = [P(X_0 = 1) \quad P(X_0 = 2) \quad P(X_0 = 3)]$$
$$= [0.25 \quad 0.40 \quad 0.0.35].$$

The transition probability matrix for the hidden Markov chain is

$$P = [p_{ij}] = \begin{array}{c|ccc}
\text{State} & 1 & 2 & 3 \\
\hline
1 & p_{11} & p_{12} & p_{13} \\
2 & p_{21} & p_{22} & p_{23} \\
3 & p_{31} & p_{32} & p_{33}
\end{array} = \begin{array}{c|ccc}
X_n \backslash X_{n+1} & 1 & 2 & 3 \\
\hline
1 & 0.24 & 0.56 & 0.20 \\
2 & 0.38 & 0.22 & 0.40 \\
3 & 0.28 & 0.37 & 0.35
\end{array}$$

The observation symbol probability distribution matrix is

$B = [b_i(k)] =$

State	Observation Symbol c	Observation Symbol d	Observation Symbol e	Observation Symbol f
$i=1$	$P(O_n = c\|X_n = 1)$	$P(O_n = d\|X_n = 1)$	$P(O_n = e\|X_n = 1)$	$P(O_n = f\|X_n = 1)$
$i=2$	$P(O_n = c\|X_n = 2)$	$P(O_n = d\|X_n = 2)$	$P(O_n = e\|X_n = 2)$	$P(O_n = f\|X_n = 2)$
$i=3$	$P(O_n = c\|X_n = 3)$	$P(O_n = d\|X_n = 3)$	$P(O_n = e\|X_n = 3)$	$P(O_n = f\|X_n = 3)$

State	Observation Symbol c	Observation Symbol d	Observation Symbol e	Observation Symbol f
$B=[b_i(k)]=$ $i=1$	$b_1(c)$	$b_1(d)$	$b_1(e)$	$b_1(f)$
$i=2$	$b_2(c)$	$b_2(d)$	$b_2(e)$	$b_2(f)$
$i=3$	$b_3(c)$	$b_3(d)$	$b_3(e)$	$b_3(f)$

State	Observation Symbol c	Observation Symbol d	Observation Symbol e	Observation Symbol f
$B=[b_i(k)]=$ $i=1$	0.30	0.18	0.20	0.32
$i=2$	0.26	0.34	0.12	0.28
$i=3$	0.16	0.40	0.34	0.10

(a) Consider the particular sequence, $O=\{O_0,O_1\}=\{f,d\}$, of $M+1=2$ observations. Use exhaustive enumeration to solve HMM problem 1 of Section 6.2.4 by calculating $P(f,d|\lambda)$.

(b) Use the forward procedure to solve HMM problem 1 of Section 6.2.4 by calculating $P(f,d|\lambda)$.

(c) Use the Viterbi algorithm to solve HMM problem 2 of Section 6.2.4 by calculating $P(X_0,X_1; f, d|\lambda)$, where $X=\{X_0,X_1\}$ is the state sequence which maximizes the probability of having generated the given observation sequence, $O=\{O_0,O_1\}=\{f,d\}$.

References

1. Grassmann, W. K., Taksar, M. I., and Heyman, D. P., Regenerative analysis and steady state distributions for Markov chains, *Op. Res.*, 33, 1107, 1985.
2. Kemeny, J. G. and Snell, J. L., *Finite Markov Chains*, Van Nostrand, Princeton, NJ, 1960. Reprinted by Springer-Verlag, New York, 1976.
3. Kohlas, J., Numerical computation of mean passage times and absorption probabilities in Markov and semi-Markov models, *Zeitschrift für Op. Res.*, 30, A197, 1986.
4. Sheskin, T. J., A Markov chain partitioning algorithm for computing steady state probabilities, *Op. Res.*, 33, 228, 1985.
5. Rabiner, J. L., A tutorial on hidden Markov models and selected applications in speech recognition. *Proc. IEEE.* 77, 257, 1989.
6. Koski, T., *Hidden Markov Models for Bioinformatics*, Kluwer Academic Publishers, Dordrecht, 2001.
7. Ewens, W. J. and Grant, G. R., *Statistical Methods in Bioinformatics: An Introduction*, Springer, New York, 2001.

Index

Printed in the United States
by Baker & Taylor Publisher Services.

Printed in the United States
by Baker & Taylor Publisher Services